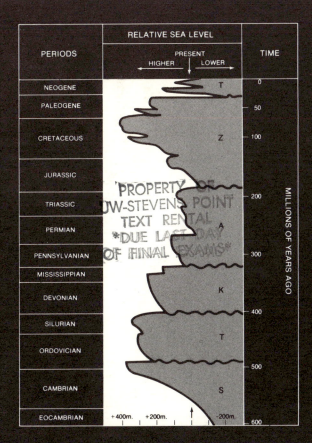

PERIODS	RELATIVE SEA LEVEL		TIME
	HIGHER ← PRESENT ↓ → LOWER		
NEOGENE			0
PALEOGENE			50
CRETACEOUS			100
JURASSIC			
TRIASSIC			200
PERMIAN			
PENNSYLVANIAN			300
MISSISSIPPIAN			
DEVONIAN			400
SILURIAN			
ORDOVICIAN			500
CAMBRIAN			
EOCAMBRIAN	+400m. +200m. -200m.		600

(Relative sea level chart labeled with letters T, Z, A, K, T, S; vertical axis MILLIONS OF YEARS AGO)

Worldwide sea-level fluctuation curve for the past 600 million years. See Fig. 3.21 for explanation.

D0162642

Explanation of symbols used on facies maps, paleogeographic and paleogeologic maps, and most cross sections in remainder of book (*Principal sources of data: Alberta Society of Petroleum Geologists, 1964*, Atlas of facies maps of western Canada; *Clark and Stearn, 1960*, Geological evolution of North America; *Douglas et al., 1963*, Petroleum possibilities of Canadian Arctic; *Eardley, 1962*, Structural geology of North America; *Kay, 1951*, North American geosynclines; *Levorsen, 1960*, Paleogeologic maps; *Martin, 1959*, American Association of Petroleum Geologists Bulletin; *Donald E. Owen, unpublished maps; Raasch et al., 1960*, Geology of the Artic; *Sloss et al., 1960*, Lithofacies maps; *Rocky Mountain Association of Geologists, 1972*, Geologic atlas of the Rocky Mountain region.)

Conglomerate
Sandstone
Graywacke
Shale
Shale and Chert
Red Beds
Carbonate Rocks (Maps) (Limestone and Dolomite)
E Evaporites (Maps)
× × Evaporites
Limestone ⎫ (cross sections)
Dolomite ⎭
225 Isotopic date
Zero thickness line (present limit of strata designated)

+ + Granitic Rocks
Basement Rocks (Granite or metamorphic)
Miscellaneous igneous plutons
V Volcanic Rocks
Prevailing wind
Paleocurrent orientations in sediment
Inferred ancient ocean bottom currents
R Reefs
E Evaporites
C Coal
Faults
Transcurrent or lateral faults
Thrust faults
Unconformity

Third Edition

EVOLUTION OF THE EARTH

Robert H. Dott, Jr.
University of Wisconsin, Madison

Roger L. Batten
American Museum of Natural History, New York
and Columbia University

Maps and Diagrams by
Randall D. Sale
Cartographic Laboratory
University of Wisconsin

McGraw-Hill Book Company

New York St. Louis San Francisco Auckland Bogotá
Hamburg Johannesburg London Madrid Mexico
Montreal New Delhi Panama Paris São Paulo
Singapore Sydney Tokyo Toronto

To our parents
and all
of our other teachers

This book was set in Palatino
by Black Dot, Inc.
The editors were Marian D. Provenzano
and Scott Amerman;
the designer was Joan E. O'Connor;
the production supervisor was Phil Galea.
Rand McNally & Company
was printer and binder.

3 4 5 6 7 8 9 0 R M R M 8 9 8 7 6 5 4 3 2 1

Library of Congress Cataloging in Publication Data

Dott, Robert H, Jr.   date
Evolution of the earth.

Includes bibliographies and index.
1. Historical geology. I. Batten, Roger Lyman,
joint author. II. Title.
QE28.3.D68   1981   551.7   80-18876
ISBN 0-07-017625-6

CONTENTS

PREFACE

Time present and time past
Are both perhaps present in time future,
And time future contained in time past.
T. S. Eliot,
Four Quartets

The third edition of *Evolution of the Earth* differs from the second in several important ways. First, we have reorganized several chapters. Introductory ones have been compressed somewhat and organic evolution is now discussed at a later point (Chapter 10) after some of the record of early life has been presented. Secondly, we have expanded the overall discussion of life at several points. An entirely new chapter (Chapter 9) entitled *Early Life and Its Patterns* has been introduced. Thirdly, discussions of physical history have been strengthened by updating the plate tectonic discussions at many points. Many new diagrams have been prepared to supplement these discussions. Fourth, a completely new *World Map Supplement* has been added near the middle of the book between Chapters 14 and 15. This contains six paleogeographic and six companion paleotectonic (lithosphere plate position) maps that can be referred to easily while reading any part of the text. Paleogeography can be compared quickly with paleotectonics for each corresponding time. We also have added supplemental items such as endpapers for easier reference while reading. Included are a complete *Geologic Time Scale*, *Phanerozoic Sea-Level Fluctuation ("Vail") Curve* (repeat of Fig. 3.21), a simplified *Classification of Igneous Rocks*, and a map of *State and Province Boundaries for North America*. Finally, we have added a Glossary.

When our first edition appeared in 1971, there was a vacuum in historical geology instruction. The long-standing approach to present a very detailed, dry descriptive relation of sequences of formation names for many areas, but with little or no unifying theme and little emphasis upon interpretation and hypothesis testing. That approach was aptly characterized by our friend L. L. Sloss as "The Roll Call of the Ages." We selected, instead, an overall chemical-and-physical as well as organic evolution of the earth as a central theme, and it was fortunate for us that the unifying plate tectonic theory appeared in time to provide more coherence to physical history. We also pioneered with the first use of color for figures in a historical geology book. Looking back, we take considerable satisfaction that our hoped-for renaissance in the treatment of earth history has been successful as evidenced both by the popularity of *Evolution of the Earth* and by the obvious influence of many of our innovations upon several other books that have appeared subsequently. The general response of our own and others' students has been most pleasing of all! Every year we receive appreciative letters from far and wide. *Evolution of the Earth* has not escaped the notice of combatants in the continuing controversy between evolutionists and creationists, either. We receive praise and condemnation, for besides being criticized for endorsing evolution, we also have been scolded for being too soft-spoken on the issue. A worse fate is to be ignored.

Our basic philosophy remains unchanged in this new edition. As we stated in the Preface to the first two editions of *Evolution of the Earth*, we are firmly committed to the proposition that introductions to the sciences should be primarily conceptual rather than informational. Students deserve introductions that reveal the logical framework of a discipline, show relations of that discipline to the totality of human knowledge, and give them some idea of what it is like to be a participant in the discipline. At the same time, there should be depth and rigor that challenge the mind. Our experience indicates that students are stimulated by constant exposure to scientific controversies, occasional spicy personal feuds, or an amusing faux pas. The student also needs to become a partner in the endless process of hypothesis testing, which is what we believe science is all about. In this book, we have tried to use these approaches.

In keeping with our strong preference for a "How do we know?" rather than "What do we know?" approach, we stress what assumptions are made by earth historians, what kinds of evidence (and tools for gathering that evidence) and what processes of reasoning and limitations of hypotheses are involved in reconstructing and interpreting the past. At the same time, we have again in this edition tried to emphasize that which is more or less unique about geology as an intellectual discipline. We feel that the tendency to neglect earth history in many introductory courses is most unfortunate, because to us it is its historical nature that sets geology apart from the more familiar nonhistorical sciences. Therefore, we have deliberately tried to draw out and to illustrate this uniqueness. As readers use the book, we hope that they will keep in mind the following three maxims, which seem to us to be of transcendent importance for all educated people: (1) new concepts of time, (2) the universality of evolutionary changes, and (3) the importance throughout time of ecological interactions between life and the physical world.

Initial encounters with the sciences involve major shifts of scales, both spatial and temporal, as well as language shifts. For the average person, these commonly are very difficult. For non-science-oriented people, apathy and even antagonism may result from an early difficulty in grasping such shifts. In geology, the initial adjustment for the lay person undoubtedly is best made in the field where the raw data lie. Ideally, most introductory geology should be taught in the field almost exclusively. For the most part, however, this is not practical, so procedures must be used to lend synthetically some reality and to induce some student involvement in the subject. For earth history, we have had success by beginning with discussions of the more familiar geologic processes that have affected humans most directly in historic times. Gradually the discussion shifts into truly geologic frames of reference. For the more science-oriented student, we recommend also simple homework problems that deal with rates of a wide variety of geologic processes. Suggested examples are contained in a supplementary *Instructor's Manual.* We have had success in using such problems not only to supplement the text, but also to provide departure points for discussions in small class and laboratory groups.

Evolution of the Earth developed from a second-semester course at the University of Wisconsin, Madison, taken mostly by first- and second-year college students representing a broad spectrum of science and nonscience backgrounds. The first two editions were used successfully in this same course, and the modifications incorporated in this new edition reflect an evolution resulting from our own experience together with the constructive criticisms of many other teachers at a variety of different colleges having widely differing requirements. We have also tried to adapt to changing times and rapid advances in the science. We believe that this new edition of *Evolution of the Earth* lends itself to greater flexibility of usage. The book assumes only a prior knowledge of general geology comparable to a high-school earth science or first-semester college physical geology course. We have retained the classical foundations of earth and life history, but certain material, especially of a chemical and physical nature that seems necessary to us for an adequate understanding of modern geology, may prove difficult for some readers. We are confident that the skillful teacher can help students through such material by drawing out the essential highlights without intimidating them. The Glossary should aid all readers, but especially those with limited backgrounds.

The historical elaboration of the development

of basic principles for interpreting earth and life history invoked in the early chapters is continued in somewhat streamlined form. First, it provides a kind of proxy for the reader's actually recapitulating discoveries and interpretations accomplished by past generations that make up the fabric of geologic principles. Second, it helps to make clear that the science is a human activity, and that the quest for the understanding of nature is an ongoing, open-ended process in which readers themselves could participate if they should so choose. Finally, the historical approach reveals the cultural relationships of the science, which are unusually rich in the case of geology. Historical geology provides a great opportunity—even obligation—to present a broad integration of diverse material and to clarify the relevance of science. Apparently this will increase, not decrease, in importance, for it is estimated that knowledge will grow fourfold by the year 2000. The first three chapters in which the historical approach is most prominently incorporated can be read rather quickly and understood by most students with a minimum of classroom elaboration if the instructor so chooses. Nonscience students especially like this material. But we are mindful that some readers became impatient with our prior five-chapter historical treatment, and so we have compressed this part and relocated the organic evolution chapter. Chapters 5 through 7 present some of the more complex principles and concepts basic to interpreting earth and life history. These are very important to the remainder of the book, and experience shows that they require classroom supplementation. Large crustal displacements through time are now sufficiently well established that it seems mandatory to use them as a framework for the analysis of the local record. Therefore, Chapters 8 and 9 present a broad background leading from mountain building through continental drift and sea-floor spreading to plate tectonics. These chapters, in short, provide the *raison d'être* for the remainder of the book and so deserve special treatment.

After these background chapters, which themselves contain abundant earth history, we prefer to bring up new important concepts, and to expand upon some already-introduced ones, within the context of the actual historical record as each becomes important to the discussion. This is our rationale for taking up organic evolution after reviewing the record of Prepaleozoic and early Paleozoic life. In this way, evolution is discussed after the need becomes apparent for a general explanation of the actual fossil record. Similarly, the principal elaboration of the concept of the geosyncline and of mountain building is reserved until the first major orogeny of the Paleozoic is treated under Ordovician history; mountain building is further elaborated several times later, too. And reefs are not discussed in detail until the middle Paleozoic, when large fossil reefs built by animals first became widespread.

In developing new emphases in historical geology, we have minimized the encyclopedic "Roll Call of the Ages" so prevalent in the past. Chapters 8 and 11 through 18 treat the actual historic record for the earth with North American data integrated still more closely with that for other continents than in the second edition. But whether a given discussion concerns North America or some other part of the world, the purpose always is to illustrate how geologists unravel and interpret the historical record more than to burden the reader with a formidable mass of factual detail and terminology. Only a handful of stratigraphic names appear in the text, and a lengthy relating of evidence and events in every corner of every continent is not presented. Instead, while a broad chronologic framework is provided throughout, chief emphasis is placed upon interpretation of physical and biological environments, reconstruction of paleogeography, and evolution. Our unifying theme throughout is that of *overall chemical evolution* of the earth! Interactions between the living and the nonliving are emphasized, as are major stratigraphic and tectonic patterns that recur in both time and space. In this way, each of the North American chapters tends to be topical as well as chronological, with only one or a few rather unique circumstances (such as evaporite deposition and organic reefs in Chapter 13) developed fully in each at the expense of more mundane, nonunique stratigraphic detail.

Factual detail that is presented is for documentation only; students should not be expected to memorize it but rather to understand the principles and logic of the historical arguments for which the facts provide essential evidence.

The concluding chapter, as in the last edition, reminds the reader of the three fundamental maxims that follow from a study of earth history, which were stated above. Brief illustrations of each are discussed, and then the extrapolation of some implications of these maxims into humanity's future are developed briefly in order to suggest some conclusions following from a study of earth history that have relevance to all members of the human race.

It is impossible to acknowledge all of the countless individuals who have contributed directly or indirectly to the completion of this book. We especially thank L. L. Sloss and W. P. Gerould for originally encouraging us to embark on the writing of the first edition. Reviews of the manuscript by Wakefield Dort, Jr., George W. Lynts, and W. Graham Heaslip have improved it materially. Colleagues who read portions of one or more of the editions, or in other ways made important specific suggestions, include C. J. Bowser, C. R. Bentley, R. F. Black, R. A. Bryson, C. W. Byers, E. N. Cameron, H. S. Chafetz, C. S. Clay, L. M. Cline, C. Craddock, K. A. W. Crook, I. W. D. Dalziel, R. H. Dott, Sr., C. E. Dutton, R. E. Garrison, L. R. Laudon, G. R. McGhee, L. J. Maher, L. G. Medaris, D. M. Mickelson, J. R. Moore, L. C. Pray, J. D. Ryan, R. Schaeffer, and R. S. Steinhart. We single out especially C. W. Byers for his many constructive criticisms. Some others who made unpublished material available or who provided important insights through general discussions include, in addition to the above, R. J. Adie, R. T. Bell, S. A. Born, W. C. Bradley, S. Bridson, A. V. Carozzi, D. L. Clark, T. A. Cohen, W. Compston, Grant Cottam, B. Dennis, R. M. Gates, M. Kay, T. S. Laudon, O. Loucks, M. W. McElhinny, R. P. Meyer, A. A. Meyerhoff, N. D. Newell, N. A. Ostenso, Edna P. Plumstead, W. R. Reeder, R. Siegfried, K. O. Stanley, R. Stauffer, P. R. Vogt, and J. W. Wells. R. D. Sale, of the University of Wisconsin Cartographic Laboratory, prepared line drawings, thus maintaining the prize-winning quality of the illustrations. Sources of drawings and photographs are cited in the figure captions. Agencies that were especially helpful in providing photographs were the American Museum of Natural History, Exxon Production Research Co., Geological Survey of Canada, Geotech Marine Sciences Division of Teledyne Exploration Co., Manned Space Flight Center and Jet Propulsion Laboratory of the National Aeronautic and Space Agency, Marathon Oil Co., Shell Oil Corp., Smithsonian Oceanographic Sorting Center, United States Geological Survey, and U.S. National Museum. Karen E. Dott drew a few illustrations, and Cynthia E. Dott assisted in the assembly and labeling of illustrations. Lu Batten created the index. We are also pleased to acknowledge support of various research projects by the National Science Foundation, especially its Office of Polar Programs, and the Graduate School of the University of Wisconsin, Madison, which allowed one or both of us to visit and to study several foreign areas, including Antarctica. Indirectly, these opportunities have enhanced greatly the final product. Lastly, we again recognize the apparent patience of our families.

And I have written books on the soul,
Proving absurd all written hitherto,
And putting us to ignorance again.
  (So said Robert Browning's
  skeptical philosopher Cleon.)

Robert H. Dott, Jr.
Roger L. Batten

1

TERRESTRIAL CHANGE

There's nothing constant in the
universe,
All ebb and flow, and every
shape that's born
Bears in its womb the seeds of change.

Ovid,
Metamorphoses, XV (A.D. 8)

FIGURE 1.1
Fossilized human foot-
prints in volcanic ash
near Managua, Nica-
ragua, formed by peo-
ple running from an
erupting volcano
2,000–5,000 years ago.
*(Photo by F. B. Richard-
son, furnished by Howel
Williams.)*

1

It was in August of 1883 that a small island named Krakatoa, located between Java and Sumatra in Indonesia, disappeared in what was long considered the most violent explosion ever witnessed by humanity. A 300-meter-(1,000-foot-) deep hole was formed in the bottom of the sea where the island had stood. The explosion was heard 5,000 kilometers (3,000 miles) away. Volcanic ash, which was blown 80 kilometers high and fell on ships 2,000 kilometers from the island, darkened the sky for 2 days (Fig. 1.2). Huge *tsunamis*, waves 30 meters high, crashed upon surrounding islands, drowning 35,000 people and destroying 300 towns.

Another geologic catastrophe recorded in history books was the mischief of Vesuvius volcano in Italy in A.D. 79. Earthquakes warned of eruptions, and many people fled from the Roman city of Pompeii at the foot of the mountain near present-day Naples. Those who lingered were suddenly killed by poisonous volcanic gases and then immediately buried with their houses, furniture, and even pets beneath volcanic ash and mudflows.

A volcanic event of still greater magnitude occurred in the eastern Mediterranean region about 1500 B.C., but its story has only recently been worked out. Santorin, or Thera, is a crescent-shaped island between Greece and Crete (Fig. 1.3). It is a volcanic caldera, that is, a ring enclosing a submerged crater 400 meters deep—a puny remnant of a once lofty mountain. (Crater Lake in Oregon is also a caldera, but it stands well above sea level.) Many small eruptions have occurred at Thera within historic times—the last in 1920 and 1950—but old lava flows and thick volcanic ash deposits tell of a more violent earlier history. Recent sampling by oceanographers of deep-sea sediments in the eastern Mediterranean has revealed widespread buried ash layers that become thicker toward

Thera, their obvious source. Dating indicates that the last very great eruption occurred about 1500 B.C. The material erupted was *three or four times greater* than from Krakatoa.

Thera's eruptions are of special interest not only for their geologic magnitude but even more so because of their possible historical implications. The earliest European civilization—called Minoan—began on Crete only 125 kilometers (75 miles) south of Thera, and a well-preserved Minoan settlement has recently been excavated on Thera itself. But the Cretan civilization collapsed suddenly about 1450 B.C. for reasons that have eluded archaeologists for generations. Greek archaeologist Spyridon Marinatos postulated in the 1930s that an eruption of Thera may have ended the Minoan culture. Thera's cataclysmic eruptions would have generated tsuna-

FIGURE 1.2

Krakatoa volcano and areas of audible noise and significant ash fall from its eruption of 1883. *(After F. M. Bullard, 1976, Volcanoes of the Earth; by permission University of Texas Press.)*

FIGURE 1.3

Location of volcano Thera relative to historically prominent areas of the eastern Mediterranean. Shaded zone represents probable fallout pattern of volcanic ash.

mis, waves perhaps as much as 100 meters high, which could reach northern Crete without warning in a mere 20 minutes. As around Krakatoa, such waves could do severe damage, and the Minoan capital of Knossos is known to have been virtually leveled at about this time. Prevailing winds blew great volumes of ash southeastward. Most of this ended up on the bottom of the Mediterranean, but some also fell on eastern Crete (Fig. 1.3).

Although Thera alone does not seem the whole cause of Minoan decline, it could have inspired some of the accounts of ancient havoc in the eastern Mediterranean, for example reports of periods of floods and darkness when "the sun appeared in the sky like the moon." Could Thera have caused the days of darkness of the Old Testament? Could Thera's submergence have inspired the myth of the sinking of Atlantis? Doubtless some of these claims are overly enthusiastic.

The only real certainty in this world is *change*! Upon even casual examination we can see natural changes taking place on the face of the earth today. Volcanoes are converted to shoals,

or vice versa, and memories from history lessons recall other past dramatic changes. Changes in life on earth also are apparent, and presumed life crises have figured in literature, for example the biblical "Noah's Flood." People always have been deeply impressed—sometimes even rudely inconvenienced—by "catastrophic" geologic events, whereas more subtle, slower changes, even though ultimately more profound, have gone almost unnoticed. Such typically human preoccupation has strongly influenced the development of geological concepts, particularly as they bear upon the history of the earth. Therefore, it is profitable to relate a few examples of important geological processes that have acted during the past 5,000 years or so of recorded human history. Consideration of these changes in the context of a familiar span of time may provide the best basis for approaching matters of greater antiquity in our planet's diary.

CATASTROPHIC GEOLOGIC EVENTS IN HUMAN HISTORY

Another interesting example of geologic events in recorded history can be found in the famous Old Testament story of Sodom and Gomorrah in the Book of Genesis. Recent findings suggest the probable location of these two "cities of sin" under the south end of the Dead Sea. Genesis records that when they were destroyed: "Brimstone and fire rained upon the cities . . . and the smoke of the land went up like the smoke of the furnace." It is easily seen from this graphic description how such an event would make a great impression upon early Mediterranean civilizations of approximately 2000 B.C. Viewing the Jordan Valley today, one can readily understand the impact for it is an inhospitable region. Modern scholars are finding the Bible to be an amazingly accurate historical document; so let us look at what is known of the geology of this area to see if we can discover some verification and explanation of this famous catastrophic event interpreted by the Hebrews as retribution. First, we find that the Jordan Valley is a long, straight-sided depression bounded by plateaus

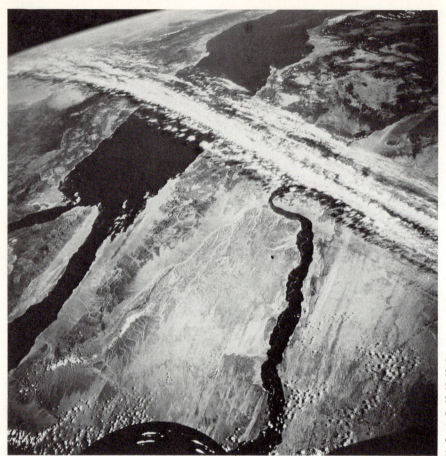

FIGURE 1.4
Gemini spacecraft photo showing prominent linear clouds over northern Egypt (looking southeast across Sinai Peninsula and Red Sea). Jordan Valley and Dead Sea lie just beyond left-center edge of view; heavily vegetated Nile Valley lies right of center. *(Courtesy NASA; photo no. 66-63533.)*

and extending south to join the larger Red Sea depression (Fig. 1.4). Second, we would be impressed with evidence of present and past hot spring activity. Evidence of still greater past heat is borne out by lava flows. Straightness of the valley walls, as well as the hot springs and lavas themselves, indicate the presence of great cracks or faults penetrating deep into the earth. When we discover that earthquakes are common, the whole story becomes clear. Severe tremors accompanying faulting must have resulted in slight southward tilting of the Dead Sea fault block, causing the sea to flood the cities. Hot sulfur springs probably became unusually active in the Jordan Valley as a result of the same tremors.

One of the most volcanically active regions in the world is Central America, and here, too, there have been several interesting cases of human involvement in geologic change. Perfectly preserved human footprints in fine-grained volcanic mud deposits occur near Managua, Nicaragua (Fig. 1.1). The muds obviously were sticky when people hastened across the Managuan plains fleeing from nearby volcanic eruptions sometime between 2,000 and 5,000 years ago. Fine ash was falling continually, and buried the prints for posterity. Animals, too, lumbered over the ash; they included bison and some possible domesticated animals. Volcanic eruptions continue to inconvenience people. Take, for example, the 1973 eruption of Heimaey volcano in Iceland, the lava and ash from which buried many houses (Fig. 1.5).

Equally dramatic in recent times have been very devastating earthquakes such as have occurred in San Francisco in 1906 (Fig 1.6), near Yellowstone Park in 1959, and in Alaska on

FIGURE 1.5
Basaltic lava flow encroaching upon houses in Heimaey, Iceland, during a 1973 eruption. *(Courtesy Campbell Craddock.)*

Good Friday, 1964. Tremors are also commonly associated with volcanism, as we have already seen. A very close link has been established by observation for many years in Hawaii, where eruptions are benevolently forewarned by tremors. Though forgotten today, the violent earthquake and tsunamis that destroyed Lisbon and were felt all over western Europe and northwestern Africa on All Saints Day, 1755, made far greater impact on people's minds than any of these other ones. The timing of that catastrophe deeply impressed superstitious Europeans.

That earth tremors are the rule is indicated by the fact that annually about 800,000 are recorded worldwide but not felt by people. At least 10,000 people are killed each year by earthquakes.

SUBTLE GEOLOGIC EVENTS IN HUMAN HISTORY

CLIMATE

Human civilization appeared more or less simultaneously along the lower Nile River in Egypt, in the Tigris and Euphrates valleys in

FIGURE 1.6
With curious irony, this statue of famous geologist Louis Agassiz (see Chap. 18) was toppled from its mounting at Stanford University during the 1906 San Francisco earthquake. It has since been remounted, little the worse for its Humpty-Dumpty-like fall. *(From Lawson, et al., 1908, The California earthquake of 1906, Carnegie Institution of Washington.)*

present Iraq, on the Indus delta of northwest India, and in northeast China. The eastern Mediterranean–Middle East region, about which we know the most, is today a harsh, arid country that seems an unlikely cradle for agricultural societies. Yet when the great Ice Age glaciers covered northern Europe, this region enjoyed a climate that was comfortably warm and more humid than now. Climatologists believe that an atmospheric high-pressure cell then centered over the ice sheet pushed moist westerly winds south of the Mediterranean Sea. Knowing this, we should not be surprised that early humans spread across the area and here developed their culture rapidly. After the great glaciers retreated, the Middle East climate harshened drastically. By 5000 to 3000 B.C., major cultural centers were restricted to the large, through-flowing river systems, such as the Nile, Tigris-Euphrates, and Indus, that were more or less independent of the increasing drought. It is from just such areas that our earliest written records have come. They tell of already highly developed agricultural systems with elaborate irrigated plantations and advanced urban centers.

Other interesting slow geologic changes have also occurred during human history. Between 1890 and 1950, measurements at widely scattered seaports demonstrated that sea level had been rising over the entire earth an average of 1.2 millimeters per year. This means a total rise of about 70 millimeters (about 3 inches), certainly not devasting in 60 years. But if long continued, this might be of some concern in the future, especially on an earth rapidly becoming crowded by a staggering population boom. Why should sea level be changing at all? If this could be answered, then the probable impact on coastal human habitation might be better assessed. We do not have far to look for the explanation. Over approximately the same 60-year interval, the climate was warming. For example, winters had been much less severe than in the "good old days" of great-grandfather's youth when people skated safely across a completely frozen lower Hudson River from New York City to New Jersey. As the general climate warmed from 1890 to about

1950, by from 0.5 to 5°C (depending upon latitude), glaciers in high mountains over most of the earth retreated gradually, in some cases as much as 16 kilometers (10 miles). The world's largest ice masses, those of Greenland and Antarctica, apparently retreated slightly at their margins, too. The obvious effect of melting so much ice would be to raise sea level, as has been the case.

GLACIERS AND SEA LEVEL

If we assume that melting will continue indefinitely and then gaze into the crystal ball, we can make some startling predictions. From estimates of the total volume of water still locked up in glacial ice, it is calculated that sea level would be raised through complete melting by *at least 50 meters more above its present level*. This would, of course, flood most coastal areas where many of the world's largest concentrations of civilized activity are found. In North America, Washington, D.C., would become a seaport, while New York, Miami, New Orleans, Houston, and Seattle all would be submerged like Sodom and Gomorrah. Memphis, now 560 kilometers from the sea, would become a major port, and Washington, D.C., lies at a critical elevation so that it would suffer very interesting consequences. Most of the government buildings are located on Potomac River lowlands less than 30 meters above present sea level. The Pentagon, only 11 meters above sea level at its base, would become useless. The Washington Monument, whose base has an elevation of 9 meters, would show just its top above water as a useful navigational marker.

While we are at it, we should examine the opposite effect of glaciation, namely the worldwide lowering of sea level during maximum glacial advances. Intuition alone could tell you what must have happened. The last glacial maximum occurred about 20,000 to 25,000 years ago, and from evidence of submerged ancient shoreline features, it is inferred that sea level must have been approximately 100 meters (roughly 300 feet) lower than at present. Several geologists have carefully studied the changes of sea level during the past 20,000

FIGURE 1.7

Eastern Mediterranean–Mesopotamian region showing the Fertile Crescent (white), centers of early agrarian civilizations, and the known extent of the post-Paleolithic marine deposits of the Tigris-Euphrates valley.

years, and some have attempted to link these with historical records. They suggest that an average rise of 100 centimeters per century occurred from about 17,000 up to 6,000 years ago. About 4000 B.C., rapid rise culminated, and all subsequent rise has averaged only 12 to 15 centimeters per century. Clearly, the last phase of rapid rise may have caused flooding of early Bronze Age coastal settlements, and it is certainly interesting to find reference to a deluge in the legends of many separate ancient cultures including Greek, Babylonian, Hindu, and Hebrew.

About 1925, a 3-meter-thick clay layer with marine shells was discovered beneath Ur, one of the world's most ancient cities located in the lower Euphrates valley and known to date back at least to 3000 or 4000 B.C. (Fig. 1.7). Paleolithic human artifacts underlie the clay, thus dating the oceanic incursion at between 4000 and about 8000 B.C. Similar buried clay, though thinner, was then found to underlie a large area of the Tigris-Euphrates valleys. Seemingly, proof of the ancient flood had been discovered at last! It is interesting to note the close correspondence of the age of the clay established archaeologically with the culmination of postglacial rapid rise of sea level established geologically. In 3000 B.C., Ur stood at the head of the Persian Gulf; so it is possible that rising sea level rapidly flooded the lower valley until the rivers could extend their deposits southward again by sedimentation. Since 3000 B.C. it is known that the Tigris-Euphrates delta has advanced southward nearly 175 kilometers (100 miles); today it advances an astounding 25 meters per year.

Could recent small oscillations of sea level have given rise in the Mediterranean–Persian Gulf region to the many ancient Flood legends? If the biblical story were correct, there should have been simultaneous wetting of all other great ancient coastal civilizations. But it has not been possible to establish any synchroneity of deluges reported in the many ancient traditions. Probably these events resulted from periodic local river floods, which caused devastation of the great cultural centers concentrated on low-lying deltas of large rivers. Even today the Tigris-Euphrates delta is subject to frequent floods, and annually two-thirds of Bangladesh, on the immense Ganges delta, disappears beneath several meters of water during the monsoon season. Both regions are low and swampy for hundreds of miles inland.

For the past 6,000 years, maximum vertical fluctuations of only about 3 meters above or below present sea level are indicated. Even when sea-level rises did occur, they could not have inundated the whole then-known world, as tradition would have us believe, but only low coastal areas. On the other hand, as we have seen already, a slight warming of climate for the next 1,000 years or so might result in an additional rise of at least 50 meters. Perhaps we should look still further ahead and prepare instead for cooler times again, as our present "interglacial" stage may last only about 10,000 more years, at which time large ice sheets may begin forming and sea level falling once more.

CRUSTAL WARPING

Much of the low European coastline has been flooded slowly at least since the 1600s when tidal records were first made (Fig. 1.8). This is to be expected from what we have just seen. In central Scandinavia, on the other hand, an awareness developed about 1700 that the coastline there had *retreated*, suggesting an apparent fall of sea level of more than 1 meter in the previous century. It had been assumed that the earth's crust was fixed until, in 1765, a Finnish surveyor proved that the northern Baltic coastline was changing more in the north than in the south. He concluded that the crust must be rising differentially rather than sea level falling.

Amsterdam records indicate that sea level is apparently rising 20 centimeters (8 inches) per century there. But this is about twice the rate deduced for the worldwide effect of glacial retreat. How can we reconcile this discrepancy? Where the apparent rate of rise is too large, we are forced to conclude that there also has been some subsidence (sinking) of the land, and,

conversely, where the apparent rate of rise of sea level is too small or nonexistent, the coastal land also must have risen as sea level has risen. Therefore, study of any coast may need to take into consideration several factors; another cause of oceanic deluge—namely local subsidence of the earth's crust—must be added to those already discussed.

FIGURE 1.8
Relative crustal subsidence (-) and uplift (+) in northwestern Europe in the 10,000 years since the last glacial advance. Numbers indicate known rates of present coastal movements in millimeters per year. (*After R. Fairbridge, 1960,*The changing level of the sea; *copyright © by Scientific American, Inc. All rights reserved.*)

There is no better place in the world to illustrate the interplay of land and sea-level changes than northwestern Europe. A long history of careful measurements shows that, as the North Sea region has subsided to flood the low countries, most of interior central Europe as well as the northern two-thirds of Great Britain and Scandinavia have been rising at a geologically rapid rate (Fig. 1.8). Borings in the Netherlands indicate that old shoreline sediments, known from their fossils to have been deposited during the Ice Age roughly 1 million years ago, now lie 2,000 meters below present sea level (i.e., beneath 2,000 meters of younger deposits). The crust of the earth here has subsided about 2,000 meters in 1 million years, or an average of about 2 millimeters per year. Measurements in central Scandinavia indicate that parts of the North Sea and Baltic coasts are now rising as much as 10 to 11 millimeters per year. The maximum total rise has been 250 meters (825 feet) since the retreat of glacial ice from the region about 10,000 years ago. The average apparent rate of rise, then, has been about 2.5 meters per century, an astoundingly rapid rate as we shall soon see—and much greater, notice, than the simultaneous rate of subsidence of the nearby low countries (Fig. 1.8).

Submarine topography and dredgings indicate conclusively that much of the North Sea floor and the English Channel formed a low, swampy land connecting Britain with the Continent at the end of the last glaciation about 10,000 years ago. Glacial ice had then melted entirely from Britain and must have been shrinking rapidly from Scandinavia. Although sea level had begun rising rapidly, it did not flood this region until 6,000 or 7,000 years ago, when, you will recall, we estimate that the present general level was achieved. Easy migration of plants, animals, and early humans was possible across this area from the Continent to Britain for at least 3,000 years. As low areas were flooded, former ice cap centers of Scandinavia and Britain have risen by *isostatic rebound* in response to removal of the great load of ice that had weighed down the crust there. This clear cause-and-effect is duplicated for North America, where measurements prove that most of

east-central Canada is also rising in response to deglaciation. This is one of our strongest confirmations of the general plasticity of large parts of the earth's interior. To compensate for the rise of Scandinavia and Britain, we conclude that some subcrustal material must flow slowly under these areas, and lateral migration from beneath the North Sea may account for the subsidence there. Subsequently, we shall explore the historical implications of the tendency for a balancing of levels of different parts of the outer crust of the earth in this manner.

HUMAN BEINGS AS GEOLOGIC AGENTS

Human beings sometimes have contributed to geologic change, usually unwittingly. It is well known that historic Venice is sinking at a present rate of 6 to 10 millimeters per year. At the most, 1.5 millimeters is due to worldwide sea-level rise, as we have seen, and only 2 millimeters or less can be attributed to regional *crustal subsidence* (Fig. 1.8). What, then, explains the alarming flooding of this precious city? It has been shown that withdrawal of ground water lowered the water table 20 meters (65 feet) beneath a nearby industrial center during the last 30 years. Thus about 80 percent of the sinking of Venice is caused by rapid removal of water by humans, which has caused compaction of sediments beneath the area (that is, the squeezing and shrinkage under the weight of overlying material as fluids were removed). Pumping of water recently has been reduced, and happily the sinking has slowed.

Near Los Angeles, California, people have for 30 years been encouraging encroachment of the sea. The second largest oil field in California extends beneath the coast under Long Beach harbor. This area is the only satisfactory harbor in the greater Los Angeles region; therefore, a large naval base, an automobile assembly and shipping facility, and other industrial activities are concentrated along the waterfront. Over the years, extraction of petroleum and natural gas from incompletely cemented sandstones more than 1,000 meters below the surface has lowered the pressure in the pores of the strata and allowed the sediment grains to compact; subsidence has resulted. An area 8 kilometers (5 miles) in diameter with subsidence amounting to as much as 6 meters in 23 years is centered almost directly under the coastal facilities. You can imagine the consternation of irate naval and industrial landowners! To stem further subsidence, petroleum companies today replace extracted petroleum with water pumped into the sandstones at a rate equal to that of extraction of the oil.

People have acted the role of a geologic agent in this same region in other ways. First they persist in completely stripping vegetation from hillsides for new suburban housing tracts, which encourages landsliding during heavy rains. Fires caused by people denude the mountain slopes, further aggravating the erosion problem. Housing developments also continue to be erected across known active faults and old landslides. Human activities have had a more subtle effect upon southern California beaches. Prior to a great wartime building boom of the 1940s, rivers continually replenished the beach sands, which were slowly swept southward along the coast to Newport, south of Long Beach, there to be flushed into deep water via a submarine canyon. Metropolitan expansion brought complete usage of normal runoff waters so that rivers are completely dry except during rare winter downpours. Furthermore, through much of their lengths, the rivers now have artificial cement channels, which deprive them of considerable sand that would otherwise be added from their banks. The beaches have been starved of sand, and so winter storm waves are taking their toll of the shore and some choice homes. This calamity has necessitated building elaborate sand-drift traps and expensive barging and pumping to refeed the beaches.

Alarming depletion of natural water supplies in the arid Southwest, and necessity for transporting water hundreds of miles at fantastic expense is another story of humans versus nature. Most ground water now being extracted in such areas fell as rain thousands of years ago, and is being withdrawn more rapidly than it is being replenished. As pressure on finite resources increases, other regions must also take notice, for many people today are living precariously on borrowed water—borrowed from the past without any adequate reinvestment program for the future's mushrooming thirsty population.

TIME AND RATES OF GEOLOGIC PROCESSES—HOW FAST IS FAST?

GEOLOGIC TIME

Having reviewed examples of different geologic processes active within the brief span of human recorded residence upon our little planet, we shall now begin to investigate the matter of geologic time. Here we step into a new frame of reference for most people. Time is to the geologist much as the immensity of space is to the astronomer. Five thousand years of recorded human history seems a long time, and if we tell you that humans first appeared between 1.5 and 2 million years ago, they will seem quite respectably antique. What would you think, though, when we tell you that *the earth is almost 3,000 times older?*

Because experience shows that realization of the magnitude of geologic time is so new to the uninitiated, geologists find that analogy with a more familiar measure of time is helpful. Our favorite is to compare it to the 7 days of the week. If we take Monday as the first day and consider the Monday morning sun to have risen about 5 billion years ago (roughly the earth's birthday, give or take a half billion), we would find that the record of Monday through Saturday, or over three-fourths of earth history is exceedingly obscure! For example, the first recognizable fossil animals appeared a mere 600 to 700 million years ago, or early Sunday on the last day of our analogy. We can get a truer perspective when we realize that recorded human history would begin only after 11:55 P.M. Sunday. The oldest known living things, a California bristlecone pine tree 4,600 years old and a Sequoia tree almost 4,000 years old, both sprouted "minutes" later; Christ would have been born only seconds before midnight at the close of the "week." Subsequent events—such

TABLE 1.1
Comparison of rates of some geologic processes

Slowest				Fastest
Average erosion of continent	Cutting of Grand Canyon	Postglacial rise of sea level	Rise of Scandinavia	Advance of Tigris-Euphrates delta
0.03 mm per year	0.7 mm per year	5 mm per year	10 mm per year	25,000 mm per year

as the entire industrial revolution, discovery of electricity, invention of the automobile, airplane, and television—hardly seem worth mentioning on this scale.

To illustrate geologic time still another way, let us perform some simple arithmetic to discover approximate rates of additional geologic changes, particularly the so-called subtle kinds of changes.

Examples of Rates

Recall that we said sea level was approximately 100 meters lower about 20,000 years ago and has risen with minor fluctuations since. The average rate of rise would then be 0.005 meter, or 5 millimeters, per year. In other words, it would take an average of 66 years to rise 1 foot. On the other hand, looking to the future and taking the view that sea level will rise another 50 meters if existing ice caps melt, and assuming the same average rate of melting and rise as above, it should take but a mere 10,000 years to produce the final flood. Table 1.1 summarizes for comparison the rates of four other common geologic processes.

Besides the question: "How fast is fast?" there is also a question: "How frequent is frequent?" To people, river floods are among the most familiar geologic catastrophes. Moreover, the flood plains of rivers are as much a part of the total watershed systems as are the main channels. This should be obvious to even the most casual observer who lives near a large river. But people have long ignored this fact and have continued to build houses, towns, and factories on flood plains that inevitably are destined for dousings. As a wise Louisiana geologist once said: "Flood plains are for floods!"

On the carefully controlled Mississippi River, severe floods have been infrequent, and so

inhabitants have become complacent over many decades. In 1965 a combination of large ice jams and sudden melting of heavy snow produced a catastrophic flood. None of the old-timers could remember such an event, nor could they recollect stories of such from their ancestors. Quickly a hue and cry went up for governmental preventive aid—totally ignoring the inevitability of floods. Flooding provides a useful example of the limitation of human time perspectives. If floods like that of 1965 occur, on the average, only once every couple of centuries, there could be no record of them in our young folklore. Similarly, an "unprecedented" North Sea flooding of the Netherlands in 1953 was caused by meteorological circumstances so unusual that it is estimated they would occur only about once every 400 years.

Another seemingly unprecedented flood occurred in 1976 in Big Thompson Canyon, Colorado. Less than 3 years later, people had begun to reoccupy devastated canyon-bottom areas. Either they are suicidal or they are incurable gamblers, for they are playing the odds that it was a 300- or 500-year flood. Though by human standards such floods are rare events, these catastrophes must be considered frequent and normal on a geologic time scale.

An important question might be asked here. Namely, are measured rates of *uplift* and other structural changes adequate to produce great mountains over a reasonable span of geologic time, or do these important features require abnormally faster rates as might seem to be the case? Said another way, are we today witnessing a period of active mountain making in areas like California, or is this a relatively static time?

During his great round-the-world voyage in *H.M.S. Beagle*, Charles Darwin experienced several earthquakes. In 1835, following an espe-

FIGURE 1.9
Youthful mountain range along the southern California coast with a smooth wave-cut marine surface elevated within the past 50,000 to 100,000 years. Before elevation, the shoreline lay at the foot of the mountains. *(Courtesy W. C. Bradley.)*

cially severe one, he noted an uplift of the Chilean coast of about a meter, as well as nearby volcanic eruptions. He reasoned that a rise of a few feet per shock might produce, over geologic time, whole mountain ranges (Fig. 1.9). Figure 1.10 is a simple graph projecting some rates of measured uplift in several parts of California, obviously a relatively active region. This graph was prepared by an American geologist to show what changes of elevation might be expected 200,000 years in the future, a very short time geologically. Obviously we must conclude that present, known rates of uplift can indeed produce mountains in a relatively short time.

Different Ways of Growing and Changing
In the preceding arithmetic gymnastics, we have talked only of average rates of change and for simplicity, therefore, we have assumed constant rates over long times. Figure 1.10 illustrates a graph showing this type of constant change, which plots as a· straight line (thus termed "linear change"). However, we know that most natural processes do not proceed at constant rates. In nature linear change is quite the exception. For example, erosion of a newly uplifted mountain range would tend to proceed fastest early and then slow down after uplift ceased, because stream gradients would flatten. Growth in most animals slows as adulthood is approached. If you invest money, you want it to earn interest at an increasing rate through time, but, conversely, if you borrow money, you hope to pay out interest at a decreasing rate as you pay back the principal.

We hear a great deal about the human population "explosion," and a glance at Fig. 1.11 reveals why. *Homo sapiens* has multiplied in much the same way as you computed com-

FIGURE 1.10
Present known rates of uplift in southern California projected into the future at a constant (linear) rate. (*Adapted from J. Gilluly, 1949; with permission of Geological Society of America.*)

pound interest in elementary mathematics. In other words, population grows at an ever-increasing (non-linear) rate. We see that human beings in their first million years did not increase their total numbers greatly. It must have taken hundreds of thousands of years to break the first million mark. Records exist only for the past few centuries, but it is clear that the first documented doubling of population took two centuries (1650–1856); the second doubling a little less than one century; and today it doubles in only 30 years! The present rate is equivalent to adding the population of Britain annually. By the year 2000, people will number around 8 billion—double the present count!

All the changes discussed up to this point move in one general direction (increase of size, decrease of elevation, etc.); that is, they are nonrepeating or irregular changes. Many changes in nature, however, are repeating ones. Commonly these repeat with a more or less regular time period, as for example the fluctuations of the tides and changes of seasons. Such changes are called cyclic or rhythmic phenomena.

CATASTROPHIC VERSUS UNIFORM VIEWS OF CHANGE AND THE AGE OF THE EARTH

Ancient peoples quite understandably were impressed only by rapid and violent geologic processes. As a consequence, a view of the earth developed very early and gradually earned for itself the name *catastrophism*. By this view, which dominated human thought until 1850, most changes occurred suddenly, rapidly, and

devastatingly. The most famous example of such a supposed catastrophe was, of course, the great Deluge of the Old Testament. Others included volcanic eruptions at Pompeii and Thera and earthquakes at Lisbon and Jamaica.

A parallel idea consistent with catastrophism was a universally accepted belief that the earth

FIGURE 1.11
Growth curve for human population of the world, example of a nonconstant (nonlinear) rate of change (data from United Nations estimates).

was but a few thousand years old. This conception is revealed in Shakespeare's *As You Like It*, written around 1600, where reference is made to the age of the earth being about 6,000 years. In 1654 a more definite pronouncement came from obscure Anglican Archbishop Ussher, who announced with great certainty that, based upon his analysis of the Scriptures, the world had been created in the year 4004 B.C. on the 26th of October at nine o'clock in the morning! Some years later, another authority fixed an equally precise date for the great Deluge, November 18, 2349 B.C. At that time, practically all leading intellectuals believed implicitly in a literal interpretation of the Bible, and so it is not surprising to find that a 6,000-year age for the earth was gospel for fully 200 years.

In the seventeenth century, the entire Scriptural account of the Creation was a point of departure for science. Humans were considered early comers who lived for a time in paradise. As punishment for their sins, the Deluge was sent to throw the earth's surface into chaos. Countless creatures suddenly were buried in sediments, violent waves and currents scoured out valleys and piled up rocks to form mountains, and the earth was wracked by earthquakes and volcanic eruptions. After the Flood, the mountains began to be worn down and valleys to be filled.

When geology emerged as a vigorous intellectual pursuit around 1800, an alternate concept of earth change began to develop. By the new notion, much as we tried to show with some of our arithmetic above, it was held that the same common processes now observed operating upon the earth were largely responsible for all past changes imprinted in the rock record. A noncatastrophic or more *uniform view* of change thus emerged, which required only a limited role for exceptionally sudden events like earthquakes, volcanic eruptions, floods, and avalanches. Annually the world experiences about 150 major earthquakes, 20 volcanic eruptions, and countless floods. None of these violent events can be considered unusual. Today we envision the earth as dynamic, ever-changing, and evolving under the influences of many complex physical, chemical, and biological processes. These processes compete with one another constantly to produce a situation tending toward some balance among them, but it is always being modified and kept off balance by new changes. The resulting modifications are in part nonrecurring, or irregular, changes, and in part regular, recurring, or cyclic, in character.

A corollary of the uniform change concept, which had great impact upon society in the 1800s, was its inescapable implication that the earth must, in fact, be much more than a few thousand years old in order to allow enough time for all observable ancient changes to have occurred in noncatastrophic ways. We begin to see here the seeds of a social controversy whose repercussions affected our entire civilization with the age of the earth and the nature of geologic changes as principal points of disagreement. The controversy was brought to a head largely by the findings of geology in the early nineteenth century. In the next chapter we shall examine this philosophical crisis as we start to develop the first principles upon which to base our interpretation of the fascinating and perplexing diary of our evolving planet.

The earth is not finished, but is now being, and will forevermore be re-made. (C. R. Van Hise, 1898)

Key Concepts

500-year flood
isostatic rebound
rates of change:
   constant (linear)
   nonconstant (nonlinear)
   nonrepeating (irregular)
   repeating (cyclic or not)
subsidence
tsunami waves
uplift

Readings

Alaska's Good Friday earthquake, March 27, 1964: Washington, U.S. Geological Survey, Circular 491.
Archaeological Services of Greece, 1969, International scientific congress on the volcano Thera: Athens.

Bullard, F. M., 1976, Volcanoes of the earth: Austin, Univ. of Texas Press.

Doumas, C. (editor), 1978, Thera and the Aegean world: Warminster, Aris and Phillips.

Fairbridge, R. W., 1960, The changing level of the sea: Scientific American, v. 202, May 1960, pp. 70–79.

Man against volcano: The eruption of Heimaey, Vestmann Islands, Iceland: Washington, U.S. Geological Survey INF-75-22.

Ojakangas, R., and Darby, D., 1976, The earth and its history: New York, McGraw-Hill. (Paperback)

Vitaliano, D. B., 1973, Legends of the earth—Their geologic origin: Bloomington, Indiana University Press.

2

FLOODS, FOSSILS, AND HERESIES

Is There "No Vestige of a Beginning— No Prospect of an End"?

These shells are the greatest and most lasting monuments of antiquity, which in all probability, will far antedate all the most ancient monuments of the world, even the pyramids, obelisks, mummys, heiroglyphicks, and coins . . . Nor will there be wanting media or criteria of chronology which may give us some account even of the time when (they formed).

Robert Hooke (1703)

The mind seemed to grow giddy by looking so far into the abyss of time.

John Playfair (1805, at Siccar Point)

FIGURE 2.1
Seventeenth-century illustrations of fossil ammonoids (*Cornua ammonis* or "snake-stones") prepared by Robert Hooke. *(From Posthumous works of Robert Hooke, 1703.)*

17

2

Until the late eighteenth century, the biblical Genesis was so perfectly rationalized with geologic facts that only a handful of people questioned the conventional geology, and they did so at no small risk. For example, it was the adding up of ages of successive generations recorded in the Bible that formed the basis for the generally accepted 6,000-year age for the earth. Even up to 1800, agreement of biblical revelation with reason seemed adequate to most people. The presence of abundant fossil marine creatures in strata strewn over the land offered proof of the Flood. The earth was thought to have been a paradise with a mild, uniform climate and a flat surface before the sinful fall of humanity. Under such conditions, it was not surprising that Adam and his progeny each lived to be more than 900 years old according to the Old Testament. But after the Flood, conditions were quite different. Formation of mountains by the Deluge caused in turn a harshening of climate. As a result, people did not live so long and they now had to work much harder to win an existence. Moreover, erosion of the land implied a decay of the earth that would culminate in the end of the world as foretold in certain biblical passages. Martin Luther even predicted that the ultimate catastrophe would occur by 1560!

How do important new ideas develop? Why are they sometimes quickly accepted to produce "breakthroughs," while at other times are attacked or lost? Answers to these questions are, of course, complex, and relate to the whole social, economic, and intellectual history of humanity. For example, the old idea that the earth lay at the center of the universe was accepted and tenaciously held for 14 centuries because it seemed to fit the common experiences of apparent rising and setting of the sun and rotation of stars in the heavens. It also neatly fit

the popular assumption that humanity, and therefore earth, lay at the center of everything. Though Columbus set out to circumnavigate the earth and prove it to be spherical, he actually did neither. Vikings had beaten him to America 500 years earlier. But the Greeks even before that had deduced that the earth was not flat, for it cast a circular shadow on the moon during lunar eclipses, the Flat Earth Society notwithstanding. Moreover, the curvature, radius, and circumference of the planet had been determined with suprising accuracy 2,000 years before Columbus!

Realization came very slowly that explanations must be based upon detailed evidence from the earth itself. Geology's awakening began with revelations about fossils and strata in 1669. But it was a slow awakening. One of the most difficult issues was a search for some mechanism either of equilibrium or rejuvenation of the earth to provide a habitat for life in the face of obvious decay of the landscape. A first step was to be a reinterpretation of the Deluge as a corrective redesign of the Creation.

Being an invention of humans, science is a special kind of social institution, which therefore shares the agonies of cultural change. Social, political, economic, and religious disorders of the Renaissance affected science markedly, and not only philosophically. For example, colonialism and wider trade created needs for better navigation, timekeeping, and map making, which stimulated astronomy and mathematics. Geology was affected through mining, quarrying, and clay production. Also the study of earth history received a boost from the great interest in the past generated by the discovery and excavation of buried Pompeii in the eighteenth century and the publicizing of Egyptian, Mayan, and other ancient civilizations in the early nineteenth century.

FIGURE 2.2
A well-preserved ant in Pleistocene amber, an unusual mode of preservation. (*Courtesy American Museum of Natural History.*)

WHAT IS A FOSSIL?

EARLY QUESTIONS

Among ancient peoples there were many controversies concerning geologic matters. Besides supposed catastrophic floods, probably the most important things that puzzled the ancients were fossils. What is a fossil? The word means simply "something dug up." but, as generally used, a *fossil* is any recognizable evidence of preexisting life. Shells of old animals, bones, plant structures such as petrified tree trunks, impressions of plant leaves, soft worms, or jellyfish, and even tracks, burrows, and fecal material formed by animals qualify. Fossils are preserved in many different ways (see Fig. 2.2).

Some of the first ideas about fossils came from Greece. Large bones were interpreted as relics of a former race of heroic human giants, but apparent oceanic organisms hundreds of feet above sea level and miles inland posed weightier questions. Had the sea receded or had these objects "grown" in the rock much like mineral crystals? Perhaps, somehow, fish and other animals had crawled into cracks in the rock and died, then were converted into stone by mysterious "vapors." In this view, it was supposed that fossils were much younger than the rocks in which they occurred. Still another quaint idea was that fossils were "figured stones" created in

the rocks by "plastic forces." Centuries later, some people thought that they grew in the rocks from seeds. It even was suggested that they grew from fish spawn caught in cracks in the rocks during the great Flood.

Four major questions about fossils were posed by these early speculations: (1) Are fossils really organic remains? (2) How did they get into the rocks? (3) When did they get there? (4) How did they become "petrified"?

AN ARTIST SAW THE WAY

The most important early record of a careful interpretation of fossils was that left by the Italian genius Leonardo da Vinci, one of the most original natural philosophers of the early Renaissance. About A.D. 1500 he recognized that fossil shells in north Italy represented ancient marine life even though found in strata exposed many miles from the nearest seashore (Fig. 2.3). In opposition to the popular view that fossils had been washed in by the biblical Deluge, he argued that clams could not travel hundreds of kilometers in 40 days. Many of the shells certainly could not have been washed inland great distances because they were too fragile. He also pointed out that the associations or assemblages of different kinds of fossils

FIGURE 2.3
Fossiliferous and barren strata such as those studied by Leonardo da Vinci and Nicholas Steno in Italy.

Sandstone with rare abraded forms

Shale with buried oyster reefs

Broken shells in pebby sandstone

Abundant varied shells in shale

Barren sandstone

Delicately ornamented shells in shale
(small and large forms mixed)

found in these ancient strata were still intact and resembled living communities of organisms that he observed at the coast. Leonardo further noted that there were many distinct layers that were fossil-rich and that these were separated by completely barren, unfossiliferous ones (Fig. 2.3). This suggested to him, by reasoning from seasonal flooding of rivers, that there were many events recorded rather than a single, worldwide deluge.

NICHOLAS STENO

In the middle seventeenth century, another important man, Nicholas Steno, also made some clear observations about fossils, and his writings were widely and quickly circulated beginning in 1669.

Steno, like Leonardo, felt that fossils formed with the rocks slowly, bit by bit. Steno was a careful observer and based conclusions upon those things that he actually observed in the rocks of Italy. He recognized that shell-bearing strata beneath the site of ancient Rome, which were used in constructing its buildings, must be older than the city; that is, more than 3,000 years. But Steno clung to the belief that earth history was not much longer than human history, and went awry in interpreting large mammal bones in central Italy as the remains of Hannibal's famous elephants; he missed their age by at least a million years!

STENO'S PRINCIPLES

Besides correctly interpreting fossils, Steno drew some even more important conclusions about the strata in which they occur. The result was his statement of the most basic principles for analysis of earth history. Steno showed great insight into the significance of individual strata; for example he recognized that particles would settle from a fluid in proportion to their relative weights or mass, the larger ones first and so on. Any changes in size of particles would cause development of horizontal layering or *stratification*, the most conspicuous single property of sedimentary rocks (Fig. 2.4). He also appreciat-

ed the importance of solution and precipitation of chemically soluble sedimentary materials. In short, he recognized the tendency for uniformity of texture and composition in individual strata, and he inferred that the characteristics of different strata reflected changes of such conditions as temperature, wind, currents, or storms.

Finally, and most significantly, sedimentary strata consisting of particles and shells could not all have existed from the beginning of time (i.e., all be the same age), but must have been deposited particle by particle and layer by layer,

FIGURE 2.4

Conspicuous stratification, the most characteristic feature of sedimentary rocks, interpreted fully by Nicholas Steno in the seventeenth century (sandstones and thin shales of late Paleozoic Itarare Formation, Brazil). *(Courtesy J. C. Crowell.)*

FIGURE 2.5
View in eastern Grand Canyon illustrating Steno's principles. A discordance (called angular unconformity) marked
by line separates older tilted strata (Eocambrian) from
younger (Cambrian) ones. *(Courtesy E. D. McKee.)*

one on top of another. Therefore, in a sequence
with many layers of stratified rocks, a given
layer must be older than any overlying layers.
We now call this the principle of superposition
of strata, and, though it seems obvious, it is the
crux of interpretation of geologic history from
the rock record. Steno was the first to state
formally this and two other principles, which
made him one of the first men to recognize that
strata contain a decipherable chronological record of earth history, the analysis of which
today falls under the heading of stratigraphy.

Steno's principles provide the ultimate bases
of practically all interpretation of earth history,
and so their importance can hardly be overemphasized even though they may seem obvious
today. Briefly stated, they are

1 *Principle of superposition:* in any succession of
strata not severely deformed, the oldest stratum
lies at the bottom, with successively younger
ones above. This is the basis of relative ages of
all strata and their contained fossils.

2 *Principle of original horizontality:* because sedimentary particles settle from fluids under gravitational influence, stratification originally must
be horizontal; steeply inclined strata, therefore,
have suffered subsequent disturbance.
3 *Principle of original lateral continuity:* strata
originally extended in all directions until they
thinned to zero or terminated against the edges
of their original area (or basin) of deposition.

The Grand Canyon illustrates all three principles (Fig. 2.5). First, the upper three-fourths of
the canyon walls expose horizontal strata undisturbed for a long time. At the bottom of the
canyon, however, one can see tilted strata that,
by the second principle, must have been originally horizontal. They are separated from the
horizontal sequence by a discontinuity or break
in the rock sequence, called an *unconformity*.
From the first principle, we see that the oldest
stratum respectively in either sequence must lie
at the bottom. Furthermore, we conclude that
the tilted sequence as a whole is older than the
flat sequence, for it underlies the latter. Finally,
the third principle is illustrated by the fact that
each stratum exposed on the far wall of the
canyon has its clear counterpart on the nearer

walls, and it requires little imagination to see that each was continuous across the present canyon area prior to cutting of the gorge by the Colorado River. Although here the original lateral continuity of strata is obvious, in most areas it is less so. Yet, everywhere that we see the eroded or broken edges of strata exposed, we know that once they were more extensive laterally than they appear to be today. This becomes of paramount importance in trying to restore ancient conditions on earth from a much-eroded and fragmentary stratified record.

UTILITY OF FOSSILS

ROBERT HOOKE'S OVERLOOKED HYPOTHESIS

About the same time that Steno's writings appeared, famous British scientist Robert Hooke also argued that fossils were truly organic, and joined other challengers of the conventional notion that all had been emplaced by the single biblical Flood. Hooke studied and meticulously illustrated fossil shells (Fig. 2.1), making extensive use of the microscope, which had recently been invented. Most significant was Hooke's suggestion that fossils might be useful for making chronologic comparison of rocks of similar age, much as old Roman coins were used to date human historical events in Europe (see quotation at the beginning of the chapter). He speculated that species had a fixed "life span," for many of the fossils he studied had no known living counterparts. This was one of the earliest hints of extinction of species, it long having been assumed that all life was created about 6,000 years ago, and is still living. But extinctions would be evidence of imperfections in the Creation. It was nearly a century before his idea actually was tested and proven valid.

Geology had not yet arrived at a stage in which very much attention was being given to detailed study of individual strata. Until the late eighteenth century, natural scientists were preoccupied with grander problems. Combat still continued with the Deluge enthusiasts, known as *diluvialists*, who held such interesting ideas as the suggestion that the Flood had first dissolved all antediluvian matter, except fossils, and then reprecipitated the sediments in which the fossils became encased as they settled out of the turbulent waters. But where were the remains of men killed by the Flood? This question was of sufficient moment that a Swiss naturalist, J. Scheuchzer, in 1709 excitedly interpreted a giant salamander skeleton as the fossil remains of a man drowned by the Deluge. First things first—clearly science was not yet ready to *use* fossils! Slowly, doubts arose over the geologic importance of the Flood, and gradually it became regarded as too brief to have altered the earth's surface significantly.

FOSSILS AND GEOLOGIC MAPPING

By the mid-eighteenth century, a great deal of fossil collecting had been accomplished, and the mapping of earth materials was beginning. As early as 1723, an English naturalist (John Woodward) suggested an identity or correlation of certain strata on the European mainland with those in Britain on the basis of "great numbers of shells and other productions of the sea." In 1746 one of the earliest crude maps, showing mineral and fossil localities in part of France and Britain (Fig. 2.6), was produced by J. E. Guettard. Guettard showed that, like their modern counterparts, fossil shells commonly have others attached to them, may have holes bored through them, and may tend to be broken and worn.

Just before 1800, a civil engineer named William Smith was actively involved in land surveying in England. From visits to mines, he early recognized a widespread regularity in the rock succession, which already had been demonstrated locally in coal mines (Fig. 2.7). Smith finally proved the enormous utility of fossils. In

FIGURE 2.6
Guettard's early mineralogic map of France and England, forerunner of true geologic maps (prepared with assistance from A. L. Lavoisier). Note symbols for rock types, ores, coal; shaded oval band represents the chalk-bearing strata. (*From* Histoire de L'Academie Royal des Sciences, *Paris, 1746.*)

EAST   PLATE VI   WEST

A Section of the Strata at Chatsworth, from the Old Park to the River Derwent.

John Whitehurst del.

FIGURE 2.7

An early cross section showing the stratigraphic sequence and faulting in a coal field near Newcastle, England. The author had a quaint explanation of the tilting and faulting shown: subterranean heat caused expansion and cracking of the earth; sea water then leaked downward to be converted to steam, resulting in sudden convulsions of the crust, climaxed by the biblical Deluge. *(From J. Whitehurst, 1778, An inquiry into the original state and formation of the earth.)*

1796 he wrote of the "wonderful order and regularity with which nature has disposed of these singular productions (fossils) and assigned to each its class and its peculiar Stratum." Culmination of his effort was preparation of the first geologic map of high quality, some sheets of which were completed before 1800.

Working in a humid region where rock outcrops are rare, artificial excavations and mines were of great importance in helping Smith to form his ideas. As a practicing engineer, he was engaged in extensive canal excavations in southern England, which were intended to increase the transport capabilities for the rapidly expanding British commercial empire in the early industrial revolution. Smith traced and mapped different strata according to their color and mineral composition as well as distinctive fossils. As his mapping extended across England (Fig. 2.8), he could recognize most of the strata over long distances by their fossils, much as one might recognize bygone eras from the distinctive styles of old coins, bottles, or cans in the layers of different metropolitan rubbish heaps.

Smith's map and its accompanying table of strata completed in 1815 was a cartographic masterpiece. He was careful in collecting, locating, and marking his fossils, a most important precaution if they were to be used for identifying particular strata. Smith's prowess as a careful observer is exemplified by his distinction of fossils still in their original place from badly abraded ones that would be less reliable for tracing individual strata because they had been transported and mixed after death. The map immediately proved useful, not only for planning canals, quarries, and mines, but also as a guide to soils. Its author long since had recognized that soil develops from underlying bedrock, and, therefore, reflects some characteristics of the rock. Smith gradually was able to establish a detailed sequence of strata, which then enabled him to predict distribution of different rock bodies in new, unmapped country.

Meanwhile in the Paris region, the principle of fossil correlation and its usefulness in geologic mapping was discovered simultaneously. Young A. L. Lavoisier, a great chemist-to-be, by 1780 had become aware of some order to fossil occurrences. He recognized different assemblages of fossils associated with different rock types and reasoned that these associations reflected environmental conditions, a most advanced idea for that time. Two other Frenchmen named Georges Cuvier and Alexandre Brongniart, however, developed these ideas more fully and proved that around Paris there is a definite relationship between fossil occurrences and the succession of the sedimentary deposits in which they occur (Fig. 2.8). Like Smith, they also developed an essentially modern geologic map (published 1811) showing the distribution of various rock divisions based largely upon the similarity of fossils found in the particular strata. Cuvier wrote:

> These fossils are generally the same in corresponding beds, and present tolerably marked differences of species from one group of beds to another. It is a method of recognition which up to the present has never deceived us.

The simultaneous, independent Smith-Cuvier discovery that strata could be distinguished by their fossils was a "scientific breakthrough." After the giant step was taken around 1810, the use of fossils was accepted quickly. The subdivision, tracing, and mapping of European strata went forth at a great pace with stimulation from the industrial revolution.

Index Fossils and the Principle of Fossil Succession

The first construction of geologic maps was based largely upon the ability to compare and trace individual strata using their unique fossils to help identify them, in other words, to correlate rocks with similar fossils. Paleontologic *correlation* involves the most important new principles developed after those contributed by

FIGURE 2.8

Index map of Europe showing areas of the Smith and Cuvier-Brongniart geologic maps, as well as some localities referred to in Chaps. 2 and 3.

Steno. All these together form the stratigraphic basis for analysis of earth history.

From the combined work of Cuvier, Brongniart, and Smith, two important and closely related principles have come to us. The first of these is generally known as *fossil succession*, which forms a sort of corollary to the important principle of superposition given us by Steno, and discussed above. The principle of fossil succession states that, in a succession of strata containing fossils, obviously the lowest fossils are oldest. This is simply a direct extension of Steno's principle. The other important tenet coming from their work may be called the *principle of fossil correlation*, namely that like assemblages of fossils are of like age, and

FIGURE 2.9
The use of fossils for correlation across the English Channel of strata studied by Smith in England and Cuvier and Brongniart in France. Fossils x, y, and z are useful index fossils, but clam c is not, because it ranges through the entire sequence.

therefore strata containing them are of like age (Fig. 2.9). Application of the principle involves what we now call *index fossils*. An index fossil is one that is particularly useful for correlation of strata; that is, it is an index to a particular stratum or group of strata.

What constitutes an index fossil, and do not all fossils qualify as such? For a fossil to be considered a stratigraphic index, it must be: (1) easily recognized (i.e., unique); (2) widespread in occurrence; and (3) restricted to a very limited thickness of strata (i.e., the organism represented was short-lived). To have utility, all these conditions must be met, but not all fossils meet these criteria. Many are difficult to distinguish from others, some are not wide-spread, and many were too long-lived; they occur through great thicknesses of strata without change (for example, fossil *c*, Fig. 2.9). Note that, as Hooke guessed 100 years before Smith and Cuvier, an index fossil has special qualities in common with archaeological artifacts like old Roman coins, which can be used to date and correlate human historical events.

What makes correlation with index fossils actually work? Neither Smith nor Cuvier formally set down all the logic in Hooke's suggestion, which they rediscovered and proved, but today we can do this in retrospect. The proven utility of an index fossil requires these conditions: first, that once a species became extinct it never reappeared; and second, that no two species are identical. The method, as Cuvier observed, "up to the present has never deceived us." And this is still true after nearly 200 years of additional experience!

It was Leonardo and Steno and a few other early workers whom we have mentioned who first pointed the way for the science by insisting upon consulting the earth itself for the answers to its riddles, unencumbered by dogmas, superstitions, and ill-founded assumptions. They were among the first to oppose blind acceptance of preachings that were in no way verifiable, that is, whose validity could not be tested by actual observation of nature. Such intellectual crusading against orthodox dogmatism was a risky business, for it landed several men in prison and cost at least one his life.

EXPLANATIONS OF CHANGE AMONG FOSSILS

Smith felt little compulsion to explain why unique fossils occurred in certain strata. He simply accepted the happy fact and used it. Cuvier, on the other hand, was very much concerned with the problem of fossil succession, and he developed an enormously influential theory to explain it. His great stature as a biologist and the relative harmony of his theory with theology earned wide acclaim.

Cuvier already had proved by comparative anatomy that some fossil and living elephants, though similar, belonged to different species. Then he discovered other fossil bones that belonged to creatures completely different from any known living forms. He had thus proved that some past inhabitants of earth had vanished completely; he had discovered the first clear examples of extinction of organisms.

From their studies of strata of the Paris Basin, Cuvier and Brongniart correctly concluded that there were a number of major unconformities or interruptions in the sequence. As one can see from the column of strata in Fig. 2.10, there are alternations of sandstones, shales, and limestones. In addition, there are several zones of

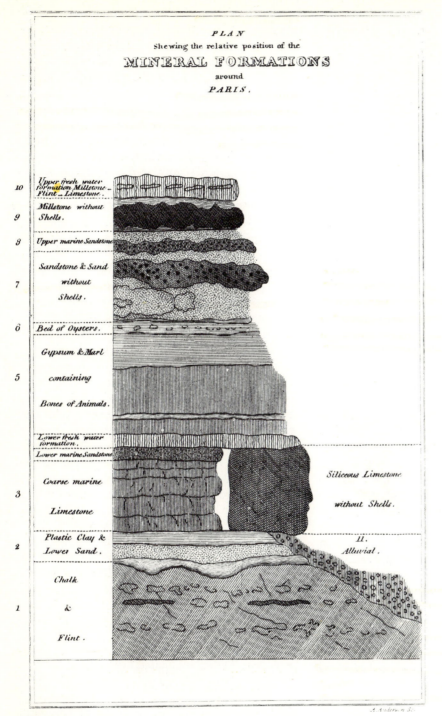

FIGURE 2.10
Stratigraphic columnar section of strata studied by Cuvier and Brongniart around Paris. It is interesting to note that in the translation of this work from the French, some English editor added what is obviously Napoleon Bonaparte's profile to the cliff for unit 3. (*From G. Cuvier,* Essay on the theory of the earth; *English translation, 1818.*)

ancient soil with fragments of plants in them as well as several breaks in deposition marked by concentrations of wood, shell fragments, and small pebbles. Obviously the record of sedimentation is not a smooth, continuous one, but tells a story of interruptions of deposition accompanied by abrupt changes of fossil assemblages. Cuvier's theory of fossil succession assumed that portions of assemblages represented in successive groups of strata had become extinct before succeeding strata were deposited. He interpreted the evidence to indicate many wholesale catastrophic extinctions of assemblages of millions of organisms caused by violent oscillations of the sea. Land organisms were thought to show effects of the catastrophes even more dramatically than marine ones. Cuvier noted that fossils in successively younger strata were more like modern organisms, successive extinctions having eliminated many now-unknown species so that, with each catastrophe, life moved toward the ultimate "perfection" of the modern scene.

Although today Cuvier's hypothesis is regarded as inadequate, nonetheless it is important to realize that it was a reasonable conclusion based both upon local evidence and its seeming accord with popular theological assumptions. Recall from Chap. 1 that ancient peoples understandably had regarded most major changes in the earth as produced catastrophically by unknown agents. This prejudice was woven early into the fabric of Judeo-Christian traditions, had become dogma, and was thereby perpetuated. Against this background, and on the authority of scholars such as Cuvier, a distinct philosophy called *catastrophism* matured in the early nineteenth century. About 1850, Alcide D'Orbigny modified the hypothesis by postulating the creation of new species following each of 27 catastrophes. The complete theory became known as *catastrophism and special creations*. It said, in effect, that there was no connection in life between the index fossils of successive strata.

The alternative to Cuvier's concept of catastrophism, which was a theory of extinctions to explain fossil succession, was that fossils in younger strata are descendants from older groups of organisms by change from one species to another. Even Cuvier noted a general tendency for younger fossils to be progressively more like living organisms, but insisted it was due solely to selective extinctions. Other geologists also noted such change; for example, another Frenchman (Giraud-Soulavie) as early as 1780 subdivided strata in five "ages" based upon the gross characteristics of their fossil assemblages. Rocks of the "first age" included no forms analogous to living ones, the "second" included both extinct and living marine forms, the "third epoch" had shells of modern types only, the "fourth" was plant-bearing, and the "fifth" was distinguished by mammal remains. This was one of the first attempts to subdivide the stratigraphic rock record into broad divisions based upon changes in the development of life. It is a precursor to a geologic time scale, for it implies an irreversible or unidirectional trend of development—a systematic change through time—*evolution*!

Organic evolution, the alternative to Cuvier's catastrophism, was not new in Cuvier's time. The general concept had been suggested at least 100 years earlier, and two of Cuvier's own colleagues were ardent evolutionists who continually feuded with him. Apparently no urgent "need" was felt for such a theory, however, until more factual evidence had accumulated and the intellectual climate began to change near the end of the 1700s.[1] This important idea

---

[1] Feuding about evolution seems perpetual. Mark Twain's acid pen had this to say about a century ago: "It was foreseen that man would have to have the oyster. . . . Very well, you cannot make an oyster out of whole cloth, you must make the oyster's ancestor first. This is not done in a day. You must make a vast variety of invertebrates, to start with—belemnites, trilobites, Jebusites, Amalekites, and that sort of fry, and put them to soak in a primary sea, and wait and see what will happen. Some will be a disappointment—the belemnites, the Ammonites, and such; they will be failures, they will die out and become extinct, in the course of the nineteen million years covered by the experiment, but all is not lost, for the Amalekites will fetch the homestake; they will develop gradually into encrinites, and stalictites, and blatherskites, and one thing and another as the mighty ages creep on . . . the Archaean and the Cambrian Periods . . . and at last the first grand stage in the preparation of the world for man stands completed, the oyster is done. An oyster has hardly any more reasoning power than a scientist has; and so it is reasonably certain that this one jumped to the conclusion that the nineteen million years was a preparation for *him*; but that would be just like an oyster." (From the essay *Was the World Made for Man?*)

is discussed extensively in Chap. 10 after other essential background has been developed.

THE FOSSIL RECORD

From Steno's time onward, geologists quickly recognized that fossils were the remains of once living organisms by making comparisons (where possible) with living organisms. A fossil clam shell is not that much different from a living clam in general appearance. By the nineteenth century, cataloging and describing of fossils had reached the point where it was possible to correlate strata with some precision, to begin to construct a picture of the complex changes in the history of life, to reconstruct past climates, and to document evolution.

The most common and important fossils are the hard parts of organisms—shells, bones, teeth—for the simple reason that soft tissues usually decay and disappear rapidly following death. Hard parts may remain on the surface for some time, but the faster they become buried by sediment, the better preserved they are. For example, shells lying on the beach or in shallow water are moved back and forth by waves and tides; soon they become broken up and eventually are ground up into small particles. Indeed, many lime sands and muds in the tropics are almost completely made up of such fragments. In time, as more sediments pile up, pressure and solution cause the grains to be cemented together, forming limestone (Fig. 2.11).

Not all organisms have had hard parts. These are preserved only due to unusual burials, and are important because they give us clues otherwise not available. Frequently, in volcanic ashes or in black muds, where oxygen is absent, leaves, jellyfish, insects, and many other organisms may leave impressions, or form carbonized films after the organic molecules are reduced to the nonvolatile element carbon. Spores and pollen of plants are common in many rocks and have become very useful tools in assaying past climates and ecologic zones, as well as being used even for intercontinental correlation.

Trace fossils (indirect evidences of life) include a large variety of objects or marks made by

FIGURE 2.11

Microscopic photograph of a limestone composed of abraded and sorted fossil skeletal debris (grains average about 0.5 millimeter). Such clastic limestones are common (Lower Pennsylvanian, northern Nevada; *by* R. H. Dott, Jr.).

organisms. Tracks and trails left by animals tell us about the nature of their locomotion and habits. Burrows, coprolites (fossilized fecal matter), and various surface or subsurface marks aid in estimating depth of water, softness of sediments before lithification, water circulation, and clues to the types of organisms living in the water and the food they ate (see Figs. 2.12 and 16.1).

The fossil record has provided us with a vast, but incomplete, library of information to help interpret earth history. This record is being continually improved by new discoveries, but there are some built-in biases that probably can never be overcome. For example, it has been estimated that of 3,000 plant and animal species living in and around modern coral reefs, fewer than 75 species are recognizable after death, and most of these have mineralized hard parts or skeletons.

Another major type of bias is the availability of fossiliferous rocks and their chance location. Hence, much more has been known about the fossil record in the Northern Hemisphere because of larger populations and more geologists. Additionally, over 75 percent of the record on continents represents land or shallow-sea environments, with only a few rare deeper water deposits. During the past several decades, deep-sea sediments have been extensively

FIGURE 2.12
The trace fossil *Zoophycus* on a stratification plane in Pennsylvanian sandstone, Texas. It was formed by a burrowing animal that systematically mined its way through the sediment in concentric arcs to extract organic nutrients. *(Courtesy J. Bourgeois.)*

cored, and vital data have been obtained to reconstruct the history of the ocean basins. However, most of the fossils found in the cores are single-celled forms, thus presenting a bias. Further, the oldest deep-water fossils found date back only 130 to 150 million years; thus there is an all-too-short record of deep-water events.

Population density relative to size of a given organism is yet another form of bias. Microscopic organisms in the oceans are present in astronomical numbers, partly because of reproductive frequency and larger numbers of progeny compared to dinosaurs or elephants, which have very small populations and produce fewer offspring. Hence, the chance of finding a dinosaur is far less than that of finding a plant spore. A very common bias relates to the random availability of rock outcrops, which may be ephemeral. Some outcrops yield vast numbers of fossils and are long lasting, but many disappear due to formation of soil, grassing over, or abandoning of quarries, which fill with water.

FIRST UNIFIED HYPOTHESES OF THE EARTH

THE COSMOGONISTS

The history of knowledge has been characterized by periodic formulation of hypotheses that explained most of the facts available at a given time. Science is a process of continuous refinement and testing of such explanations. Hypotheses, such as the different explanations of index fossils, inevitably have been colored by the temperaments, experiences, and prejudices of their advocates, which makes them all the more interesting to study.

Several Renaissance thinkers attempted to formulate general, all-inclusive hypotheses (called cosmogonies) to explain the origin and development of earth, life, and the entire universe. Though riddled with speculation, these early explanations had a long-lasting influence. It was assumed that the earth originally was hot and glowed like the sun. Subsequently, as it cooled, a primitive hard crust formed, and the water and atmosphere became segregated according to relative densities. The interior had a fiery core surrounded concentrically by other materials.

BUFFON'S BREAK WITH GENESIS

One of the most fertile eighteenth-century thinkers in natural science was the Frenchman G. L. de Buffon, who published beginning in 1749 a 34-volume work entitled *Histoire Naturelle*. Accepting the idea of a molten origin, Buffon suggested that planets originated through detachment of hot portions of the sun by collision or near-collision with a large comet, an idea doubtless derived from great interest in comets stimulated by studies of their periods beginning 50 years earlier (see Fig. 5.1). Subsequent earth history consisted of cooling through six distinct epochs totaling 75,000 years. He was one of the first to question the literal significance of the Six Days of Creation and to postulate such a great age for the earth. "A year is to God as a thousand years to man." Buffon represented a major turning point in the history of geologic thought by his rejection of Scripture as a source of geologic insight and his extension of the supposed age of the earth. Even more importantly, he rejected catastrophes, believing instead that ordinary processes now operating could explain geologic phenomena.

FIRST TRUE GEOLOGIC CHRONOLOGY

Ideas commonly have arisen independently in different minds thousands of miles apart. Such was the case with the first concepts of natural subdivisions within the rocks of the earth's crust. Whereas Buffon's Epochs were largely speculative, several contemporaries working in Germany, Scandinavia, Italy, Switzerland, and England were recognizing and naming local natural rock divisions based upon evidence observed in the field. In the English coal fields as early as 1719, a succession of strata was recognized and illustrated in detailed cross sections. Two Germans in the middle eighteenth century also discerned an order in the vertical succession of rocks. J. Lehmann distinguished stratified, gently tilted fossiliferous deposits (or *Flötzgebirge*) from more primitive, unfossiliferous, steeply tilted ones cut by dikes and ore veins (material congealed from fluids that were injected along fractures). He believed that all had formed in the sea, but that *primitive* ones

were deposited in disorder during the early Creation while the *Flötzgebirge* formed during the Deluge. A third group of deposits postdated the Deluge and were formed by accidents of nature such as earthquakes, landslides, and volcanic eruptions. He assumed that these divisions matched the epochs of Genesis. In preparing early geologic maps and cross sections, G. Füchsel observed that certain strata are characterized everywhere by land plants and coal, while others contain only marine fossils. He originated the modern concept that the strata of a given *formation* reflect similar formative processes and environments. Both men assumed that lithologic differences among formations reflected worldwide fluctuations of the sea.

In 1759 an Italian authority on mining and mineralogy, Giovanni Arduino, formally distinguished in northern Italy: *Primitive Mountains* (unfossiliferous schists—altered sedimentary rocks—and veins of the high Alpine mountain core), *Secondary Mountains* (with limestone and shale containing abundant fossils), *Tertiary Mountains* (richly fossiliferous clay, sand, and limestone of the low hills), and finally the youngest *Volcanic Rocks*. This scheme became the first formal stratigraphic standard for subdivision of the rocks of the earth's upper crust. Being the simplest hypothesis, a universality of historical development of the entire earth's surface long had been assumed, and it already was recognized that fossiliferous marine strata similar to those of Italy were widespread over much of Europe. Therefore, it was natural for Arduino's chronology to be applied throughout Europe and America. It was destined to have a profound influence upon the progress of geology.

NEPTUNISM

A. G. WERNER

In 1787, one of the most influential men in all of the history of geology, Abraham Gottlob Werner, published a general theory of the origin and rock sequence of the crust. Most predecessors had postulated a former universal ocean, and believed that most rocks were products

TABLE 2.1

Standard geologic column as conceived by 1800

Alluvial rocks (also Tertiary of Arduino)	Relatively loose gravel, sand, peat, and some limestones, sandstones, and shale
Secondary (*Flötz*) rocks	Sandstone, limestone, gypsum, salt, coal, basalt, obsidian
Transition rocks	Hard graywacke, slate, and some limestone with first organic fossils; chiefly chemical, but with first mechanical deposits
Primitive (or Primary) rocks	Chemically formed original surface of earth (granite, schist, gneiss, serpentine, etc., with no organic fossil remains)

thereof. Since Viking times, it was known that the shoreline of the northern Baltic Sea was retreating, for new rocks kept appearing; it was natural to suppose this reflected a worldwide change. Mountains were assumed by many to be original irregularities of a chaotic early Creation, and the succession of strata was presumed to reflect subsequent gradual diminution of the sea and emergence of land. Because of the great emphasis upon the sea for explaining all of the crust, this "theory" earned the apt nickname *neptunism* (for Neptune, Roman god of the sea) (Fig. 2.13).

Werner adopted the early neptunian concept of the earth, and amplified it through a more detailed accounting of the chronological succession of strata. But, unlike some early neptunians, Werner was not a catastrophist. He regarded the earth as much older than humanity, and did not bother correlating stratigraphic history with Scripture. Primitive rocks, such as granite and schist, were regarded as unfossiliferous and entirely of chemical origin (Table 2.1). Because they appear in the high axes of mountain ranges, they must be the earliest rocks precipitated from the sea. As the water subsided, fossiliferous Transition rocks were deposited. These included the first mechanically formed (i.e., fragmental) impure sandstones, called graywackes, as well as some chemical limestone. The inclination of these strata (Fig. 2.13) was assumed to be due to initial deposition on the irregular Primitive crustal surface or to collapse into subterranean caverns. Lower topographically were *Flötz* (flat) rocks, later referred to as the Secondary. These comprised sandstones, limestones, gypsum, rock salt, coal, and basalt. Finally, in the lowlands and valley bottoms, the youngest or Alluvial deposits were formed as the sea receded to its present position; they are nonmarine. (Arduino's Tertiary and Volcanic were not recognized by Werner as separate divisions.) Werner's grouping of rocks by supposed relative age soon became almost universally accepted as a standard for worldwide comparison. He had succeeded in mobilizing, at last, a conscious scientific desire to study the history of the earth.

FIGURE 2.13

Early neptunian concept of deposition of strata by a receding, universal ocean. Note that as drawn here from early descriptions of the neptunian theory, positions of the strata seem to defy superposition; that is, successively younger ones do not everywhere directly overlie older. Later neptunians, such as A. G. Werner, envisioned the older strata as dipping and flattening beneath younger ones away from the mountains.

FIGURE 2.14

Auvergne volcanoes, south-central France. Note the perfect volcanic cones and hummocky-surfaced lava flows extending out from two of the craters near the center. (*After G. P. Scrope, 1825,* Considerations on volcanoes.)

THE HEATED BASALT CONTROVERSY

The neptunian scheme suffered from several obvious shortcomings, such as a gigantic water-disposal problem. Wernerians ignored many of these, even while proudly announcing that they rejected hypothesis and speculation and built their entire case upon irrefutable facts, a pious claim pressed with equal vigor by the opposition.

Most difficult and controversial was the origin of *basalt*, a dark, fine-grained rock considered today to have crystallized from hot, molten lava. To Werner, volcanoes all were recent and had no great importance in the history of the earth in spite of dramatic evidence marshalled by the previous generation of antineptunian Italian geologists, who thought that all land was formed initially by volcanic eruptions such as they themselves had seen build new islands in the Mediterranean. Werner, instead, endorsed an old idea that volcanic eruptions, when they do occur, originate from combustion of buried coal seams; therefore volcanoes could have occurred only after deposition of the great Secondary coal layers. He contended stubbornly to his last day that basalt interstratified with sediments had been precipitated from the universal ocean; where it looked like lava, it had been fused by combustion of adjacent coals.

It now seems astounding that most geologists could seriously entertain the neptunian origin of basalt, but it is even more surprising to realize that two Frenchmen had published volcanic interpretations of famous basalts in central France (see Fig. 2.8) before Werner first published his theory (Fig. 2.14)! Together with the Italians, those men founded a different school of geologic thought that came to be called vulcanism (for Vulcan, god of fire), to which three of Werner's own students soon were converted. Subsequently, volcanic rocks were found interstratified with Secondary and Primitive deposits, proving that volcanism had occurred throughout all of earth history.

PLUTONISM—THE BEGINNING OF MODERN GEOLOGY

JAMES HUTTON

In the last quarter of the eighteenth century, geology finally was coming of age; the name itself came into general use then. It is a curious coincidence that the two most influential early geologists—and of opposite persuasions—lived and wrote at the same time. The great Scottish innovator, James Hutton, was a gentleman farmer and geologist. Farming generated an interest in the earth. Hutton was obsessed with discovering a mechanism for maintenance of the environment for land plants and animals in the face of obvious destructive forces acting to wear away the landscape—a dilemma noted before. First announcement of his revolutionary theory was in 1785 when Werner's influence was at its peak. Eventually Hutton's theory was to provide the foundation of modern geology.

AN OLD DYNAMIC EARTH

Hutton early recognized that rocks exposed today to the atmosphere tend to decay and produce gravel and soil. He noted by analogy that many rocks contain debris derived from older rocks, which apparently had decayed similarly in the past. Moreover, he appreciated that counterparts of ancient sedimentary rocks are still forming, chiefly in the sea. These observations led him to a cyclic view of earth change with construction of new products neatly balancing destruction of old. Such tendency toward *equilibrium* among dynamic natural processes today is described as a physical system in a *steady state*. Energy is expended, but the system appears generally the same at any time; a beach is a modern example, for it looks essentially the same day-to-day in spite of constant pounding by the surf. A prevailing view in Hutton's day—at the beginning of the industrial revolution—was that an external intelligence had set all of nature in motion like some gigantic machine. The earth was like a great heat "engine," with volcanoes acting as safety valves, but it was also a cyclical "machine" such as a clock whose pendulum swings to and fro.

Hutton's greatest contribution was the original full appreciation that the earth is internally dynamic and ever-changing in sharp opposition to the Wernerian view that rocks had formed on a rather static solid earth foundation. This was really the central conflict with neptunism! To Hutton, it was absurd that steeply inclined Transition and Secondary strata in mountains were originally deposited in such positions. He reasoned instead that these strata had been tilted and crumpled later by internal earth forces.

Hutton also attacked catastrophism. He recognized that valleys were cut by rivers rather than the sloshings of great oceanic floods; he objected to the belief that "everything must be deluged," and found, instead, modern earth processes quite capable, given enough time, of having produced the record of the past.

Hutton saw that, if we assume that the earth is only 6,000 years old, we *must* subscribe to catastrophic earth change to explain geologic facts. The fresh condition of a 2,000-year-old Roman wall in northern England, however, convinced him that ordinary geologic processes—by human standards—act slowly. Because rocks seemed to reveal only the results of ordinary processes like those visible and acting at the present, a great deal more time must have been required for the almost imperceptible changes to have accomplished great geologic work. "What more can we require? Nothing but time," said Hutton.

"NOTHING IN THE STRICT SENSE PRIMITIVE"

Hutton's first quarrel with neptunians was over the latters' assumed uniqueness of the different chronologic rock divisions, especially the Primitive. Such was counter to Hutton's own observations as well as to his concept of an earth in dynamic equilibrium and acted upon by the same (i.e., uniform) processes through all time. Presence of pebbles and sand within Scottish schists (altered or metamorphosed sediments of diverse kinds) otherwise regarded as Primitive attested to the existence of still older rocks somewhere from which those fragments had been eroded. So-called Primitive rocks were not entirely of chemical origin, as the neptunians claimed; neither were they unique in lacking fossils. Seemingly there was "no vestige of a beginning, no prospect of an end."

BASALT

Hutton was impressed early with the importance of subterranean heat, whose existence was inferred from hot springs and volcanoes. Also he interpreted clinkerlike coal seams transected by basaltic dikes as having been baked. He studied a well-exposed basalt sill interlayered with sediments, and found clear evidence both of its hot origin and forcible intrusion after the sediments had formed.

Hutton's igneous interpretations of basalt and natural rock glass were reinforced experimentally in 1792 by a friend (Sir James Hall), who

Ideal Transverse Section of a Mountain range.

FIGURE 2.15

The plutonist concept initiated by Hutton of upheaval of mountains, folding of strata, and intrusion of granites, all as results of internal heat. Note the discordance between the upper (Secondary) strata and the more contorted schists. (*After G. P. Scrope, 1825,* Considerations on volcanoes.)

melted basalt between 800 and 1200°C, only to produce glass with rapid cooling, but basalt again with slow cooling.

GRANITE AND MINERAL VEINS

Hutton knew that mineral veins occur in areas of Primitive granites and schists (*granite* is a coarsely crystalline, light-colored igneous rock that crystallized slowly below the earth's surface). In 1764 he observed a peculiarly textured granite in northern Scotland and thought that it, as well as the complex minerals in metallic veins, could have formed only through crystallization from hot fluid material. Sea water could not hold all the different compounds simultaneously in solution, therefore veins and dikes were not simple oceanic precipitates in open cracks as the neptunians preached. Influenced by the work of a prominent chemist friend, Hutton shrewdly inferred that great pressure deep in the earth would affect chemical reactions markedly; compounds that are gaseous at high temperature and low pressure (such as those expelled to the atmosphere from volcanoes) would not be lost as gases from hot solutions under high pressure at depth. Moreover, there could be intense heat without fire at such depths. Hutton's interpretation of a hot origin for granite eventually displaced the neptunian view. Great appeal for subterranean heat to produce basalt, granite, and mineral veins

earned for his theory the nickname *plutonism* (for Pluto, god of the lower, infernal world).

EVIDENCE OF UPHEAVAL OF MOUNTAINS

Hutton proposed that heat also consolidated sediments, and then caused upheaval of rocks to form mountains by thermal expansion of the crust (Fig. 2.15). Granites, veins, and basalts were assumed to have formed at such times. He deduced that dikes of granite cutting across or intruding older rocks must exist. Having predicted their ultimate discovery, he scoured Scotland for examples and found many such dikes transecting schists in central and western Scotland. His elation could not have been greater had they been veins of solid gold.

Hutton also discovered what he had predicted and long had sought, namely a great unconformity in the stratified rock record (Fig. 2.16). He suspected that the contact between Secondary and older rocks might reveal clear evidence of great upheaval of mountains from the ancient sea floor. In all three exposures that he discovered, pebbles worn from older rocks occurred just above the unconformity. In Hutton's examples, the dip of strata below the unconformity was steep, and from Steno's principle of original horizontality, one concludes that severe upheaval had tilted the rocks and formed mountains, which were then worn away; their remnants were later covered by younger, nearly flat strata.

The unconformity, together with intrusive granite dikes, proved conclusively that mountains are neither static bumps left from a primeval earth surface nor relicts of the Deluge, but were formed by repeated dynamic convulsions of the crust (Fig. 2.17).

FIGURE 2.16
Unconformity with basal "puddingstone" or conglomerate along the River Jed, south of Edinburgh, Scotland, the second such example discovered by James Hutton in 1787. (*From* Theory of the earth, *1795.*)

FIGURE 2.17
Diagrammatic representation of Hutton's concept of cyclic deposition, upheavals, intrusion of granites, and development of unconformities to form a continuous progression of new landscapes born from the wastes of old.

CHARLES LYELL'S UNIFORMITY OF NATURE

GEOLOGY'S GREATEST BOOK

Before Hutton's unifying theory and the nearly simultaneous development of principles of correlation and geologic mapping by Smith and Cuvier, modern geology as a unified science really did not exist. These developments finally provided the link among very diverse studies, and provided both method and purpose. Evidence of subsequent rapid growth of geology is provided by the founding of the first geological society in London (1807) and the first governmental geological survey in Great Britain (1835).

In spite of these breakthroughs around the beginning of the nineteenth century, geologic efforts were still somewhat diffused. Many people active in earth studies either were not immediately aware of the significance of Hutton, Smith, and Cuvier, or were unwilling to abandon neptunism. For example, a Cambridge University professor told his class, "There is enough water in the crystals in my museum to extinguish all the fires of the plutonists." Such water, which formed vapor bubbles as the minerals crystallized, was proof enough for the good professor that basalt had precipitated from sea water. Moreover, Hutton was a poor writer and received bad reviews because of his anticatastrophist preachings (Fig. 2.18). Particularly, his contention of great antiquity of the earth expressed in the subtitle to this chapter was like a red flag to a bull. In the eyes of many, no science was valid if it did not support and clarify Scriptures, which Hutton had refrained from doing, even though he was firmly convinced of some grand design or wisdom in nature.

A persuasive synthesis of mushrooming factual data with a noncatastrophic, Huttonian interpretation was supplied by British geologist, Charles Lyell, in a great work entitled *Principles of Geology*, which was revised 11 times between 1830 and 1872. Lyell, who was born the same year that Hutton died, produced one of those rare books of almost unprecedented impact. He painstakingly illustrated the concept of uniformity of nature through time. He traveled extensively in Europe and North America and

FIGURE 2.18
Caricature drawn in 1787 of ". . . the eminent old geologist, Dr. James Hutton, rather astonished at the shapes which his favorite rocks have suddenly taken"—the faces of several antagonists. *(From John Kay, 1842, Edinburgh portraits.)*

was able to show overwhelmingly that geologic processes observed today can be assumed to have operated in the past. As an antidote for Cuvier's catastrophism, Lyell adopted a steady-state view of the entire earth more extreme than that of Hutton. He believed staunchly that the general intensity of processes and conditions had varied hardly at all through time. Of course some violent events like earthquakes occur, *but no more frequently in one epoch than in another*. If one continent were large, another simultaneously was small; if the climate here were cold, it was balanced there by hot conditions. By such rationalizing, he envisioned that conditions remained essentially constant through time for the earth as a whole. To Lyell, his mission was "freeing the science from Moses." To acknowledge any irregular variations of intensity was to

FIGURE 2.19
One of many noncatastrophic, historic geological changes documented by Charles Lyell. The ancient village of Eccles in Norfolk, England, was beginning to be attacked by the North Sea around 1600. By 1839 coastal dunes had buried all but the top of the tower (left), but by 1862 wave and wind attack had exposed the ruins again (right). (*After C. Lyell,* Principles of geology; *10th ed. 1867, v. 1, pp. 514–515.*)

leave the door dangerously ajar for catastrophism. Therefore, only uniformly repeating (i.e., steady-state) changes were permissible.

The uniformity doctrine actually was older than Lyell or even Hutton, for it is implicit in earlier writings of Leonardo da Vinci, Buffon, and a Russian named Lomonosov. But Lyell most successfully interpreted and publicized it for society at large. His importance is underscored by the fact that early American geologists were ardent neptunian catastrophists ignorant of Hutton's theory until Lyell began publicizing the new geology, as well as by his impact upon theology and Victorian writers. Moreover, *Principles of Geology* had a great influence upon Lyell's friend, Charles Darwin. Lyell, himself, however, had great difficulty in abandoning a fixed species concept, for to accept change of one species into another in an irreversible line of descent seemed to him contradictory to his unbending devotion to a uniform, cyclical earth. Evolution and a strict steady state are incompatible.

> I could get along very well if it were not for those geologists. I hear the clink of their hammers at the end of every bible verse. (John Ruskin 1851)

THE DOCTRINE OF UNIFORMITY TODAY

The uniformity of nature, or uniformitarianism, must be examined carefully because it is so basic to studying earth history and so much controversy has surrounded it. First, let us stress that uniformity is an *assumption* about nature—a doctrine rather than a logically proven law. Hutton himself said that "the uniformity of nature, even if not strictly true, is necessary for our clear conception of the system of nature" (1788).

Much confusion exists about the uniformity doctrine, even among scientists, but most especially among neocatastrophists (e.g., Immanuel Velikovsky, author of *Worlds in Collision*) and antievolutionists, who now seem to be increasingly vocal. The uninitiated still today interpret uniformity as implying that the earth always has been *exactly the same*—a Lyellian influence. An old cliché that the *present is the key to the past* is a misunderstanding by many, too, for the present earth is unique in terms of climate, topography, and life. In fact, the *past can be a key to the present*, too, for the historical record provides a perspective against which to compare the present. It is just common sense to compare both ways, for the present provides such a short observational time span compared to geologic time.

How, then, is uniformity to be regarded today? Only a static earth could be completely unchanging, yet ours clearly is dynamic. Lyell allowed some change, for that is what the *Principles* was all about (Fig. 2.19). His changes, however, were orderly, cyclic ones confined within narrow limits. But a Lyellian steady-state dynamic earth would defy basic laws of physics. This contradiction was recognized by the nineteenth-century British physicist Lord Kelvin, who challenged strict uniformitarianism as perpetual motion—an earth machine that never ran down was a physical absurdity! Kelvin reasoned that the energy reservoir of the entire solar system must have been greater in the past and was gradually being dissipated. His position implied significant differences in past conditions, thus a noncyclic, evolutionary view of the earth.

After Lyell and Kelvin squabbled, geologists became increasingly aware that the intensity of geologic processes had, indeed, varied greatly through time.[2] Moreover, in support of Kelvin, it was clear that conditions on earth had changed irreversibly so that in certain respects different epochs of geologic time were unique—history was not strictly uniform as Lyell had contended. Examples of major noncyclic events include the change from an early oxygen-free to an oxygen-rich atmosphere; the origin of life under conditions that no longer exist; evidence of meteorite impacts on all planetary bodies in the distant past far more intense than today; and, most familiar of all, the complete extinction of large groups of organisms such as the great dinosaurs.

Lyell's strict uniformitarianism was reasonable because it was the simplest assumption to make about past change, but we should not forget that his extreme position must have seemed tactically necessary to counter catastrophism. Later textbooks have tended to distort

[2]Consider the following anecdote: "One is told that a certain boy grows one-tenth of an inch in a year, and that he is four feet high. The geologist then argues ten years to grow one inch, 480 years for four feet, therefore the boy is 480 years old. A person who knows more about the boy in question comes forward to inform us that when he was younger, he grew much faster, and that, after all, his age is only ten years. The geologist replies, it may be so. Let me see, one-tenth of an inch in a year, on the doctrine of uniformity, one inch in 10 years. No, you are clearly mistaken; 10 years is far too short a time. The organic processes of growth could not possibly have occurred in this time. I grant you my own calculation is not rigorously exact; let us say then our boy is but 400 years of age." (W. J. Sollas, 1877, On evolution in geology: *The Geological Magazine*, v. 4, pp. 1–7.)

the uniformity controversy by polarizing the issue too much. Even in the nineteenth century, there was a broad range of positions between Lyell and Cuvier with good arguments supporting all of them. We can recognize at least four historic variations of the uniformity controversy: (1) natural versus supernatural causes; (2) presently observable causes versus processes no longer operating (e.g., origin of life); (3) gradual versus sudden (catastrophic) change; and (4) perfectly repeating (steady-state) conditions versus nonrepeating, irreversible evolutionary development.

Because of all the controversy and many different shades of so-called uniformitarianism, we feel that this term should be relegated to the historic artifact of Lyell's extreme position. Today we envision neither a violently catastrophic nor a rigidly uniform earth, but rather an evolutionary one that has changed through an irreversible chain of cumulative historic events and that is still changing today. The only assumption geologists now make is that physical and chemical laws are constant, which is properly called *actualism* (an older French term than uniformitarianism). The following very modern statement of actualism made by Hutton's friend John Playfair in 1802 makes one wonder why the Lyellian controversy ever arose:

Amid all the revolutions of the globe the economy of Nature has been uniform, and her laws are the only thing that have resisted the general movement. The rivers and the rocks, the seas, and the continents have been changed in all their parts; but the laws which describe those changes, and the rules to which they are subject, have remained invariably the same.

By inductive reasoning and analogy, the study of geologic processes acting today provides us with powerful clues to their past action, but we do *not* assume that those processes always acted with the same rates and intensities—contrary to what our critics say we believe! Confusion about what is meant by "catastrophic" processes and a lack of appreciation of the vastness of geologic time feeds the continuing controversy over uniformity in nature. Geologists today routinely accept sudden, violent, and even certain unique events as perfectly consistent with contemporary earth theory. We feel that catastrophic should refer only to postulated supernatural events, and that only by substituting actualism for the ambiguous uniformitarianism can misconceptions be minimized.

THE SCIENTIFIC METHODOLOGY OF GEOLOGY

HISTORICAL SCIENCE

Physics and chemistry are concerned almost exclusively with phenomena controlled by presumably universal natural systems that are nonhistorical, that is, *independent of the time at which they operate*. Geology, astronomy, and biology, on the other hand, are more historical. When a geologist focuses only upon present processes and configurations of earth materials, he or she is an applied physicist or chemist. But when beginning to interpret a series of past events, a geologist becomes a unique historical scientist. While assuming all physical, chemical, and biological theory, reconstruction and explanation of history has become the chief goal. Moreover, prediction of results from known causes (deductive reasoning), so important in nonhistorical science, is replaced in geology by the interpretation of ancient causes from their historical results (inductive reasoning).

SCIENTIFIC EXPLANATIONS

Only the disciplined intellectual assault called science has a good track record of success in understanding nature. Science consists simply of the formulation and testing of hypotheses based upon observational evidence; experiments are important where applicable, but their function is merely to simplify observation by imposing controlled conditions. Scale models of objects or processes are familiar and of obvious value, say, for studying transport of sand by water. Experimentation with scale models has limitations, especially if very long spans of time or an excessive number of factors are involved in the process being modeled. In historical

sciences, we commonly use modern phenomena as models of ancient events, in which cases a natural modern analogue (e.g., an entire river) is a proxy for an unnatural experiment. For large and complex systems we let nature do the "experiment" while we observe and measure.

Contrary to a popular textbook myth, it is rare for a scientist to make many observations without already having a tentative hypothesis in mind to test. Also, many brilliant breakthroughs have resulted from accidental discoveries or intuitive flashes based upon skimpy evidence. For example, Hutton had formulated his theory and deduced from it the existence of intrusive dikes and uncomformities long before he had seen either feature.

The average person may not appreciate the differences between scientific proof and absolute proof in logic, which is the difference between *probability* and *certainty*. Science deals with probabilities, that is, one hypothesis seems for the moment more correct than another. Therefore, scientific explanation is never-ending, for it produces only an approximate working model of nature as we think we understand it at a particular time. In geology our explanations are less certain than in, say, physics because of the enormous complexity of the earth and the very incomplete record we have of its history.

As a series of observations leads to many interrelated hypotheses, it becomes necessary periodically to establish broad simplifications, either unifying theoretical laws or empirical generalizations. Newton's famous laws of motion provide clear examples, and ones which can be expressed in concise, unambiguous mathematical language. Empirical stratigraphic generalizations developed by Smith and Cuvier, on the other hand, cannot be so expressed, but their significance is no less.

In this book we shall regard *speculation* as conjecture—simply opinion or guess—for which there is little or no evidence. A *hypothesis*, however, is a logical, but tentative, explanation of verifiable phenomena based upon evidence. More than one hypothesis may appear equally capable of explaining a given set of facts; then further evidence for testing of the alternate hypotheses is needed to identify the most probable one.

As a rule, the simplest hypothesis that explains the largest number of facts is preferred. But this doctrine of simplicity—known also as the doctrine of least astonishment—must not be overworked, for there is no assurance that the simplest explanations will always be correct; many things in nature are not simple. A *theory* is distinguished here as an elegant and well-verified explanation that supersedes or encompasses a number of separate hypotheses. In science, theories finally may be elevated to the status of *laws* if rigorously tested and found to be invariable; a law, in other words, is a theory about which there is no longer any reasonable doubt. Formally speaking, there are many hypotheses in geology, few theories, and no exclusively geologic laws. The laws of physics, chemistry, and biology, however, provide the ultimate bases for understanding all geologic phenomena.

MULTIPLE WORKING HYPOTHESES

Most geologic phenomena involve many interacting factors, and it is commonly impossible to restrict the number of factors under consideration or to evaluate each and every factor precisely. Moreover, the record of evidence is fragmentary. As a result, it is often impossible (at least temporarily) to discover the decisive evidence to test a tentative hypothesis. Because of this difficulty, an American geologist, T. C. Chamberlin, in 1890 advocated a procedure of inquiry he called the method of multiple working hypotheses. He pointed out that if one constructs only a single explanation, one may become too strongly wooed by it, and unwittingly seek only the facts to support it or bend the hypothesis to fit new facts. Moreover, absence of an alternative explanation is no assurance that the truth has indeed been discovered. Multiple alternative hypotheses divide our affection, suggest new tests, and expose more facets of the problem. Because his method leads to a high degree of objectivity, Chamberlin commended it to all walks of life. As an ideal, the method has much to recommend it, but in practice most scientists—including Chamberlin himself—have tended to push their pet ideas to the exclusion of others. The results have not been so bad at that.

SUMMARY

By now the central position of fossils in the history of geology should be clear, as well as the importance of fossils among the tools used for interpreting earth history. Cuvier even contended that we owe directly to fossils the beginnings of any real theories of the earth, for until they were studied, there was no reason to suspect distinct, successive events in the development of our planet.

Fossils are studied by geologists for three chief reasons. *First,* they are direct records of ancient life and illustrate, at least partially, the development or evolutionary change of that life through time. Put another way, they are results of organic evolution; also related to evolution is the fossil evidence of extinction of ancient life forms.

Second, fossils are important records of past environments; in fact they are among our most important tools for interpreting ancient climates and geographies. Of special importance is the proof they provide that large parts of continents have been submerged, not once but dozens of times. Fossils also provided early evidence suggesting drastic past climatic changes. Hooke believed that fossils proved that England once had a tropical climate, which is now well accepted.

Third, fossils have a great utilitarian value to geologists in the correlation of strata, as was predicted by Hooke but demonstrated by Smith, Cuvier, and Brongniart. Fossil correlation required careful collection and description of index fossils and an appreciation of Steno's principle of superposition.

It is important to remember that the stratigraphic utility of fossils is quite independent of questions about "What is a fossil?" or of the controversy between catastrophism and organic evolution, which explains theoretically *why* index fossils are useful. The fossils could as well be nuts and bolts of distinct shapes, sizes, or colors as far as their utility alone is concerned; anything unique about any particular stratum is potentially useful for correlation! For example, varieties of grapes should work, for French wine districts show a marked relationship to soil and bedrock types; the Champagne region lies on chalk (Fig. 2.6).

In this chapter we have tried to illustrate something of the complex way in which geologic thought developed. One of the most important early ideas about the earth was the deluge concept, which, as we noted in Chap. 1, appears in many ancient traditions. Its ultimate origin is lost in antiquity, but it is perfectly natural that ancient Greeks would take marine fossils high in their mountains to be proof of submergence. Jews, followed by Christians, adopted this idea, renamed the event Noah's Flood, and the single-deluge origin of all fossils became entrenched for centuries.

When the true origin and utility of fossils was proved about 1800, that knowledge was immediately and widely applied, yet the philosophical showdown over catastrophism was delayed for half a century—even long after most thinkers had abandoned the single-deluge tradition. A final break with medieval attitudes was agonizing, as shown by Cuvier's stubborn position as late as 1831:

> The thread of the operations is broken, the march of nature is changed; and not one of her agents now at work would have sufficed to have effected her ancient works. (G. Cuvier, *Discourse on the Revolution of the Globe*)

Early Italian geologists reasoned from their experience that all land came into being through volcanism. Neptunians at the same time argued, in effect, that the earth's crust was structurally static. Mountains either were inherited vestiges of an early molten, chaotic period of the Creation, or were formed by catastrophic waves and currents during the Noachian Deluge. Practically all rocks, including basalt, were seen as deposited from a receding, universal ocean. Gradually, French and Italian vulcanists were persuasive in removing basalt from Werner's list of oceanic precipitates; the mere 6,000-year age of the earth was challenged by Buffon and Hutton, and the Flood became relegated to a brief, young event of little geologic importance. Moreover, scientific explanations due to the passage of comets through the solar system were being offered for the Flood as well as for the origin of the earth itself.

Finally, at the end of the eighteenth century, modern geology began as Hutton exposed the fallacies of the neptunian arguments. As the first plutonist, he fully recognized the dynamic nature of the earth's interior. At the same time, Hutton laid the groundwork for later overthrow of catastrophism by Lyell and Darwin, through his insistence that present observable processes, given enough time, could have produced all the geologic record.

Three chief doctrines of geologic change have prevailed. Catastrophism invoked sudden, violent upheavals and floods produced by unknowable causes not now operating. Actualism assumed a uniformity of causes through time but considered that their rates, intensities, and locations have varied. Strict uniformitarianism, on the other hand, assumed that processes in the past were constant both in kind and intensity (or rate). While the Lyellian doctrine of uniformitarianism in its strict form is no longer acceptable, the assumption of uniformity of natural laws (actualism) is mandatory for rational historical analysis.

Actualism provides a connecting thread between the present and past and allows us to reconstruct events never witnessed by humans. Comparative or analogical reasoning both among ancient situations and between ancient and modern ones must be employed very extensively. This is a two-way process, for in some cases the past is a key to better understanding of the present. History provides a powerful test of geological hypotheses; if historical evidence demands that a certain thing occurred, then it must be possible even if difficult to conceive by existing theory. Failure of current physical or chemical theory to explain some phenomenon is hardly adequate reason to deny its possibility, though this commonly has been done. That which *has* happened *can* happen!

Galileo's telescope had stretched human minds to consider an infinity of space and other possible worlds. In the eighteenth century human minds had discovered an infinity of time and history. The most important aspect of history is that it never exactly repeats itself. Historical events involve so many complex factors that they are at least in small degree unique events, therefore the probability of duplicating exactly a known complex historical series of events at another time becomes very improbable. Individual phenomena, to be sure, have recurred many times, but sequentially related events are not repeatable *in exactly the same way*. It follows that the earth is a dynamic, evolutionary planet, not a cyclical steady-state one as Lyell thought.

The important changes in geologic thought noted here were occurring within the context of a more general change of attitude of Europeans toward nature. The pessimistic view of a wrathful God and a miserable, decaying world doomed to complete destruction was being replaced by more optimistic images of a benign deity and a world and human race both capable of rejuvenation. Ideas of equilibrium and rebirth rather than hopeless decay were inspired by analogies with the life cycle. Hutton now offered the first geologic cycle—a kind of physiology for the earth. It provided a concrete means of continual terrestrial rejuvenation whose result would be an equilibrium environment for life.

Key Concepts

actualism
basalt
catastrophism
correlation
cosmogonist
diluvialist
dynamic earth
ecologic zone
empirical generalizations
equilibrium
evolutionary earth
formation
fossil
fossil succession
granite
historical science
index fossil
neptunism
organic evolution
original horizontality
original lateral continuity
plutonism
population density
Primary, Secondary, Tertiary
scientific laws
steady-state earth
stratification
superposition
trace fossils
unconformity
uniformitarianism

Readings

Adams, F. D., 1938, The birth and development of the geological sciences: New York, Dover. (Paperback, Dover, 1954)

Albritton, C. C., Jr., 1967, Uniformity and simplicity: Geological Society of America Special Paper 89.

Clark, D. L., 1968, Fossils, paleontology, and evolution: Dubuque, W. C. Brown. (Paperback)

Cuvier, G., 1818, Essay on the theory of the earth, with mineralogical notes by Professor Jameson and observations on the geology of North America by Samuel L. Mitchill: New York, Kirk & Mercein.

Davies, G. L., 1969, The earth in decay: New York, American Elsevier.

Geikie, A., 1905, The founders of geology (2d ed.): New York, Macmillan. (Paperback, Dover, 1962)

Gillispie, C. C., 1951, Genesis and geology: Cambridge, Harvard. (Paperback, Harper Torchbooks, 1959)

Greene, J. C., 1959, The death of Adam: Ames, Iowa State Univ. Press, chaps. 1–3. (Paperback, Mentor Books, 1961)

Hooykaas, R., 1963, The principle of uniformity in geology, biology, and theology: Leiden, Brill.

Hutton, J., 1795, Theory of the earth with proof and illustrations: London, Cadell & Davies.

Lyell, C., 1830–1872, Principles of geology (14 eds.): London, J. Murray.

McAlester, A. L., 1968, The history of life: Englewood Cliffs, Prentice-Hall. (Paperback)

Nicholson, M. H., 1963, Mountain gloom and mountain glory: New York, Norton. (Paperback)

Playfair, J., 1802, Illustrations of the Huttonian theory: London, Cadell & Davies.

Porter, R., 1977, The making of geology—earth science in Britain 1660–1850: Cambridge, Cambridge University Press.

Rudwick, M. J. S., 1972, The meaning of fossils: Episodes in the history of palaeontology: London and New York, MacDonald-Elsevier.

Schneer, C. J., ed., 1969, Toward a history of geology: Cambridge, M.I.T.

Simpson, G. G., 1963, Historical science, in The fabric of geology: Geological Society of America.

Warshofsky, F., 1977, Doomsday—The science of catastrophe: New York, Readers Digest Press.

Woodford, A. O., 1971, Catastrophism and evolution: Journal of Geological Education, v. 19, pp. 229–231.

Von Zittel, K. A., 1901, History of geology and palaeontology to the end of the nineteenth century: London, Walter Scott.

Specimen Nº 1. Scratched
by a Glacier Thirty three
Thousand three hundred
& thirty three Years before
the Creation

Scratched by a cart
wheel on Waterloo
Bridge, the
day before
yesterday

Prodigious Glacial Scratches

Scratched by T. Sopwith

The Rectilinear Course of these
Grooves corresponds with the
motions of an IMMENSE
BODY the momentum of which
does not allow it to change its
Course upon Slight Resistances

COSTUME of the GLACIERS

3

THE RELATIVE GEOLOGIC TIME SCALE AND MODERN STRATIGRAPHIC PRINCIPLES

If history always repeats itself, how come there's so much to learn?

(Anonymous Schoolboy)

Historians explain the past and economists predict the future. Thus, only the present is confusing.

(Anonymous)

FIGURE 3.1
Cartoon of W. E. Buckland, noted British catastrophist, who steadfastly defended the role of the biblical Flood in producing many geologic features. Finally, by 1840, he concurred that glaciers had produced most of the features attributed to the Deluge. *(From Gordon, 1894,* Life and correspondence of William Buckland.)

45

3

After publication of Smith's and Cuvier's first geologic maps, stratigraphic studies went forth at a rapid pace. Though the Wernerian chronology (see Table 2.1) continued to be used as a general standard of reference for supposed relative age, local names for strata became more and more numerous. Naming of distinctive local rock bodies was a natural by-product of mining and mapping. The names developed unconsciously as a shorthand, and reflected geographic localities or peculiar rock types (Fig. 3.2). At first such names were used informally and only locally, but as studies were extended, the lateral continuity of certain distinctive strata became apparent, and it was natural that some names were extended more widely. At the same time, fossils were being collected and studied more and more, and strata with similar assemblages were correlated from widely separated areas even where complete physical continuity between them could not be observed because of discontinuous outcrops. Names for strata became extended still farther from the local areas where originally applied, and the geologic map of all western Europe developed.

As the more important named divisions became extended widely in Britain and northwestern Europe, relationships among different groups of strata gradually became clearer through application of the principle of superposition. A group of Secondary strata named Juras for the Jura Mountains of France and Switzerland were found to overlie another group named Trias in central Germany and to underlie a third group in France named Cretaceous (Fig. 2.8). By application of Steno's principle of original lateral continuity, the Cretaceous name was extended from France across to England (see Fig. 2.6). Meanwhile, the coal-bearing strata of Britain were named Carboniferous, and correlation allowed extension of that name to the Continent.

In a rather haphazard, trial-and-error fashion, old rock divisions became subdivided into clearly traceable, named groups of related strata called formations. The names applied to each group, once arranged in their proper vertical succession according to superposition, provided the basis of a more detailed and practical chronology that gradually displaced the older scheme.

There was a period of overlap of out-of-date concepts with the growing new chronology. For example, although the idea of a single major deluge had yielded to growing contrary geologic evidence, many people still clung to a nearly universal but brief, recent Flood, and it became commonplace to classify strata as antediluvial and postdiluvial (before and after the Flood). Such a flood was thought essential especially to account for widespread and large erratic boulders scattered over northern Europe (Fig. 3.3). The boulders and associated clay or drift (called Diluvium) were assumed to have been drifted in by icebergs. Many prominent geologists—the diluvialists—staunchly held this belief, which retained overtones of catastrophism. It was not displaced by the glacial theory until after 1840 (Fig. 3.1), and as late as 1829 the formal name Quaternary was proposed in good Wernerian tradition to include these young deposits. The occurrence of primitive Paleolithic human artifacts with such deposits led Lyell to conclude that humans had existed through most of Quaternary time.

THE MODERN RELATIVE TIME SCALE

At first there was no thought of building a new systematic time scale. But there developed the need to classify and organize material into some manageable, orderly form; otherwise the study

FIGURE 3.2

Exposures of Silurian-aged Ludlow Rock (foreground) named for Ludlow castle and town (background) in southwestern England. The castle, seat of Tudor royalty, was built between A.D. 1200 and 1400 of Old Red Sandstone, which also crops out nearby. (*After W. B. N. Berry, 1968,* Growth of a prehistoric time scale, *by permission of W. H. Freeman Co.; copyright © 1968, modified from R. I. Murchison, 1839,* The Silurian System.)

FIGURE 3.3

Distribution of the drift or Diluvium containing erratic boulders associated with unsorted "boulder clay" over Europe, and long considered to have been drifted in by icebergs during the Noachian Flood. After 1840 it was attributed to glacial transport. (*Adapted from L. J. Wills, 1951,* Palaeogeographical atlas; *by permission of Blackie and Son Ltd.*)

of strata would be chaos. About 1830 a conscious effort was undertaken to name formally the entire European succession. Two British geologists, Adam Sedgwick and Roderick Murchison, decided that the old Transition rocks had been neglected. Therefore, they addressed themselves to the task of studying them in Wales.

In 1835 two divisions were named, the older the Cambrian (for Cambria, ancient Roman name for Wales) and the younger the Silurian (for an ancient Welsh tribe, the Silures). Sedgwick and Murchison next extended their work southward to Devonshire. There they encountered unfamiliar strata with marine fossils judged to be intermediate between Silurian and Carboniferous ones, thus perhaps contemporaneous with the nonmarine Old Red Sandstone that lies in a similar stratigraphic position in Wales. A new, third division was named Devonian, and soon thereafter deposits of nonmarine (Old Red) type were found interstratified with some marine Devonian layers in South Wales, confirming correlation with the Old Red Sandstone (Fig. 3.4).

Sedgwick recognized need for a still more

formal stratigraphic classification with different levels of subdivisions based upon more rational criteria than those of the archaic scheme. He therefore proposed the concept of very large divisions based upon the gross characteristics of fossils that were by now rather well known. Each of the large divisions would include a number of smaller subdivisions. For example he proposed the Paleozoic Era (meaning "early or old life") to comprise the Cambrian, Silurian, and Devonian divisions. The Paleozoic was to include all the divisions from the oldest recognizable fossil animals through all the divisions dominated by invertebrate animals of related types. We now know that abundant and highly organized invertebrate fossil animals first appear in Cambrian strata, though primitive ancestral forms occur in lithologically similar Eocambrian ones. The Prepaleozoic has been given many different names such as Azoic ("lacking life") and simply Precambrian, which is the most commonly used term today. In formal usage we prefer Prepaleozoic, although we also use Precambrian out of long habit.

The largest time scale divisions, based solely upon fossil life, are now called eras of geologic time. The divisions of strata subdivided and named according to specific European strata and their contained fossils are called systems. Today we must distinguish between the tangible physical rock record, the systems, and abstract time. Thus the Cambrian System of rocks was formed during the Cambrian Period of time. In Table 3.1 we show the relative geologic time scale with eras subdivided into periods, and for the younger periods, into still smaller divisions called epochs.

Development of the geologic time scale was anything but an orderly affair. One of geology's many stormy feuds arose over the division and naming of strata. Sedgwick and Murchison, who were fast friends during the 1830s, later fell out bitterly over definition of the boundary between the Cambrian and Silurian Systems. As their separate mappings in Wales drew closer together, it became apparent that each had included one particular group of rocks in "his" system. Sedgwick's topmost Cambrian overlapped Murchison's lowermost Silurian (Fig.

FIGURE 3.4

Relations of the nonmarine Old Red Sandstone facies of Wales to marine Devonian facies of Devonshire as inferred by Sedgwick and Murchison. Intertonguing of marine and nonmarine deposits proved the Devonian age of the Old Red and also that different environments of deposition existed simultaneously.

TABLE 3.1
The modern relative geological time scale compared with the archaic scale (see also end papers)

Archaic scale as applied in Britain	Modern scale		
	Eras	Periods or Systems	Epochs or Series
Quaternary (1829)* — Alluvium / Diluvium	Cenozoic (Recent Life) (1841)	Neogene (1853)	Holocene (or Recent) (1885) (1833) / Pleistocene (1839) / Pliocene (1833) / Miocene (1833)
Tertiary (1759)		Paleogene (1866)	Oligocene (1854) / Eocene (1833) / Paleocene (1874)
	Mesozoic (Middle Life) (1841)	Cretaceous (1822) / Jurassic (1795) / Triassic (1834)	
Secondary (1759)	Paleozoic (Ancient Life) (1838)	Permian (1841) / Carboniferous (1822)	North America: ⎰ Pennsylvanian (1891) ⎱ Mississippian (1870)
(Old Red Sandstone)		Devonian (1837) / Silurian (1835)	
Transition (of Werner) (1786)		Ordovician (1879) / Cambrian (1835) / "Eocambrian" (informal; see Chap. 11)	
Primitive or Primary (1759)	Prepaleozoic or Precambrian (Local subdivisions are used, but their worldwide correlation is difficult; further discussion appears in Chap. 8.)		

*Dates indicate when divisions were named.

3.5). What to do? These otherwise proper English gentlemen ceased speaking to each other and the problem was unresolvable during their lifetimes. Much later a compromise was struck by removing the rocks in question and erecting a new system called Ordovician (Table 3.1). Besides its amusement value, this incident carries an important message about the difficulty of stratigraphic subdivision, especially in the early days.

While feuding erupted in Wales, several epochs of the old Tertiary division, by then incorporated as a system, were defined and named (Table 3.1) on the basis of relative percent-ages of living species of organisms represented among Tertiary fossils in the Mediterranean region. Soon the Mesozoic Era ("middle life") and Cenozoic Era ("new life") were named. The Permian System was named next, and, with the illogical retention of archaic Tertiary and Quaternary as systems of the Cenozoic Era, the time scale was essentially complete. Attempts have been made to replace Tertiary and Quaternary (Table 3.1), but the terms still appear.

Discussion of the time scale's development has been included to underscore that it "grew like Topsy" over a wide region. Nonetheless, it

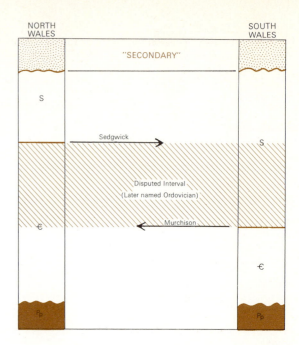

FIGURE 3.5
Geology of Wales (left), where Sedgwick and Murchison worked and feuded over correlation of Cambrian and Silurian strata (right). The Old Red Sandstone marks the base of the archaic Secondary division (see Table 3.1).

has evolved into an organized, workable scheme of classification well known even outside the walls of geology.[1] In the process of mapping and correlation of strata from their "home" or *type* areas, many problems were encountered, some of which are still argued heatedly today. But building of the scale illustrates the success of application of the principles of superposition, original horizontality, original lateral continuity, and of similar fossil assemblages.

Most of the systems were established in areas of dominantly marine, fossiliferous deposits, which provide admirable standards (called *type sections*) for comparison of many equivalent-aged strata over the world. A special problem

was presented by nonmarine deposits such as the Old Red Sandstone. As shown in Fig. 3.4, the tracing of thin, datable marine layers, called *tongues*, into the nonmarine deposits was required for dating the latter.

Final proof of the inadequacy of Werner's theory of the earth also came with the new chronology. Werner's Primitive rocks were shown to include Paleozoic, Mesozoic, and even some Cenozoic ones. His Transition ended with the Silurian in Britain, Carboniferous in Germany, and the Eocene in the Alps. Clearly, lithology, structure, and metamorphism were not, after all, reliable indices of age, though this important axiom was slow to be fully recognized.

LOCAL ROCK STRATIGRAPHIC UNITS

THE FORMATION

The most basic unit of stratigraphy is the *formation*, first defined in Germany about 1770. The original notion of a distinctive series of strata that apparently originated through the

---

[1]In *Life on the Mississippi (1883)*, Mark Twain wrote of an old St. Louis hotel: "The Southern was a good hotel, and we could have had a comfortable time there . . . true, the billiard-tables were of the Old Silurian Period, and the cues and balls of the Post-Pliocene; but there was refreshment in this, not discomfort; for there are rest and healing in the contemplation of antiquities.''

FIGURE 3.6
Designation of formations by rock type is clear in A. It is not so in B because of subtle gradations of color, nor in C because of intimate interstratification of two lithologies. In the latter two cases, some arbitrary division must be chosen. In C, scale is important, for a single thin sandstone layer hardly has utility as a formation. (Dots represent sandstone, dashes are shale, and brick pattern is limestone.)

"same formative processes" is still valid 200 years later. Formations must be distinctive in appearance in order to be easily recognizable. Ideally they are named for type localities where they are typically displayed.

Many practical problems arise in defining formations. Designation of upper and lower limits or contacts is especially difficult as shown in Fig. 3.6. Obviously the definition of a formation is somewhat arbitrary and is influenced by scale factors. A stratum is the thinnest rock layer observable; formations include more than one stratum, and systems contain many formations.

Characteristics chosen to define a formation may include one or more of the following types: (1) composition of mineral grains; (2) color (reflecting some chemical property); (3) textural properties (size of grains, etc.); (4) thickness and geometry of stratification; (5) overall character of any organic remains; and (6) outcrop character. The sum total of such characteristics is conveniently referred to as the *lithology* of a formation (from "lithos," meaning rock).

SEDIMENTARY FACIES

So far nothing has been said about the mutual age or time relations of formations, or of their lateral relationships. For many years after both neptunism and catastrophism had been laid to rest, there still persisted a belief in universal patterns to the history of the entire earth's surface. It was natural to assume that obvious

stratigraphic divisions being named in Europe—and destined to become the standard systems—must be universally present and of the same age worldwide (i.e., synchronous). The finding of similar fossils in most respective systems over much of Europe seemed ample evidence. This concept of universality was also implicit in even the earliest definition of a formation. Thus a kind of layer-cake arrangement was conceived with each formation assumed to extend indefinitely laterally without change. This premise had been, of course, the cornerstone of Werner's time scale, and it was not easily given up because it was the simplest possible concept.

It is not clear when geologists first began to suspect that distinctive, locally named formations could not extend unchanged indefinitely. The earliest clear records occur in French writings, which show a recognition that similarity of fossils in similar sedimentary rocks might reflect environmental factors even more than strict age equivalence.

In 1789 the French chemist A. Lavoisier published several diagrams and text showing the relation of different adjacent sedimentary environments and the effects upon both sediments and organisms expected from changes of relative land and sea levels (Fig. 3.7). Using comparative reasoning, he also illustrated strata around Paris that seemed to reflect such changes in the past. Lavoisier and his teacher (Rouelle) appear to have been the first to appreciate fully the importance of adjacent environments as well as changes through time upon developments of sedimentary deposits and organisms. Shallow, nearshore marine sediments tend to be coarser and contain organisms adapted to rough water, whereas contemporaneous offshore, deeper, and quieter marine sediments are finer and contain delicate bottom-dwelling organisms together with floating and swimming forms. Thus products of each environment have unique characteristics even though they accumulate contemporaneously and grade imperceptibly into one another. Such advanced ideas were not widely known outside France, however.

Recall that Sedgwick and Murchison, after

FIGURE 3.7

A. Lavoisier's diagram of the relationship of coarse littoral (*Bancs Littoraux*) and finer pelagic (*Bancs Pelagiens*) sediments to the northern French coastline. Lavoisier recognized that gravel could be moved only by waves near shore, whereas finer sediments are carried into deeper water. He also recognized that distinctive organisms inhabited each environment. But if sea level rose, flooding the land (*la Mer montante*), both littoral and pelagic sediments would migrate landward. Conversely, if sea level fell (*la Mer descendante*), they would shift seaward. (*From A. Lavoisier, 1789, Memoires d'Academie Royal Sciences.*)

different environments. Yet, full impact of this conclusion was slow in developing. It did not happen until 1838 when a perfectly exposed example of lateral changes in both lithology and fossils within a single stratum was described in Switzerland.

To fully characterize strata, it now became necessary to describe not only vertical but also laborious studies, determined that the Old Red Sandstone of South Wales was equivalent in age to strata originally defined as their Devonian System (Fig. 3.4). The former, with plant and fish remains, reflected environments markedly different from the marine condition of the latter; yet strata formed in those contrasting environments were of the same age.

Analogies between features such as ripple marks in the Old Red Sandstone with counterparts forming in modern sediments (Fig. 3.8) produced some of the earliest environmental interpretations of sedimentary rocks. Simple application of the principle that present processes provide keys to the origins of ancient features would seem to indicate at once that many different sediment types inevitably must have formed simultaneously side by side in

FIGURE 3.8

Modern ripple marks exposed at low tide, Sapelo Island, Georgia. The straight, sharp, symmetrical ripple crests are typical of those formed by waves in shallow water. Note also holes formed by burrowing organisms. (*By R. H. Dott, Jr.*)

FIGURE 3.9
Sedimentary facies around a hypothetical island showing tendency for coarser sediments to occur in strongly agitated, nearshore environments. Bottom-dwelling organisms in the different environments differ considerably, also. (The upper surface of the diagram is a map of most recent bottom-sediment types showing lateral variations only at a moment in time; sides are cross sections showing facies relationships through time.)

lateral variations, for which the term *sedimentary facies* was adopted. Facies variations are changes in the total aspect of strata, which may be rather subtle in nature (e.g., a gradual increase in the ratio of sand to mud). A particular three-dimensional body of sediment—a facies—grades laterally by some statistical change in its properties to another, adjacent facies. The facies concept relates sediments and sedimentary rocks to their environments of deposition. A map of modern ocean-bottom sediments or of animal communities reveals the reality of variations of products of adjacent, synchronous environments (Fig. 3.9). On land today—at the same time—we find arctic, temperate, and tropical environments of markedly different characteristics side by side.

TRANSGRESSION AND REGRESSION BY THE SEA

An example from the historical period in the Netherlands shows some implications of facies variations through time. As indicated in Chap. 1, the North Sea coast of the Low Countries has been submerging since the last glacial advance. Whether submergence was due to worldwide rise of sea level or to land subsidence (or both), the sedimentary result would be essentially the same. Bore holes have been dug in Holland for several centuries, producing an accurately dated Pleistocene and Holocene succession (Fig. 3.10). Advance of the sea over the land, termed *transgression*, caused a continuous shift of environments and their sedimentary and biologic products landward. Retreat of the sea from a land area, called *regression*, has the opposite effects. The result is a more or less continuous, nearly flat layer of sandy nearshore deposits and finer offshore ones as shown. The relation between these two sandy and muddy facies is complex, but it involves a seaward gradation along any single datum of like age (such as the present sea floor), from sand to mud.

If these deposits are preserved and lithified, what should future geologists designate as formations? Defined on physical, chemical, and organic characteristics, the most obvious choices would be a sandstone formation and a shale formation. But note that these two formations would *in part grade laterally into one another*, that is, they are not in simple superposition even though one tends to lie mostly above the other. Some lateral projections of sand, called tongues, penetrate into the generally shaly facies and vice versa. A sand tongue may reflect a period of violent storms, which moved sand farther offshore than normal.

It follows that neither formation is of exactly the same age everywhere along the cross section. Each is older at the seaward than at the landward end because of the progressive transgression by the sea. On the other hand, at any point some part of one is the same age as part of the other. Clearly, then, formations cannot be thought of as either: (1) laterally indefinite and unchanging, or (2) exactly one age throughout their entire extent. Rock divisions must be defined relatively locally and more or less independent of time. After they have been defined and mapped carefully, then it may become possible to evaluate fully their time or

FIGURE 3.10

A historic example of transgression on the Netherlands coast, showing landward shift of facies dated by archaeological and carbon 14 evidence. Note migration of characteristic molluscan animals with the sandy facies; the fish, being a swimming form, was independent of the bottom environment and might be found in either facies. (Angle of slope is exaggerated.)

age relationships. However, such evaluation requires a great deal of information on a regional scale.

Two basic types of facies patterns must be distinguished in order to reveal past changes of land and sea areas. A *transgressive facies pattern* is illustrated by Fig. 3.10 (see also Fig. 3.7). Transgressive facies generally (1) reflect shrinkage of land area; (2) are preceded by an erosional unconformity; (3) show a landward shift of adjacent sedimentary facies through time; and (4) are reflected by deposits that tend to become finer upward at any one geographic locality. A *regressive facies pattern*, on the other hand, generally results from a relative fall of sea level or a rise of land level. It has opposite characteristics to those of the transgressive pattern, thus regressive facies (1) reflect en-

largement of land area; (2) have an erosional unconformity at the top; (3) show a seaward shift of the different sedimentary facies through time; and (4) are reflected by deposits that tend to become coarser upward at any one locality. Figure 3.11 illustrates an example of regressive facies due to the recent rise of northern Scandinavia as a result of deglaciation (compare Fig. 1.8). Other important examples include the exposure of land bridges between continents or islands by the worldwide fall of sea level during the last glaciation.

About 100 years ago, German geologist Johannes Walther studied the relationships of modern sediment types (facies) to their environments. He noted that inevitably environments shift position through geologic time, and as they do so, their respective sedimentary facies will succeed one another in vertical depositional sequences as already noted. But the succession cannot be random, for only the facies characteristic of directly adjacent environments can succeed one another. Then it was noted, as an important general consequence of this fact, that the vertical progression of facies will be essentially the same as corresponding lateral facies changes. Therefore, in a *transgressive facies pattern*, just as the sequence becomes finer upward (characteristic 4), the finer facies also spreads

FIGURE 3.11

A historic example in the northern Baltic Sea of regressive facies, that is, the seaward shift of different sediment types as environments shifted southward (left) during the past 14,000 years due to the rise of Scandinavia since the melting of glaciers.

FIGURE 3.12
Contrasting effects of sea-level fluctuations on low versus steep coastlines due to upwarping of the ocean basin. Roughly half of the total continental surface today lies within a few hundred meters of sea level; thus a modest sea-level change produces a profound change in land area. Many such modifications have occurred in the past.

laterally in the direction of transgression (see Fig. 3.10). Conversely, just as a *regressive sequence* becomes coarser upward, the coarser facies also spreads laterally in the direction of regression (see Fig. 3.11). This important linking of vertical and lateral facies changes is called *Walther's principle*, and it is of fundamental importance in interpreting ancient geography (paleogeography).

CAUSES AND CONSEQUENCES OF TRANSGRESSION AND REGRESSION

As was hinted in Chap. 1, large-scale changes of sea and land levels have been extremely important in earth history. What causes could produce such colossal changes? Worldwide sea-level changes may result from fluctuating continental glaciation or large-scale warpings of deep ocean basins. Worldwide regressions during Pleistocene glacial episodes were very important to early humans, having allowed them to migrate from Siberia to America and from Europe to Britain. The relief of the land will dictate enormous differences in the magnitude of shoreline movements during transgression or regression. Low, flat land, such as the Netherlands, would be inundated or drained very widely by but a slight relative change of sea level, whereas a bold coastline such as that of California would be only locally affected (Fig. 3.12).

Local changes such as mountainous uplifts or crustal subsidence can result in local transgression or regression as well. But, if more than one cause occurs simultaneously, the effects along a given coast will be more complex, and may even give us some surprises. For example, transgression in the Netherlands has been accentuated by subsidence of the earth's crust at the same time that postglacial sea level has been rising worldwide (Fig. 3.10). Conversely, the Baltic Sea coast simultaneously has experienced regression in spite of the postglacial sea-level rise. This is because of the more rapid uplift of Scandinavia (Fig. 3.11).

There are other complications that may result from relative rates of sedimentation and erosion along a coastline. Very rapid sedimentation due to uplift of a distant inland area or a climatic change that accelerates erosion may cause the seaward retreat of a shoreline independent of structural warping of the crust or worldwide changes of sea level. For example, the rapid advance of the Tigris-Euphrates delta noted in Chap. 1 has caused the northern shore of the Gulf of Arabia to retreat from the land toward the gulf 175 kilometers (100 miles) in the past 5,000 years, even as sea level has risen about 100 meters worldwide (Fig. 3.13). The delta advance resulted from accelerated sediment transport due in large part to human agricultural practices far inland (see also Fig. 1.7).

Ideally we would like to be able to distinguish worldwide causes of transgressions and regressions from more local changes along individual shorelines due to sedimentation, erosion, or local crustal warping, but such a distinction

FIGURE 3.13

Advance of the Tigris-Euphrates delta 175 kilometers into the Gulf of Arabia during the past 3,000 years in spite of the worldwide rise of sea level of about 4 meters during that same period. Thus, in spite of the worldwide postglacial transgression, rapid sedimentation has caused local regression here. Not to true scale (compare Fig. 1.7).

FIGURE 3.14

The relation of a preserved rock record with discontinuities and the abstract time continuum corresponding to that rock record plus its unconformities. A tangible rock record exists only for the shaded portion at the right, while the blank areas (hiatus), represent unconformity intervals.

often cannot be made. Obviously from the fragmentary evidence preserved in ancient rocks, we cannot be very hopeful of determining the actual or absolute cause of every transgressive or regressive facies pattern; we see only an apparent sea-level change reflected. A great deal of continentwide or even worldwide information is required to test and reject possible alternate explanations. We shall illustrate many examples of these important interpretations subsequently.

MODERN STRATIGRAPHIC CLASSIFICATION

ROCKS VERSUS TIME

As was indicated early in this chapter, it has become mandatory to make clear distinctions between abstract geologic time, which is continuous, and the actual rock record, which is riddled with unconformities or "omissions" of varying magnitude. Yet all that we are to know of the historical continuum must be gleaned from the imperfect rock record. A unit of relative geologic time has meaning only in terms of rocks formed during that time. We have distinguished the Cambrian System of rocks from the Cambrian Period of time. The rocks of this system in Wales, where they were first studied and named, provide a world standard for comparison by correlation of rocks anywhere else, which, on the basis of similar fossils, are

judged to have formed during that same period of the Paleozoic Era.

A pure time division, such as the period, must include the time represented by any and all discontinuities as well as the actual preserved rocks in the corresponding universal rock division, the system. Figure 3.14 illustrates this difference. It follows that if elsewhere a more complete sequence of Cambrian rocks were found with fewer and smaller unconformities and more fossils, it would provide a better world standard of reference than does that of Wales. While this is true, the original European System standards are now so firmly established by long usage that such changes are rarely made.

Most classifications contain different levels of subdivision. A simplification of the most generally accepted hierarchy of stratigraphic subdivisions is as follows:

Relative time divisions	Equivalent universal rock (or time-rock) divisions
Era	Erathem
Period	System
Epoch	Series
Age	Stage

The above rock divisions are termed universal because they comprise at least continentwide if

FIGURE 3.15

Significance of contrasting rates of evolution and rates of environmental shift (transgression). The brachiopod *Lingula* in the sandy facies evolved very slowly and is a poor index fossil; it has migrated with its shifting sandy environment and is unchanged biologically after millions of years. Cephalopods found in the muddy facies, however, were swimming forms free of the bottom environment. They also evolved rapidly, therefore their species are admirable index fossils for times 1, 2, and 3.

not worldwide standards as opposed to the purely local rock units such as formations. Though the rocks of Wales that define the Cambrian System have certain peculiar lithologic characteristics, all rocks considered of like age elsewhere regardless of their lithology are also referred to the Cambrian System. Comparison among these is made only in terms of age equivalence.

BIOSTRATIGRAPHIC CONCEPTS

Identity of age of strata, especially over long distances, has been established largely by application of the principle of fossil correlation. The subdivisions of any newly studied stratigraphic sequence are correlated with the standard universal rock divisions on the basis of similar fossils. When Charles Walcott, a geologic pioneer in western America, identified thick strata in southeast California as Cambrian a century ago, he was performing a *correlation* with Wales based upon index fossil assemblages. Such correlations assume that similar evolutionary stages of development were reached simultaneously by particular organisms in all parts of the world. Experience suggests that, at the level of precision of correlation so far achieved, this assumption is generally valid. The facies concept, however, makes it clear that evolution

through time is not the only change operating, for environmental differences in space also must be considered as factors affecting distribution of fossils, as emphasized in the preceding discussion of sedimentary facies.

Environmental changes are, by and large, relatively short-term affairs, and so are not very significant when considering fossil assemblages in the large universal rock divisions. However, for smaller time divisions, the vagaries of local environmental influences must be considered, for environmental changes may have been more rapid than evolutionary ones. Moreover, only certain groups of organisms—those which evolved very rapidly—can provide reliable correlations for short time intervals. Figure 3.15 shows how differing rates of evolution may compare with environmental change. Several related aspects of index fossil assemblages enter into their use for correlation. A widespread biologic datum; that is, a restricted thickness of strata characterized by a distinctive index fossil, is commonly termed a *fossil zone* (Fig. 3.16).

Because sedimentary environment may influence bottom-dwelling organisms so profoundly, such creatures may be of little value in correlation of strata of adjacent different facies. Many fossil types are notoriously restricted to one or a few lithologies ("facies fossils"). For example, certain clam species today burrow only in beach sands, while others live only in muddy tidal flats. These would be less-than-perfect index fossils because of their narrow preference for a particular sediment type. It follows, then, that the best index fossil is one which lived more or less independent of the bottom environment where sediments form. Obviously these would be floating or swimming organisms (Fig. 3.17). Fortunately, there are several such groups that

LOCALITY 1   LOCALITY 2   LOCALITY 3

A B C   A B C   A B C

LOWER TRIASSIC SERIES

TOTAL KNOWN RANGE of A

FIGURE 3.16
Correlation using three different index fossils. (Brown patterns indicate ranges and relative abundances for each species.) In practice, assemblages are more useful than a single species, but close attention must be paid to overlapping stratigraphic ranges of index fossils. Note also that the range and the maximum development (widest brown pattern) of a single species vary slightly from place to place. This imposes a lower limit of resolution to fossil correlations beyond which it is fruitless to attempt further refinement.

FIGURE 3.17
Relations of organisms to environments and sedimentary facies. Bottom-dwelling organisms tend to be restricted to but one facies, whereas floating and swimming organisms are independent of facies, therefore the latter tend to be the very best index fossils for correlation among different facies. Volcanic ash layers also provide valuable time datums for correlation among facies.

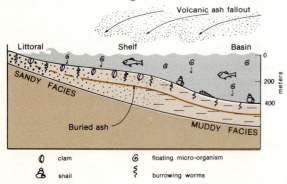

Volcanic ash fallout

Littoral   Shelf   Basin

SANDY FACIES

Buried ash   MUDDY FACIES

0
200   meters
400

0 clam   6 floating micro-organism
& snail   ξ burrowing worms

also evolved rapidly and that therefore serve well for correlation even between different sedimentary facies (Fig. 3.18). Figures 3.16 and 3.19 show in practice how index fossil zones together with lithology are used for correlation and mapping.

UNCONFORMITIES, THE UNIVERSAL TIME SCALE, AND ROCK SEQUENCES

Unconformities, like rock units, can be traced and mapped to establish their physical continuity, and they also can be studied from the standpoint of age. Continuity and age are established by mapping the relationships of strata and fossils immediately above and below the unconformable contact. It generally turns out that, like some formations, an unconformity surface varies somewhat in age from place to place. More importantly, the total time interval represented by the discontinuity may vary greatly; at one place there may be far more rock record missing than at another point along the same surface. An unconformity may even disappear laterally into a continuous, unbroken, conformable sequence of strata (Fig. 3.20). Un-

Formation contacts extrapolated where concealed between outcrops

FIGURE 3.19
Three-dimensional relationships of formations and index fossil zones; top surface shows a geologic map constructed by establishing physical continuity of formations between isolated outcrops. Age correlation by fossil zones shows that formation 3 is synchronous everywhere, but formations 1 and 2 vary in age due to lateral facies changes. (Compare with Figs. 3.15 and 3.10.)

FIGURE 3.18
Lateral view of a conodont. Conodonts are small anatomical parts of uncertain origin, but were probably supports for the digestive system of an extinct, floating organism. Evolutionary structural changes occurred rapidly, and conodonts are widespread in different rock types, so they are ideal index fossils. This conodont is from the Permian of Texas (magnified 110 times). *(Courtesy D. L. Clark.)*

conformities, then, show important lateral and vertical differences equally as important as those of rock units.

Today geologists tend to distinguish different types of unconformities (Fig. 3.21). What is more important than names is that some unconformities represent profound upheaval and deep erosion, as in mountain making, whereas others represent little or no structural disturbance, only erosion or nondeposition. But even the subtle ones may represent long time intervals.

Increasingly, geologists recognize thick, laterally extensive rock units called *sequences* that include many formations as packages bounded by exceptionally profound regional unconformities. No restricted time connotation (synchroneity) is attached to the unconformable sequence

boundaries; in fact they are known to be of varying age from place to place. *Sequences* constitute an additional type of stratigraphic division—a regional rock unit, but not a universal time division with strict age connotations; conventional period or era boundaries are ignored in defining them. Certain unconformities are judged to be the most important ones among many. Inevitably not all geologists agree entirely on the validity of the suggested sequence boundaries, but as is reflected in the

FIGURE 3.20
Variations in unconformities. Note that one type may change laterally to another. *Nonconformities* are underlain by igneous or metamorphic rocks; *disconformities* show little or no angular discordance.

FIGURE 3.21
Worldwide sea-level fluctuation curve for the past 600 million years. Note two maxima (high levels) about 500 and 75 m.y. ago, and three minima (low levels) about 600 m.y., 200 m.y., and the present. Maximum fluctuations were about 350 meters above and about 150 meters below present level. Letters are abbreviations for unconformity-bounded sequences in central North America (see Fig. 12.5). *(After Sloss, 1963 and Vail et al. 1977).*

organization of the chapters on North American history, a few major unconformities serve to punctuate the stratigraphic record in a meaningful way (Fig. 3.21).

Recent work based upon detailed comparisons of intercontinental subsurface stratigraphic data, especially by the petroleum industry, suggests that these particular unconformities reflect large global fluctuations of sea level due to glacial and large-scale structural events like the warping of an entire ocean basin (Fig. 3.21). Thus the *sequences* bounded by these great unconformities are thought to represent "global cycles" of sea-level fluctuation with variable

periods on the order of 10 to 80 million years. The magnitude of apparent vertical changes of sea level is staggering; the suggested maximum rise was about 350 meters (1,150 feet) above present level (near end of Cretaceous) and the maximum fall was about 200 meters (650 feet) below present (late Paleogene). (For comparison, Pleistocene fluctuations were about 200 meters). Clearly, such large worldwide fluctuations should have had profound effects upon the stratigraphic record of shallow marine and nonmarine regions all over the earth. We shall want to be alert for evidence of this as we study the historical record more fully.

APPLICATIONS OF BASIC STRATIGRAPHIC CONCEPTS

HUTTON'S SCOTLAND

Now we can illustrate and at the same time begin to interpret stratigraphy in a modern sense using Great Britain as an example. Figure 3.22 shows a cross section through Scotland, illustrating most of the important features first studied by Hutton. A great unconformity can be seen beneath the Devonian Old Red Sandstone. Rocks below are Silurian and older, and so it is apparent that sometime in the later Silurian, and perhaps early Devonian, profound mountain building occurred in a belt running northeast through Scotland; it also affected Wales and western Scandinavia. Southeast of this old disturbed belt the Old Red Sandstone is more nearly conformable with the Silurian.

The fact and age of the mountain-building event is clear on the southeast (right) end of the Scottish cross section from the superpositional relations of the Old Red Sandstone over the angular unconformity. But to the north (left) of the Highland Boundary Fault, things are not so definite. Severe upheaval, metamorphism, and igneous activity occurred, but when? Did these changes happen at the same time that Hutton's unconformity developed? Downfaulted Old Red Sandstone at the far left suggests that the upheaval was, at least in part, pre-Devonian. A rough clue to magnitude of events, though not

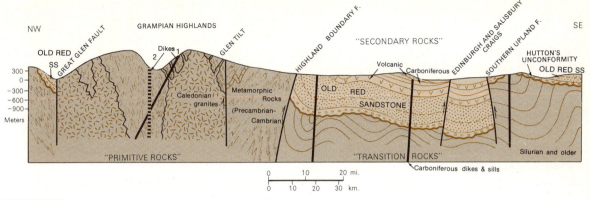

FIGURE 3.22

Cross section showing structural relationships among the rocks studied by Hutton in Scotland (vertical scale exaggerated to show details; see Fig. 3.23 for location). *(Adapted from* Geological map of Great Britain, *1957.)*

to precise age, is provided by the presence of strongly metamorphosed rocks beneath the pre-Old Red unconformity. Exposure of such rocks at an old erosion surface implies deep erosion of the crust to expose rocks that formed in a relatively high-pressure and high-temperature environment.

RELATIVE AGE OF IGNEOUS BODIES

How can we narrow the age of all the other metamorphic rocks and granites of the Grampian Highlands where Old Red Sandstone is absent? Some important principles exist for deciphering the relative age of igneous bodies, and if such bodies in turn can be related in age to fossiliferous sediments, then the igneous rocks, too, can be related to the geologic time scale. It is self-evident, as emphasized by Hutton, that an *intrusive igneous body must be younger than all rocks intruded by it.* We may call this the *principle of cross-cutting relationships,* which is a kind of extension to intrusive igneous rocks of the principle of superposition. Commonly it is possible to establish relative age of several intrusive masses by mutually cross-cutting relations among them. Thus at the left end of Fig. 3.22, the granite is clearly younger than the metamorphic rocks, in which a few Cambrian fossils occur. The first dikes that

Hutton studied were of this granite. But the granite is older than basaltic dike 1, which is in turn older than dike 2. These relations are purely relative and say nothing about how much younger each successive rock is than another. Is it to be measured in minutes, days, years, or what? It is a reasonable hypothesis that at least the granite, and possibly also the metamorphism, date from a little before Hutton's pre-Old Red Sandstone unconformity, which developed about the end of the Silurian Period. Similarly, it is tempting to correlate dike 1 with basalt sills and dikes studied by Hutton at Edinburgh, which were intruded into the Old Red Sandstone (see Chap. 2).

RELATIVE AGE FROM INCLUDED FRAGMENTS

But why not correlate basalt dike 2 of the Highlands with these instead? Just how different in age are these two Highland dikes? Some new lines of evidence are needed to answer these questions. What we require, clearly, is some definitive younger age limit for the igneous and metamorphic rocks. If the Old Red sediments rested unconformably upon the granite, dating of granite emplacement would be simple by superposition, but this is not the case. We can seek indirect evidence for relative age in conglomerates of the Old Red itself, for coarse fragments provide direct evidence of the character of rocks exposed by erosion while they were accumulating. An important additional principle elucidated by Hutton is that *rocks represented as fragments in another rock body must be older than the containing rock*. This could be called the *principle of included fragments*. Note that it applies as well to pieces (inclusions) of foreign rocks in igneous intrusions as well as to grains or pebbles in sediments.

In our example, we find pebbles in the Old Red that are identical to granites in the Grampian Highlands. Therefore, our argument is tightened, and a great deal more confidence can be attached to our hypothesis. Hutton's original intrusive granites and his unconformities turn out to be closely related, and they constitute the principal evidence of one of the earth's greatest mountain-building events much as Hutton envisaged it. This event has since been named the Caledonian orogeny (Caledonia was the Roman name for ancient Scotland; orogeny means mountain building).

RELATIVE AGE OF FAULTS

Let us return momentarily to the cross section (Fig. 3.22) and reconsider the basalt dikes. Further evidence near Edinburgh indicates that these dikes and related lavas are of Carboniferous age. It is reasonable, but by no means provable, that basaltic dike 1 in the Highlands is of like age. Dike 2 may be only slightly younger or it may be much younger. Without other evidence from beyond the confines of this cross section we can say no more, but in western Scotland there is a swarm of Cenozoic basaltic dikes, which probably are correlative with dike 2. We should also note that the faults shown can be dated as having been active after the Caledonian orogeny and chiefly after the Devonian. There is no evidence to narrow their age further. We see then that, like igneous intrusions, *faults are younger than all rocks that they displace and older than the oldest rock that unconformably overlaps them*. Strictly speaking, however, this only dates the last fault movements; many faults have been active for very long periods, even hundreds of millions of years, and thus other

indirect evidence may be required to date their first movements.

REGIONAL ANALYSIS

On a regional scale, we can see other important relationships in strata of the same age as those just considered. Figure 3.22 shows the unconformable overlap of the Devonian Old Red Sandstone across deformed Silurian and older rocks in Scotland. Figure 3.23 shows a sedimentary *facies map* (A), and a *restored cross section* (B), for the Devonian of Europe. The map distinguishes present distribution of Devonian rocks, and, by applying the *principle of original lateral continuity* to the facies patterns, suggests how much more extensive the Devonian facies probably were (diagonal patterns). Cross sections, such as Fig. 3.23B, that show vertical variations among rocks are important complements to facies maps. Geologists use many kinds of maps and cross sections to illustrate geometric relationships of rocks, but those of Fig. 3.23, unlike more familiar ones, ignore (or restore) post-Devonian deformation and erosion.

From Fig. 3.23 we see that mountain building profoundly disturbed the northwest margin of Europe as a whole during Silurian and Devonian time, while the region to the southeast remained more stable. This contrast is reflected by: (1) more and greater unconformities at the west; (2) coarser and thicker nonmarine sediments at the west; (3) finer, thinner, more marine deposits at the east. Mountainous lands in Britain and Norway shed immense volumes of coarse debris eastward due to vigorous erosion; greatest thicknesses are nearest the mountains (Fig. 3.23B). Uplift of the mountains was spasmodic, for a major unconformity divides the Old Red Sandstone at the left end of the section. Clearly the time represented by unconformities is greatest nearest the old mountains, as is the contrast between rocks above and below them. Eastward the same unconformities become more subtle; some even disappear completely within *conformable* sequences.

B - Restored stratigraphic cross section (vertical scale exaggerated; earth's curvature ignored)

FIGURE 3.23

Devonian facies of Europe. *A (above)*: Facies map shows areas of preserved strata as well as inferred restorations where erosion has removed them (diagonal pattern). *B (below)*: Restored cross section showing facies, thickness, and the variations in unconformities. Effects of Caledonian mountain-building event shows at left. *(Adapted from R. Brinkmann, 1960, Geologic evolution of Europe; by permission of Ferdinand Enke Verlag; and L. J. Wills, 1951, Palaeogeographical atlas; by permission of Blackie and Son Ltd.)*

SUMMARY

In this chapter we have seen how the relative geologic time scale developed in northern Europe. Geology now had both a viable unifying theory and new methodology. Evolution of the time scale illustrates the application and refinement of the basic stratigraphic principles of Steno, Smith, and Cuvier. The relative geological time scale provides a standard of reference for rocks throughout the world. But relative geologic age generally must be established by fossils and by determining superpositional relations of strata containing them. Hutton illustrated other important means of determining relative age. For igneous rocks, cross-cutting intrusive relations and the unconformable superposition of younger, datable sediments provide bases for their relative dating. Composition of conglomerate pebbles and of included fragments in igneous rocks also provide evidence for relative age. In subsequent chapters, we shall apply these principles and methods to North America.

Extension of mapping and of stratigraphic knowledge inevitably confronted geologists with the realization that formations must change in character laterally in response to differences of ancient sedimentary environmental influences, which imposes restrictions on correlation using index fossil assemblages. Some fossils, in fact, occur exclusively in only one facies type. Sedimentary facies patterns reflect many influences, but especially important are transgression and regression by the sea due to relative vertical changes of land and sea levels. Sedimentary facies provide the chief clues for deciphering major changes of paleogeography such as the changes of land-sea relationships. The concept of facies is probably the most difficult one in this chapter, but a clear understanding of it is of utmost importance.

Today we require a rather elaborate stratigraphic classification scheme that makes special allowance for distinction of rock units having tangible spatial dimensions (systems, series, etc.) from abstract time (era, period, epoch, etc.). The classification also has two levels for rock units, the universal (worldwide) standard rock units, such as the system, and the local rock units, such as the formation. Many discontinuities occur in the rock record; therefore we have an incomplete record of earth history. It is now apparent that local rock units may vary in age from place to place. Similarly, unconformities are recognized to vary in age and magnitude of time represented from place to place. Recently it has proved useful, especially for continental- and intercontinental-scale comparisons, to subdivide the stratigraphic record at especially widespread unconformities regardless of their positions within the standard time scale. The large rock units lying between such unconformities, called *sequences*, appear to reflect major worldwide fluctuations of sea level ("global marine cycles"). While at this point they doubtless seem to you like an unwanted added complication, sequences seem to offer promise of a great simplification in the study of world stratigraphy. It has been an elusive hope of geologists ever since the neptunians that some kind of worldwide patterns might exist in the stratigraphic record. After many bitter disillusionments with earlier panaceas, it appears at last that, indeed, there may be some nearly universal punctuations in an otherwise enormously complex record.

Key Concepts

conformable
cross-cutting age relations
facies map
facies tongue
formation
included fragments
index fossil zone
lithology
relative time scale
restored cross section
rock sequences
sedimentary facies:
   regressive
   transgressive
time versus rock divisions
type section
Walther's principle

Readings

Ager, D. V., 1973, The nature of the stratigraphical record: New York, Wiley-Halsted.

Bennison, G. M., and Wright, A. E., 1969, The geologic history of the British Isles: New York, St. Martin's.

Berry, W. B. N., 1968, Growth of a prehistoric time scale: San Francisco, Freeman. (Paperback)

Brinkmann, R., 1960, Geologic evolution of Europe: New York, Hafner. (English translation)

Carozzi, A. V., 1965, Lavoisier's fundamental contribution to stratigraphy: The Ohio Journal of Science, v. 65, pp. 72–85.

Cohee, G. V., Glaessner, M. F., and Hedberg, H. D., 1978, Contributions to the geologic time scale: Tulsa, American Association of Petroleum Geologists.

Eicher, D. L., 1968, Geologic time: Englewood Cliffs, Prentice-Hall. (Paperback)

Geikie, A., 1905, The founders of geology (2d ed.): New York, Macmillan. (Paperback, Dover, 1962)

Gignoux, M., 1955, Stratigraphic geology: San Francisco, Freeman. (Emphasis on Europe)

Harbaugh, J. W., 1968, Stratigraphy and geologic time: Dubuque, W. C. Brown. (Paperback)

Krumbein, W. C., and Sloss, L. L., 1963, Stratigraphy and sedimentation (2d ed.): San Francisco, Freeman.

Levin, H. L., 1978, The earth through time: Philadelphia, Saunders.

Matthews, R. K., 1974, Dynamic stratigraphy: Englewood Cliffs, Prentice-Hall.

Payton, C. E., et al., 1977, Seismic stratigraphy: Tulsa, American Association of Petroleum Geologists, Memoir 26.

Rutten, M. G., 1969, The geology of western Europe: Amsterdam, Elsevier.

Woodford, A. O., 1965, Historical geology: San Francisco, Freeman.

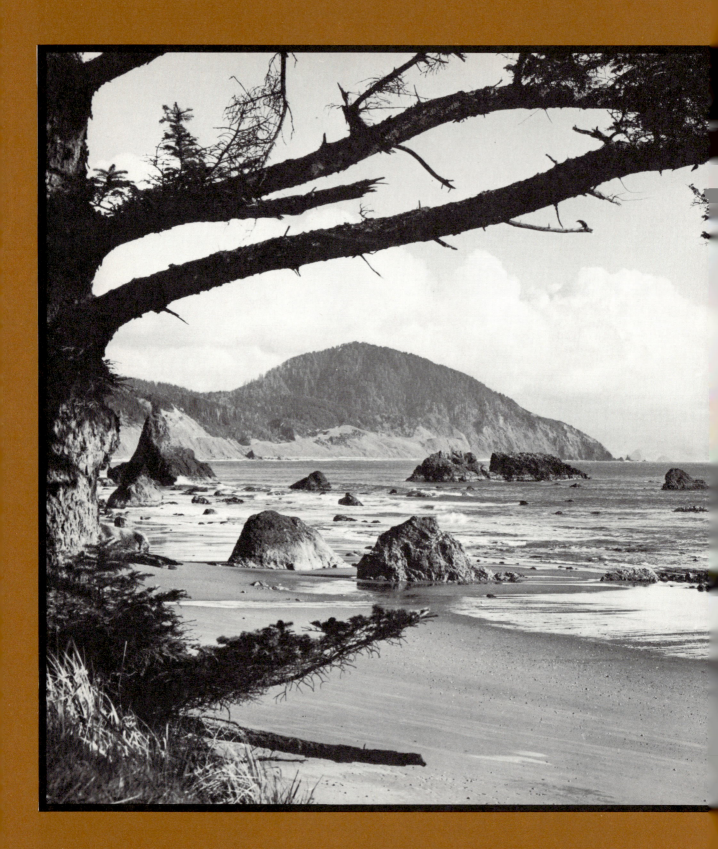

4

THE NUMERICAL DATING OF THE EARTH

Some drill and bore
The solid earth,
and from the strata there
Extract a register, by which we learn
That He who made it,
and revealed its date to Moses,
was mistaken in its age!

William Cowper (Late eighteenth
century)

4

Evolution is one of the most powerful of man's ideas, for it has revolutionized our way of thinking about natural and even social phenomena. Among great books, Darwin's *Origin of Species* probably ranks second only to the Bible in its impact on Western thought. After 1859, the basic concept of evolution, or change from one form into another through time, was also applied beyond biology. For example, studies of heat energy in the nineteenth century led physicists toward an evolutionary view of the physical realm. Today, science views nature as a whole as evolutionary, thus constantly changing in a series of linked but unique (nonrepeating) historic events.

It appears that the entire earth is the result of one grand chemical evolution from a heterogeneous, formless, lifeless beginning to a highly organized, complex, inhabited world. The transmutation of new isotopes—different species of a single chemical element—from radioactive parents, a subject of this chapter, clearly represents inorganic evolution. Buffon's hypothesis of epochs of cooling of the earth and of life development mentioned in Chap. 2 represents the first crude evolutionary conception of earth history. Basic to any idea of the origin and subsequent history of the planet is the question of the total age of the earth, which even Buffon greatly underestimated. In this chapter the different schemes for so-called absolute dating are examined as a prelude to our study of the sequential evolution of earth as a whole. In short, we need a calendar against which we can measure time numerically rather than in relative terms only.

It is difficult to overstress the revolutionary impact on modern geology of isotopic dating of minerals. Not only has it stretched our conception of the total age of the earth by nearly 1,000

times, but it has allowed at last the subdivision and long-range—even intercontinental—correlation of Prepaleozoic rocks, which represent 75 percent of earth history. It also has given more precision to the subdivision of the younger record, especially in complex mountain belts where igneous and metamorphic rocks are widespread. In addition, isotopic dating allows the exciting possibility of evaluating numerically the rates of processes acting in the past, adding a desirable degree of quantitative rigor.

EARLY ATTEMPTS TO DATE THE EARTH

You will recall that, on the basis of the Scriptures, it was long ago argued that the earth was only 5,658 years old, having been formed in 4004 B.C. Buffon suggested in 1760 that it was 75,000 years old. Hutton, Lyell, and their followers reasoned from the implications of their uniformitarianism and the observable rates of geologic processes that the earth must be far older than that. Finally, Charles Darwin needed as much time as possible for the evolution of the amazing array of life forms known by 1859; millions of years seemed indicated before the development of human beings.

Near the end of the nineteenth century, a British geologist estimated that all the time since the beginning of the Cambrian amounted to about 75 million years. He based his conclusion upon the maximum known thickness of strata of that age span multiplied by an assumed average rate of sedimentation derived from study of modern depositional rates. Use of the accumulation rate of salt in the sea was attempted by an Irish scientist named Joly in 1899. Using the then-known average rate of delivery of salt to the sea by rivers, he found that it would have taken about 100 million years to develop the present salinity. Obviously neither of these "ages" could have been the total age of the earth, as their authors realized.

All the early attempts at assessing the age of the earth in terms of numerical time involved a host of shaky assumptions. For example, if we were to choose an element other than sodium or chlorine to estimate the age of sea water, we would find a different apparent age because each element accumulates at a different rate. Today it is impossible to accept the simple extrapolation back in time of present rates of practically any processes. It is partly because of the absurdities that arise from such simple extrapolations that we emphasized in Chap. 2 that processes had not been perfectly uniform in intensity through time. Nonetheless, the early attempts to date the earth showed ingenuity and an evolution of thought in one direction, namely toward older and older estimates for our planet's birthdate. By the middle of the last century, most geologists thought in terms of a total age on the order of several hundred million years, but they had no numerical method to prove this hunch.

KELVIN'S DATING OF THE EARTH

In 1846 the great physicist Lord Kelvin launched a 50-year combat with geologists over the earth's age and the nature of its historical development. This was an outgrowth of his displeasure with strict uniformitarianism because it seemed inconsistent with physical principles as noted in Chap. 2. A completely uniform, unchanging earth was, to him, impossible. That temperature increases with depth in the earth was known, and was taken as proof that the earth is losing heat from its interior. Either the earth must be cooling after an initial very hot stage, or else it contains a continuing internal source of heat energy. Lyell appealed to chemical reactions in the interior to produce heat in an endless cycle allowing for a steady-state earth. Kelvin considered Lyell's untiring heat engine to be perpetual motion, thus unac-

ceptable. He discounted any renewing internal heat source and accepted the nearly universal assumption that the earth was originally molten. Therefore heat was being dissipated through time, and the earth could not be unchanging as long as this expenditure of energy continued. He had, in fact, endorsed an evolutionary view of the earth that eventually would displace Lyell's conception.

From deep mines an approximate rate (or gradient) of the downward increase of temperature was known. Using this gradient (which was about three times too great), Kelvin extrapolated the apparent rate of cooling of the earth backward to a time when the earth presumably was molten; this should be close to the total age of the earth, he reasoned. Kelvin's calculations satisfied him that the earth is between only 20 and 30 million years old. His arguments seemed flawless, for no other terrestrial or solar heat sources were known. In the eyes of most scientists Kelvin had won his campaign decisively, and left the geologists reeling. They were reluctant to accept such a modest age, but were incapable of mustering a rigorous counteroffensive until the American geologist T. C. Chamberlin in 1899 challenged Kelvin's assumption that the earth had begun as a molten body. He introduced an important new hypothesis to be discussed in the next chapter, which proposed that the planets formed by the accretion or collecting together of cold, solid chunks of matter. According to Chamberlin, then, the earth must have heated up sometime after its initial formation through some internal process quite independent of solar heat.

RADIOACTIVITY

The last great breakthrough, which completed the stage for modern historical geology, occurred in 1896 in the Paris laboratory of physicist Henri Becquerel. The Frenchman and assistants, the Curies, recognized the phenomenon of radioactivity through the exposure of photographic film tightly sealed from light but held next to a uranium-bearing mineral. Some kind of radiation invisible to the eye had exposed that film. Two hitherto unknown chemical elements were soon discovered, radium and polonium, which form from uranium by changes in the atomic nucleus that involved unknown types of radiation. Demonstration that many elements have several nuclear species called *isotopes* followed quickly. Some of these species are unstable; that is, their atoms are not permanent as had been assumed for all atoms. The nuclear species (isotopes) of a single element differ in having variations in the number of neutrons in the atomic nucleus. Yet, isotopic species of any given element have similar chemical properties, and it is for this reason that isotopes went undetected for so long.

Isotopes with unstable nuclei (parent isotopes) undergo spontaneous change—or mutation—until a permanent, stable configuration is reached (daughter isotopes). This process of change is called *radioactive decay*, and it results in one or more of three types of *emanations* or emissions from the nucleus at a fixed average rate characteristic of any particular isotope. It was such emanations that exposed Becquerel's film. The emanations were originally named *alpha* (which proved to be a helium nucleus, i.e., a helium atom stripped of electrons), *beta* (a free electron), and *gamma rays* (similar to x-rays). Radioactive decay can be written like a chemical reaction in the following manner:

Unstable parent isotope→stable daughter isotope + nuclear emanations + heat

Isotopes are designated by a number representing the total sum of neutrons and protons in the nucleus, called *mass number*, which is generally written at the top left of the chemical symbol for the respective element. (This number differs from the more familiar *atomic number*, which designates only the number of protons). Emissions from the nucleus of a *radiogenic* isotope change the identity of that isotope by increasing or decreasing the number of neutrons in the nucleus (Fig. 4.2). A brief example may help. When the uranium isotope with mass number 238 (^{238}U) decays, an alpha particle (He+) is emitted from its atomic nucleus. This has the effect of reducing both the number of neutrons and number of protons in the nucleus by 2 each, therefore the resulting new isotope has a mass

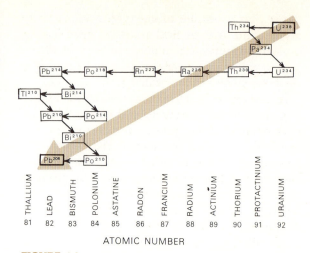

ATOMIC NUMBER

FIGURE 4.2

Diagrammatic portrayal of the decay history of the uranium 238 series. Half-lives for the intermediate daughter products vary from less than 1 second (polonium 214) to 1,622 years (radium 226). *(From* How old is the earth? *by P. M. Hurley. Copyright © 1959 by Educational Services, Inc. Reprinted by permission of Doubleday and Company, Inc.)*

number that has been reduced by 4. This new isotope is thorium 234 (^{234}Th), which is itself unstable. By emission of a beta particle (an electron), it in turn decays to protactinium 234 (^{234}Pa), and so on through a complex series of intermediate, unstable daughter isotopes ultimately to stable lead with mass number 206 (^{206}Pb), as shown in Fig. 4.2. Thus radium and polonium discovered in Becquerel's laboratory proved to be but intermediate daughter isotopes in the complex ^{238}U decay series. Uranium and thorium isotopes have the most complex series, whereas most isotopes used for dating involve simpler, direct decay from parent to daughter.

Further details of changes within the nucleus during decay are not essential to understanding the isotopic dating of minerals and rocks, therefore they are not treated further here. What is most important to realize is that the occurrence of an individual emanation is a statistical event. That is, neither the identity of the particular atom that will decay next nor the exact time at which the event will occur can be predicted absolutely. What we can predict is what fraction of the total parent atoms present will decay in a given period of time. Observations with devices

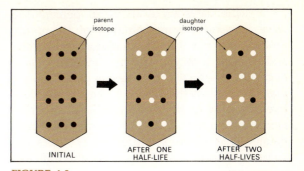

like a Geiger counter of many emanations from many atoms of a particular nuclear species over an extended time period provide a *statistical average rate of activity or decay*. In the laboratory, neither heating nor cooling, nor changes in pressure or chemical state have affected in any detectable way the average rate of spontaneous decay. Because the rate apparently cannot be artificially changed, it is assumed that it always has been uniform for a given isotope. If true, then decay of unstable isotopes would seem to be the only change in nature that we dare consider to have a constant rate. A few workers have even questioned this.

FIGURE 4.3

Diagrammatic portrayal of decay of a radioactive isotope within a mineral crystal at the time of initial crystallization. Note the changing ratio of parent to daughter isotope after one and then two half-life time intervals.

FIRST ISOTOPIC DATING OF MINERALS

Further knowledge of radioactivity came with breathtaking rapidity after the turn of the present century. The British physicist E. Rutherford, after counting radium emanations for many days, reasoned that *total emission activity* was proportional to the number of unstable parent atoms still present. This meant that emission must decrease in some regular fashion through time, and the decay could be expressed mathematically.

In 1905 a Yale chemist, Bertram Boltwood, suggested that lead was among the disintegration products of uranium. This is an ironic twist of the alchemists' ancient dream of making precious gold from lead; instead, apparently common, ugly lead was formed from a rare and almost equally precious element. In 1907 Boltwood, following a suggestion by Rutherford, developed the idea that a *decay series* such as that of uranium to lead could provide a means of dating the time of crystallization of minerals that contained a radioactive or radiogenic element. Boltwood reasoned that in unaltered minerals

> of equal ages, a constant proportion must exist between the amount of each disintegration product and the amount of parent substance . . . and . . . the proportion of each disintegration product with respect to the parent substance must be greater in those minerals which are the older.

From chemical analyses of a number of uranium-bearing minerals from localities whose relative geologic ages were known, he calculated average lead-uranium ratios and discovered a very significant pattern. Without exception, the proportion of lead was progressively greater for geologically older localities (Fig. 4.3). Boltwood felt that this systematic relationship confirmed his inference that lead was a product of uranium decay. Rutherford also had shown that helium was a product of uranium decay. Boltwood found helium gas in the same minerals that were analyzed for lead and uranium, and he inferred that indeed it was formed as another product of the decay of uranium. Attempts to date minerals by their helium-uranium ratios have met with little success, however, because helium, being a gas, escapes readily from the mineral crystals, causing results that are "too young."

To compute the ages of minerals, Boltwood needed to know the rate of decay of uranium, which was so slow that it had not yet been measured accurately. So he estimated it indirectly from the known rate for radium, a rapidly decaying intermediate member of the uranium series (Fig. 4.2). He then calculated the ages of minerals from their lead-uranium ratios as:

$$\text{Age} = \frac{\text{amount of daughter isotope (Pb)}}{\text{amount of parent isotope (U)}} \times 10^{10} \text{ yrs}$$

Boltwood's results for 10 localities on 3 continents ranged from 410 to 2,200 million years. In reality, his dates were about 20 percent too large because the mathematics of the decay process was not fully understood until 1910, and because, unknown to him, more than one decay series—each with differing decay rates—was involved in the minerals that he studied. Even allowing for this error, the oldest apparent age listed by Boltwood was nearly 100 times greater than Kelvin's figure and 10 times the wildest guess of any geologist for the total age of the earth!

THE CONCEPT OF DECAY

To analyze many different isotopes whose rates of decay vary widely, it is convenient to develop a generalized expression applicable to all decay series. This also should help to understand better the principles of geologic dating by the decay phenomenon. Remember that what is dated is the *time of crystallization* of a mineral; that is, the time of incorporation of an unstable isotope into that mineral (not the time of origin of that isotope in the universe, which in most cases was earlier than the origin of the earth). It is assumed that once a mineral crystallized, a closed chemical system was formed such that any daughter product now present in the mineral was formed only from the decay of the original unstable parent isotope therein (Fig. 4.3). It is like the closed system of an hourglass, which entraps sand grains. Those grains in the

top of the glass are like parent isotopes, whereas those falling to the bottom are like daughter isotopes; the ratio between them at any moment is a measure of time elapsed.

Dating is a matter of determining the ratio of parent to daughter atoms (Fig. 4.3) and then calculating a mineral's age by multiplying by the decay rate of the parent. More atoms decay early in the history of a mineral, and emanations then decrease in a regular way.[1] Figure 4.4 shows a decay-curve graph with time expressed in half-life units. The *half-life*, which is the time required for one-half of a given amount of any particular unstable nuclear species to decay, is a more convenient expression of rate for purposes of illustration than is the decay constant (λ). Half-life can be better envisioned by considering an analogy. Imagine a glass full of water to represent all the parent isotopes in some mineral crystal. After one half-life elapses, you drink half a glass of water. Then after one more half-life passes, and finding your thirst still unquenched, you drink half of what remains. After a third half-life elapses, you drink half of the remainder, and so on. The decrease of water in the glass is like the decrease of parent isotopes as decay proceeds in a mineral. By using half-life as a time-scale unit, our curve applies to the decay of any isotope regardless of the fact that actual half-lives vary enormously,

---

[1] $\text{Age} = \dfrac{\text{amount of daughter isotope}}{\text{amount of parent isotope}} \times \dfrac{1}{\lambda}$

Being exponential, decay can be conveniently expressed in logarithmic form such that:

$$\lambda t = \ln \frac{N^0}{N^t}$$

where λ = decay constant
 N^0 = number atoms of parent isotope at time zero (crystallization)
 N^t = number atoms of parent after time t of decay
 ln = natural logarithm (base 2.78)

In other words, the time of decay is proportional to the natural logarithm of the ratio of N^0 to N^t. It is convenient to have some fixed unit of decay time that would express concisely the decay characteristic of a particular isotope. Such a unit is the half-life ($t_{1/2}$), i.e., the time required for one-half an original amount of a nuclear species to decay. At one half-life, the N^0 to N^t ratio equals 2, therefore:

$$\lambda t_{1/2} = \ln 2$$

$$t_{1/2} = \frac{\ln 2}{\lambda} = \frac{0.693}{\lambda}$$

FIGURE 4.4
Simple arithmetic plot of a universal isotopic decay curve. After one half-life has elapsed, 50 percent of the original parent isotope remains; after two half-lives, half of that, or 25 percent, and so on.

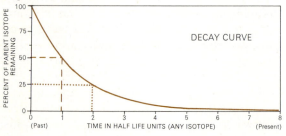

TABLE 4.1
Principal decay series used for mineral and total-rock dating

Parent isotope	Half-life	Ultimate stable product	Effective age range
Rubidium 87	47 billion years (b.y.)	Strontium 87	>100 million years (m.y.)
Thorium 232	13.9 b.y.	Lead 208	>200 m.y.
Uranium 238	4.5 b.y.	Lead 206	>100 m.y.
Potassium 40	1.3 b.y.	Argon 40	>100,000 years
Uranium 235	0.71 b.y.	Lead 207	>100 m.y.
Carbon 14	5,730 years	Nitrogen 14	0–60,000 years

as shown in Table 4.1. Undoubtedly some short-half-life isotopes contained in the early earth have so completely decayed that we cannot detect them; yet their daughter products must remain. For example, two isotopes of the gas xenon are thought to be decay products respectively of iodine 129 and plutonium 244, which do not occur naturally on earth (although they have been produced by nuclear explosions). Their half-lives are so short that, if they were present in the early earth, as is supposed, after only a few hundreds of millions of years they had diminished to unmeasurably small amounts.

Weathering or other chemical modifications of minerals, particularly heating during metamorphism, may disturb the original parent-daughter ratio. This obscures the true age of crystallization of the mineral, generally producing an apparent age that is "too young." In interpreting the geologic significance of isotopic dates, one must be alert for such *resetting of isotopic clocks*, as we shall see in later discussions.

MODERN ISOTOPIC DATING

THE DECAY SERIES USED

During the middle twentieth century, "absolute" or isotopic dating of minerals increased tremendously. By 1930, uranium-lead dating had progressed to the point that it could be said that the end of the Cambrian Period was about twice as old as the Permian. In 1934 the first isotopic time scale was published, and geology textbooks were beginning to recognize isotopic dating. Discovery about 1940 of the rubidium 87–strontium 87 (^{87}Rb–^{87}Sr) series and the proof in 1948 that radiogenic potassium 40 decays to the inert gas argon 40 (^{40}K–^{40}Ar) initiated the use of other isotopes as geologic clocks in addition to those of well-known uranium and thorium. The most important decay series in current use are shown in Table 4.1. The differences in rates of decay impose certain limitations on the relative usefulness of each series for minerals of different relative ages. Carbon 14, for example, decays so rapidly that after about 80,000 years there is not enough of the parent left to be measured accurately with present analytical techniques. Therefore, this isotope is useful only for late Pleistocene and Holocene times; it is especially useful in archeology, climatology, and studies of the movement of ground water and ocean circulation. Until recently, the range from about 100,000 years to 1 million years ago was the least accessible, but refined K–Ar techniques, the use of certain intermediate daughter products of uranium, and fission-track dating (described below) have helped to close this gap.

The long-half-life isotopes are of no use for dating minerals less than about 100 million years because no perceptible change will have occurred in such a short time. Uranium, thorium, and rubidium are especially applicable to Prepaleozoic rocks (i.e., over 700 million years). The potassium-argon series is slightly less useful for very old rocks. This is not because of its decay rate, but because the ultimate stable daughter product argon, being a gas, is lost from crystals more easily than other daughter products. Argon leakage is a particularly serious problem if a mineral has been reheated after initial crystallization through burial or metamorphism to temperatures exceeding 150°C. In short, potassium-bearing minerals can have their isotopic clocks reset more readily than certain other minerals. Dating metamorphic events, however, may be just as important as dating the original crystallization of a rock. In fact there is growing awareness that isotopic dates, strictly speaking, are dates of cooling of minerals down to some critical temperature at which daughter isotopes become fixed in mineral structures. This date appears in many cases to be geologically identical with the date of initial crystallization of the minerals, but there are important exceptions. Resetting of isotopic clocks is generally detected by studying at least two isotope series that have responded differently during a rock's history.

RATIOS AMONG DAUGHTER ISOTOPES

Dating by comparison of ratios of only daughter products is being used increasingly. Comparison of different daughter lead isotopes, such as ^{206}Pb and ^{207}Pb (as in Table 4.2), provides some of the most accurate rock dates possible for the entire uranium series. Because these daughter isotopes result from series with different decay rates, proportions of the lead isotopes change through time. Their changing ratio provides a special isotopic clock.

Isotope ratios for lead minerals (such as galena, PbS) also are used for dating by comparing the proportions of different lead isotopes

TABLE 4.2

Concordant isotopic ages from two different decay series for uraninite and from the ^{207}Pb–^{206}Pb ratio from a Prepaleozoic pegmatite dike, Black Hills, South Dakota *

Isotope ratios used	Age
^{206}Pb:^{238}U	1,580 m.y.
^{207}Pb:^{235}U	1,600 m.y.
^{207}Pb:^{206}Pb	1,630 m.y.

*After Wetherill et al., 1956, *Geochimica et Cosmochimica Acta*, v. 9.

present. One isotope, ^{204}Pb, is not known to be forming today by any radioactive decay process now operating on earth. Significantly, it is more abundant in old lead minerals, whereas geologically younger minerals contain progressively more of the *radiogenic lead* isotopes produced by uranium and thorium decay (Table 4.1). In other words, as decay has created radiogenic lead, some of that lead has been freed from its parent uranium-bearing minerals by melting or weathering, and has become recombined with other leads in later-formed lead minerals. Such dilution of the earth's original nonradiogenic ^{204}Pb has increased through geologic time in direct

FIGURE 4.5

Use of daughter lead isotopes for dating. The ratios of each of three radiogenic lead isotopes to nonradiogenic lead 204 (which has no known parent on earth) have all changed through time, but at different rates. Comparison of such ratios for a mineral provides a means of dating. Also extrapolation of the solid curves derived from many lead minerals of differing ages back to *primordial ratios* found in meteorites (see text) provides an estimate of total age of the earth. (*Based upon work of Nier, Patterson, and Tilton.*)

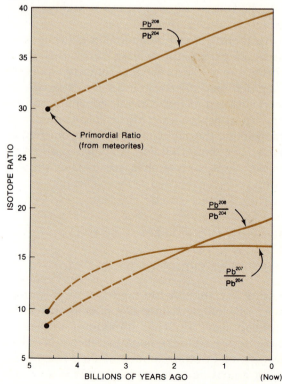

proportion to the continuous generation of new radiogenic lead isotopes through decay. The systematic change in the ratios among the lead species in younger and younger minerals provides a special isotopic clock because each species is formed at a different rate (Fig. 4.5).

An ingenious method has been developed for dating Pleistocene deep-sea clays using the ratio of protactinium 231 (half-life 34,300 years) to thorium 230 (half-life 80,000 years). Both are intermediate daughter products of uranium decay as are thorium 230 and uranium 234, which also are used (Fig. 4.2). All these isotopes are taken from sea water by clay minerals, and their ratios in sediments become a function only of the time since absorption from the water. This method allows dating of sediments as much as 175,000 years old or double the range of carbon 14 dating.

CARBON 14—A SPECIAL CASE

The value of carbon 14 for dating was discovered in 1947. Unlike most other isotopes used for dating, an unstable isotope of carbon is produced in the upper atmosphere. High-energy cosmic particles shatter the nuclei of oxygen and nitrogen atoms there, releasing neutrons, which in turn produce ^{14}C when they collide with other nitrogen atoms (Fig. 4.6). Unstable ^{14}C decays eventually to stable nitrogen 14. Most of the ^{14}C produced is oxidized to carbon dioxide (CO_2), some of which enters the hydrosphere. Production of ^{14}C in the atmosphere is assumed to be at a nearly constant rate; therefore ^{14}C in the atmosphere should have long since reached a steady state. That is, a constant balance should exist between its production and its decay such that the ratio of ^{14}C to stable carbon is assumed to be essentially constant through time.

From either the air or water, organisms acquire small amounts of ^{14}C along with the stable isotopes ^{12}C and ^{13}C. After death no new carbon is added, so that over time the ^{14}C content decreases progressively by decay. Any dead organic matter, such as wood, charcoal, some bone, human artifacts such as cloth, and also shells and young limestones are all potentially

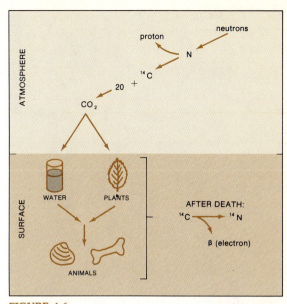

FIGURE 4.6

Constant generation of carbon 14 in the upper atmosphere by cosmic particle bombardment that produces neutrons, which convert nitrogen atoms to ^{14}C by expelling a proton. Some ^{14}C is ultimately incorporated into organisms, remains of which can be dated.

datable (Fig. 4.6). The number of ^{14}C and ^{14}N atoms present is so small that direct measurement is impossible except with sophisticated nuclear reactors. Instead, dating is generally done by use of a radiation counter to compare the *radiation activity* (i.e., the level of beta emissions) of the specimen with that of a standard modern sample whose activity is known very accurately. The number of emissions or "counts" in a given time period is proportional to the number of atoms of ^{14}C present; the older the specimen, the less its activity level. The rather short ^{14}C half-life of 5,730 years, together with the limit of sensitivity of counting devices, imposes an upper limit of measurable age of about 70,000 to 80,000 years by conventional methods. A new approach using a reactor to count actual atom ratios may push the limit back further. Since the industrial revolution, combustion and atmospheric nuclear testing have altered the ^{14}C content of the total carbon reservoir artificially, which has created problems for maintaining reliable modern standard samples.

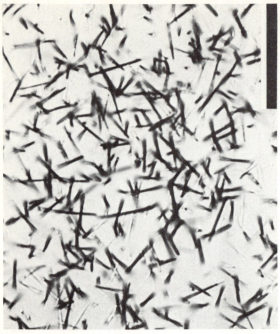

FIGURE 4.7
Fission tracks in a zircon crystal after etching. Each track is the result of spontaneous fission of an atom of uranium 238. The number of tracks within a given area is a function of age. (Black bar at upper right is 0.02 mm long). *(Courtesy J. D. Macdougall.)*

FISSION-TRACK DATING

The newest method of isotopic dating is based not upon spontaneous radioactive emissions, but upon a rarer effect of decay characteristic of a few isotopes, namely *spontaneous fission* of an atomic nucleus. In uranium 238, the principal fissionable isotope in nature, for approximately every 2 million decay emissions of alpha particles (helium nuclei), one atom undergoes fission instead. Its atomic nucleus splits, forming two energetic particles that repel one another. As these particles speed apart, they strip electrons from among adjacent atoms. The havoc produced leaves minute imperfection lines in minerals and glasses. When etched, these become visible under a microscope as the tracks of fleeing nuclear fragments, thus *fission tracks* (Fig. 4.7). One can count the density of tracks to obtain an estimate of the number of atoms that have already undergone fission, and then determine by conventional analytical procedures the

amount of ^{238}U still present. In principle, we can then calculate the age, as with other types of dating, from the ratio of original total ^{238}U atoms to the number of unfissioned atoms still remaining, times the known average rate of fission events for this isotope.

Fission-track dating is relatively simple and inexpensive, can be used for very small samples, and is almost as precise as other dating techniques. Unlike other methods, it can be applied to earth, lunar, or meteoritic materials of practically any age. Its initial successes were in dating archeological and young geological materials ranging from several hundreds to a few millions of years in age, that least-accessible range between ordinary carbon 14 and potassium-argon dating. Manmade glass colored by uranium-rich minerals as young as 30 years has been dated as has pottery as young as 700 years. Fission-track dating has a bright future.

DISCORDANT DATES

Daughter isotopes are usually trapped within the crystal structures of minerals in which they originated. But metamorphic recrystallization of a rock tends to purge minerals of many daughter products formed therein, because daughter isotopes are chemically dissimilar from their parent species, and so may not form true chemical bonds in the mineral. Helium and argon may diffuse completely out of the original rock. Other daughter products, however, will become incorporated in new minerals, while remaining parent isotopes begin a new decay history in purged crystals (Fig. 4.8). The latter grains will yield isotopic dates of the metamorphic, clock-resetting event, whereas if none of the premetamorphic daughter products were lost from the rock as a whole during its recrystallization, the parent-daughter ratio for the *total rock* still will indicate the time of original formation of the rock (Fig. 4.8). In refined dating today, several isotopic dates are determined for the total rock and for several different minerals; cross checks with different isotope series also are desirable. In this manner, the maximum possible can be learned about a rock's isotopic history.

FIGURE 4.8
The effect of metamorphism in redistributing daughter isotopes (circles) in a rock that crystallized originally 1,000 million years ago (1); (2) after 500 million years, some parent isotopes (dots) in a feldspar crystal had decayed; (3) metamorphism 480 million years ago drove the daughter atoms out of the crystal, but they were retained in the surrounding rock; (4) today dating of the feldspar would reveal the metamorphic event, while a total-rock date would reveal the original crystallization 1,000 million years ago.

Fission-track clocks also may be reset because moderate heating tends to cause healing of the tracks. Thermal energy excites atoms sufficiently to allow them to recover from fission damage and restore order within crystals. The result will always be ages that are "too young" (as with K–Ar resetting). As a consequence, fission-track dates commonly are discordant with other isotopic dates, but this can generally be detected and used to advantage in discerning complex rock histories. The tracks themselves may give a clue to resetting by showing two distinct groups, one of older, partially healed, short tracks and another of younger, normal-length ones formed after a heating event.

THE ACCURACY OF ISOTOPIC DATES

Several factors limit the accuracy of "absolute" isotopic dates. First, the statistical nature of the decay process itself means that, even under the best of conditions, only average decay rates are determinable. This introduces a small uncertainty, though for very old minerals it is negligible. A second source of uncertainty originates in the laboratory analyses of the isotope ratios. The mass spectrograph provides by far the most sensitive analyses of isotope abundance, with from ±0.2 to 2.0 percent accuracy possible.

Because of analytical limitations, isotopic dates generally are reported with an uncertainty figure expressed, such as 100 ± 5 million years. This means that analytical limitations do not allow one to say more than that the age lies between 95 and 105 million years. Under ideal conditions of rock freshness, analytical care, and relatively old material (with accurately measurable amounts of daughter material present), the analytical error for isotopic dates can be as low as ±2 percent. For a mineral 10 million years old, that would be only ±0.2 million years; for one 100 million years old, a mere ±2.0 million years; for one 1,000 million years old, ±20 million years. It is important to realize that some uncertainty exists for every isotopic date, but in an introductory book such as this, the uncertainty factor is of little consequence, and so it will be omitted from most dates quoted.

THE ISOTOPIC TIME SCALE

Most of the datable minerals containing radioactive elements originate in igneous rocks; therefore the bulk of isotopic dating must be confined to such rocks. Detrital minerals containing unstable nuclear species occur in sediments, to be sure, but to date them is not to date the time of sedimentation but rather the original time of crystallization of the mineral in a parent igneous (or metamorphic) rock or mineral vein prior to erosion, transport, and deposition in a sediment.

Because the relative geologic time scale is based upon fossiliferous sedimentary rocks, obviously it is urgent somehow to establish numerical dates for the sediments. Direct isotopic dates of sedimentary rocks are possible for only a few minerals that contain unstable species *and that crystallize in the environment of deposition*. Such a mineral is *glauconite*, a green micalike silicate mineral containing potassium ($FeKSi_2O_6$). It apparently forms only in marine environments, where deposition is slow. The potassium in glauconite consists in part of ^{40}K; thus, the mineral can be dated by the K–Ar method. Potassium-bearing feldspars form in certain sediments soon after deposition, and these, too, are potentially datable. Unfortunately neither glauconite nor such feldspars are very common, and they give rather unreliable results. The uranium method has been used for black shale containing minute amounts of

TABLE 4.3
The numerical isotopic time scale*

Millions of years		Era	Period	Epoch	
0	Phanerozoic time	Cenozoic Era	Neogene Period		0
					23
			Paleogene Period		
					65
100		Mesozoic Era	Cretaceous Period	Late	
					100
				Early	
					135
			Jurassic Period		
					192
200			Triassic Period		
					230
		Paleozoic Era	Permian Period		
					290
300			Carboniferous	Pennsylvanian Period	
					320
				Mississippian Period	
					350
400			Devonian Period		
					410
			Silurian Period		
					435
			Ordovician Period		
					485
500			Cambrian Period	Late	
					505
				Middle	
					530
				Early	
					560
600			"Eocambrian Period"		
700		Prepaleozoic time			700+
800	Rare, primitive animal fossils				
3,500	Oldest known fossil algae and bacteria				
4,000	Oldest dated rocks				
4,500	Meteorite dates / Earliest organization of definite mineral phases				
5,000	Approximate origin of protoplanet				

*Modified from J. L. Kulp, 1961, *Science*, v. 133; W. B. Harland, 1964 and 1971, *The Phanerozoic time scale:* London, Geological Society; and Cohee et al., 1978, *The Geologic Time Scale:* American Assoc. of Petroleum Geologists.

uranium-bearing minerals associated with carbon compounds. The uranium compounds formed in the sediments, and so they provide dates of sedimentation. Carbon 14 can be used in certain very young sediments.

Because most dating must be done on igneous or metamorphic rocks, it is necessary to relate such ages indirectly to the relative geologic time scale, and thus to establish a numerical age scale (Table 4.3). Here the basic principles of stratigraphy come into play again. A volcanic rock interstratified with fossiliferous sediments and containing datable minerals clearly is of the same geologic age as those sediments, providing an ideal point on the isotopic time scale. An intrusive igneous body, however, must be younger than all rocks through which it cuts. If the latter are fossiliferous strata, we have an older geologic age limit for the intrusion. If, perchance, younger fossiliferous sedimentary rocks rest unconformably upon the intrusion, or at least contain pebbles derived from it, then a maximum geologic age limit can also be established (Fig. 4.9), for in either case the sediments must be younger than the intrusion. If the age difference between the older and younger fossiliferous sediments is not great, the intrusion is closely dated in the relative geologic scale, and its isotopic date provides another precise control point on the numerical time scale. The scale is constantly being refined, and techniques for analysis of smaller and smaller amounts of isotopes are being developed continually.

Isotopic dating has value for the field geologist in the dating and correlation of local rock bodies where fossil evidence is poor or lacking. This is simply reversing the process used for originally establishing points on the isotopic scale. Because of cost this sort of use has received little application as yet, and commonly the different rocks in a limited area under study are so close in age that very great sensitivity is required to discriminate them isotopically.

MEMORABLE DATES

At last we can inquire intelligently of our planet's true birthdate. The chemical composition of the solar system must be relevant to its

FIGURE 4.9

Scottish rocks studied by Hutton showing how key dates are established for an isotopic time scale. The granite is post-Ordovician and pre-Devonian, thus is of Silurian age. A date for it establishes the Silurian Period as about 425 million years old. The date for a lava flow within the Old Red Sandstone establishes that part of the Devonian Period is about 370 million years old.

age as well as to its mode of origin. It is thought that most of the natural chemical elements in the universe evolved through thermonuclear reactions in stars; additional ones also developed outside the stars (as in the earth) through radioactive decay of unstable isotopes. The chemical elements in our part of the universe apparently evolved from the lightest and simplest (hydrogen, 75 percent; and helium, 24 percent) to the heaviest and rarest ones (such as uranium and lead). Presence still in the earth of moderate-half-life isotopes such as ^{40}K ($t_{1/2} = 1.3$ b.y.) places an upper limit on the formation of the earth. If the earth were more than 6 or 7 billion years old, all of those would have decayed.

No individual isotopic date represents the time of origin of a parent isotope, for most of them originated before the earth was formed (the majority of elements in the solar system probably are about 7 to 10 billion years old). The dates instead indicate times of incorporation of a given isotope into a mineral crystal. Isotopes had been decaying ever since their extraterrestrial origin, but only after incorporation into the closed system of a crystal could the decay products be trapped with the parent isotope to provide a measurable isotopic ratio useful for dating (Fig. 4.3). Uranium, lead, and rubidium dating have revealed that igneous rocks as old as 3 to 3.5 billion years occur in South Africa and northwestern Russia near Finland, and over 3.7 to 3.8 billion years in southwestern Greenland and in Minnesota. Thus some of the outer crust of continents is more than 3.5 billion years old. Some of the moon's crust appears to be of about the same age according to isotopic dating of the Apollo-mission specimens (3.5 to 4.2 billion years).

The oldest rock dates still do not represent the total age of the earth, for it must have taken the crust a while to form, as we shall see in the next chapter. An ingenious way to estimate the elusive total age was suggested through work with lead isotopes in the 1930s. Today there is about one part of "new" radiogenic lead to two parts of original "common" lead in the crust. Certain meteorites contain leads but no uranium or thorium parents; therefore the proportions of different lead species therein are assumed to have remained fixed since the meteorites formed. And lead ratios in these meteorites are taken as a standard representing the approximate primary or *primordial lead ratios* in the early earth. By comparing ratios of the four lead isotopes in varying-aged minerals, we can extrapolate back to ratios identical to the assumed primordial leads of meteorites. Such extrapolation indicates that the primordial ratio should have existed on earth about 4.6 to 4.7 billion years ago (Fig. 4.5). This date apparently represents the time when elements in the earth first became arranged into definite chemical compounds, and it is taken as the beginning of truly geologic time. It must be close to the planet's birthdate.

How much earlier the very first assembly of the earth began is more difficult to fix. Astronomers believe, from theories of stars and galaxies, that the Milky Way galaxy is about 10 billion years old, and that our sun formed about 6 billion years ago. Meteorites yield dates averaging between 4.3 and 4.6 billion years, nearly identical with that indicated by lead ratios for the initial formation of earth minerals. Moon rocks also yield dates approaching 4.0 billion years. Assuming that meteorites, as well as the planets and moons, always have been as closely linked to the sun as they are today, they can hardly be older than that star. Therefore, it appears that the earth as a solid sphere is not

more than 5.0 billion years old, and that the outer crust of continental areas started forming at least 3.8 billion years ago (Table 4.3).

The first bona fide records of fossil life come from rocks that are about 3.0 to 3.5 billion years old, which contain structures and chemical compounds formed by bacteria and possibly primitive algae. The first recognizable animal fossils are a mere 1 billion years old or so, but the first abundant animal record began only 560 million years ago; man is a mere 2 to 3 million years old. Therefore, for almost one-fourth of earth history, there is no recognized record of any life form, and only the last quarter, often referred to as Phanerozoic time, has an abundant record, particularly of index fossils useful for geologic dating and correlation of strata. If we were to proportion this book strictly according to the absolute time scale, 16 chapters should be given over to the Prepaleozoic, 1.81 to the Paleozoic, 0.6 to the Mesozoic, and only 0.25 to the Cenozoic. However, we actually do the opposite because we know the most about the latest, brief part of our planet's history.

SUMMARY

In this chapter we have seen that estimates of the age of the earth have been increasing constantly up to the present-day figure of 4.7 to 5.0 billion years. The maximum figure does not seem likely to increase very much more, however. Early dating methods, such as those based upon increasing salinity of the seas or total deposition of stratified sediments, were ingenious, but suffered from erroneous assumptions and inaccurate knowledge of true rates of the processes involved. We can never assume constant rates of most processes over long intervals of geologic time. So far, only radioactive decay shows apparent uniformity of rates, and even this has been questioned.

Lord Kelvin's calculation of age from the earth's internal heat gradient was equally ingenious and infinitely more persuasive because of the author's great stature and the compelling logic of his mathematics. But the most brilliant argument is no better than its weakest assumption! Kelvin's argument was doomed by the incorrect notion that all the earth's internal heat was residual from a hot origin. Radioactivity was discovered near the end of his life, and so he had no way of knowing that the earth contains its own internal heat-generating mechanism. Isotopic dating has exonerated Hutton and Lyell by showing the earth to be very old indeed, in fact many times older than even their wildest estimates.

From isotopic dating we can assess actual lengths of the geologic periods and derive a numerical calendar. From the results of extensive analyses, the dates of most period boundaries are now fairly well established as summarized in Table 4.3. Average length of the periods was 30 to 40 million years. The Cambrian apparently was longest (100 to 125 million years); the Silurian was the shortest (only 10 to 12 million years). Isotopic dating also allows rough comparisons of rates of change, such as mountain building, erosion, sea-level changes, and organic evolution. Throughout the book, we shall examine many events in terms of absolute, as well as relative, geologic age and rates.

Truly, isotopic dating of minerals has revolutionized geology in the twentieth century! Besides offering absolute dating, radioactivity also provides an internal source of earth heat, the chief form of terrestrial energy. Radioactive heat generation was about five times greater early in history than now because of the presence then of many more atoms of unstable isotopes, and it may have been sufficient to melt part of the interior for a time.

The discovery of radioactivity provided a dramatic example of inorganic evolution—the creation or mutation of new elemental species from older ones. Modern thermonuclear theories of stars and of the origins of the elements provide a compelling evolutionary conception for the entire solar system, as well as for the earth. This is explored more fully in the next chapter.

Key Concepts

activity, emissions
atomic nucleus
atomic number
cosmic particles
daughter-isotope-ratio dating
decay series
discordant dates
emanations (or emissions)
evolutionary earth
fission tracks
glauconite
half-life
isotopes:
   parent (unstable)
   daughter (stable)
isotopic dating
isotopic time scale
mass number
primordial lead ratios
radioactive decay
radiogenic lead
rate of decay
resetting of clocks
steady-state earth
total-rock dating

Readings

Burchfield, J. D., 1975, Lord Kelvin and the age of the earth: New York, Science History.

Cohee, G. V., et al., 1978, The geologic time scale: Tulsa, American Association of Petroleum Geologists.

Dalrymple, G. B., and Lanphere, M. A., 1969, Potassium-argon dating: San Francisco, Freeman.

Faul, H., 1966, Ages of rocks, planets, and stars: New York, McGraw-Hill. (Paperback)

Harland, W. B., 1964, The phanerozoic time scale: London, Geological Society, and 1971, Publication No. 5 (supplement).

Harper, C. T., 1973, Geochronology: Radiometric dating of rocks and minerals: Benchmark papers in geology: Stroudsburg, Dowden, Hutchinson, and Ross.

Hurley, P. M., 1959, How old is the earth?: New York, Doubleday-Anchor. (Paperback S 5)

Joly, J., 1909, Radioactivity and geology: London, Constable.

Macdougall, J. D., 1976, Fission-track dating: Scientific American, December, pp. 114–122.

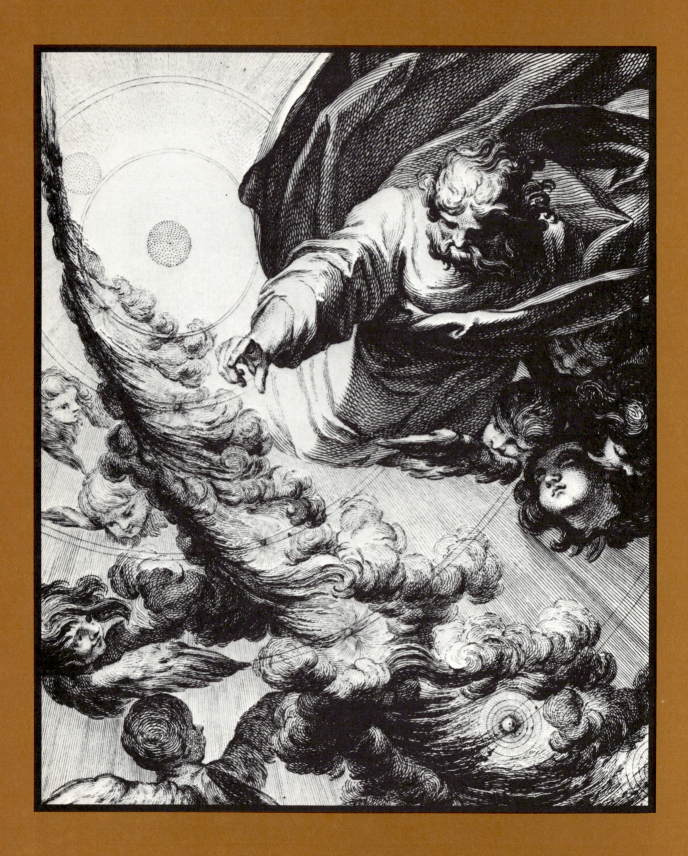

5

ORIGIN AND EARLY EVOLUTION OF THE EARTH

The eye sleeps until the mind
wakes it with a question.

Arabian Proverb

There is something fascinating about
science. One gets such wholesale
returns of conjecture out of such a
trifling investment of fact.

Mark Twain,
Life on the Mississippi

FIGURE 5.1
Artistic rendition of
Buffon's hypothesis of
planetary origin by
passage of a comet
near the sun. (*From de
Buffon, 1749,* Histoire
naturelle.)

5

Early explanations of planetary births had in common an assumption of a hot origin followed by cooling; they also assumed the sun to be considerably older than the planets. But we no longer accept the storybook pictures of the earliest earth as a fiery red ball of boiling lava shrouded in ominous steam clouds. G. L. de Buffon popularized the long-held hot origin in the eighteenth century with a vivid hypothesis that the gravitational pull of passing comets had torn away hot masses from the sun, which then cooled to form planets (Fig. 5.l).

Buffon's hypothesis was abandoned largely because space is so vast in relation to the nearly insignificant size of individual celestial bodies that collisions seemed unlikely (moreover, comets have weak gravity fields). Spatially, the planets in our solar system can be likened to a few lonely squirrels running around over the entire earth. After Buffon, several different suggestions were made also calling for condensation of the planets from a hot, gaseous cloud (nebula) surrounding the sun rather than being derived from the sun itself. But these hypotheses all had major defects. During our century, more complex hypotheses have been proposed to overcome difficulties of the earlier ones. While each improves upon others in certain respects, no fully satisfactory explanation exists even yet.

The average density of rocks exposed at the earth's surface is about 2.7 times that of water (i.e., 2.7 grams per cubic centimeter), with the very densest ones being slightly over 3.0 times. The densities of these rocks were easily measured long ago, and in the late eighteenth century a clever experiment showed that the overall bulk density of the entire earth is about 5.5 times that of water. Obviously there must be very dense material in the earth's interior. Although early cosmogonists postulated differential settling of matter according to density from an assumed chaotic early stage, resulting in a concentric arrangement of solids, liquids, and gases, determination of bulk density was the first evidence that the earth's interior is somehow zoned.

To explain the planet's origin as well as the mechanism and history of differentiation of solid earth, ocean, and atmosphere is the greatest challenge for geologists. Because it provides a unifying focus for all branches of geology and geophysics, the present chapter is devoted to this important though speculative phase of earth history. First we summarize opinions on the origin of the planets, then we shall review the evidence available for interpreting the present interior makeup of the earth, and finally we shall examine hypotheses for the origin of the crust, oceans, and atmosphere.

PROBABLE ORIGIN OF THE EARTH

A COLD BEGINNING

After 1900, a hypothetical origin of planets through aggregation of cold clouds of dust and gases became preferred. According to this scheme, much of the internal heating of the earth is presumed to have occurred subsequent to its aggregation through gravitational contraction and radioactive decay within the mass. The energy produced by such heating apparently has caused the continual disturbances seen in earthquakes, volcanoes, formation of mountains, and the like.

The first important suggestion of a cold origin of the planets was the *planetesimal hypothesis* of Chamberlin and Moulton formulated early in the twentieth century. (Chamberlin was a prominent American geologist; Moulton an astronomer.) Like Buffon's before, their hypothesis called upon close passage of another star and our sun. Pull of the intruder's gravity field presumably extracted solar gaseous material that condensed in space to form *planetesimals*. Then, unlike Buffon's hypothesis, countless numbers of these small, cold, meteoritelike particles aggregated to form planets. As the planets grew, their strengthening gravity fields attracted still more particles.

THE SOLAR NEBULA

In the mid-twentieth century, the planetesimal concept was modified by two astronomers, von Weizäcker and Kuiper. They, for the first time, tried to explain the simultaneous origin of the entire solar system in a unified *nebular hypothesis*. They considered both the differences existing among the planets as well as the recent discoveries of nuclear physics. It is important to realize that our sun is a rather ordinary star and that space contains countless ones like it. Many others probably have planets circling them, too. In trying to explain the origin of our solar system, then, we do not assume that it is unique. Indeed, we gain great insights from observations of other star systems.

The solar system is a small part of the Milky Way galaxy composed of an estimated 100 billion stars. The solar system rotates through the galaxy at 200 kilometers (120 miles) per second, and we are now in about the same relative position as in the late Paleozoic Era. Astronomers estimate that the Milky Way galaxy is about 10 to 15 billion years old. From the regularities of rotational direction and the revolution of planets in practically the same plane around the sun, as well as from the thermonuclear theory for stars, von Weizäcker and Kuiper reasoned that about 5 or 6 billion years ago a giant, disc-shaped interstellar cloud of gases and "dust" was spinning in our part of the galaxy. It probably consisted chiefly of hydrogen and helium with lesser amounts of oxygen, neon, carbon dioxide, methane, and ammonia. Rotation was induced by motion of the galaxy as a whole. Gravity would concentrate more mass at the center of the disc, and the resulting compression would raise the temperature there to perhaps several million degrees. At a critical temperature, thermonuclear reactions would begin, providing the sun's heat energy through hydrogen fusion.

Spin of the embryonic solar system would inevitably induce turbulence in the envelope of gas and dust—*the solar nebula*—surrounding the growing and heating sun. In the earlier version of the nebular hypothesis, planetesimal-like bodies tens of kilometers in diameter were thought to have condensed from the nebula first. Then, in regularly spaced eddies in the spinning nebula, cold planetesimals, dust, and gases tended to concentrate and collide rapidly to form so-called protoplanets (Fig. 5.2). Orbits of the protoplanets probably were erratic and variable, causing many collisions that resulted in fewer but larger bodies. The planetary orbits that have resulted are those that least disturb each other. (Comets are considered to be surviving planetesimals with very elliptical orbits.) Moons either may have formed in like manner as sisters in secondary eddies associated with the planetary ones, or they may have been captured later as slaves to their respective host planets. Isotopic dates for moon rocks suggest that our moon, wherever it formed, is the same age as the planet.

As the gravity fields strengthened, the protoplanets would enlarge by sweeping up still more material from the dust cloud. After perhaps several million years, the protoplanets were essentially complete, but were hundreds of times larger in diameter than now (Fig. 5.3). Although their masses were much greater, their densities would have been small, for they consisted at first primarily of hydrogen and helium with very minor amounts of the rare, heavy elements. As their gravity fields strengthened, the protoplanets would have contracted and become more dense. Apparently they also experienced a separation or *differentiation* of most heavier elements toward their centers. It was long assumed that the composition of all protoplanets initially was about the same, but because hydrogen and helium are very light, they could readily escape the modest gravity fields of the smaller planets. Accordingly, the four inner planets, like Earth (densities 5.1 to 5.5), presumably lost far more of their lighter

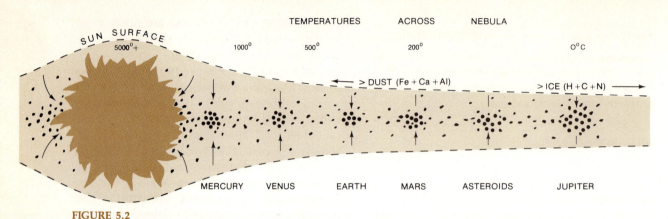

TEMPERATURES   ACROSS   NEBULA

SUN SURFACE 5000°+   1000°   500°   200°   0°C

> DUST (Fe + Ca + Al)   > ICE (H + C + N) →

MERCURY   VENUS   EARTH   MARS   ASTEROIDS   JUPITER

FIGURE 5.2

Cross section of spinning disc-shaped nebular cloud showing formation of the inner protoplanets by concentration of planetesimal bodies. Temperatures shown are temperatures of initial condensation in different parts of the nebula estimated from physical chemical theory applied to the mineral phases observed in meteorites. (Not drawn to scale.)

elements than did the larger, outer ones (densities 0.71 to 2.47). Further modification of compositions presumably resulted from a relative depletion of the light gases from the inner solar system by the sun's radiation (the "solar wind") literally blowing them and remaining interplanetary nebular gases to the outer part of the system, where they were swept up by the large planets, or escaped completely.

More recently it has been argued that the planetesimals and protoplanets accreted from the hot solar nebula with initially different compositions. Sophisticated chemical theory provides an accurate idea of the probable temperatures and order of condensation of different compounds from a hot nebula to form planetesimals. (The nebula was heated by its own gravitational contraction, remember.) The most common types of meteorites provide clues to composition of the original planetesimals (Table 5.1). The most refractory elements, calcium, aluminum, and titanium, would condense into chemical compounds in the range 1500 to 1300°C. Elements such as iron, nickel, cobalt, magnesium, and silicon would condense and form compounds in the range 1300 to 1000°C, while most remaining elements would condense between 1000 and 300°C; water would condense below 100°C. Today the relative iron content of the planets decreases with distance from the sun, suggesting that within the solar nebula there was a temperature gradient such that more iron condensed in the hotter neighborhood of the sun (Fig. 5.2). Moreover, the lesser density of the larger outer planets may be due in part to their iron occurring in a more oxidized state, because iron oxide minerals are less dense than pure iron.

Some geochemists believe that there is too much Fe and Ni left in the earth's mantle for one-stage accretion and differentiation of the

A. INITIAL ACCRETION

B. CONTRACTION AND DIFFERENTIATION

FIGURE 5.3

Hypothetical stages in the evolution of the early earth from a large, low-density, homogeneously heterogeneous protoplanet to a smaller, denser, internally differentiated planet. Note loss of light hydrogen (H) and helium (He) to space.

TABLE 5.1

Comparison of estimated abundances of principal elements in the earth and in meteorites*

Major elements	Earth's crust, %	Total earth, %	Meteorites, av. %
Iron (Fe)	5.6	35.0	29.0
Oxygen (O)	45.0	28.0	32.0
Magnesium (Mg)	2.0	17.0	12.0
Silicon (Si)	28.0	13.0	16.0
Sulfur (S)	0.03	2.7	2.1
Nickel (Ni)	0.007	2.7	1.6
Calcium (Ca)	4.2	0.61	1.3
Aluminum (Al)	8.2	0.44	1.4
Cobalt (Co)	0.002	0.20	0.12
Sodium (Na)	2.4	0.14	0.60
Manganese (Mn)	0.09	0.09	0.21
Potassium (K)	2.1	0.07	0.15
Titanium (Ti)	0.57	0.04	0.13
Phosphorus (P)	0.10	0.03	0.11
Chromium (Cr)	0.01	0.01	0.34

*Adapted from B. Mason, 1966, *Principles of geochemistry* (3d ed.): New York, Wiley.

protoplanet (Fig. 5.3) to be adequate. They propose instead a two-stage accretion by which about 80 percent of the earth first accumulated and the core differentiated as described above. Then, after the nebula had cooled somewhat, the outer 20 percent was accreted to form the upper mantle rich in the more volatile elements and carbon, but still containing considerable iron.

Some moons, and possibly even early planets, probably collided and broke up to form meteors, which career through space and occasionally fall into planetary or lunar gravity fields. The hundreds of asteroids may represent orbiting debris from a planet that was disrupted (or prevented from aggregating?) by Jupiter's powerful gravity field. Many meteors today probably come from the asteroid belt. Dating of moon rocks suggests extreme meteor bombardment until about 4 billion years ago with a great reduction since.

The large, low-density, outer (Jovian) planets presumably condensed from the cool outer portion of the nebula wherein ices of water (H_2O), ammonia (NH_3), and methane (CH_4)

were prevalent. These planets are notably depleted in the heavier elements such as iron (Fig. 5.2). Apparently condensation was most efficient in the neighborhoods of Jupiter and Saturn so that growth was fastest and greatest there.

Either variation of the nebular hypothesis explains simultaneously the formation of sun and planets and accounts for regular spacing and concentric orbits, as well as for the direction of rotation and revolution of most planets. Coincidence of isotopic dates both for meteorites and the initial mineralogic organization of the earth and moon also is explained. In short, because it accounts for many seemingly diverse phenomena and provides simultaneous origin of sun and planets *as part of one great evolutionary process*, we believe some variation of this hypothesis to be the best available working model of the solar system. Table 5.2 summarizes the probable time scale for solar system evolution from the initial contraction of an interstellar cloud to the end of the protoplanet stage about 4 billion years ago. But further analysis of this fascinating subject belongs to astronomy.

MOONS

Planetary satellites probably formed from the nebula like the protoplanets, but did this occur within their present orbits, or were they captured later by the planets from erratic orbits? On

TABLE 5.2

Possible time scale for early solar system evolution during the first 1 billion years beginning with contraction of an interstellar cloud. Time zero is assumed to have been about 5 billion years ago.*

0	Spinning nebula of dust, gas, and ices.
10,000	Protosun separating from nebular cloud.
100,000	Asteroid-sized planetesimals accreting from nebula. Jupiter and Saturn forming in outer nebula.
1,000,000	Inner four terrestrial planets forming from nebula. Separation of cores of terrestrial planets.
1,000,000,000 years	Last planetesimals swept up and large lunar craters formed by last major impacts.

*Adapted from J. A. Wood, 1979, *The solar system:* Englewood Cliffs, N. J., Prentice-Hall.

FIGURE 5.4
Footprints on the moon. Mount Hadley rises 4,400 meters above the dust-covered plain. Note apparent planar structures on the mountainside inclined downward to the left, possibly indicating an outcrop of rocks tilted by meteor impact or volcanic eruption. *(Apollo 15 surface photo; NASA Photo AS15-87-11849. Courtesy Paul D. Lowman and National Aeronautics and Space Administration.)*

the other hand, were they split off from planets? All origins seem possible. Our own moon (Fig. 5.4) is almost a planet itself, being nearly as large as Mercury and having a differentiated crust, mantle, and possibly a small core; it is as old, too. The low (maria) areas of the moon have rocks most like terrestrial basalt, but they are much richer in titanium, zirconium, yttrium, and chromium, whereas they are relatively depleted in sodium (Na), potassium (K), and rubidium (Rb). Their densities are 3.1 to 3.5. The lunar highlands contain less-dense rocks rich in plagioclase feldspar (anorthositic gabbro). These lunar "continents" are rich in aluminum like the earth's continents, but are depleted in sodium, potassium, silicon (Si), and rubidium. Apparently the moon's core is poor in iron and nickel, but its gabbroic crust is six to eight times thicker than the earth's crust. Judging from heat-flow measurements, the moon is two to three times richer in isotopes of uranium (U) and thorium (Th). Isotopic dates of different moon rocks indicate extensive melting of its surface to a depth of at least 100 kilometers around 4.5 to 4.6 billion years ago followed by differentiation of its interior between 4.4 to 4.5 billion years ago. Meteor bombardment and cratering were very intense until about 4 billion years ago. The mare basalts filled the large craters mainly between 3.0 and 4.0 billion years ago. For at least the last 2.5 billion years, apparently the moon has been nearly static. Just how long it has been a slave to the earth is unclear.

Much less is known about the satellites of other planets, of course, but recent space probes

FIGURE 5.5
Jupiter viewed from the Voyager 1 spacecraft when it was 28.4 million kilometers from the planet, February 5, 1979. The dark moon Io can be seen near the right edge of Jupiter's disc and bright Europa to the right. Colorful bands with giant turbulent eddies, especially the Great Red Spot (several times the earth's diameter), are clearly visible in Jupiter's atmosphere. Jupiter's larger satellites and the Great Red Spot were first observed by Galileo with his crude telescope 400 years ago. (*NASA photo P-21083, courtesy Jet Propulsion Laboratory.*)

have given some tantalizing hints about a few of Jupiter's 13 moons (Fig. 5.5). Ganymede, Callisto, and Europa have densities of only 2 to 3, and probably consist mostly of ice. Ganymede and Callisto have many impact craters, but Europa does not. Is its surface too soft to preserve them? Io is orange-colored, suggesting a high sulfur content. It lacks impact craters, but has a number of large *volcanic craters (calderas)*. No less than six apparent plumes from active eruptions were photographed by Voyager 2 in 1979! Very likely the volcanoes erupt pure sulfur magmas and gaseous sodium. The eruptions may be caused by a strong gravitational tug-of-war between Jupiter and one or two of the other satellites with Io caught in the middle (Fig. 5.5).

DISTRIBUTION OF THE ELEMENTS

Before proceeding further, we must compare the overall chemical composition of the earth with that of the solar system as a whole. Only about one-quarter of the one-hundred-odd known elements are at all common, over 95 percent of the universe being composed of only the two lightest, hydrogen and helium. In a general way, the heavier elements tend to be progressively less abundant. With the development of nuclear physics in the 1930s, it was concluded that most of the elements evolved successively from hydrogen through a series of complex thermonuclear reactions in stars. Only very large celestial bodies develop sufficient internal gravitational pressure (and temperature) to trigger such reactions and form stars, which is why the puny planets are just planets.

We must emphasize that practically all the atoms of various elements making up our earth were formed long before the protoplanets. A relatively few, however, have evolved within the earth. These are the daughter products of radioactive decay of initially incorporated unstable isotopes. Only 20 elements can be called major constituents of the earth, and some of these are really not very conspicuous. To the 15 elements listed in Table 5.1 must be added hydrogen (H), carbon (C), nitrogen (N), neon (Ne), and argon (Ar). Helium (He), though second only to hydrogen in the universe, is almost totally missing from earth by loss to space. But why is there any hydrogen, which is still lighter? Because much was trapped by combining with oxygen (O) to make relatively heavy water molecules (H_2O). Clearly the proto-earth underwent some changes of bulk composition and more profound physical changes as it shrank in size and became much more dense under the influence of its own gravity field (Fig. 5.3).

Space-probe data suggest that Mars and Venus have crusts like the earth, and strongly support differentiation of these bodies as well. What differences in composition of the solid planetary bodies exist probably reflect relative positions within the solar nebula (Fig. 5.2). Their atmospheres differ considerably, however. Venus has an atmospheric pressure 90 times that of Earth's, and its atmosphere is composed of 97 percent carbon dioxide (CO_2) and <0.1 percent hydrogen (H_2); water and free oxygen (O_2) are essentially nonexistent. Mars has a much less dense atmosphere, with nitrogen (N_2) and CO_2 dominant, and water vapor and oxygen only a fraction as abundant as in Earth's atmosphere. Assuming that all of the inner four (terrestrial) planets underwent differentiation similar to that of the earth, then their atmospheres were also derived by the lighter elements escaping from their interiors. Subsequently, the lightest elements escaped to space, with the smallest planets and moons losing the most. For the remaining gases, chemical changes have produced other differences among the atmospheres. For example, whereas most of Venus' CO_2 still remains in its atmosphere, most of the earth's is stored in carbonate rocks ($CaCO_3$) and by organisms in the biosphere. As we shall see, the role of organisms in modifying the composition of the earth's surface has been profound.

PHYSICAL NATURE OF THE EARTH'S INTERIOR

ANALOGIES DRAWN FROM METEORITES

A major clue about the earth's interior came from, of all things, meteorites. In 1873 an American geologist named J. D. Dana studied the mineral composition of various types of meteorites and suggested that the interior of the earth might be made up of similar materials. If so, it would account nicely for the bulk density of the earth, for meteorites are much denser than crustal rocks. The assumed kinship of meteorites at least with the inner planets is strengthened by very similar ratios of certain elements in both the earth and meteorites (Table 5.1).

TABLE 5.3
The major zones of the earth[*]

	Thickness or radius, km	Volume, 10^{27} cm³	Volume, %	Mean density, g/cm³	Mass, 10^{27} g	Mass, %
Atmosphere	(Outer atmospheric limit indefinite)			0.12×10^{-8}	0.000005	0.00009
Hydrosphere	3.80 (av.)	0.00137	0.1	1.03	0.00141	0.024
Crust	30	0.015	1.4	2.8	0.043	0.7
Mantle	2870	0.892	82.3	4.5	4.056	67.8
Core	3471	0.175	16.2	10.7	1.876	31.5
Total earth	6371	1.083	100.0	5.52	5.976	100.0

[*]After B. Mason, 1966, *Principles of geochemistry* (3d ed.): New York, Wiley.

There are two major types of meteorites, metallic iron-nickel ones (25 percent) and non-metallic or stony ones (75 percent). While their combined, overall composition approaches that of the total earth, it shows at least five times as much iron and only three-fourths as much oxygen and silicon as are found in crustal rocks (Table 5.1). Why should there be different kinds of meteorites, and why are they so different from the accessible earth's crust? According to the protoplanet hypothesis, all embryonic planets underwent gravitational collapse simultaneously, and presumably density differentiation was universal among them; recent studies of the moon, Venus, and Mars confirm it. If one or more bodies disintegrated later, fragments both of very dense core material (presumably metallic) and less dense outer material (stony) would have been cast into space as meteorites.

The prevalent interpretation of meteorite origin is that metallic meteorites do represent, as Dana suggested a century ago, materials like the earth's interior. The similarity of isotopic dates for initial organization of both meteorites and earth as noted in Chap. 4 strengthens the comparison.

SEISMOLOGICAL EVIDENCE OF INTERNAL STRUCTURE

Below the deepest mines and wells, we have compelling, though indirect, geophysical evidence of internal density zonation. Results from seismology, especially, led to recognition of the zones in the earth's interior shown in Fig. 5.6 and in Table 5.3.

A near-surface discontinuity in earthquake shock-wave transmission is named the *Mohorovicic discontinuity or Moho*; the velocity of wave propagation accelerates below it. The Moho generally has been taken as the base of the earth's crust. It lies an average of 5 kilometers below the sea floors and 33 kilometers below continental surfaces, but is as deep as 70 kilometers under some portions of continents (Fig. 5.7). But a zone of relatively low seismic velocity and low rigidity—probably due to partial melting—is present below the Moho within the upper mantle (between 60 and 250 kilometers deep). Apparently it is of great importance in major structural adjustments. Indeed, it is now realized that this *low-velocity zone* is more fundamental than the Moho. Everything above this zone is today termed *lithosphere* ("rocky sphere") of which the crust and Moho are part (Fig. 5.7). A second great seismic discontinuity at about 2,900 kilometers below the surface separates the solid mantle from the liquid outer core (Fig. 5.6).

FURTHER EVIDENCE ABOUT THE CORE FROM THE MAGNETIC FIELD

The earth's magnetic field must be related to overall earth physics, and it has had a history which, as we shall see later, has a profound

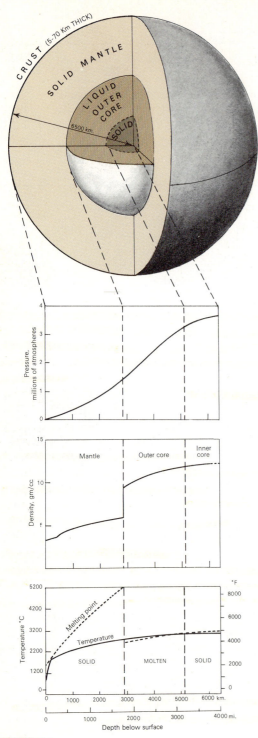

FIGURE 5.6

The earth's interior. (*Adapted from A. N. Strahler, 1960, Physical geography, with permission of John Wiley and Sons; and A. N. Strahler, 1963, The earth sciences, with permission of Harper & Row, Publishers.*)

bearing upon the interpretation of the history of the crust. About A.D. 1600 the field was recognized as similar to one surrounding a gigantic two-poled bar magnet aligned with the earth's axis. Precise observations indicated that the field was not completely fixed; the poles were found to migrate in a crudely circular path spanning approximately 20 degrees of longitude in five centuries. Also, it has been shown recently that the field has reversed its polarity many times. During the last 5 million years, it has reversed 10 times, but the period of reversals has varied considerably, as we shall see in Chap. 6.

The cause of the magnetic field has not been fully explained, though interesting hypotheses have been proposed. Several features must be explained: (1) the field has two poles located near the geographic poles; (2) it shows irregular variations both in position and polarity; (3) these variations bear no relation to the crust, and so must have their origin deep in the earth.

The most widely accepted view, which was first suggested by the Frenchman Ampere in 1820, is that internal electric currents produce a field much like that formed around a wire transmitting a current. The assumed electrical currents require some driving mechanism to maintain them, however. An iron-nickel-rich core (similar to metallic meteorites) would be a good electrical conductor, and a fluid outer part of such a core would allow mechanical motion of electrical charges. Physicist W. M. Elsasser suggested in 1939 that interaction of motion and electrical currents in the outer core could generate and sustain the magnetic field (Fig. 5.8). Near coincidence of the field and spin-axis orientations is consistent with this explanation. Elsasser assumed that the outer core is an electrical conductor and is in motion due to heat convection therein. Recently it has been suggested that wobble (precession) of the earth's rotational axis, coupled with the Coriolis effect (see p. 249), may drive the dynamo instead. In any case, it is well known that if a conductor moves within a magnetic field, an electric current is generated within it. An analogy can be made with a conventional electric generator. The earth's core is the conductor moving within an electromagnetic earth field; it is as though

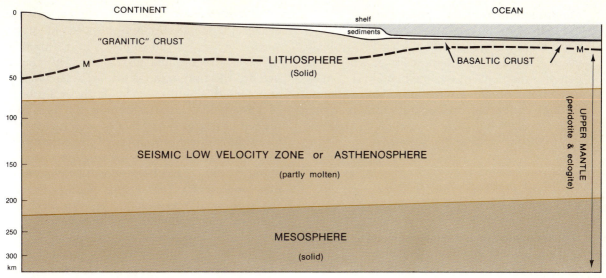

FIGURE 5.7

Structure of the upper 300 kilometers of the earth showing relations among the crust, upper mantle, and lithosphere. The M seismic discontinuity (or Moho) long chosen to define the crust-mantle boundary now seems to have been overemphasized. Although M probably represents a compositional contrast (see Fig. 5.10), the contrast between lithosphere and the low-seismic velocity zone is physically more profound. The zonation shown is based almost entirely upon seismic velocities; compare Table 5.3 for other properties. (Not to exact scale.)

FIGURE 5.8

Cutaway diagram illustrating Elsasser's dynamo theory of the earth's magnetic field. Differential rotation of solid mantle (faster) and fluid core (slower) presumably induces eastward flow of electrically charged, ferromagnetic material in the outer core. Flow of electrical charges induces a strong field (large curved arrows), which is maintained by the earth's rotation. Large anomalies may be caused by complex eddies in the outer core (small curved black arrows).

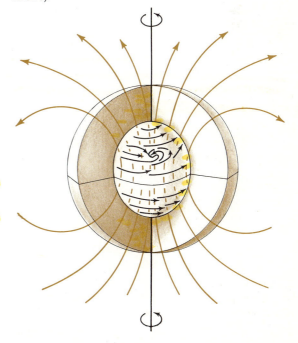

the current being generated were fed back through the electromagnets to maintain the field.

According to Elsasser's *dynamo hypothesis*, the field results directly from core motions, and rotation of the earth affects both orientation and strength of the field. Core motions, which otherwise might be random in direction, are preferentially oriented by the earth's spin, producing a strong field whose axis on the average over time closely parallels the spin axis. The reversals of polarity are not fully understood, but may result from instabilities of motion within the fluid outer core. Apparently the reversals are completed suddenly in a geologic sense.

The field shields living organisms from damaging cosmic radiation. That the field is old is suggested by presence of a record of shallow marine life on earth for over 3,000 million years and of land life for at least 400 million years.

Measurements of remanent magnetism in very old rocks indicate approximately constant average intensity of the field for at least 2,700 million years. There is even a hypothesis of accelerated genetic mutations on the one hand and of mass extinctions of life on the other by radiation during temporary weakening accompanying pole reversals, but more compelling evidence is needed for us to take this seriously.

CHEMICAL COMPOSITION OF THE DEEP INTERIOR

Besides analogies with meteorite compositions and evidence from the magnetic field, there is other evidence that can be brought to bear upon the important question of composition of the earth's interior. First we assume that the interior consists chiefly of relatively common elements (see Table 5.1) arranged in mineral phases that would yield observed seismic velocities and densities at inferred temperatures and pressures. Observation of heat flow from the interior indicates that radioactive isotopes cannot now be very concentrated below the crust, or the earth would be much hotter. Finally, analogies long have been drawn from the densities and seismic properties of known surface-rock materials and meteorites in order to satisfy conditions at depth.

From the sum total of all evidence, it was reasoned years ago that the bulk composition of the crust beneath oceans is close to that of *basalt* (density 2.9), continental crust is close to that of *granitic rocks* (density 2.7), and the upper mantle seems to be most closely comparable with the unusual *ultramafic rocks* ("entirely mafic"; density about 3.3), which occur locally within structurally complex mountain belts. Ultramafic rocks such as *peridotite* consist almost exclusively of dark, so-called *mafic minerals* (ma-, magnesium, and -fic, ferric or iron), such as the pyroxenes, olivine, and garnet, and lack feldspar and quartz, so are relatively poor in potassium and silica. They are similar to the nonmetallic (stony) meteorites.

An ultramafic mantle of density 3.3 to 4.5 still would not satisfy the overall earth density of 5.5. Therefore, the core must be much denser because it makes up only 16 percent of the earth's volume; it must be 10 to 12 times as dense as water (Table 5.3)! Iron-nickel meteorites are the only natural materials known that approach this figure, and so partly by a process of elimination, composition of the core was judged to be much like that of such meteorites. Only the seismic properties of the core can be measured, and they give information primarily about rigidity. It has been shown by comparing observed density and seismic characteristics of the core with experimental results for many chemical elements that the transition metals (e.g., iron, nickel, cobalt) satisfy all known conditions. Of these, only iron and nickel are abundant enough in nature to form the major part of the core. But if the core were pure iron-nickel, it would be more dense and would have greater seismic velocities than are observed. Therefore, some lighter element must be present to form an iron-nickel alloy. About 10 percent sulfur would best satisfy geophysical observations, and would also explain why the rest of the earth seems to be greatly depleted in sulfur.

Large masses of peridotite are seen in many mountain belts where they have been exposed by erosion following profound structural movements, which carried them up many kilometers from the mantle. These dense, silica-poor rocks provide the most direct evidence of the make-up of the upper mantle. Unusual small bodies of another ultramafic rock called *kimberlite* also represent samples of the mantle punched up through the crust by explosive gases. Kimberlite generally contains blocks of peridotite and may contain the form of pure carbon called *diamond*, as it does at Kimberley, South Africa, where the rock was named. Like all of the other minerals characteristic of the ultramafic clan, diamond is very dense as well as hard, which reflects the high-pressure environment of the mantle. On the other hand, the less-dense form of carbon called *graphite* (as in your "lead" pencil) forms near the surface at much lower pressures than does diamond.

Volcanic eruptions in oceanic areas such as the Hawaiian Islands also provide samples of the mantle, but due to partial melting, their composition does not directly reflect that of the mantle. Most of the oceanic lavas originate at depths of around 40–60 kilometers, which is the depth range of earthquakes that accompany the eruption episodes. But those lavas yield olivine-rich basalt, which is more siliceous than peridotite. Some of the basalts, however, contain inclusions of ordinary peridotite brought up intact. How is it that lavas can be derived from mantle peridotite yet have a different composition? A complex chemical fractionation of elements called *partial melting* preferentially gives rise to basaltic magma that is richer in silica than the average mantle (see the end papers). It is variations of pressure, temperature, and water content in the area of melting that account for this partial melting, further discussion of which is beyond the scope of this chapter.

THE DAWN OF EARTH HISTORY

CHEMICAL AND THERMAL EVOLUTION

Let us now explore the implications of our model of the earth's interior in terms of a probable thermal history and overall chemical evolution. An originally molten planet certainly would differentiate readily according to relative weight of the elements as it began cooling, but how could a relatively cold aggregation of planetesimals do so? Initial heating by gravitational contraction of the protoplanet would raise the temperature at the center to about 1000°C, and the early earth must have had about five times as much radioactive heat production as now (Fig. 5.9), which would have raised the temperature another 2000°C or so. A third source of heat would be large meteorite impacts, for which the moon provides abundant evidence prior to 4 billion years ago.

Assuming an original bulk composition for the earth somewhat like that of chondritic stony meteorites (Table 5.1), American geophysicist Francis Birch postulated that, within its first 500 million years or so, the protoplanet's deep

FIGURE 5.9
Estimates of changing heat production from major radioactive isotopes since the origin of the earth. Note that most heat results from decay of ^{235}U and ^{40}K. (*Adapted from B. Mason, 1966,* Principles of geochemistry *(3d ed.); by permission of John Wiley & Sons, Inc.*)

interior heated up to or near the melting point of iron and nickel (2000°C at depths of 800–1,000 kilometers). These two elements presumably separated and sank to form the core between 5.0 and 4.7 billion years ago, the latter time being fixed by the earliest isotopic dates we have for the organization of distinct mineral phases both in the earth and in meteorites. Lighter silicate minerals formed the mantle; separation would have been greatly facilitated by enormous heat energy released by the sinking of iron to the core. Persistence of a liquid outer core beneath a solid mantle may seem puzzling at first, but magnesium- and iron-rich silicate minerals typical of the mantle have relatively higher melting points than iron at any pressure and temperature combination. Present temperature in the outer core is estimated to be near 3000°C, that is, above the melting point of core material but below that of mantle silicate minerals at that depth (Fig. 5.6). Recent estimates of the temperature necessary to melt pure iron and nickel in the early earth seem impossibly high, but the presence of sulfur with iron would lower the melting temperature by almost 1000°C. This makes the major differentiation of the earth possible within a more plausible temperature range.

Is the earth as a whole today heating up, cooling off, or in a thermal steady state? Volcanoes, hot springs, and high temperatures in deep mines and wells provide proof that heat is still flowing from the interior, and although much of the original heat-producing radioactive decay has ceased (Fig. 5.9), insulation by the crust and mantle impedes heat dissipation so much that the interior may not yet have cooled very much. In any event, the interior is still evolving. Consequences of this chemical and thermal evolution are important, for the heat energy causes structural disturbances suffered by the earth. We shall see that mountain building, the most conspicuous of such disturbances, has gone on throughout earth history. Eventually structural turmoil must decline, as it did long ago on the moon, when the thermal energy reservoir is exhausted. But we cannot judge where in its energy dissipation history the earth is today.

ORIGIN OF THE CRUST

We now turn our attention to the outer crust of the earth, which is of chief concern for the remainder of the book. Origin of the crust must fit into the overall chemical segregation of the earth by further separation of light mineral phases from original mantle material. It was suggested in 1928 by N. L. Bowen that the crust of continents, whose average properties approach that of granite, originated by chemical

FIGURE 5.10

Present distribution of major chemical elements and also uranium, thorium, helium, and argon, which reveals the elemental differentiation of the atmosphere, sea water, and crust from the earth's mantle. For each division, elements are listed in approximately decreasing abundance from left to right. (Sea-water depth is greatly exaggerated to accommodate lettering.)

fractionation from a mantle with properties closer to olivine-rich basalt or stony meteorites. It is theoretically possible by *fractional crystallization* and concentration of relatively lighter elements from a basaltic magma to distill off a granitic residue representing 7 or 8 percent of the original magma. The entire granitic crust totals volumetrically only about 0.01 percent of the present mantle! According to this concept, then, only a relatively small, light-element fraction of a "basaltic" mantle need have been extracted to make the continents, which suggests that a great deal more fractionation is still possible.

As the mantle differentiates, relatively light silicon, oxygen, aluminum, potassium, sodium, calcium, carbon, nitrogen, hydrogen, and helium, and lesser amounts of other elements, rise to the surface to form the crust, sea water, and atmosphere (Fig. 5.10). The most familiar rock-forming minerals are produced in the proportions shown in Table 5.4. Atomic weight is not the only factor to determine their ultimate residence, however, because atomic size and electrical charge are more important for heavy elements such as uranium, thorium, and the rare earth elements. They are too large to fit into closely packed crystal structures of dense sili-

TABLE 5.4

Proportions of commonest rock-forming minerals in the crust*

Alkali feldspar (KAlSi$_3$O$_8$ and NaAlSi$_3$O$_8$)	31.0%	Total feldspar 60.2%
Plagioclase feldspar (NaAlSi$_3$O$_8$↔CaAl$_2$Si$_2$O$_8$)	29.2	
Quartz (SiO$_2$)	12.4	
Pyroxenes (Ca [Mg, Fe] Si$_2$O$_6$)	12.0	
Iron and titanium oxides (magnetite, hematite, ilmenite)	4.1	
Biotite mica (complex K, Mg, Fe, Al, Ti hydroxyfluo silicate)	3.8	
Olivine ([Fe, Mg]$_2$ SiO$_4$)	2.6	
Muscovite (complex K, Al hydroxy-fluo silicate)	1.4	
Other minerals	3.5	
	100.0	

*After L. H. Ahrens, 1965, *Distribution of the elements in our planet:* New York, McGraw-Hill.

cate minerals found in ultramafic rocks, but are comparable in size to potassium or calcium. It is not surprising, therefore, to find them preferentially concentrated in the more open structures of the minerals found in crustal rocks. Everything that we have learned about the distribution of elements in the earth, meteorites, the moon, and other planets points to a wholesale chemical differentiation of lighter crust from a heavier interior.

If the entire earth had as much radioactive material as the continents, the globe would be entirely molten. Even though not exactly abundant in granites, the 0.0006 percent uranium there is 10 times as much as in basalt and 1,000 times more than in ultramafic rocks! Isotopic dating of continental rocks suggests that continental crust, for whatever reasons, did not begin to be stable until about 3.8 to 3.9 billion years ago, or nearly 1 billion years after separation of the core and mantle. On the other hand, the entire surface may never have been molten; probably crust formed, was assimilated and re-formed repeatedly. With roughly five times present radioactivity, apparently there was ample heat to keep the crust unstable through almost continuous volcanic activity. But for at least the past 3.5 billion years, large portions of continental crust have persisted.

ORIGIN AND EVOLUTION OF THE ATMOSPHERE AND OF SEA WATER

THE ATMOSPHERE

Origin of the atmosphere and oceans also should fit into the great chemical differentiation of the earth. Even the origin of life seemingly was linked with differentiation, for it required certain unique chemical characteristics of the early atmosphere and oceans. Once formed, life influenced the further development of both. The atmosphere has undergone important changes through time, and is evolutionary like the solid earth. To investigate the possible nature of its development, we compare our present atmospheric composition with that of a large, outer planet, Jupiter, which apparently retained original gases from the solar nebula, and with meteorites, which are regarded as resembling the early protoplanets in composition. From Table 5.5 it is clear that these compositions are all markedly different, and so no obvious explanation of the earth's atmosphere is immediately forthcoming therefrom. In the 1950s, some suggested that the earth's atmosphere may have evolved by complex chemical changes from a Jupiter-like primitive atmosphere to its present composition. Such an origin is no longer favored, however, due to chemical reasons that we cannot develop here.

Two major hypotheses exist to explain the evolution of the present atmosphere from a primeval one, and there is no reason why both mechanisms could not have contributed. There is one common feature, namely the assumption that considerable free hydrogen and helium have been lost to space because of their very small atomic masses (Fig. 5.2). What hydrogen remained was mostly oxidized to form sea water. The greatest single problem facing any hypothesis of atmospheric development is to explain the abundance of free oxygen, which is missing from both Jupiter and the meteorites (Table 5.5).

TABLE 5.5
Comparison of present atmosphere with the gases of other planetary bodies and volcanoes. Gases are listed in approximate decreasing order of abundance.

Earth's present atmosphere			Jupiter's atmosphere	Meteorites (average)	Volcanoes (average)	Geysers and fumaroles
Major:						
Nitrogen	(78%)	Stable	Methane	Carbon dioxide	Water Vapor (73%)	Water vapor (99%)
Oxygen	(21%)	Unstable (reacts with Fe and C)	Ammonia	Carbon monoxide	Carbon dioxide (12%)	Hydrogen
Argon	(0.9%)	Stable	Hydrogen	Hydrogen	Sulfur dioxide	Methane
Carbon dioxide	(0.03%)	Unstable (reacts with silicates)	Helium	Nitrogen	Nitrogen	Hydrochloric vapor
Water vapor	(Variable)	Unstable (reacts freely)	Neon	Sulfur dioxide	Sulfur trioxide	Hydrofluoric vapor
Minor:	(Traces only)					
				Methane*	Carbon monoxide	Carbon dioxide
Ozone		Unstable (dissociates)				
Neon		Stable		Nitrous oxide*	Hydrogen	Hydrogen sulfide
Helium		Stable		Carbon disulfide*	Argon	Ammonia
Krypton		Stable		Benzene*	Chlorine	Argon
Xenon		Stable		Toluene*		Nitrogen
Hydrogen		Unstable (reacts with oxygen to form water)		Naphthalene*		Carbon monoxide
				Anthracene*		
Methane		Unstable (dissociates)				

*Only present in nonmetallic carbonaceous chondrites.

OUTGASSING HYPOTHESIS

The generally accepted explanation, which was first urged by geologist W. W. Rubey in 1951, invokes an origin of most atmospheric gases from the interior by gaseous transfer (outgassing) to the surface through igneous processes. It had been suggested in 1910 that the traces of atmospheric helium had originated by radioactive decay of uranium, and in 1937 it was suggested that atmospheric argon, which is surprisingly abundant compared with the other rare gases, was derived similarly from decay of potassium 40 in the earth. Both the present abundance of argon 40 and the decay rate of potassium 40 are in good agreement with this hypothesis.

Volcanoes and hot springs are known to expel steam, carbon dioxide, nitrogen, and carbon monoxide (Table 5.5). By assuming *outgassing* to be a normal part of overall density differentiation, we could explain all the atmospheric nitrogen, carbon dioxide, helium, argon, and water vapor. In addition, the overwhelming preponderance of steam in the expelled gases provides a ready source of sea water. The greatest single problem facing any hypothesis of atmospheric development is to explain the abundance of free oxygen, which is missing from Jupiter, the meteorites, and volcanic gases (Table 5.5). It has long been assumed that the early atmosphere did not have free oxygen, and two hypotheses have been proposed to explain its appearance.

PHOTOCHEMICAL DISSOCIATION HYPOTHESIS

The first hypothesis assumes an early atmosphere like that of Jupiter today, with methane, ammonia, and some water vapor. The atmosphere would be devoid of any free oxygen and, therefore, of free *ozone* (O_3), a special form of oxygen gas composed of three instead of two oxygen atoms. Ozone filters out most of the lethal, short-wavelength ultraviolet radiation from the sun, making the lands habitable (Fig. 5.11). An ozone layer occurs at the upper part of the oxygen-molecule-rich lower atmosphere (about 15 to 30 kilometers). It is manufactured in the atmosphere by high-energy ultraviolet radiation from the sun. Oxygen molecules (O_2) are broken, and the stray oxygen atoms then combine with other molecules to form ozone (O_3). Ozone is unstable, however, and so the third atom soon goes astray to combine with other strays to form O_2 again. The process is a steady-state one; that is, ozone is constantly forming at the same rate it breaks down, thus maintaining a nearly constant amount. High-energy ultraviolet (uv) radiation can trigger other photochemical or light-induced reactions. Such reactions are known to occur in the upper atmosphere today and presumably were more common throughout the atmosphere before ozone accumulated. It is estimated the ozone would begin to form when the oxygen content reached about 10 percent of its present abundance. Changes resulting from such reactions in the primitive atmosphere might have been:

1 Dissociation of primeval water vapor into hydrogen and oxygen with most hydrogen escaping to space:

$$2H_2O + uv \ light \ energy \rightarrow 2H_2 \uparrow + O_2$$

2 Newly freed oxygen reacted with methane to form carbon dioxide and more water vapor:

$$CH_4 + 2O_2 \rightarrow CO_2 + 2H_2O$$

3 Oxygen also would react with ammonia to form nitrogen and water:

$$4NH_3 + 3O_2 \rightarrow 2N_2 + 6H_2O$$

4 After all CH_4 and NH_3 were converted to CO_2 and N_2, *then* free O_2 could accumulate as further dissociation of water vapor occurred.

In this way, the present nitrogen–carbon dioxide–oxygen atmosphere might have formed.

PHOTOSYNTHESIS

The alternate source of free oxygen is plant

FIGURE 5.11
Profile of the atmosphere, showing stratification, types of radiation, radiation shields (ionosphere and ozone layers), and surface-atmosphere interchanges.

photosynthesis. There is a close balance between carbon in sedimentary rocks and free oxygen in the atmosphere, which seems to confirm that most O_2 has been released by the breakdown of CO_2 molecules through photosynthesis with C temporarily stored in organisms and sediments. Any subsequent oxidation of that carbon, as by burning or respiration, returns CO_2 to the atmosphere. Today photosynthesis is universally regarded as the major source of free oxygen. Primitive plants (blue-green algae) had appeared by 3.5 billion years ago, but undeniable evidence of a first strongly oxidizing atmosphere occurs in rocks formed only 1 to 1.5 billion years ago. The evidence lies in widely distributed metal oxide minerals, especially of iron. Therefore, plant proliferation and accumulation of much free oxygen may have taken as long as 2 billion years. Once free oxygen was present, ozone could form and provide an ultraviolet shield for further life development. There is an interrelation between organisms and the atmosphere; ozone provides a shield for all land life, and yet apparently organisms themselves produced the oxygen from which it forms.

PROBABLE EVOLUTION OF THE ATMOSPHERE

A plausible atmospheric evolution is summarized in Fig. 5.12. If the atmosphere is indeed evolutionary, then we must inquire if it is still changing, and if so, at what rate. Although organisms generally are remarkably adaptable, it is difficult to conceive of drastic changes of atmospheric or oceanic composition since the appearance of highly organized fossil marine animal forms nearly a billion years ago. Complex interactions between life and its total ecologic environment provide a strong argument for relative stability of the atmosphere and oceans for at least the past billion years or so. Moreover, sedimentary rocks of late Prepaleozoic age are not different from those of younger ages, suggesting that the chemistry of the seas and atmosphere has not changed drastically since.

Important exchanges occur between life forms and the atmosphere. Bacteria are important in overall cycling of nitrogen between earth and atmosphere; some release it, while others fix it in nitrogenous compounds. Plant photosynthesis in general requires carbon dioxide and releases oxygen, whereas animal respiration consumes oxygen and releases carbon dioxide. Carbon is temporarily removed and stored in coal, and carbon dioxide is similarly stored in large volumes of carbonate rocks (limestone and dolomite). It is estimated that over 600 times as much CO_2 is so stored as there is now in the atmosphere, hydrosphere, and biosphere combined. Oxygen is used in large quantities in the oxidation of minerals at the crust surface, a process of major importance for at least the past 1.5 to 2 billion years.

Humans have produced recent changes in the atmosphere, especially through additions of fuel-combustion products. Pollution eventually could cause a general climatic change by altering the transparency of the atmosphere to incoming solar radiation and to outgoing radiation from the earth's surface. Once started, such changes may be difficult to reverse; therefore, atmospheric pollution poses serious threats to life. Humans have the capability to upset within a few centuries an ecologic equilibrium between life and earth that has existed for at least 2 billion years!

ORIGIN OF SEA WATER

Sea water is not difficult to explain, because our hypotheses for the atmosphere also provide abundant water. The origin of the oceans becomes largely a question of the beginning of and

FIGURE 5.12

Probable evolution of composition of the earth's atmosphere. (*Adapted from H. D. Holland, 1964, in The origin of the atmosphere and oceans; by permission of John Wiley & Sons, Inc.*)

the rate of accumulation of water and of dissolved salts.

Following the outgassing hypothesis, rate of accumulation of sea water would be tied directly to atmospheric production and, therefore, to chemical fractionation of the solid earth. Rubey reasoned that the volume of sea water has grown in direct proportion to the increase in volume of continental crust through time. Did atmosphere and seas (and crust) accumulate slowly at a more or less uniform rate, or did they accumulate early and rapidly? A rough index might be obtained from the rate of release of helium and argon 40 to the atmosphere, assuming the outgassing hypothesis to be correct. If water were released at a comparable rate, it would suggest that the oceans accumulated over a long period of time, but more rapidly early in history when greater abundance of radioactive elements produced about five times as much heat as now. Furthermore, as we shall see subsequently, there is isotopic evidence that most of the present area of granitic (continental) crust already existed at least 2 billion years ago. Therefore, most of the outgassing of the atmosphere and sea water also should have been completed by then.

The oceans undergo exchanges of chemical materials with the solid earth, atmosphere, and life. Gases such as oxygen and carbon dioxide dissolve in sea water in proportion to their abundance in the atmosphere, helping to stabilize the composition of both. Some of the salts dissolved in sea water presumably resulted directly from outgassing, but chemical weathering of rocks provides large volumes of salts, too. Salts such as calcium and potassium are reprecipitated in marine sediments so that proportions among dissolved material in sea water are very different from those in streams. Important subtractions include carbonate rocks and layers of ancient rock salt, but these represent only temporary "bank deposits" from which withdrawals can be made later by erosion. In spite of such changes, the overall chemical content of the seas seems to have been rather constant. Today there is a kind of chemical steady state among oceans, atmosphere, rock weathering, and life.

THE ORIGIN OF LIFE

EARLY CONCERNS

Ever since humans began to think, we have wondered where we came from and how all living things came to be as they are. Before the theory of evolution was developed, people thought in terms of multiple origins. They observed that higher animals mated to produce offspring, though the exact process of conception eluded understanding until the invention of the microscope. One important early concept was that, because the female produced the offspring, only she was involved with heredity. With the development of the microscope, the sperm was discovered, which led to the curious idea that the sperm contained a complete miniature adult, and the female provided only the environment for growth of the fetus. A more vexing problem was that of small living things that seemed to spring to life from inorganic matter in a putrefying environment. Experiments by Pasteur on putrefaction, along with the use of the microscope, demonstrated that even the lowest forms of life, like larger ones, produce offspring by transferring a portion of their cell nuclei.

Thus, prior to the microscope, people confused reproduction with the actual origin of life. During the Middle Ages, many scholars were preoccupied with the idea of *spontaneous generation*, believing that putrefaction somehow produced a miraculous metamorphosis of nonliving to living matter. Father Athanasius Kircher (1602–80), professor of science in the College of Rome, for example, divided animals into two groups, one that reproduced sexually and another that formed continuously by spontaneous generation. "It is obviously pointless to give these latter forms a place in the already encumbered ark," he wrote in discussing the passenger list for Noah's voyage. Even Buffon believed that organic molecules, released by putrefaction, came together to form simple organisms. It was Pasteur who proved, in the early 1860s, once and for all, by his famous sealed-flasks experiment, that there is no spontaneous generation.

Pasteur was asked by the wine growers of southern France to find out why their wines were turning to vinegar. He believed that airborne microbes may have infected the wine and then converted it. He designed balloon-shaped flasks made with long, drawn out wavy necks. The airborne bacteria entering the flasks were caught in the bends of the necks, and the liquid, sterile nutrients he had placed in the chamber of the flasks remained clear and sterile. Other flasks, with the necks broken off, containing the same sterile liquid nutrients rapidly putrefied as the entering bacteria colonized the nutrients. Thus he showed that putrefaction caused the wine to go bad and, more importantly, that it was the product of an infection of bacteria and not newly created life ("spontaneous generation").

CONDITIONS FOR LIFE

Until the middle of this century, there was very little information to provide a clear picture of the origin of life. With the work by Rubey and others on the origin and nature of the early atmosphere and the discovery of the nature of DNA (deoxyribonucleic acid), we now have a basis for speculating rationally on the beginnings of life. The notion that life originated as a natural—perhaps inevitable—consequence of chemical conditions on the nonliving early earth first was developed in the 1920s. The earth contains abundant carbon (with remarkable combining properties), hydrogen, oxygen, and nitrogen, the major elements present in all organisms. Moreover, it is well endowed with the universal solvent, water, which, by virtue of the earth's critical temperature relation to the sun, exists largely in the liquid state. No other planet in our solar system now has all of these conditions. Perhaps Pangloss, in Voltaire's famous story Candide, was correct that this is, after all, the "best of all possible worlds." The

modern biochemical theory calls for arrangement of the basic elements of life into more and more complex molecules until, finally, true organisms developed.

We have seen that current views on the nature of the early (anaerobic) atmosphere favor one poor in free oxygen but rich in water vapor and compounds containing hydrogen, carbon, and nitrogen. Whichever hypothesis proves correct for the subsequent changes to produce our present nitrogen–carbon dioxide–oxygen atmosphere (aerobic), or how quickly this occurred, the change was intimately associated with the origin of life.

All life shares one common feature: the presence in the nucleus of each cell of the complex molecule DNA, which, because of its composition and structure, serves as a code-carrier with the complete information for the specific entity of which it is a part. Importantly, DNA-borne information serves as a chemical template to initiate biochemical reactions that will allow the living entity, whether a diatom or an elephant, to replicate every single part of itself during growth. Much is now known about the construction of this remarkable molecule of life and its function. For example, it is estimated that the human body contains a total of about a yard of DNA in all its cell nuclei. The constituents of the DNA molecule can be arranged in 4×10^{109} different ways, which accounts for all the many characteristics of different life forms.

DNA itself, while it is the information carrier which specifies the making of proteins, does not directly take part in the actual formation of the replication. This is accomplished by the messenger RNA (ribonucleic acid), which carries the genetic information to the protein formers. The formation of proteins requires that raw material obtained from outside the cell be converted into simple nutrients such as glucose, and from these building blocks amino acids are formed. Energy for this system is induced by ATP (adenosinetriphosphate), which, like DNA and RNA, is found in all living cells.

The key to the origin of DNA, and thus life itself, is in the natural synthesis of amino acids, which are the materials from which proteins are made. To oversimplify, it is a matter of linking more and more chains of molecules (polymerizing) to construct the complex DNA double helix. The historical sequence could have been the origin of amino acids, then the chemical linkage of amino acids into larger molecules (proteinoids), and next the formation of proteins and of *nucleic acids* from nucleotides (four different nucleotides compose DNA and RNA). At the protein stage, it was necessary to develop a chemically stable environment to assure the next stage leading to DNA, whose replicating function marks the level we call life. But DNA is not self-sustaining; therefore development of an enclosed cell was essential to ensure the development of life. It should be kept in mind that it was a far greater leap from small amino acid molecules to the first cell capable of self-replication than from that first cell to the organism that is now reading this page! Much experimental evidence is available to help us construct hypotheses to explain this sequence. For a full discussion of the biochemical details, see L. E. Orgel, 1973.

THE PRODUCTION OF AMINO ACIDS

The prime factors necessary for the production of amino acids are an elevated temperature, a high-energy source, and the appropriate chemical elements. An experiment was performed in 1953 by Miller and Urey at the University of Chicago that demonstrated one way in which amino acids could have been produced. An electrical discharge was introduced into a chamber containing methane (CH_4), ammonia (NH_3), hydrogen, and water, all of which were heated to 100°C. After a week, four amino acids common to protein were synthesized. Since 1953 a series of experiments has been conducted using different energy sources, including ultraviolet light and other high-energy radiation, applied to very high concentrations of basic gases or chemicals in a hot-water environment. The most recent experiments suggest that if methane and ammonia were dominant in the early atmosphere, the action of solar radiation could lead to the formation of the poisonous gas hydrogen cyanide (CHN), urea, and amino acids. In fact, laboratory syntheses have been

made under plausible primitive earth conditions for nearly all complex organic molecules essential to DNA, basic enzyme systems, and to photosynthesis. Recently there has emerged additional evidence for the abiologic production of amino acids when it was discovered that chondritic meteorites found in Antarctic ice contained a number of amino acids (produced in outer space) that could be used in protein formation. We are unsure exactly what the environment was that allowed the first amino acids to form, but we feel that perhaps they could have developed in pools of water near volcanoes where elevated temperature would favor synthesis.

SYNTHESIS OF LARGER MOLECULES AND CELLS

However amino acids originated, nearly all theorists believe that for a time they formed continuously and concentrated in water. Some feel that this occurred in open seas, while others feel it more likely that concentration occurred in pools of hot water. Dr. Sidney Fox of the University of Florida feels that synthesis of the larger molecules could occur only if amino acids were greatly concentrated and then dried. Fox splashed dried amino acids with water, and formed simple proteins or proteinoids. While relatively small amino acid molecules and proteinoids probably formed rapidly, formation of larger molecules such as enzymes would be very improbable events under most circumstances. What was required was some additional factor to increase the probability of catalyzing these far more complex molecules. Such conditions could include hot acidic water, lightning discharge, or ultraviolet radiation.

Of even greater importance was some mechanism for transferral of information. Only when this became possible could replication and selection begin. One of the most intriguing suggestions is that clay minerals or other crystalline materials may have acted as template surfaces on which complex, large organic molecules were synthesized. The crystal structures of clays tend to have minor imperfections that might have influenced the polymerization of organic molecules adsorbed to clay surfaces in aqueous solutions. These irregularities among different clay particles might constitute a kind of "information" that could influence the synthesis of slightly differing organic molecules, providing a kind of primitive genetic system. Some of these variable molecules would be more stable and would tend to have greater survival value.

Sidney Fox and the Russian biochemist A. I. Oparin have shown how the next step to cells might have occurred. In the experiment in which Fox created proteinoids, amino acids in a hot, dry state were splashed with water. Instant polymerization took place to form proteinoids bound together in a double-walled, spherical, cell-like envelope (termed microspheres). These have been seen to divide into smaller spheres, not unlike the process of cell division; they could be called precells. Structures very similar to these have been found in banded iron deposits in western Australia (about 2.7 billion years old). Fox's experiments point the direction toward possible formation of even more complex proteins within their own protective spheres! The next step would be synthesis of DNA.

Up to this point, it was essential to have an anaerobic environment, for if free oxygen were present, the organic molecules would be destroyed quickly by oxidation. With the formation of a cell membrane, protection could be afforded for the appearance of anaerobic bacteria (although it is likely that they were preceded by a viruslike intermediary). The cell membrane is a complex structure that was crucial to form before the process of photosynthesis could occur. Anaerobic single-celled plants probably also appeared very early. When the free oxygen–carbon dioxide–nitrogen atmosphere developed, no more amino acids could be produced outside a living organic system, and probably none are forming today. A most important danger confronting early life was the presence of ultraviolet radiation in lethal quantities. However, organisms would have been afforded protection in a moderate depth of water or by surviving in the shade of rocks.

The primitive anaerobic plants produced carbon dioxide as a metabolic by-product. Some plants such as sulfur bacteria may have been the precursors of plants which began synthesizing chlorophyll and produced oxygen. The blue-green algae were contributing free oxygen, but this oxygen was probably locked up by other elements such as iron during the early Prepaleozoic. An oxygen-rich atmosphere did not form until later.

THE DEVELOPMENT OF PHOTOSYNTHESIS

From the chemical analysis of rocks and other data, we know that an oxidizing atmosphere developed by about 2 billion years ago. For example, in most of the Prepaleozoic outcrop areas of the world, red, oxidized iron-bearing sediments first appear in the stratigraphic column. (see p. 176). The change from the original atmosphere was gradual and was accompanied by the formation of an ozone layer high above the earth (although at first it was probably close to the surface), which reduced the amount of lethal ultraviolet radiation reaching the earth. This dominance of free oxygen was almost certainly produced by plants.

The anaerobic plants living during the time of the primitive atmosphere were *heterotrophs*— that is, they could not manufacture their own food, but rather consumed the inorganically formed amino acids and other organic molecules existing in their own environment. Only when photosynthesis developed were plants able to manufacture their own food from inorganic elements (becoming *autotrophs*), which later (including today) provided the basic food eaten by all heterotrophs—including humans.

This became possible through the development of photosynthesis, which provides an energy source for biological utilization of solar energy to manufacture food (i.e., stored chemical energy as organic molecules). It is the chief way that inorganic matter can be converted to food. The basic equation is

$$CO_2 + H_2O + light \xrightarrow{chlorophyll} (CH_2O) + O_2$$

The synthesis of the complex chemical chlorophyll[1] is the key to the process of photosynthesis. *Fermentation* is another mode of energy production, which does not require light; it is more primitive than photosynthesis.

The exact manner in which this changeover from anaerobic to aerobic metabolism took place is not known, and it may have taken millions of years for chlorophyll to be formed and free oxygen put into the atmosphere by plants. From fossil evidence, it appears that by the time the oldest known (Australia) blue-green filamentous algae appeared nearly 3.5 billion years ago, chlorophyll was present, in part substantiated by the presence of some chemicals which are usually found in aerobic organisms. These chemicals generally result from the alteration of an alcohol portion of chlorophyll during decomposition. If this is true, the chlorophyll-producing plant fossils are in rocks that are among the oldest known.

However chlorophyll did develop, it is found in plant cells in minute bodies called chloroplasts. Interestingly, chloroplasts, along with other structures called organelles within each cell, may be living fossils, because microbiologists recently have found that they are able to live independently and thus may be symbionts (separate organisms living within another organism for mutual benefit). The new plants were not, at first, the main contributors of free oxygen to the atmosphere, but as they became more common, they gradually took over as chief suppliers of free oxygen.

In spite of the lithologic and paleontologic evidence of an oxidizing atmosphere about 2 billion years ago, there is no evidence that animals had evolved by that time. In fact, it took over 1 billion more years for them to appear. Animals probably evolved in plant communities as single-celled mobile forms. The first animals in the fossil record were already diverse and possessed complex features, thus indicating a profound and unfortunate gap in the record.

---

[1]Chlorophyll is composed of porphyrine-based cyclic compounds and has the generalized formula $C_{55}H_{72}N_4Mg$.

SUMMARY

The discovery of radioactivity provided a dramatic example of inorganic evolution—the creation of new elemental species from older ones. Modern thermonuclear theories of stars and of the origins of the elements, together with the protoplanet hypothesis, provide a compelling, unifying evolutionary conception for the entire solar system. Although the differences among the planetary bodies defy any simple explanation, nonetheless it is generally believed that the planets and their moons formed as part of the overall evolution of the solar system, by condensation from a solar nebula; that is, a gaseous cloud surrounding the sun.

There is much geophysical and geological evidence of concentric arrangement of material within the earth. We have seen impressive evidence that, starting as a homogeneous, accreted cold protoplanet about 5 billion years ago, the earth contracted and heated internally—primarily by radioactivity. Density differentiation produced a very dense core and intermediate-density mantle (by about 4.5 b.y. ago), less dense crust (beginning no later than 3.9 b.y. ago), and least dense sea water and atmosphere (approaching their present character at least 2 b.y. ago). The earth still is undergoing chemical evolution; energy continues to be dissipated by volcanoes, earthquakes, hot springs, and uplift and subsidence of the crust.

The crust probably formed by repeated selective melting (fusion) and rise of lighter elements from the mantle. We now know that the low-seismic-velocity zone in the upper mantle is a much more fundamental boundary than the Moho (base of crust). In the next two chapters we shall see the significance of the lithosphere above that low-velocity zone. In this section we have considered the origin of sea water, but origin of the oceans is a twofold problem. We have not yet explained the ocean basins. In the next chapter we shall assess that problem and suggest alternate hypotheses proposed to explain the distribution of both oceanic and continental crust, which constitute a thin top veneer on lithosphere plates.

Biochemists believe that on the early, chemically evolving earth, given carbon, nitrogen, and oxygen and hydrogen combined as liquid water, and sources of heat and other energy, the development of the simpler organic compounds, such as the amino acids, was virtually inevitable. Synthesis of more complex molecules, such as the nucleic acids and enzymes, must have required very special conditions. Once DNA and a protective cell membrane developed, true life was formed and natural selection could begin. We see that already an impressive interaction was occurring between the living and nonliving realms during the first three-quarters of the earth's evolution. Apparently the origin of life was possible only in an oxygen-free environment, but the later development of respirative organisms required the oxygen generated by earliest creatures. Animals are not only indebted to plants for their food, but also for the oxygen that they breathe.

Key Concepts

aerobic versus anaerobic
amino acids
autotrophs versus heterotrophs
chemical differentiation
core-mantle separation
dynamo hypothesis
eclogite
fermentation
fractional crystallization
granitic rocks
impact craters
kimberlite
lithosphere
low-seismic-velocity zone
mafic minerals
mare basalts
metallic meteorites
Moho (Mohorovicic discontinuity)
nucleic acids (DNA and RNA)
outgassing
partial melting
peridotite
phase change
photochemical dissociation
photosynthesis
planetesimals
protoplanets
solar nebula
spontaneous generation
stony meteorites (chondrites)
ultramafic rocks
volcanic craters (calderas)

Readings

Ahrens, L. H., 1965, Distribution of the elements in our planet: New York, McGraw-Hill. (Paperback)

Bernal, J. D., 1967, The origin of life: London, Weidenfeld & Nicolson.

Birch, F., 1965, Speculations on the earth's thermal history: Bulletin of the Geological Society of America, v. 76, pp. 133–154.

Brancazio, P. J., and Cameron, A. G. W., eds., 1964, The origin and evolution of the atmospheres and oceans: New York, Wiley.

Cloud, P. E., Jr., 1968, Atmospheric and hydrospheric evolution on the primitive earth: Science, v. 160, pp. 729–736.

Elsasser, W. M., 1958, The earth as a dynamo: Scientific American, May, pp. 1–6.

Gaskell, T. F., ed., 1967, The earth's mantle: New York, Academic.

Jacobs, J. A., 1963, The earth's core and geomagnetism: Oxford, Pergamon.

————, Russell, R. D., and Wilson, J. T., 1973, Physics and geology (2d ed.): New York, McGraw-Hill.

Kopal, Z., 1973, The solar system: Oxford, Oxford Univ. Press.

Lowman, P. D., Jr., 1972, The geological evolution of the moon: Journal of Geology, v. 80, pp. 125–166.

McLaughlin, D. B., 1965, The origin of the earth, in Kay, M., and Colbert, E. H., Stratigraphy and life history: New York, Wiley, chap. 27.

Mason, B., 1966, Principles of geochemistry (3d ed.): New York, Wiley.

————, and Melson, W. G., 1970, The lunar rocks: New York, Wiley-Interscience.

Orgel, L. E., 1973, The origins of life: molecules and natural selection: New York, Wiley.

Press, F., and Siever, R., 1974, Earth: San Francisco, Freeman.

Ringwood, A. E., 1975, Composition and petrology of the earth's mantle: New York, McGraw-Hill.

Robertson, E. C., ed., 1972, The nature of the solid earth: dedicated to Francis Birch: New York, McGraw-Hill.

Rosenburg, G. D., and Runcorn, S. K., 1975, Growth rhythms and the history of the earth's rotation: New York, Wiley.

Schopf, J. W., Haugh, B. N., Molnar, E., and Satterthwait, D. E., 1973, On the development of metaphytes and metazoans: Journal of Paleontology, v. 47, pp. 1–9. (Contains bibliography of recent papers on Prepaleozoic life)

Wood, J. A., 1979, The solar system: Englewood Cliffs, Prentice-Hall. (Paperback, Foundations of Earth Science Series)

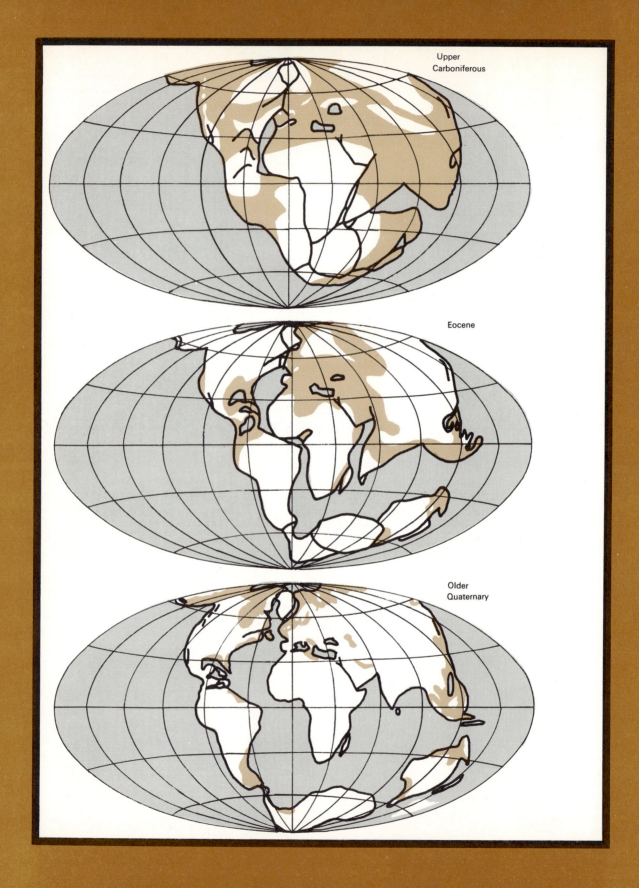

Upper
Carboniferous

Eocene

Older
Quaternary

6

MOUNTAIN BUILDING AND DRIFTING CONTINENTS

What matters it how far we go?
His scaly friend replied,
There is another shore, you know,
Upon the other side.

The Mock Turtle,
Alice in Wonderland

FIGURE 6.1
Famous reconstruction of the continents by Alfred Wegener, showing three inferred stages of continental drift. Brown represents extensions of the sea at the respective times. (*Adapted from Wegener, 1929,* Die Enstehung der kontinente und Ozeane; *by permission of F. Vieweg and Sons.*)

Evidence for an overall chemical differentiation of the earth suggests that the crust might once have been uniform over the entire earth, even though it is not so today. As we have seen, relatively low-density, outer continental crustal material apparently represents the ultimate chemical distillate for the solid earth. Apparently it formed by a sweating out of some fraction of the earth's light elements (see Fig. 5.10). But was the rate of continental accumulation constant? Once formed, have continents remained fixed in position, or have they moved? How the crust has changed through time is of great interest in geology. Therefore, in this chapter we shall formulate working hypotheses to explain it. Through the remainder of the book, we shall test them against known historical and structural evidence.

There are two fundamentally different groups of hypotheses. One group postulates an early, uniform, complete continental-type crust, with the present discontinuous distribution of continents having arisen through one of several different mechanisms. The second group postulates an original, uniform oceanic-type crust, with later growth of continental crust at the expense of the the oceanic through time.

One variation of the latter calls for growth of a few huge supercontinents followed by fragmentation and separation (Fig. 6.1). In this chapter, then, we shall examine two separate issues: *first*, was continental or oceanic crust the more primitive, and *second*, regardless of how they were formed, have continents remained fixed in their positions? Also we shall review the problem of maintaining continents in the face of erosion, the making of mountains, and the fascinating theory of continental drift.

AN EARLY, COMPLETE CONTINENTAL CRUST?

The *continental assimilation* hypothesis assumed that the ocean basins were areas of "oceanization" of large portions of an original, all-encompassing continental crust by addition from below of elements such as magnesium, iron, and calcium, or perhaps just the intrusion of countless basaltic dikes. The resulting more dense mass then subsided to form the basins. The chief objection to this hypothesis is that it runs counter to the general chemical and density differentiation by which it appears that the least dense crust has formed *from* more dense material, not the reverse (see Fig. 5.10). Nonetheless, certain geologists favor some variation of this explanation. One of the oldest hypotheses is that of George Darwin (son of Charles Darwin), who, a century ago, postulated the *extraction of the moon* to form the Pacific Ocean basin. He suggested that the earth's interior was liquid and that originally there was a uniform continental crust. The sun's gravitational pull presumably set up tidal oscillations in the crust. Bulging of the crust increased until a large chunk of crust and upper mantle was thrown out and launched into orbit around the earth. The mass reshaped into a sphere with a core of mantlelike material and a crust of continentlike material, yielding a bulk lunar density of 3.3 to 3.4. Meanwhile, the earth's remaining crust became fragmented and redistributed by profound stresses resulting from this "catastrophic" event, but leaving the Pacific Ocean basin as a scar. Darwin's idea was abandoned by most geologists when the solid nature of most of the earth's interior was demonstrated by seismology.

An alternative explanation of the crust, which seems even more incredible, is *expansion of the earth*, first seriously suggested about 1925. This idea derived from the simple observation that present continental crust could cover a sphere with about half the surface area of the present earth. But a large expansion is difficult to imagine, for no mechanism is now known that could produce it.

AN ORIGINAL, COMPLETE OCEANIC CRUST?

The growth or *accretion of continents* at the expense of original oceanic crust by gradual fractionation and concentration of potassium, aluminum, silicon, oxygen, sodium, and other elements is an old hypothesis with much to recommend it, as we shall see. It fits overall chemical differentiation of the earth (see Fig. 5.10), but unlike the idea of an early, complete continental crust, the process of crustal fractionation would be extended over much of geologic time. Traditionally, advocates of the accretion hypothesis have assumed that continents accumulated in complex patterns where they are now located, but subsequent rearrangement by displacements is also compatible with accretion.

PERMANENCY OF POSITIONS?

It is natural (and uniformitarian) to assume that continents have always been located in the same places with respect to each other and to intervening ocean basins, regardless of how their relative sizes may have changed. Long ago, workers observed the curious parallelism of outlines of the opposing coasts of the South Atlantic, suggestive of separated pieces of a jigsaw puzzle. In 1859 the first map fitting South America against Africa was published. Then in the early part of our century, the concept of large-scale rearrangement of continents or *continental drift* began to receive considerable attention, especially through the writings of a German meteorologist, Alfred Wegener (Fig. 6.1).

Today, some form of continental separation is widely acclaimed by most geologists, though still loudly disclaimed by some. A recent modification of continental displacement is the suggestion of *spreading sea floors*. Presumably, oceanic crust moves laterally outward from ocean ridges, where new crust forms, and the continents are rafted along as if on conveyor belts. According to this view of crustal structure, the earth's surface today is subdivided into several immense plates that are actively spreading apart along oceanic ridges, while being telescoped together either along deep oceanic trenches or in mountain ranges like the Himalayas. The boundaries and patterns of movements of such plates presumably have changed through time in response to profound subcrustal changes. This revolutionary theory called *plate tectonics* is outlined fully in the next chapter after some essential background has been developed.

STRUCTURE OF THE CRUST

WHAT ARE CONTINENTS?

As we pointed out in Chap. 5, the earth is a huge energy system, and earth history records the expenditure or dissipation of energy. Understanding of the evolution of the earth's crust is a combined problem of chemical and dynamic changes. We have examined the chemical aspect of the problem in some detail. Now we shall review the large-scale structural features of the earth's crust as they exist today, for we must know the product of evolution before we can ask intelligently how it evolved. Geologically speaking, continents extend beyond their present coastlines to depths between 200 and 2,000 meters below sea level. We have shown already that sea level is transitory; frequently it has risen to cause shallow flooding of almost the entire present land area, while, conversely, it has also fallen well below its present level. Because sea level is an elusive datum, a more fundamental geologic boundary is needed. Most of the earth's surface lies at two levels. One-third averages about 300 meters above sea level (continents) and the other two-thirds (the ocean basins) nearly 5,000 meters below. The principal topographic break between the two lies at the continental shelf edge (−200 meters). For the illustrations in this book, we have shown the edge of the present continents as the outer edge of the continental shelves because the thickness and composition of the crust is essentially ''continental'' to the shelf edge.

HOW DO CONTINENTS PERSIST?

Why should there be two prominent topographic levels with almost negligible intermediate ones? The *principle of isostasy* first proposed 100 years ago seems to provide the answer. Based upon gravity measurements, it was reasoned that high mountains are buoyed up at depth more or less hydrostatically according to the principle of Archimedes. This buoyancy is thought to arise in the same way as the support of an iceberg by water displaced by the great mass of ice below the water surface. Isostasy postulates a tendency for buoyant equilibrium among the larger units of the earth's crust according to their total mass. Thinner and/or more-dense units (like ocean basins) stand lower, while thicker and/or less-dense ones (like continents and mountains) stand higher. Furthermore, if a large load of sediment or ice were added, the crust would sink, whereas if some rock or ice were removed, the crust would rise.

From seismic evidence, it is clear that continental crust is from six to eight times thicker than oceanic crust (30 to 40 kilometers versus 5 kilometers), and is slightly less dense (2.7 versus 2.8). Sure enough, thicker and less-dense crust does stand higher than thinner, more-dense crust. Indeed, isostasy predicts that the elevation of any large segment of the crust is directly proportional to the thickness of the crust. But more dramatic confirmation of isostasy comes from measured subsidence of the crust of nearly 200 millimeters (8 inches) under the load of impounded water after the building of Hoover Dam in 1935, and from the evidence of the depression of the crust by huge Pleistocene ice sheets followed by measurable uplift or

FIGURE 6.2

Two different possible mechanisms of isostatic uplift and subsidence of the crust. *A:* Lateral plastic flow of material in the mantle. *B:* Physical phase changes at the base of the crust (M discontinuity) from more- to less-dense material, or vice versa (change to lesser density thickens and raises crust, while change to greater density thins and depresses crust to maintain isostatic equilibrium).

rebound of 250 meters after melting (see Fig. 1.8).

THE MECHANISM OF ISOSTASY

How can large segments of the earth's crust rise and sink to maintain isostatic equilibrium? Either the mantle must be plastic enough to flow from or to different areas in response to changes in crust level above, or the crust and upper mantle must be physically transformed to more- or less-dense forms called *phases*, thus changing their mass-volume relations (Fig. 6.2), or both. In a mountainous area, erosion tends to thin the crust from above, and isostasy dictates that, to maintain equilibrium, such crust should rise, just as melting of glaciers caused the crust to rise. Eventually, erosion and isostatic uplift would reach gravitational equilibrium beyond which, barring new structural disturbances of the crust, no further change will occur. When this is achieved, mountainous terrain will have been reduced to a plain. If, on the other hand, the lower crust or uppermost mantle then were

FIGURE 6.3

Deformation by creep revealed microscopically in a natural ultramafic mantle rock (left) and in ultramafic rock deformed experimentally at 1100°C (right). In both cases, large olivine and pyroxene grains (darker) have been flattened and broken down into smaller grains by creep. (Photographs taken through polarizing microscope; each area is 2 millimeters long.) *(After Kirby and Raleigh, 1973, Tectonophysics, v. 19, p. 179.)*

to become more dense (say, due to cooling), such mass change would cause the overlying crust to sink. There is evidence to suggest that this may occur at the base of the crust (Fig. 6.2).

Can mantle material flow laterally? Many of the physical properties of materials deep in the earth are strange to us at the surface. Most solids, including rocks, show different types of strain, depending upon pressure, temperature, fluids present, and rate of stress. A sudden shock may cause brittle rupture, as has occurred along faults, but a slowly applied stress (particularly at great pressure and temperature) may produce in the same material a kind of plastic flow without any rupturing. An example of such dual properties is provided by familiar Silly Putty. At room temperatures it has elastic properties such that it can be bounced like a rubber ball. It will break in a brittle fashion when suddenly stretched; yet if deformed slowly, it flows plastically like taffy. Most rock materials both in the deep crust and upper mantle appear to show similar properties. Below 700 kilometers, however, no earthquake foci are recorded; pressure and temperature must be high enough for entirely plastic behavior. Be-

cause plastic flow is so dependent upon rate of stress, it is apparent that over the vastness of geologic time, very large-scale deep flow is entirely possible. Based upon microscopic studies of old mantle rocks now at the surface and of samples deformed at high pressure in the laboratory, flow or creep apparently occurs by minute slips both within mineral crystals and between grains, accompanied by some recrystallization (Fig. 6.3). Especially important for flow is the zone of relatively low seismic velocity, a less-rigid or "weak" zone of lower viscosity in the upper mantle between roughly 60 and 200 kilometers—well below the crust (Fig. 5.7). We might expect any major lateral movements in the outer earth—thus isostatic compensation—to be concentrated there.

In summary, structural disturbances of the crust, and isostatic adjustments to mass changes in the crust and upper mantle, profoundly affect relative land and sea level. The stratigraphic record is the indirect result of isostatic responses to deformation and erosion coupled with the influences of mountain building, climate, and biologic processes.

PRESENT STRUCTURAL FEATURES

Besides the two most obvious topographic entities, the continents and the deep-ocean floors, called *abyssal plains*, important second-order structural divisions of the crust are also prominent. Average structural behavior of a large area of the crust through a long time interval is conveniently termed its *tectonic* character (tectonic derives from a root meaning "to build"). The most stable tectonic subdivision of continents is the *craton*, which has subdued topography and lies near sea level (Table 6.1). Tectonic disturbance of this stable division is primarily by broad, gentle warping.

Most abyssal plains, formerly thought to be as stable as cratons, are divided by oceanic rises or ridges (Fig. 6.4). Rises are long, high prominences that are volcanically active, show greater-than-average heat flow from the interior, and suffer many shallow-focus earthquakes (0 to 100 kilometers deep). Their summits are very irregular and locally form midoceanic volcanic islands such as Iceland. Composition of the lavas on the rises is largely that of typical

TABLE 6.1
Major structural or tectonic elements of the crust

Relatively stable:	Continental cratons Oceanic abyssal plains
Relatively unstable:	Oceanic rises or ridges Mobile tectonic belts (as now seen): 1. *Marginal* (between continental cratons and ocean basins) 2. *Intercratonic* (between neighboring cratons) 3. *Interoceanic* (island arc-trench systems

oceanic basalt. The study of ocean rises is recent, and has produced some revolutionary insights into global tectonics. As noted above, new oceanic crust appears to be added in the rises, causing lateral sea-floor spreading and compression against continents; continents may even be displaced in this manner. But more of this later.

In many ways, the most interesting and

FIGURE 6.4
Major tectonic features of the earth's crust for the past 1 billion years. Note the asymmetric distribution of continents and complexity of mobile belts with respect to continental cratons. Note also parallelism of the opposing Atlantic coastlines. (*Bartholomew's Nordic Projection; with permission of John Bartholomew and Sons, Edinburgh.*)

MAJOR WORLD TECTONIC FEATURES

☐ Abyssal Plains   ▨ Continental Cratons   Mobile Belts   Oceanic Rises ("Ridges")

TABLE 6.2

Contrasts between ocean ridges and mobile tectonic belts including modern volcanic or magmatic arcs

	Ocean ridges	Mobile belts
Seismicity	Shallow only (0 to −70 km)	Shallow and deep (to −700 km)
Volcanism	Chiefly basaltic	Chiefly andesitic
Plutonism	Gabbro or basalt dikes	Granitic batholiths
Structure	Extensional (rifted)	Compressional (complex folds and thrust faults)
Metamorphism	Minor local thermal effects	Major high-grade regional metamorphism (gneiss, schist, etc.)
Heat flow*	High (2.0-3.5×10^{-6} cal/cm²-sec)	Low in trenches (0.99×10^{-6} cal/cm²-sec)

*Heat flow is measured in calories of heat per square centimeter of surface area per second of time. Average flow for continental cratons is 1.4×10^{-6} (or 0.0000014) cal/cm²-sec, while that for abyssal plains is 1.3×10^{-6} cal/cm²-sec.

baffling of all major tectonic features are the long, narrow, and typically arc-shaped zones of greatest structural instability referred to as *mobile tectonic belts* (Fig. 6.4). Such belts are the loci of volcanoes, granitic batholiths, regional metamorphism, and frequent earthquakes (including all deep-focus ones from −300 to −700 kilometers). The oceanic trenches show less-than-average heat flow. Severe structural disturbances include folding and all types of faulting, but especially distinctive is overthrusting (the shoving of huge slabs of rock over one another). Characteristic mobile-belt disturbances collectively are called *orogenesis* ("mountain building"); thus such zones are also called *orogenic belts*. Mobile belts and ocean ridges are contrasted in Table 6.2.

We can classify three species of mobile or orogenic belts based upon present spatial relations to cratons and ocean basins (Table 6.1). The western North American Cordilleran and South American Andean mountain complex is a marginal belt; the Himalayan Ranges make up an intercratonic belt; whereas many of the arcuate volcanic island and deep-sea trench systems (as in the western Pacific) are interoceanic mobile belts with many of the same structural features as are found in other mobile belts. We must be careful, however, not to assume simplistically that the ancient belts were always situated exactly as they are now. If continents have in fact moved together and apart, then we should not be surprised to find that what is now an intercratonic belt began as a marginal one, and what is now marginal was at some time intercratonic.

Figure 6.4 shows that the present patterns of mobile belts and cratons are very complex. They are, in fact, so asymmetrically distributed and irregular in relation to each other as almost to defy explanation. Clearly, whatever causes mobile belts does not always discriminate neatly between crust types; yet, one naturally is inclined to seek some general simplifying hypothesis for all. As we shall see in Chap. 8, presence of granitic batholiths and extreme regional metamorphism in old mobile belts led, just before the present century, to beliefs that it was in such belts that new continental (i.e., "granitic") material was generated. Therefore, conventional wisdom has long held that the marginal mobile belts are sites of lateral growth or accretion of new continental crust outward from old nuclei—the cratons. But the role of nonmarginal belts (intercratonic and interoceanic) was not at all clear.

The distribution of modern earthquakes and volcanoes indicates that mobile belts and ocean rises must play the major roles in the dissipation of the great energy generated in our planet. We have taken pains in preceding chapters to emphasize that the earth is structurally dynamic, and we shall concentrate henceforth upon seeking an explanation of the relations of these major active tectonic features to overall crustal evolution.

THE GEOSYNCLINAL CONCEPT

Beginning in 1857, American geologist James Hall of New York laid the groundwork for one of

FIGURE 6.5
One of the first accurate cross sections of a mountain range showing folds in strata in the Appalachian chain. *(From Rogers and Rogers in* Transactions of American Geologists and Naturalists for 1840–1842, *pp. 474–531.)*

geology's early generalizations. For many years, geologists in both Europe and America had been speculating about the crumpled nature of strata seen in mountain ranges (Fig. 6.5). Hall noted that, in the Appalachian Mountains—today classed as a mobile belt—the strata of the Paleozoic Systems are not only more deformed but are also 10 times thicker than are their counterparts in the Mississippi Valley region farther west—today classed as part of the craton. He suggested a simple cause-and-effect such that the greater load of sediments in the Appalachian belt simply depressed the crust as deposition occurred (Fig. 6.6). Finally, the crust could bend down no farther, and so it failed and the strata crumpled.

In 1873 another leading American geologist, J.D. Dana, coined the term *geosynclinal* (large or "earth syncline", that is, a "downfold") for the old zone of thick strata. On isostatic grounds, he rejected Hall's simple explanation of the great subsidence by sediment-loading alone because less-dense sediments could not depress isostatically more-dense crust nearly so much as was required by the great thickness of strata. The geosyncline was inferred by Dana to be a result—not a cause—of the fundamental structural instability of mountain belts. Dana argued that buckling of the crust beneath geosynclines was the primary cause, first, of subsidence and thick sedimentation, and, ultimately, of mountain building (Fig. 6.6).

European geologists quickly adopted the basic concept of thick sediments as the ancestors of mountains. But, unlike the Americans, who believed that sedimentation always kept the subsiding geosyncline full of shallow-marine and nonmarine deposits, the Europeans believed that geosynclines began as deep furrows on the sea floor and that much of the thick sediment contained therein was of deep-water origin (Fig. 6.6). By 1930 the Europeans had gone farther in suggesting a close similarity between geosynclines and deep-marine trenches and volcanic islands of the Pacific. This began an important series of proposed links between ancient mobile belts and the modern island arc-trench systems (Fig. 6.4). From these early ideas developed one of the great unifying concepts in geology, which brought for the first time a recognition that the stratigraphy and structure of mountain belts are closely related, and that, in turn, such belts may generate new continental crust.

FIGURE 6.6
Contrasting explanations of geosynclines by the Americans Hall and Dana and by European geologists. (Gray is sea water; arrows show Dana's compression idea.)

GEOSYNCLINE, MOBILE BELT, AND MOUNTAIN BUILDING

For our purposes, we shall define a geosyncline as a great elongate belt of relatively thick strata, reflecting greater subsidence of a portion of the earth's crust relative to adjacent stable regions. As definitions go, this is passable, but it also is essential to recognize that the Hall-Dana conception of a geosyncline was linked closely with the formation of mountains. Ancient geosynclines characteristically show tightly compressed folds, various types of faults (including low-angle thrust faults), and most contain granitic batholiths and regionally metamorphosed rocks. Thus geosynclines have developed within structurally mobile belts. But is a geosyncline simply the same thing as a mobile belt? Emphatically not, for *mobile belt* is a structural term summarizing a fundamental tectonic behavior of parts of the crust. *Geosyncline* is primarily a stratigraphic concept summarizing thickness of strata lying within long, structurally mobile zones as contrasted with those deposited in more stable (cratonic or abyssal) regions. The problem is that there are mobile belts, such as the modern interoceanic island arc-trench systems, with many of the same structural qualities as ancient mountain belts, *but with very little sedimentation*. A geosyncline, then, is any mobile belt that received a great accumulation of strata; as we think of geosynclines, they simply represent sediment-filled mobile belts.[1]

Several factors seem required to produce a geosyncline. We feel, like Dana, that great structural mobility is the most fundamental condition in such belts. Formation of a geosyncline appears to depend, *first*, upon a degree of

structural mobility to depress a long zone of crust, and *second*, upon availability of a large source of sediments to fill subsiding troughs within a mobile belt. Modern interoceanic belts (Fig. 6.4) qualify in the first but fail in the second requirement; they are *potential* geosynclines only—in other words, unfilled mobile belts.

As the study of mountain belts progressed, it became apparent in the early twentieth century that their development involved all facets of geology. Besides early recognition that granitic batholiths were formed in mobile belts, it was also realized that volcanic rocks make up a sizable proportion of the stratified geosynclinal sequences. Therefore volcanism somehow must play an important role in the evolution of such belts, which further linked modern volcanic arcs with ancient mobile belts as the Europeans had suggested in the 1930s. A "geosynclinal cycle" was developed by 1940, which claimed that crustal evolution has proceeded by (1) crustal downbuckling and geosynclinal accumulation of sediments and volcanic material, (2) orogenic upheaval accompanied by metamorphism and formation of batholiths, which provide new continental crust, and (3) erosion that finally produces isostatic equilibrium and stability. This "cycle" was much oversimplified, as we shall see. We now believe, for example, that mobile belts destroy at least as much oceanic crust as they may generate of new granitic crust (see next chapter).

VERTICAL VERSUS HORIZONTAL TECTONICS

The observation in mountain belts by 1800 of tightly folded strata and the recognition of great overthrust faults in southern Germany in 1826 naturally cried out for explanation. Plutonists such as James Hutton had explained the folding as wrinkling due to a simple vertical uplift of

[1]Unfortunately, geosyncline has acquired almost as many meanings as there have been geologists using the term. One is reminded of Humpty Dumpty's pronouncement in Lewis Carroll's *Through the Looking Glass*: "When I use a word, it means just what I choose it to mean—neither more nor less." The term probably has little value today.

FIGURE 6.7
Deformational structures like those of mountain ranges produced in a laboratory scale model by closing the vise with the screw at the right. Layers consisted of materials like loose sand, clay, and lead shot. *(From Bailey Willis, 1902, Maryland Geological Survey, v. IV.)*

mountain ranges by the rise of molten granite into them; the superficial strata either were pushed or they slid to the sides (see Fig. 2.15). Beginning about 1825, the crumpling was envisioned by many as the result of viselike compression within the mountain belts (Fig. 6.7), which implied a lateral shortening of the crust. But by 1900 an intense debate had developed between those who believed that *lateral squeezing* and those who believed that only *vertical upwarping* was required to form mountains. Most geologists (like Dana) thought that the folds and thrust faults proved that the earth's circumference had been shortened across mountains, but several physicists and a few geologists (following Hall) argued instead that isostatic upwarping could create topographic mountains and also cause the superficial sediments to deform under the influence of gravity without disturbing the more rigid basement rocks beneath (Fig. 6.8). It would be like a rug wrinkling when pushed across a floor. One of the reasons for this suggestion was the apparent weakness of the rocks involved in thrust faults. It seemed easier to imagine them sliding than being pushed for several kilometers. According to this gravitational sliding hypothesis, only the superficial strata should be deformed; but ac-

cording to the lateral compression hypothesis, the rigid basement rocks also should be faulted and deformed. This provides a test of the hypotheses. Figure 6.9 shows these two contrasting situations for the same region. While gravity sliding has played an important role in the formation of some thrust faults (Fig. 6.9-upper), practically all mountain belts show significant deformation of their basement rocks (as in Fig. 6.9-lower). Therefore, it is generally believed today that mountain belts are primarily the results of profound lateral compression along relatively narrow zones; clearly, immense

FIGURE 6.8
Downwarping and geosynclinal sedimentation followed by isostatic upwarping causing folding and thrusting by gravity tectonics (i.e., sliding laterally of the superficial sedimentary sequences). According to this hypothesis of mobile-belt structures, the basement rocks were not involved in deformation and no lateral shortening of earth circumference has occurred (compare Fig. 6.9).

Subsidence and Sedimentation

Uplift and Sliding

décollement

FIGURE 6.9

Two contrasting interpretations of different parts of the Jura Mountains of the French-Swiss border. *Upper:* Slippage of strata over the basement along a flat shear surface (*décollement*), requiring no crustal shortening. (*From Buxtorf, 1908,* Naturelle Carte Geologique Suisse.) *Lower:* Faulting of the basement as well as superficial strata along steep faults, which does require some lateral shortening of the crust. (*From Aubert, 1949; by permission of* Geologische Rundschau.) Apparently both interpretations are at least partly correct.

forces must be required to accomplish such impressive work. At the same time, the crust as a whole tends to maintain isostatic equilibrium over large regions.

A SHRIVELING EARTH AND CRUSTAL SHORTENING

Late in the nineteenth century, a relatively simple and appealing explanation of geosynclìnes and mountains was offered by Dana. From the almost universal assumption that the earth was cooling, it followed that the globe should be contracting. But a cold, rigid crustal rind must yield by bending and rupture as the interior shrinks beneath it. The wrinkled skin of a drying apple provided a favorite analogy. Where else should adjustments be concentrated but at boundaries between continents and oceans? North American mountain belts lie precisely at such boundaries and enclose the entire continent, so are marginal mobile belts (Fig. 6.4). Shrinkage of the earth would cause compression of the cold crust, which seemed amply evidenced in the folded structures of

mountains. By measuring folds, it was estimated that the crust had been shortened a few hundred kilometers in each mountain belt. First, downbending at the continent margin occurred as oceanic crust pushed against the continent, and then thick geosynclinal strata accumulated there (Fig. 6.6). Further compressive stress—like a closing vise (Fig. 6.7)—exceeded the strength of the rocks, causing rupture and intense, accordionlike squeezing and uplift of the geosynclinal strata to form mountains. *Voilà!* Here was a unified concept of geosyncline and mountain formation. Unfortunately, it was wrong in at least one basic assumption. There is no evidence that the earth as a whole is contracting; its advocates knew nothing yet of radioactive heating. Moreover, as we have seen, most of the world's mobile belts are not as neatly marginal to the continental cratons as they are in North America and as was implied in the contraction hypothesis (Fig. 6.4).

CONTINENTAL DRIFT AS A CAUSE OF MOUNTAIN BUILDING—FIRST OUTRAGEOUS HYPOTHESIS

The general idea of large-scale continental displacements began to be discussed seriously in 1908, when F. B. Taylor, an American geologist, suggested that drifting of the continents had caused wrinkling of the crust to produce all the great Cenozoic mountain systems (Fig. 6.10). Presumably the leading edges of the moving continents depressed the oceanic crust ahead to

FIGURE 6.10
F. B. Taylor's and A. Holmes's conceptions of formation of Cenozoic mountains by continental drifting. Note parallelism of mid-Atlantic submarine ridge (gray) to opposing continent margins. *(Adapted from Taylor, 1910, Bulletin of the Geological Society of America; Holmes, 1929, Mineralogical Magazine.)*

form troughs in which thick sediments could accumulate to form geosynclines. Further movement then compressed and upheaved the geosynclines to produce the present lofty mountains rimming the Pacific. Upheaval of the Alpine-Himalayan chain carried the process a step further by culminating in the direct collision of continents (Fig. 6.10). For a mechanism to drive his continents, Taylor speculated that catastrophic tidal action was induced in the solid earth by capture of the moon during Cretaceous time. Although his explanation is unacceptable, Taylor had given birth to the important new idea of linking mountain building and continental displacements.

The first detailed reconstruction of the continents was attempted almost immediately by another American, H. Baker. He postulated an original, single landmass or supercontinent that suddenly split to form the present Arctic and Atlantic Oceans at the end of Miocene time (Fig. 6.11). Baker was at least as imaginative as

Taylor in devising a cause. He speculated that variations in their orbits brought Earth and Venus close enough together to produce such

FIGURE 6.11
H. Baker's replacement globe reconstruction of continents prior to supposed separation. Note rotations and squeezing together of islands and peninsulas to improve fit. *(After Baker, 1912; with permission of Michigan Academy of Sciences.)*

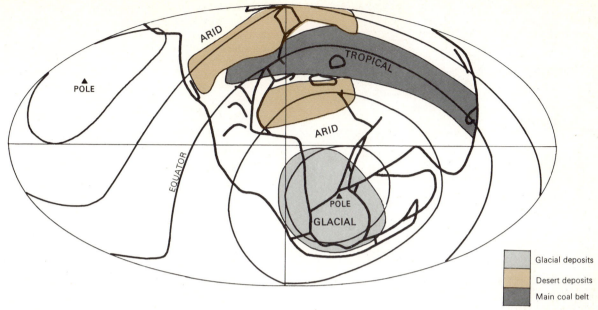

Glacial deposits

Desert deposits

Main coal belt

FIGURE 6.12

Wegener's reconstruction of the Permian continents and paleoclimatic zones inferred from peculiar rock assemblages. Based upon paleoclimatic data, paleoequator and pole positions were then deduced; they are remarkably similar to positions indicated by paleomagnetic data four decades later. Because map was drawn on present coordinates, paleolatitudes seem distorted. (*After Koppen and Wegener, 1924,* Die klimate der Geologischen Vorzeit; *by permission of G. Borntraeger Co.)*

severe tidal distortion that a large portion of original continental crust was torn from the Pacific to form the moon (à la George Darwin in 1882; see p. 110), and the remaining continental crust ruptured and slipped toward the Pacific void. There is, however, no record in Cenozoic strata of a catastrophe of such magnitude; indeed, the last major displacements of continents began in early Mesozoic time.

Others soon began to drift, the most important of whom was a German meteorologist, Alfred Wegener. Wegener drew upon evidence from geology, geophysics, biology, and climatology in developing the first complete and influential statement of the drift theory in 1912. He believed that drifting apart of the continents occurred over a long period during the Mesozoic and early Cenozoic Eras (Fig. 6.1). He became the first to attempt reconstruction of a former

supercontinent by fitting edges of continental shelves rather than present coastlines, a more geologically realistic approach. Besides the parallelism of continental margins, Wegener appealed chiefly to apparent paleoclimatic indicators in late Paleozoic rocks, including glacial, desert, and tropical rain-forest deposits, to reconstruct Permian climatic zones (Fig. 6.12). A South African geologist, A. L. Du Toit, completely consumed by drift fever, began in 1921 to strengthen the concept with a more detailed assessment of supporting geologic and paleontologic evidence than Wegener had offered. Near-identity of many fossils, on the now-widely-separated continents also lent support to drift. How could land plants with large seeds incapable of wind transport have been dispersed across wide oceans? And how could heavy land-dwelling reptiles have swum across? Even the earthworm has been used to support drift. It is said that the modern worms of Madagascar are more like those of India than of nearby Africa, and those of eastern North America supposedly are more like those of Europe than those of the American Pacific coast.

The first phase of the drift theory culminated in 1937 with publication of Du Toit's *Our Wandering Continents*, in which he scorched the

Rafting

Island Stepping Stones

Isthmian Links

Continental Drift

FIGURE 6.13
Different ways that paleontologists have tried to explain similar species on continents now widely separated by oceans. Charles Darwin once remarked that paleontologists seem to "build land bridges (or isthmian links) as easy as a chef makes pancakes." (*Drawings by J. C. Holden; from R. S. Dietz, 1967,* Sea Frontiers, *v. 13, pp. 66–82; used with permission.*)

conservative orthodox antidrifters (mostly American) with an attack upon their "groundless" worship of the dogma of fixed continents and ocean basins. He ridiculed, for example, the compromise of inventing countless narrow land or island bridges between continents to explain the undeniable similarities of many fossil and modern land organisms on opposite sides of the Atlantic (Fig. 6.13). But more of Du Toit later.

THERMAL CONVECTION —PANACEA FOR MOUNTAIN BUILDING AND DRIFT?

Wegener believed that, in spite of apparent high viscosity of subcrustal material, small forces acting over very long periods of time could cause that material to yield and allow crustal blocks to flow slowly across the upper mantle. Though it did (and still does) sound incredible, we have learned to be cautious about shouting "impossible." As noted above, modern knowledge of material behavior proves that slow plastic flow and creep do indeed occur. The question then becomes: "What force might have caused the drifting?" Wegener considered centrifugal effects of the earth's spin, tidal effects in the solid earth, and wobble of the axis as possible contributors, but all these were rejected by physicists as inadequate.

The possibility of thermal convection within the earth was suggested as early as 1839, and in 1881 it was proposed as a possible mechanism for localizing volcanoes and also for producing mountains by dragging and wrinkling the crust laterally. In 1928, a brilliant Scottish geologist, Arthur Holmes, who very early became interested in the implications of radioactivity, presented the first serious suggestion of thermal

FIGURE 6.14

The convection-current mechanism for drifting continents and forming new ocean basins as conceived by Arthur Holmes in 1928. *Upper:* Convection begins to stretch an overlying continental block. *Lower:* Block ruptured and two fragments move apart. *(From A. Holmes, 1965, Physical geology (rev. ed.), Ronald Press, copyright by Thos. Nelson and Sons, Ltd., 1944; used with permission.)*

convection in the mantle as a possible driving mechanism both for mountain building and for continental drift (Fig. 6.14). The basic idea is simple. If more heat is generated in one portion of the deep mantle due to irregular distributions of radioactive isotopes, there would tend to be a very slow plastic flow of the hotter material upward. As it rose, it would cool and flow laterally—presumably beneath the crust—and finally, when cooler than the average mantle below, it would sink. Thus a slow convective overturn of the mantle was proposed like the convective circulation of warm air in a room heated by a radiator. Holmes then assumed that where the convective flow was upward (hot), it might raise and even rupture the crust; where it was lateral, it might drag the crust along conveyor-belt fashion (see Fig. 6.14); where mantle motion became downward (cool), the light continental crust, being isostatically buoyant like a cork, would refuse to sink, and so would buckle up to form mountains (Fig. 6.15).

Gravity measurements by Dutch geophysicist Vening Meinesz in the Indonesian region around 1930 soon suggested that the crust beneath the deep-sea trench of that modern interoceanic mobile belt was being forcibly buckled down by some great compressive force (the vise again). This apparent huge crustal downbuckle under a modern mobile belt spurred interest in thermal convection cells as a possible mechanism first to form a geosyncline

and eventually to upheave a mountain system. In the late thirties, American David Griggs performed ingenious scale-model experiments, which seemed to show that Holmes's convection could indeed produce mountains (Fig. 6.15). From then to the present day—although not a proven fact even yet—mantle convection in some form has been the favorite mechanism both for mountain making and—if one were a drifter—for rearranging continents.

MORE RECENT DOGMAS

After a hot controversy over drift during the 1930s, the matter cooled in the 1940s, with geologists around the world tending to throw their lot with either the pros or the cons and getting on to other tasks. Former president of

FIGURE 6.15

Laboratory scale models of effects of possible thermal convection currents in the mantle. Asymmetric downbuckling and thrust faulting seemingly like the Appalachian belt was produced by rotating only one drum in the viscous substratum ("mantle") and bilaterally symmetrical buckling and thrusting like the Alps was produced with two rotating drums or convection cells. *(From D. Griggs, 1939, American Journal of Science, v. 237, pp. 611–650; by permission.)*

the American Philosophical Society, W.B. Scott, expressed the prevalent American view in the 1920s by describing the drift theory as "utter, damned rot!" It was un-American to be "soft" on drift even as late as 1960. Reasons for strong opposition were twofold: *First*, the idea of such dramatic shiftings of continents, and seemingly only once in history, seemed counter to the narrow uniformitarian doctrine. Certainly continents had changed, but it seemed more uniform for them to do it in place. Yet if they had to move, then they "should have" moved uniformly (i.e., continuously), or at least several times; a single great jump smacked of catastrophism! *Second*, no theoretical mechanism was known that could drive the large but very thin continents—slabs 100 times longer than they are thick—over the earth's surface. Geophysicists asked how could weak continental rocks be pushed over equally weak oceanic ones? It seemed like trying to push huge sheets of wet tissue paper. Zealots like Du Toit, however, kept the faith and "accepted the inescapable deduction from the wealth of geological evidence available to the unfettered mind." They were not overwhelmed by the protestations of the physicists, because they remembered all too well the error of Lord Kelvin's mathematical arguments about the age of the earth (see Chap. 4). Here is an important illustration of divergent attitudes. The field observer's seat-of-the-pants feeling is that if "something *did* happen, it *can* happen," whereas the theoretician tends to disbelieve anything as dramatic as drift unless he can conceive of a sound mechanism and can describe it with equations. As in any other game, each side wins some and loses some.

PALEOMAGNETISM

Rebirth of interest in drift had to await new approaches, which came—ironically—largely through physics in the 1950s and 1960s. The study of rock magnetism gave the first new impetus. Physicists had begun measuring the magnetic properties of rocks about 1850 and had discovered that many young lavas showed magnetization parallel to that of the earth's

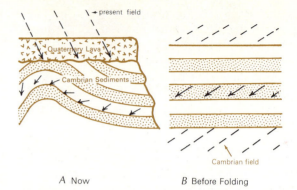

A Now   B Before Folding

FIGURE 6.16
A: Orientation of remanent magnetism with respect to the present magnetic field in two different-aged rocks. *B:* Restoration of the Cambrian magnetic field orientation after correcting for folding.

present field. About 1900, studies of ancient bricks and pottery showed that magnetization acquired during firing had been retained for 2,000 years and contrasts of orientation in lavas, dikes, and adjacent baked sediments with orientations in the unbaked older sediments showed that igneous rocks tend to acquire and retain magnetic orientations parallel to the earth's field at the time of their cooling (Fig. 6.16). Further studies showed that the magnetization is stable indefinitely unless the rock becomes heated to about 600°C, at which temperature the magnetization is lost. Consistency of rock magnetism is demonstrated by identical orientations in several different rock types of the same age in a local area.

In the late 1920s a French physicist, Mercanton, suggested that because orientation of the magnetic field now bears a close relation to the earth's rotational axis, it might be possible to test the theory of continental drift with the magnetic characteristics of certain ancient rocks. This fertile suggestion was not followed up until after World War II, when rock paleomagnetic data began to be gathered on a large scale, initially by a group of British physicists.

Several minerals (chiefly certain oxides of iron) have magnetic properties that make them *susceptible* to a magnetic field. As such minerals crystallize in a cooling magma, they tend to be oriented parallel with the earth's magnetic field.

If such minerals are deposited gently in sediments, they will tend to assume a similar orientation. Thus, rocks, as well as pottery, containing these minerals should today retain some evidence in their *remanent magnetism* of the orientation of the magnetic field when and where they were formed unless later heated. If perchance, the field shifted its position through time relative to the present position of such rocks on the earth's surface (or vice versa), then we might find a record of such shifts "fossilized" in the rocks. We could, in fact, study the history of changes of relative orientation (and also of changes of polarity) of the magnetic field, and at least one branch of geophysics would gain a historical dimension that it had lacked before.

Paleomagnetic studies based upon the above early findings have been conducted at an ever-increasing pace since 1950. The data show undeniable evidence of great changes in the magnetic field relative to present geography, but the results are still difficult to interpret. If we assume that the field was always two-poled as it is today and that it owes its orientation to the spin of the earth (see Fig. 5.8), *then the two magnetic poles should always have been more or less coincident with the geographic poles of the earth's spin axis*. If these assumptions are correct, then determination of ancient magnetic pole positions from carefully studied rock specimens should also reveal the approximate geographic paleopole and paleoequator positions. Figure 6.17 shows the orientation of magnetization in a modern rock in relation to the earth's present field (left) and in an ancient rock (right). The right diagram illustrates how different the *relative* orientation of the ancient field must have been compared to the present field. Both the *inclination angle* (with the horizontal) and *declination angle* (or direction seen in a horizontal plane relative to a present north-south direction) of the remanent magnetism for a rock are determined in the laboratory using a very sensitive magnetometer. The orientation of the specimen in the field must be established and corrections of orientation must be made if the rocks have been tilted after they formed (Fig. 6.16). A steep inclination of our "fossil magnets" indicates

PRESENT FIELD   PALEOMAGNETIC FIELD

FIGURE 6.17

Relationship of earth's magnetic field to the orientation of observed remanent magnetism in rocks (heavy arrow), showing the basis for restoration of the paleomagnetic field and rotational axis relative to an observation station. At right, field has to be shifted in a relative sense to conform to paleomagnetic data for the ancient rock (heavy arrow). Note that magnetic field curves (thin black arrows) are inclined steeply near poles, but are nearly parallel to the earth's surface (flat) near equator.

formation of the rock at high latitude (Fig. 6.17), whereas a very low angle indicates a location near the equator. Although paleomagnetism does not provide an unambiguous longitudinal location, what a powerful paleogeographic tool this offers for determining the latitudinal positions of an area through time! (Fig. 6.18). Declination angles for several widely separated samples *of the same age*, together with other clues, may help to indicate the paleopole position, as shown in Fig. 6.19.

At first, paleomagnetic data were gathered chiefly from Europe and North America. As older and older rocks from both continents were studied, it appeared that the relative positions of these two continents with respect to the magnetic field, and therefore presumably also the rotational axis, had changed markedly. But it was not a random change, for the apparent North Pole seems to have been in the mid-Pacific Ocean 1 billion years ago and has migrated or wandered to its present position, which it reached in mid-Cenozoic time (about 25 million years ago). By 1955 it was concluded that somehow the earth's rotational axis had shifted with respect to the crust so that continents indeed had occupied different latitudes through

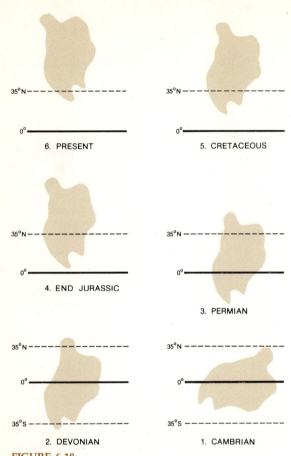

35°N -----

0° -----

6. PRESENT

35°N -----

0° -----

5. CRETACEOUS

35°N -----

0° -----

4. END JURASSIC

35°N -----

0° -----

3. PERMIAN

35°N -----

0° -----

35°S -----

2. DEVONIAN

35°N -----

0° -----

35°S -----

1. CAMBRIAN

FIGURE 6.18
Different positions of North America relative to the equator from Cambrian (1) time to the present (6) according to paleomagnetic evidence. Note the progressive counterclockwise rotation and northward shift through time. Subsequent paleogeographic maps compare geologic and paleontologic data bearing upon paleolatitude and paleoclimate with these paleomagnetic restorations.

wandering alone seemed adequate, for *the apparent ancient pole positions should have been the same for all continents at any given time in the past if only the poles (and magnetic field) were moving relative to the crust.* At first, sampling was so sparse that little confidence could be placed in the results, but geophysicists concluded that, if the magnetic field had always been of the same sort as today (see Fig. 5.8), then they must revive some hypothesis of continental displacements in order to explain their data, no matter how painful that might be. *Seemingly the continents had to have moved relative to each other.* This was a curious irony, for recall that previously it was physicists who most vehemently protested against drift because no adequate physical mechanism seemed available. But now the tables were turned, as expressed eloquently by American geophysicist Walter Munk at

FIGURE 6.19
The method of restoration of continental positions of the past using paleomagnetic data. The declination angle (between a "fossil magnet" shown by arrow and a north line) points to a former magnetic pole position. Continent positions must be adjusted until ancient pole positions for two continents coincide as shown in upper diagram. Outlines of continental margins also help, of course.

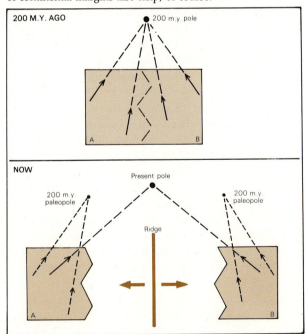

200 M.Y. AGO   ● 200 m.y. pole

A   B

NOW   Present pole

200 m.y.   200 m.y.
paleopole   paleopole

Ridge

A   B

time (Fig. 6.18). After 300 years of speculation, drastic changes of latitude now had gained favor.

Although at first it was assumed that the magnetic field had shifted relative to a fixed crust, comparison of *apparent pole wandering* or migration paths for two or more continents revealed unexpected discrepancies. They did not agree, and only movements of continents could make them do so (Fig. 6.19). Assuming that the field always had been a simple two-poled one, as today, then no hypothesis of polar the height of renewed interest in polar wandering:[2]

> In this controversy between physicists and geologists, the physicists, it would seem, have come out second best! They gave decisive reasons why polar wandering could not be true when it was weakly supported by paleoclimatic evidence, and now that rather convincing paleomagnetic (i.e., physical) evidence has been discovered, they find equally decisive reasons why it could not have been otherwise.

Paleomagnetic studies, as envisioned by their originator, not only provided a compelling test of the drift theory, but once displacements seemed proven, paleomagnetism also furnished an independent, nongeologic tool for reconstructing the predrift continents and for helping to date drift history (Fig. 6.19). Most resulting reconstructions show impressive consistency with independent geologic evidence. For example the paleomagnetic reconstructions of the continents are almost identical with those old ones of Wegener and Du Toit (Fig. 6.1), which were based upon quite different evidence. In Chap. 7 we shall study a second important phenomenon of rock magnetism, namely magnetic anomalies peculiar to the ocean floor.

EXPANSION OF THE EARTH—THE MOST OUTRAGEOUS HYPOTHESES

Expansion of the earth was first suggested in the nineteenth century to explain parallelism of Atlantic shorelines. In 1933 it was revived, and reconstructions were made on small globes to achieve a hypothetical uniform covering of continental crust (Fig. 6.20). A change of more than 2,000 kilometers of radius was required! Although this figure is staggering, expansion does have some intuitive appeal in that it allows for an originally uniform continental crustal covering, which would imply a more complete early chemical differentiation of the earth than

[2]*Nature*, 1956, v. 177, pp. 553–554.

4.0   **0.2**   **0.0**

BILLION YEARS AGO

FIGURE 6.20
Hypothesis of an expanding earth originally covered uniformly by continental crust. Here all continental crust is considered primitive, while oceanic crust formed in scars produced by radial expansion. (*Adapted from Hilgenberg 1933, Von Wachsenden Erdball; Carey, 1959; drawn by D. E. Owen.*)

does the present complex configuration.

Large-scale earth expansion seemed so outrageous and speculative that it was not taken seriously until about 1955, when the revival of interest in continental displacements began. It was suggested at that time that continental outlines, abruptly terminated mobile belts, similar stratigraphic and paleontologic sequences on widely separated continents, and oceanic ridges all might be explained as well by radial expansion of the earth as by lateral continental displacement. Even the new data favoring sea-floor spreading seem compatible with this different mechanism for expanding the oceans. Paleomagnetic pole restorations, however, cannot be reconciled as well with expansion as with lateral displacements of the crust. Calculations from paleomagnetic data suggest no more than a few hundred kilometers of expansion—if any. If expansion has occurred, note that sea water would have to have been added at a rate just sufficient to keep the enlarging ocean basins filled, for the stratigraphic record shows shallow marine deposits on continents throughout most of geologic time.

What possible mechanism could cause expansion of the earth? The first thought is heat from radioactive decay. Most materials expand when heated, but calculations show that the maximum change of earth radius by this mechanism would be only a few tens of kilometers. Based upon the expanding universe theory, it was suggested in 1937 that the gravitational constant (G) may be decreasing through time, and that the earth might expand as a result of decreasing G. Physicists estimate that 100-kilometer radial expansion might be possible, but as much as 2,000 kilometers seems impossible. The order of permissible expansion would allow for approximately a 5 or 6 percent increase of area, which falls far short of explaining complete displacement of continents. An additional consequence of decaying G and expansion of all planetary bodies, physicists tell us, is that the sun should have burned its mass so quickly that it would already be in the red giant stage, which it is not. Also, heat flow from the interior per unit of surface area of an early, small earth would have made the surface temperature too great for life, yet we have a clear fossil record extending back to within 1 billion years of the estimated origin or the earth. Finally, we would expect all mineral grains in rocks to show profound cracking due to such expansion, but they do not.

SUMMARY

In this chapter we have outlined a number of hypotheses to explain large-scale earth structure. The crust probably formed by the selective rise of lighter elements from the mantle. The two fundamental types of crust, continental and oceanic, have an irregular distribution today, which long has puzzled geologists. A basic question is which type is the more primitive. Hypotheses tend to assume either that an early continental crust covered the entire earth or that continental crust has developed slowly at the expense of oceanic crust. Regardless of the original nature of continental crust, its known existence for 3 billion years seems to require, in light of isostasy, rejuvenation of continental material from below in order to maintain topographic continents in the face of erosion. Although continental crust has suffered repeated

structural disturbances and erosion, average elevation of the continental plateau has remained close to sea level, at least for the past 2.5 billion years. The generation of andesitic rocks today in the active volcanic arcs, and the presence of huge granitic batholiths in the deeply eroded ancient mountain zones seem to indicate that the mobile tectonic belts have played a major role in the generation of continental-type crust through most of earth history.

Together with the mobile belts, which include the modern volcanic arc-trench systems, ocean ridges are the most tectonically active features. Between these more mobile zones lie the much larger stable divisions, continental cratons and submerged abyssal plains. Geosynclines represent extremely thick sedimentary and volcanic rocks that accumulated within subsiding portions of mobile belts. Early workers believed that a geosyncline was required somehow to produce mountains, but it is now evident that geosynclines are the results of mobile-belt processes, not their cause. The subsidence that allows thick sediment accumulations may be caused at least in part by any of the following: (1) sediment loading of the crust; (2) compression of the crust causing it to buckle downward; (3) removal of large volumes of magma from below to form lavas; or (4) change of the lower crust to a more-dense phase resulting in isostatic sinking. Of these, structural downbuckling seems to be the principal cause of subsidence. The geosynclinal concept is probably not very useful anymore, although it was enormously influential for a century.

Half a century ago the earth was assumed to be contracting as it cooled from a hot origin. Contraction presumably buckled the crust at the edges of continents, geosynclines were formed, and ultimately these were upheaved into mountains. Because the earth apparently has not cooled in any such simple way, this appealing hypothesis must be rejected. Conversely, it has been claimed that the earth may be expanding, but the permissible amount seems trivial. It is also claimed that the continents are not fixed in position, but have been displaced at least once. Under certain conditions, the outer earth behaves in a brittle fashion to stress, as evidenced by faulting; yet, under other conditions (particularly high pressure and temperature and very slowly applied stress), the lower crust and upper mantle behave plastically. Importance of geologic time is apparent, for a stress acting slowly over immense time can produce effects that would be impossible if the same stress acted briefly. In this way it now seems possible to explain the drifting of continents as the passive upper portions of much thicker plates. Heat convection in the earth's mantle seems the most plausible cause of the motions of those plates. Where two are compressed together, mobile belts are formed; where they move apart, ocean ridges are formed.

After continental drift was proposed as a mountain-building mechanism in the early twentieth century, the idea lacked supporters until paleontologic and stratigraphic evidence favoring drift was supplemented dramatically in the 1950s with the discovery of paleomagnetic evidence. In addition to supporting large displacements of continents, paleomagnetism also provides evidence of paleolatitude that is of great general value in interpreting paleogeography and paleoclimatology. In the next chapter, we shall see that magnetic properties of the sea floor also provide clues for restoration of continents and ocean basins.

Key Concepts

continental accretion
continental drift
contraction hypothesis
cratonic
expansion hypothesis
geosyncline
gravity tectonics
heat flow
lateral compression
magnetically susceptible
mobile belt
orogenesis
paleolatitude
paleomagnetism
   declination angle
   inclination angle
physical phase change
principle of isostasy

remanent magnetism
sea-floor spreading
tectonic
thermal convection
thrust fault

Readings

Birch, F., 1968, On the possibility of large changes in the earth's volume: Physics of the Earth and Planetary Interiors, v. 1, pp. 141–147.

Blackett, P. M. S., Bullard, E., and Runcorn, S. K., eds., 1965, A symposium on continental drift: Royal Society of London.

Carey, S. W., 1976, The expanding earth: New York, Elsevier.

Dana, J. D., 1873, On some results of the earth's contraction from cooling, including a discussion of the origin of mountains, and the nature of the earth's interior: American Journal of Science, ser. 3, v. 5, pp. 423–443; v. 6, pp. 6–14; 104–115; 161–172.

Du Toit, A. L., 1937, Our wandering continents: Edinburgh, Oliver & Boyd

Gondwanaland revisited: new evidence for continental drift, 1968: Proceedings of the American Philosophical Society, v. 112, pp. 307–353.

Hall, J., 1859, The natural history of New York, Part 6—Paleontology, J.3: Albany, Van Benthuysen, Introduction, pp. 1–85.

Holmes. A., 1965, Physical geology (rev. ed.): New York, Ronald.

Jordan, P., 1971, The expanding earth: New York, Pergamon.

Marvin, U. B., 1973, Continental drift: Washington, Smithsonian Press.

McElhinny, M. W., 1973, Palaeomagnetism and plate tectonics: Cambridge, Cambridge. (Paperback edition, 1979)

Oakeshott, G. B., 1975, Volcanoes and earthquakes: geologic violence: New York, McGraw-Hill. (Paperback)

Phinny, R. A., ed., 1968, The history of the earth's crust: Princeton, Princeton.

Poole, W. H., ed., 1966, Continental margins and island arcs, An international upper mantle project report: Ottawa, Geological Survey of Canada Paper 66–15.

Scientific American, 1976, Continents adrift and continents aground: reprints from Scientific American, San Francisco, Freeman.

Wegener, A., 1966, The origin of continents and oceans: New York, Dover. (Paperback, S 1708, English translation of 4th ed., 1929)

THE DEEP-SEA FLOOR AND PLATE TECTONICS

They (plates) can't curl down; they
must curl up
To form a kind of dish
To stop the oceans spilling out
And losing all the fish

B. C. and G. C. P. King (1971)[1]

FIGURE 7.1
Deep-sea drilling vessel *Glomar Challenger* owned by Global Marine Inc. and operated by Scripps Institution of Oceanography for the National Science Foundation. The ship weighs 10,400 tons, is 400 feet long, and the derrick rises 194 feet above the waterline. It carries over 24,000 feet of 5-inch drill pipe. Drill bits can be changed and reentry can then be made into a small borehole 3 miles beneath the ship. *(Photo courtesy Scripps Institution of Oceanography.)*

[1]*Nature*, V. 232, p. 37.

7

If continental displacements are valid, then according to the drift theory, the Atlantic, Indian, and most of the Arctic Oceans did not exist prior to Mesozoic time. Clearly, historical and structural evidence from these ocean basins should shed much light upon the whole question of crustal evolution. Indeed, the deep-sea floors provide the most definitive tests of the entire concept of continental displacements.

The deep seas were first probed in 1872 when the British *Challenger* Expedition was launched. Discovery of the mid-Atlantic submarine ridge (Fig. 6.10) about 1910 seemed to support drift, because the ridge neatly parallels the opposing coastlines; perhaps it was a scar away from which the continents moved. But the greatest period of geologic investigation has been since World War II, and progress has been breathtaking. As with rock magnetism, geophysics led the assault upon the deep seas that has produced a revolution that not only supports continental drift but goes far beyond it to provide a unifying new theory of global tectonics. Most surprising was the realization that oceanic crust, rather than being the oldest on earth, actually dates back only into the Mesozoic. Most older ocean floor seemingly has been thrust down into the mantle beneath ocean trenches as American geologist Bailey Willis prophesied way back in 1907 just as a hunch. Small amounts of old sea floor also seem to be preserved as shreds within many old mountain belts as Europeans thought 80 years ago (as noted in Chap. 6).

In the 1960s, the sea-floor-spreading hypothesis suggested that new oceanic crust is constantly added through igneous activity at ocean ridges, which are sites of extension or pulling apart. Meanwhile, the sea floor is thrust beneath the trench-arc systems, which are zones of compression. Earthquake zones define about a dozen large plates covering the surface of the earth. The different relative motions among these *lithosphere plates* apparently can explain all the major tectonic features of the earth's surface, as well as the drifting of continents. The all-encompassing theory of plate tectonics has produced the greatest revolution in the earth sciences since Darwin and Lyell. The plate-tectonic revolution began just before we published our first edition of *Evolution of the Earth*, and continued at such a rapid pace that it necessitated a complete reorganization for the second edition only five years later. Plate theory has also unified the earth sciences to a degree never before possible, and it has lent an excitement to the field that we hope will be apparent as you read ahead. It appears that the evolution of the theory will continue for some time yet.

GENERAL NATURE OF THE SEA FLOOR

Until recent years, most geologists assumed that the crust beneath ocean basins was very old, topographically featureless, structurally tranquil, and essentially permanent as to position. All these assumptions appear to be incorrect, for nowhere in deep oceanic sediments have fossils older than Jurassic as yet been definitely identified. To our utter amazement the continents seem much the older! Moreover, the total thickness of sediments on the deep-sea floor is small. Even modest assumed rates of deposition suggest that much of the present deep sea could have received significant sediment for only the past 100 to 200 million years. A final definitive answer on the maximum ages of the different deep basins must await more fossil evidence from deep drilling through the sediment column (Fig. 7.1), which has already revolutionized geology. Continental drifters obviously would find it embarrassing if Cambrian

Central Bermuda Rise

FIGURE 7.2

Seismic reflection profile across North Atlantic abyssal plain showing irregular oceanic basement surface and smoothing by sedimentation. Interpretive section appears below. Note prominent "A" reflecting zone, which is of Cretaceous age here. (*After J. Ewing, et al., Science, v. 154, December 2, 1966, pp. 1126–1132; copyright 1966 by the American Association for the Advancement of Science.*)

fossils turned up in widely scattered drill holes, but this now seems very unlikely. All available evidence points to a young—that is, early Mesozoic—origin for the present ocean basins.

Precision profiles established by reflection of low-frequency sound waves from the sea floor and buried layers beneath have shown that the ocean floors are anything but smooth. Broad *oceanic ridges or rises, deep trenches, escarpments,* and countless *submerged seamounts* characterize it instead. Indeed, the pristine surface of the oceanic crust is more rugged than most continental areas, although sedimentation has served to smooth the topography in some areas by burying original irregularities (Fig. 7.2).

RATES OF CONTINENTAL EROSION AND OCEANIC SEDIMENTATION

North America is now being eroded at a rate that could level it in a mere 10 million years, or to put it another way, 10 North Americas could have been eroded since Middle Cretaceous time 100 million years ago. If all present continents were reduced to present sea level, and their refuse were spread uniformly over the abyssal plains, a layer of sediments about 300 meters

thick would result. The observed average total thickness of deep-sea sediments is only about 600 meters, or an amount equal to the erosion of the present-sized continents *only twice during the past 200 million years* for which time there is a known record in the deep-ocean basins. From this discrepancy, as well as from other evidence, it is clear that, in the past, the rate of erosion and/or the volume of land above sea level has been much less on the average than now.

The apparent youthfulness of the entire present deep-sea floors came as a great shock, for you will recall that in Chap. 5 we argued that basaltic oceanic crust seems to be a chemically more primitive type than granitic continental crust. Now we discover that not only do continents apparently contain the only record older than 200 million years, but also that there is a shortage of deep-sea sediments compared to what would be expected through long erosion of continents. Apparently the puzzle can be answered only by losses through removal both of deep-sea sediments and oceanic crust through time.

OCEANIC RISES OR RIDGES

The most striking features, especially of the Atlantic and Indian Ocean floors, are submarine rises or ridges, mentioned before (Fig. 7.3). Their symmetrical positions invite speculation that they are scars of a predrift configuration of the crust; for example, the Americas moved west while Europe and Africa moved east away

FIGURE 7.3
World's ocean-floor topography, emphasizing roughness of much of the sea floor; note ocean ridges, linear escarpments, and deep trenches. Being less masked by sediments and vegetation than continents, this topography has great structural significance.

from the mid-Atlantic ridge (see Fig. 6.10 on page 120).

The bulk of the broad ridge features is made up of basaltic lavas, as evidenced by islands along their crests, by samples dredged from their submerged portions, and from underwater photographs (Fig. 7.4). Oceanic ridges have several important characteristics that set them aside as unique, major structural features of the crust (see Table 6.2). Besides volcanism, they display great shallow seismicity beneath their axes (unlike trenches, which also have deeper focus earthquakes). Ridges are also characterized by greater-than-average flow of heat through the crust along their axes (Table 6.2). Ridges clearly are zones of release of much subcrustal thermal energy, but *in response to*

FIGURE 7.4
Submarine ellipsoidal ("pillow") lavas on the flank of South Pacific–Antarctic oceanic ridge at a depth of 2,800 meters (longitude 145°E, latitude 56°S). (Compare Figs. 8.4 and 12.2). *(Official NSF photo, USNS Eltanin, Cruise 15; courtesy Smithsonian Oceanographic Sorting Center.)*

FIGURE 7.5
Aerial view of western part of Thingvellier graben, southern Iceland, marking surface exposure of part of crestal rift of mid-Atlantic ridge; scarp at right is about 40 meters high (view looking south). Volcanic eruptions and earthquakes are the rule in Iceland (see Fig. 1.5). *(Photo by Sigurdur Thorarinsson.)*

different stress conditions than prevail beneath island arcs.

Another peculiarity shared by most ridges is a narrow depression that extends along their axes for thousands of kilometers. In Iceland, the largest exposed ridge area on earth, prominent graben structures are conspicuous across the center of the island from north to south (Fig. 7.5), and 30 active volcanoes occur along this zone. American B. C. Heezen about 1960 noted a close parallel of scale and morphology with African rift structures, and suggested a similar origin. The northern Indian Ocean ridge even passes into the continental African rifts (Fig. 7.3). Small research submarines allowed direct observation in the 1970s of many fissures along ridges. Prevalence of topographic rifts along many ridge crests, together with seismicity and volcanism, all suggest—by analogy with structures of land—that ridges are zones of extension along which the crust is being torn open. Ridge patterns may be complicated by branching. Any juncture of three separate plates is called a *triple junction*. The simplest case involves three branching ridges, but the arms of triple junctions also may be trenches or major faults.

The East Pacific Rise is unique in being asymmetrically located in its ocean basin. It extends northward to or beneath the western edge of the North American continent. Significantly, the crust and mantle there are geophysically anomalous and display evidence of unusually widespread extensional structures in the western United States, suggesting that the ridge has somehow disturbed the edge of the continent there much as the Indian Ocean ridge affected eastern Africa.

SEA-FLOOR SPREADING—ANOTHER OUTRAGEOUS HYPOTHESIS

It was submarine extension or ocean spread constantly in progress . . . Asia was therefore formed not by overthrusting, but by underthrusting. (Bailey Willis, 1907)

Recall that, in 1928, Scottish geologist Arthur Holmes presented a hypothesis of convection in the mantle as a cause of mountain building (Fig. 6.14), and in the same year, Alfred Wegener also accepted convection as one possible mechanism for continental drift. Since that time, convection has been the most widely endorsed mechanism

to explain large-scale tectonic features. In 1962, Princeton University geologist H. H. Hess proposed a bold new hypothesis that two opposing thermal convection cells rising beneath ocean ridges produce tension in the crust (thus rifting) and also cause the abnormal heat flow observed there. As rifting occurs, earthquakes are generated beneath ridges, and new crustal material is erupted volcanically at ridge axes. Hess envisioned that, finally, the slow, convective flow laterally away from ridge axes carries older oceanic crust along as if on a conveyor belt, causing the spreading of sea floors through time.

The dramatic *hypothesis of sea-floor spreading* postulates a youthful origin of the Atlantic and Indian Ocean basins (and their crusts) through disruption of former continental areas by rifting over rising convective cells followed by progressive separation of the dismembered continental fragments as juvenile oceanic crust is generated between (Fig. 7.6). Presumably, eastern Africa and Arabia are today experiencing the beginning of a new phase of such disruption in response to an assumed shift of mantle convection patterns, which brought rising cells beneath that region. It is significant that the Red Sea–Gulf of Aden rift lies at the landward end of the western Indian Ocean or Carlsburg ridge. Spreading along the extension of this ridge tore open the Aden–Red Sea rift in the crust (Figs. 7.7, 7.8). African rifts during the past 25 million years seem to typify the initial breakup of huge supercontinents—the earliest stage in continental drift.

FIGURE 7.6

The hypothesis of sea-floor spreading showing its combined explanation of ocean ridges (divergent) and arc-trench (convergent) systems. Three lithosphere plates with continents forming parts of two are shown moving over the weak low-velocity zone of the upper mantle. Magmas are produced in arcs by heating along the subduction zones where earthquake foci (dots) are concentrated in the downgoing, relatively cool slab. Magmas are generated at shallower depths under hot ridges where foci also are shallower. The 1000°C contour illustrates contrast between hot upper mantle beneath ridges and the cooler region beneath arcs.

FIGURE 7.7

Gulf of Aden and southern Red Sea from a Gemini space-craft. Note how well Africa (below) could fit against Arabia (above), suggesting that Africa was torn away to form the new ocean basin between and inaugurate a new phase of continental drifting. *(Courtesy NASA; photo No. 66-54536.)*

FIGURE 7.8

The Afar Triangle, intersection of Red Sea, Aden, and East African rift systems, and possible surface expression of a hot mantle plume. Regardless of the underlying cause, such a system of diverging rifts (sometimes called a *triple junction*) seems to initiate new continental drifting episodes. Dots indicate young volcanoes. *(Adapted in part from Levin,* The earth through time, *copyright © 1978 by W. B. Saunders; by permission of Holt, Rinehart and Winston.)*

HOT MANTLE PLUMES

Cratonic rift structures apparently represent the initial stage of continental drifting, as we have suggested. The Red Sea–Aden rift system is especially revealing because the Carlsburg spreading ridge extends from a triple junction with the Indian Ocean ridge directly into the Gulf of Aden. A recent hypothesis of continental breakup involves postulated hot *mantle plumes* rising beneath the continent (Fig. 7.8). Such plumes are thought to be reflected at the surface by volcanic centers 100 to 200 kilometers in diameter that are associated with domed uplift areas. They are not associated with magmatic arcs and may or may not be associated with ocean ridges. The volcanoes erupt unique lavas rich in the alkaline chemical elements. About 200 possible late Cenozoic plumes have been identified for the earth as a whole, with Africa being especially well endowed. The Hawaiian Islands are the best-known probable example in an oceanic plate.

The suggested evolution of rifting and continental drifting by rising hot mantle plumes is as follows. First, the uplift of a flat, brittle plate results in the formation of three radiating cracks, a triple junction; swarms of radiating igneous dikes are commonly associated. Then, as extension of the cracks continues, two arms of the trio tend to open up to form a new ocean basin. The third (or failed) arm, which is nearly perpendicular to the opening seaway, becomes a sediment-filled trough (called *aulacogen* by Russian geologists). It gradually becomes stable and is buried. The southward-trending East African rift system seems to be such an arm of the Red Sea–Aden spreading arms. While the East African rifts are tectonically active, they have not opened up as has the latter system, for they have continental crust beneath rather than oceanic. Many great rivers flow down postulated failed-arm troughs, for example the Mississippi, Rhine, and Amazon. The sediment-filled troughs commonly hold important petroleum resources.

The mantle plume idea is a most ingenious one and deserves further attention, but at the time of this writing, theory was advancing faster than the accumulation of hard facts to prove it. A cause for such plumes is especially obscure.

AGE DISTRIBUTION OF OCEANIC ISLANDS AND SEAMOUNTS

About the same time that Hess first stated his concept of sea-floor spreading, Canadian geophysicist J. Tuzo Wilson had postulated that oceanic islands, which are practically all volcanic in origin, tend to be symmetrically distributed as to relative age outward from submarine ridges. Youngest islands tend to be at or near the axes of ridges, while those more distant seemed to be progressively older. Moreover, still farther removed from ridge axes are sub-

merged volcanic mountains (seamounts) at progressively deeper levels (Fig. 7.6). Many of these were islands in the past, as evidenced by dredgings of shallow-marine fossils from their tops. Obviously the fate of oceanic islands, once they have ceased to be volcanically active, is to be more or less eroded by surf and to gradually subside to great depths. Sea-floor spreading offers a consistent explanation of Wilson's conception of island age distributions. Most islands, he said, formed at ridge axes and have existed as such as long as they remained volcanically active (generally about 20 to 30 m.y.). As spreading of the crust progresses, they move laterally, conveyor-belt fashion, become inactive, and then proceed to sink as spreading carries them farther down the ridge flanks.

CORAL ATOLLS AND FLAT-TOPPED SEAMOUNTS

In the nineteenth century, low rings of coconut-tree-studded island paradises surrounding shallow lagoons in the tropics (between 30° north and 30° south latitude) invited explanations by many naturalists. The several hypotheses advanced to explain these islands generated much controversy, but Charles Darwin in 1842 struck upon the correct interpretation. He concluded that these atolls, which are nothing but rings of low islands composed of debris washed up by waves from coral reefs just offshore, are but one stage in an evolution of typical Pacific and Indian Ocean islands, which involves gradual subsidence of the sea floor beneath them.

Darwin noted that some tropical islands consisted of a prominent volcano surrounded by a *fringing reef* at its margin with little or no water between. Others had *barrier reefs* with narrow doughnut-shaped lagoons between the reef and the central volcano. Others had a very small volcano and much wider lagoon, and still others, the true *atolls*, had a ring-shaped series of low islands with a broad, central lagoon and no volcano at all. He said that all atolls began as fringing reefs surrounding active volcanoes and that in the course of time, they gradually sank isostatically as the volcano became extinct. Be-

TABLE 7.1

Sea-floor subsidence indicated by known depths of reef carbonate rock columns*

Location	Depth drilled or dredged (meters)	Oldest age encountered
Pacific		
Bikini	725	Oligocene or Late Eocene
Eniwetok	350	Miocene
Funafuti	330	Pliocene
Mid-Pacific mountains	1,900	Mid-Cretaceous
Philippine Sea		
Kita-daito-Jima	400	Early Miocene
Western Atlantic		
Bermuda	125	Early Oligocene
Bahama Islands	5,445	Cretaceous

*After K. O. Emery et al., 1954, U.S. Geological Survey Professional Paper 260-A; E. L. Hamilton, 1956, Geological Society of America Memoir 67; and N. D. Newell, 1955, Geological Society of America Special Paper 62.

cause coral growth is most active on the outer edges of reefs, the reefs expanded outward as well as upward. More important, the coral growth had to keep pace with the subsiding volcano. If it sank too rapidly beneath the critical *lighted (photic) zone*, the reef would be killed. Eventually the now-extinct volcano would disappear beneath the sea, and the atoll would have advanced to its mature state surrounding an open lagoon.

Darwin recognized that the only way his theory could be tested was by drilling into an atoll to see what lay beneath. It was not until 1952, at Eniwetok Atoll in the course of preparations for United States' Pacific nuclear testing, that drilling operations did penetrate submerged volcanoes. Most of what lay between the surface and the volcanic basement consisted of reef limestone and reef debris. Furthermore, the Pacific reefs were found to be relatively old—going back in some cases at least to early Cenozoic time (Table 7.1). Only the Darwinian subsidence theory for atolls can account for the great depth at which submerged volcanic foundations were encountered (nearly 1,000 meters). Known sea-level changes were not nearly great enough to explain such depths. In the Bahama

Islands, drilling has penetrated reef and other rocks dating back at least to Early Cretaceous time (Table 7.1); a volcanic foundation probably underlies Cretaceous or Jurassic strata. Many Pacific reefs have been growing continuously since at least early Cenozoic time. Some West Indies reefs began in the Cretaceous, but have had an interrupted history during Cenozoic time. The Red Sea reefs span only late Cenozoic time. Some modern reefs apparently began only during the Pleistocene, especially in the West Indies.

In the western and northern Pacific basin, there are many submerged, flat-topped *volcanic seamounts*. These peculiar features (Fig. 7.6), were discovered during World War II by extensive deep-sea soundings. H. H. Hess interpreted them as oceanic volcanic mountains whose tops had been planed off by wave erosion before they subsided to depths of hundreds or thousands of meters. Proof of his interpretation came at the end of the war with realization from dredged samples that dead, submerged reefs existed on the tops of some mid-Pacific seamounts west of Hawaii now at depths of more than 2,000 meters! These reefs contain Lower Cretaceous organisms that originally lived at sea level. Obviously reef growth could not keep pace with crustal subsidence. According to sea-floor spreading, all the western Pacific seamounts are on crust that is more than 30 million years old.

MAJOR OCEANIC FRACTURE ZONES AND TRANSFORM FAULTS

Exploration in the northeastern Pacific basin following World War II revealed a group of long, east-west-trending submarine escarpments explicable only as great zones of faulting (Fig. 7.3). Such fracture zones, as they are called, have been found elsewhere, too, and seem to be characteristic of sea floors. In general they are perpendicular to ocean ridges. In the mid-Atlantic, for example, the ridge crest has been offset along such zones. The Indian Ocean floor displays the most amazing array of escarpments, suggesting wholesale fragmentation of the crust along largely north-south fractures

FIGURE 7.9

Comparison of transform faults with ordinary transcurrent or strike-slip faults. Wide brown lines represent ridge axes; A and B represent two different times for each case. *Upper:* Spreading proceeds as transforms move, so fault motion suggested by offset of ridge crest is opposite actual motion. *Lower:* Spreading ceased before faulting; therefore offset of the dead ridge does provide an accurate idea of displacement. *(Modified from Wilson, 1965, Science, v. 150, pp. 482–485.)*

that look like railroad tracks along which India traveled northward (Fig. 7.3).

With important exceptions like the great San Andreas fault of California, such long, nearly straight fractures are rare on continents. Indeed, most of the Pacific fracture zones terminate at the continental margins; thus they reflect structural processes virtually restricted to oceanic crust. That the dominant movement along these zones is lateral or *transcurrent* (that is, blocks move past each other horizontally rather than vertically) is indicated by offsets of linear magnetic anomalies, offsets of the ocean-ridge crests, and from studies of motions during earthquakes that occur along them. From the apparent offsets of the mid-Atlantic ridge near the equator, it appeared at first that the motion there was all *left-lateral*, for the north side of each fault seems to be displaced left (westward) as viewed from the south. Cumulatively, such movement might account for the great midlatitude bend of the ridge (Fig. 7.3). But some of the faults have sharp terminations at the ridge crest, and quite a different interpretation of movement became possible in light of the sea-floor-spreading hypothesis. J. Tuzo Wilson, in 1965,

conceived of a mechanism that produces what he termed *transform faults*. By this scheme, the faults formed during spreading actually have movement *opposite* to that implied by mere apparent offset of ridge crests (Fig. 7.9); lateral motion is transformed at ridge crests by spreading. Apparently such faults owe their origin to the fact that spreading rates cannot be uniform along a spreading ridge on a sphere (Fig. 7.10). Careful studies of seismic records for earthquakes along ridge fractures confirm that the motions are, indeed, transform in nature (as in Figure 7.9-upper). Apparently most great oceanic fracture zones are transform faults related to sea-floor spreading. Besides connecting ocean-ridge segments, transforms may also connect ridge and volcanic-arc segments, or two arc segments. Furthermore, they may be members of triple junctions.

FIGURE 7.10

Geometrical relationships of a spreading ridge axis, transform faults, and subduction zones on a spherical earth. The lithosphere plates A, B, and C move with respect to an imaginary pole of plate rotation or spreading rather than the earth's rotational pole. Transform faults are perpendicular to the spreading axis (i.e., parallel to imaginary lines of latitude about that pole). Rate of spreading, indicated by relative length of arrows, increases from the pole to the equator of spreading. *(Adapted from Heirtzler, 1968, Sea floor spreading, copyright © by Scientific American, Inc. All rights reserved.)*

FIGURE 7.11

"Zebra-stripe" magnetic anomalies symmetrical around axis line of the mid-Atlantic ridge southwest of Iceland (in line with graben of Fig. 7.5). Symmetrical relation of stripes to ridge axes here and elsewhere led to the hypothesis that such anomalies reflect successive reversals of polarity of the magnetic field while the sea floor was spreading away from the ridge. *(From Heirtzler et al., 1966, by permission of* Deep-Sea Research.*)*

MAGNETIC ANOMALIES AND SEA-FLOOR SPREADING

Among many new kinds of information about the sea floors was the discovery in 1961 of linear magnetic anomalies apparently unique to the abyssal oceanic crust, for the patterns terminate near the continental shelves. The anomalies first were mapped in detail in the northeastern Pacific Ocean and then across the mid-Atlantic ridge south of Iceland (Fig. 7.11). Surveys elsewhere have now provided magnetic profiles across much of the remainder of the world ocean-ridge system, and all show a fairly consistent pattern of narrow, alternating anomalies, but there are areas with none or irregular patterns. Near Iceland, the stripes showed two striking features: *first*, they parallel closely the ridge axis, and *second*, they show a remarkable bilateral symmetry such that those on one side of the axis tend to mirror those on the other (Fig. 7.11). The same symmetry has been found elsewhere (Fig. 7.12).

What do the oceanic magnetic anomalies

mean? They were mapped by ships or planes towing magnetometers along repeated traverses that crisscross the ridges. Resulting data record variations in total intensity of the magnetic field across the sea floor. At first it might seem that the stripe patterns represent alternating zones of extreme contrasts of magnetic susceptibility and remanent magnetism in the oceanic crust. But the magnitude of the anomalies would require improbable alternations of very narrow, deep blocks of nonmagnetic and very strongly magnetic materials; necessary geometry and susceptibility contrasts are so unlikely as to boggle the mind. Alternatively, the anomalies could represent polarity reversals in successive bands of rock materials of uniform magnetic susceptibility. Thus the black bands in Figs. 7.11 and 7.12 are thought to represent normal polarity relative to the present earth field, and the blank ones reversed polarity. This explanation is preferred because independent evidence of polarity changes of the earth's field have been known for many years.

As early as 1905, it was found that polarity in magnetically susceptible minerals in some late Cenozoic lavas was opposed to that of the present main field. In the 1920s and 1930s, examples of reversals were documented in Cenozoic lavas from several continents. The last reversal, which produced the present field configuration, occurred in Pleistocene time nearly 1 million years ago (Fig. 7.13). In thick sequences with many Cenozoic lavas, a number of successive reversals have been documented, and the rocks now can be dated isotopically, thus providing a geomagnetic-reversal time scale. Similarly, polarity studies have been made of deep-sea sediments from submarine cores, and the results compare closely with those for lavas (Fig. 7.13). As magnetically susceptible sedimentary minerals settled to the sea floor, they assumed a remanent magnetism imposed by the magnetic field; thus, as in lavas, they preserve clues to past polarity.

Reversals of polarity of the earth's field have been important phenomena, with universal effects on minerals forming anywhere in the world in various types of rocks over at least the past 400 or 500 million years and, we assume,

FIGURE 7.12

Ocean crust magnetic anomalies plotted as profiles of measured magnetic intensity along straight-line crossings of the South Pacific–Antarctic oceanic ridge. Note symmetry of anomalies on either side of the broad ridge-axis anomaly (upper curve is same as middle, but reversed to underscore the symmetry). Note also sharp contrasts of magnetic intensity peaks, which represent alternating "zebra-stripe" patterns of anomaly maps like Fig. 7.11. Polarity-reversal episodes recorded in crust are idealized at bottom in black-and-white bar with sediments shown above. *(From Pitman and Heirtzler,* Science, *v. 154, December 2, 1966, pp. 1166–1171; copyright 1966 by the American Association for the Advancement of Science.)*

FIGURE 7.13

Diagrammatic representation of magnetic polarity reversals observed in late Cenozoic lavas in many parts of the world (which can be dated by isotopic methods) and in deep-marine sediments dated by fossils. Fossil zones, designated by Greek letters, tend to correlate closely with reversal episodes; therefore either one provides a tool for submarine stratigraphy. Polarity studies have led to compilation of a formal magnetic-reversal time scale, part of which is shown. *(Adapted from Cox et al., 1963,* Nature, *v. 198, p. 1049; Opdyke et al.,* Science, *1966, v. 154, p. 349.)*

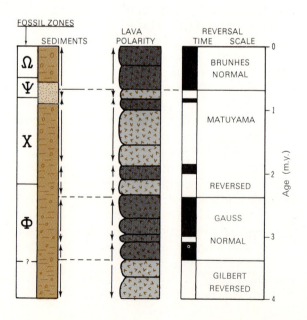

much longer. The period between reversals has varied enormously. During late Cenozoic time, for which data are by far the best, reversals have been occurring with a frequency of approximately 1 million years, although refined measurements show some periods as short as 10,000 years. Relatively short polarity episodes also characterized earlier Cenozoic time (Fig. 7.14-left), but late Mesozoic and late Paleozoic times had much less frequent reversals; early Mesozoic and middle Paleozoic times, conversely, were characterized by more reversals (Fig. 7.14-right). The actual reversing process is not understood. Apparently the poles migrate erratically, and the strength of the field varies considerably while reversing is underway; the entire change is thought to be geologically sudden, spanning only a few thousand years.

Discovery of an apparently universal polarity-reversal history for the earth's magnetic field, which is acceptable within the Elsasser dynamo theory for the field (see Fig. 5.8), represented a major recent breakthrough in geologic knowledge. It is suggested that symmetrical oceanic magnetic anomalies represent fossilized polarity episodes associated with the reversal history. If true, they would greatly strengthen the sea-floor-spreading hypothesis. The anomalies appear to record each polarity-reversal event as the sea floor was spreading; as new crust formed in the central ridge rifts and cooled to the temperature at which magnetism

becomes fixed (Curie point), it acquired and retained the polarity of the field at that time. Distance of each anomaly pair from a ridge axis would, therefore, be proportional to its age (i.e., the time since it lay at the axis). It is important to stress that the linking of the anomalies with known magnetic polarity reversals is a *hypothesis*, but it is strengthened by the fact that deep drill holes have encountered fossiliferous latest Cenozoic sediments resting upon oceanic basement on the central ridges, but progressively older strata outward toward the continents (Fig. 7.12; deep-sea drilling is still sparse, however).

SPREADING HISTORY

The oceanic magnetic anomalies have been likened by F. J. Vine to a tape recorder attached to a sea-floor conveyor belt. With D. H. Matthews in 1963, Vine proposed the relation of the anomalies to polarity reversals and spreading. These British workers showed that the anomalies can be correlated between several different ridge systems with some confidence. Vine infers that a worldwide reversal history has been "taped" with great fidelity, and by comparing anomalies outward from ridge axes (Fig. 7.12) against the known polarity-epoch time scale established on land (Figs. 7.13 and 7.14), both average rates and an estimate of duration of spreading can be determined for different ridges. Resulting apparent rates of total spreading

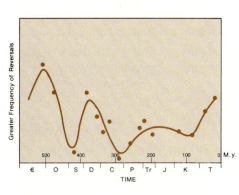

FIGURE 7.14

Left: Magnetic-reversal time scale well established for past 80 million years. *(From Heirtzler et al., 1968,* Journal of Geophysical Research, *v. 73, p. 2123; copyrighted by American Geophysical Union). Right:* Approximate relative frequency of reversals for last 600 million years. *(Simplified from McElhinny, 1971,* Science, *v. 172, pp. 157–159; copyright 1971 by the American Association for the Advancement of Science).*

Subduction zones and deep earthquakes   Transforms   Ocean ridge axes   Jurassic   Cretaceous   Paleogene   Neogene

FIGURE 7.15

Ages of the oceanic crust based upon magnetic anomalies and deep-sea drilling (numbers are ages in m.y.) (compare Fig. 7.3). Also shown are subduction zones (volcanic arc-trench systems), ocean-spreading ridges, and major transform faults. (*Adapted from Karig, 1971,* Journal of Geophysical Research, *and Pitman et al., 1974,* Geological Society of America map of age of ocean basins.)

vary from about 2 centimeters per year south of Iceland to more than 18 centimeters per year in the South Pacific, with the average being 4 to 5 centimeters per year. Knowing the reversal history from lavas on land, we can extrapolate out from ridge-crest regions, which represent only the past few million years, and estimate the total duration of spreading recorded in the anomalies across a given ocean basin (Fig. 7.15). Such analysis provides an estimate of the minimum age of the ocean basin; for examples, spreading from the mid-Atlantic ridge apparently began between 150 and 200 million years ago, from the northwestern Indian Ocean ridge between 80 and 100 million years ago, and

Australia and Antarctica did not separate until 60 million years ago. These figures match closely most independent estimates of inception of postulated continental separations. They also suggest that little if any of the present oceanic crust can be as old as the Paleozoic Era (i.e., more than 225 m.y.)! Much of the western Pacific and some of the northern Atlantic crust is Mesozoic, but most of the rest of the present oceanic crust is Cenozoic, which is confirmed by limited fossil evidence now available from the deep seas (Fig. 7.15). (Some geologists still believe, however, that, as more and deeper drilling of the sea floor proceeds, some older fossils will eventually be found).

Extrapolations of the kind just mentioned are approximations only. In reality, spreading patterns apparently vary considerably along even the same ridge. Authorities also believe that the ridge-axis spreading sites themselves shift laterally as spreading occurs; sometimes they jump suddenly. Changes in magnetic anomaly patterns and discontinuities in sediment thickness and age distribution on several ridges suggest a possible discontinuity in spreading rate about 10 million years ago (late Miocene). On the East Pacific Rise, only Pleistocene sediments have been found on the crest, whereas Pliocene and Miocene deposits occur on the flanks. A late Mesozoic period of active spreading apparently was followed by a mid-Cenozoic slackening, in turn superseded in Miocene time by the present spreading cycle.

PLATE TECTONICS

We are presented with a seemingly simple contrast of crustal strain in the major tectonic features of the crust. As noted before, ocean ridges and the volcanic arc-trench systems are overwhelmingly the most seismically active zones today (Fig. 7.16). In the middle 1960s it was argued that earthquakes define boundaries of huge, rigid plates on the outer earth. The structure of island arcs and ancient mobile belts has long suggested great zones of dominant compression, and now the oceanic ridges (together with continental rift zones) suggest extension. Such motions are confirmed by earthquake motions (Fig. 7.16). At present the broad plates between these zones are relatively stable.

An American student of ocean basins, R. S. Dietz, was an early advocate of sea-floor spreading and was one of the first to speculate (in 1963) about an all-encompassing tectonic theory based upon spreading. He suggested that the push of a spreading sea floor against a continental margin could—at least in some cases—bend down the continental margin, crumple the sediments there, and eventually upheave them; metamorphism and partial melting of the deeper sediments also were postulated, thus producing all the characteristic features of mountain belts. Dietz envisioned that new material was plastered onto continents from the ocean basins to produce gradual enlargement. The continents also could be passively swept along by spreading in conveyor-belt fashion; thus continental drift became a by-product of sea-floor spreading. Other workers soon argued that some continental margins (such as the Atlantic margin of the Americas today) are firmly linked or coupled to their adjacent spreading oceanic crust and so move quite passively with it and are not presently being deformed (Fig. 7.6). Such margins are *passive trailing edges* of continents. Only along the *active leading edges* of plates is mountain building in progress (for example, the Pacific margin of South and Central America). Also, however, oceanic crust may

FIGURE 7.16
Slip vectors from earthquake studies. Arrows show relative horizontal motion of crustal blocks. Oceanic ridges are denoted by double lines; arc-trench systems by heavy single lines. Note that motions in general appear to be away from ridges (extensions), but toward arcs (compression). (*After Isacks, Oliver, and Sykes, 1968; by permission of* Journal of Geophysical Research.)

FIGURE 7.17
Eight major lithosphere plates of today (arrows indicate inferred directions of movement of plates), The plates include the crust and from 20 to 40 kilometers of the upper mantle. Their margins are delineated by the earth's major zones of earthquakes and volcanoes; note that some plates include both oceanic and continental portions that are physically coupled together. (*Adapted from J. T. Wilson, 1968,* Proceedings of the American Philosophical Society, *v. 112, pp. 309–320; by permission of The American Philosophical Society.*)

be crushed against and under other oceanic crust, as appears to be the case in in the western Pacific volcanic island arc-trench systems (Figs. 7.16 and 7.6).

In 1968 a group of seismologists at Columbia University drew together sea-floor spreading and continental drift into a broad, unifying theory called *plate tectonics.* Based largely upon new seismic data from all over the world, they argued that the seismic zones (Fig. 7.16) define five or six large lithosphere plates comprising the earth's outer 200 kilometers or so (Fig. 7.17). Each plate seems to behave as a rigid unit—moving away from a neighbor at an ocean ridge—a *divergent plate boundary*—and toward another plate along a volcanic arc and trench—a *convergent boundary* (Figs. 7.6 and 7.16). Different plates also may touch either at triple junctions or at transform faults. Not only do motions during earthquakes as well as folds and thrust faults seen in island arcs and young mountains suggest convergence, but in the 1930s it was found that the deeper quakes (> 100 kilometers) associated with the trenches define a zone dipping 30° to 45° beneath an arc. A recently detected speedup of seismic waves beneath this inclined seismic zone indicates the presence there of a huge slab of more rigid, apparently colder rock material extending downward into the mantle to a maximum depth of 700 kilometers (Fig. 7.6). An idea from the 1930s of an inclined zone of shearing as much as 100 kilometers thick beneath volcanic arcs was thus confirmed, and, even more dramatic, the

underthrusting—called *subduction*—of a slab of oceanic material approximately 100 kilometers thick beneath the arc seemed inescapable. New high-resolution seismic profiles actually reveal the fault zone beneath some trenches. Thus, divergence of oceanic plates at ridges is indeed accompanied by subduction of the front of the plate at its convergent margin. The subducted cold material is gradually heated and probably is very slowly assimilated into the mantle below 700 kilometers. Within the upper side of the downgoing slab, friction accompanying the frequent earthquakes observed beneath arcs causes local *partial melting* of the downgoing plate and probably some adjacent mantle to give rise to the magmas erupted from volcanoes of the arc (Fig. 7.18). But these magmas are not basaltic as on the ridges. Rather, partial melting along the subduction zone (Fig. 7.18) produces magmas richer in potassium, sodium, and alumina than does melting under ridges. This is partly due to the different temperatures and pressures in the two environments (Fig. 7.6), but also water in the downgoing slab must help to mobilize these elements preferentially when heating occurs. The resulting rock is called *andesite.* The andesitic arc magmas are "continental" in average composition so that the accretion or addition of new continental rock material really is occurring in young mobile belts.

Seismology has also shown that the moving plates involve much more than the crust. So-called *lithosphere plates* are on the average about

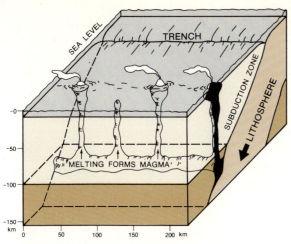

FIGURE 7.18

Formation of a volcanic arc by melting along a subduction zone. Magma forms where that zone reaches a depth of about 120 kilometers, at which a critical combination of pressure, temperature, and water content is achieved. Necessary heat is derived either from friction along the subduction zone or simply by contact of the downgoing slab with the hot asthenosphere through which it is moving. Less-dense magma rises toward the surface along conduits a few kilometers in diameter and rather uniformly spaced about 70 kilometers apart. Presumably magma is generated more or less constantly and rises about 10 to 100 meters per year. Thus any given unit volume would take no more than 10,000 years to reach the surface—a geologically lightninglike trip— *(Redrawn after P. Van Dusen in B. D. Marsh, 1979, American Scientist, v. 67, pp. 161–172.)*

60 kilometers thick, thus including the crust as a thin superficial rind as well as a portion of the upper mantle down to the low-seismic-velocity zone at about 60 kilometers (Fig. 7.6; also Fig. 5.7). That zone, recall, because of the decrease of seismic velocity within it, must be of lower strength than the plate above or the deeper mantle below. We must now think not of the crust as the fundamental physical unit of the outer earth, but rather of much thicker lithosphere plates all in motion relative to one another on a worldwide scale.

WHAT DRIVES THEM?

The driving of the plates is attributed primarily to convection—our old panacea—within or below the low-seismic-velocity zone. The two-times-greater heat flow observed at ridge crests as compared with deep-sea trenches (Table 6.2; Fig. 7.6) supports—or at least is consistent with—the convection mechanism. An illustrative analogy for plate motions due to convection is provided by lava lakes in Hawaiian craters (Fig. 7.19). Dense, chilled lava crusts form on the surface and tend to sink, allowing less dense, still-molten convecting lava to rise as ridges. Lava in these ridges becomes chilled, is pushed aside, and in turn sinks slowly into the underlying molten pool to be reassimilated.

Geophysical studies of the stress across the interiors of the relatively rigid lithosphere plates indicate that they are now experiencing compression within. The large but thin plates composed of rather weak rocks are like huge sheets of tissue paper, and so it seems impossible that they can be either simply *pushed* outward from ridges or *pulled* downward from beneath trenches (although gravity certainly must help once a plate starts down). This suggests that they are mainly being *dragged* passively along piggyback fashion as by convection within the underlying mantle (Fig. 7.20). Such dragging could produce modest compression within the main parts of the overlying plates. Rate of subduction is also a factor in plate behavior (Fig. 7.21).

From isostasy, we can readily see that a plate with a thin, oceanic crustal rind would more readily sink along a break than would a plate with a thick, light continental rind, which would make up about one-third of the total plate thickness (Fig. 7.21). The subduction of oceanic beneath continental material can be readily understood, as can the subduction of one plate with a thin oceanic rind beneath another of the same type as is happening in the western Pacific. It is argued, however, that if two plates with thick continental rinds impinge, an awesome collision will result, but neither can be subducted significantly beneath the other because of their great isostatic buoyancy.

Collisions between plate margins, especially between two continental masses, are then of great importance. Along any old mobile belt that is *presently* intercratonic, it appears that two continents collided (e.g., the Himalayas; Fig.

FIGURE 7.19

Small spreading zone defined by molten lava (light line from top to bottom) in Mauna Ulu lava crater, Hawaii, during a 1971 eruption. As fresh molten lava rises by convection, slightly older chilled lava (darker) is shoved aside to sink at some other (compressive) zone in crater (out of view). Note transform fault near middle of view and differential rate of spreading revealed by the different widths of recently chilled lava on either side of spreading line. Front edge of view is about 50 meters wide. *(Courtesy W. A. Duffield, U. S. Geological Survey Hawaiian Volcano Observatory; see Duffield, 1972,* Journal of Geophysical Research, *v. 77, pp. 2543–2555.)*

7.21). In these cases, intensely compressed and metamorphosed ultramafic rocks and basalts commonly define a *suture zone* where two plates collided. Such zones are all that is left at the surface of a former ocean that separated the two continents before collision. *Apparently these narrow, mangled zones in continents are the only relics we can expect of any sea floor older than about 200 million years!* We find, then, that *convergent plate boundaries are destructive* and that staggering areas of the earth's surface of the past have completely disappeared. It seems inescapable

that several hundred billion cubic kilometers of lithosphere have been consumed in the past 3 to 4 billion years.

GLUING TOGETHER BROKEN OLD PLATES

Because sea-floor spreading and the all-encompassing plate theory were derived from studies of the present deep seas, strictly speaking they apply only back 200 million years to the oldest existing sea floor. Positions of plates at some time within the past 200 million years can be determined with considerable accuracy by moving them together until paired magnetic anomalies match, as in Fig. 7.22. Previous plate patterns and histories however, can be inferred only from continents, and of course the evidence becomes much more scant as we proceed back through history. Plate theory tells us that all ancient mountain or mobile belts represent former convergent plate boundaries, because it is believed that the impingement of plates is the

FIGURE 7.20
Four possible mechanisms that have been suggested to drive lithosphere plates: *A:* Push from an ocean ridge; *B:* Pull of colder downgoing plate down a subduction zone; *C:* Sliding down the slopes of an ocean ridge due to gravity; *D:* Carried piggyback on convection cells. (*After R. C. Dietz and J. S. Holden, 1973, Bulletin of the American Association of Petroleum Geologists, v. 57, pp. 2290–2296; by permission*).

FIGURE 7.21
Some different plate subduction patterns: *A:* Difference of angle seems to reflect rate of subduction (steeper angle characterizes older, thicker, colder plates); *B:* Arc-continent collision due to shrinkage of an ocean; *C:* Continent-continent collision producing an intercratonic mountain belt as in the Himalayas. Other subduction patterns also are possible. (S—collision suture.)

FIGURE 7.22
Restoration of past positions of two lithosphere plates by refitting contemporaneous magnetic anomaly lines and transform faults. For each plate there is a geometric pole of rotation, whose locus is defined by lines normal to transform faults (i.e., transforms are segments of circles concentric around that pole; see Fig. 7.10). *(Suggested by J. Morgan, 1968, Journal of Geophysical Research, v. 73, p. 1962.)*

mechanism of mountain building. Therefore, we can use old mountain belts to define old convergent boundaries (Fig. 7.23), and with the help of rock magnetism and every available geologic tool for reconstructing ancient geography, we may try to restore ancient plate relationships.

The present plate pattern has lasted about 200 million years so far, and it is generally assumed that the basic lithosphere plate mechanism has operated for at least the last half of earth history (2.5 b.y.). But plate boundaries and patterns of motions have changed drastically several times. Six or seven episodes of extreme mountain building on a nearly worldwide scale probably

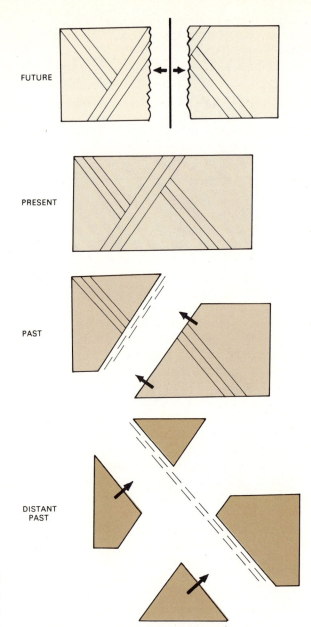

FIGURE 7.23
Delineation of ancient (especially pre-Mesozoic) lithosphere plates by old mobile belts (i.e., old convergent plate boundaries), and the step-by-step unraveling of plate interactions. This figure also illustrates the fact that continents are mosaics of old plate fragments laced by ancient convergent (mobile) zones of varying ages.

record wholesale reorganizations of plates. If so, then the average duration of a certain global plate pattern seems to be between 300 and 500 million years. Also we see that continents are mosaics assembled from pieces of plates of different ages broken apart and glued together in a crazy-quilt fashion. Small wonder that it has taken geologists 200 years to devise a truly comprehensive theory of global tectonics!

SUMMARY

The greatest surprises from postwar oceanographic work were the proof that the deep-sea floor is at most only about 200 million years old and that it is structurally very active. Only the continents contain any pre-Mesozoic historic record, whereas it was long assumed that the oceans would yield an unbroken sedimentary and paleontologic record dating back to near the beginning of the earth. Deep-sea topography, sediment distributions, and magnetic anomaly patterns first provided evidence that the sea floors might be spreading along the ocean ridges. Then in the late 1960s, seismic evidence proved not only that the sea floor is moving away from the ridges, as postulated for spreading, but also that it is moving toward the trenches. Finally, seismology also provided evidence that cold lithosphere slabs are being subducted beneath the trench-arc systems.

Plate tectonics seems to be a twentieth-century revolution in the earth sciences equal in importance to the seventeenth-century Newtonian revolution in physics and astronomy, the eighteenth-century Huttonian revolution in geology, and the nineteenth-century Darwinian revolution in biology. The plate tectonic theory was accepted with almost unprecedented rapidity in a mere 5-year period. In early 1963, a manuscript by Canadian L. W. Morley that anticipated the Vine-Matthews magnetic reversal record in ocean-floor rocks was rejected for publication, as was J. T. Wilson's first manuscript on the ages of volcanic islands. Yet, by the end of the same year, sea-floor spreading had become widely accepted in other articles. Why should any theory be adopted so rapidly?

Apparently it was because so much diverse evidence had been accumulating over a long period that the time was ripe for a new, comprehensive explanation that could be used to make successful predictions. (But, the bandwagon effect has already produced a somewhat disturbing dogmatism.)

The popularity and excitement of plate tectonics lies in the fact that it explains so much in a relatively simple, unifying manner. It allows one for the first time to envision relations of very different features scattered widely over the globe as parts of one, grand dynamic scheme on a chemically evolving planet. Ocean ridges are akin to rifts like that of East Africa, which seem to have resulted from a new phase of continental drifting. Spreading of ocean basins by addition of new crust at oceanic ridges seems also to explain the relative age distributions of many oceanic islands and atolls, submerged seamounts, the presently known distribution of deep-sea sediment ages, greater heat flow at ridge axes than at ocean margins, an increase of seismic velocity of oceanic crust away from ridges (reflecting aging and cooling of the plates), and, of course, the unique oceanic magnetic anomaly patterns. Seismic evidence from many earthquakes confirms motions of oceanic plates away from ridges, toward arcs, and laterally along transform faults that divide and offset ridge axes.

But what about mechanisms? While gravity doubtless helps pull down lithosphere slabs that are cooler and slightly more dense than the earth's interior, thermal convection in the mantle provides the most appealing chief driving mechanism for moving plates as on conveyor belts. Experimental evidence noted in the last chapter proves that rock materials like that of the earth's mantle can flow in a plastic fashion when subjected to slow stresses at high temperature and pressure. Sea-floor spreading rates suggest that such deep flow must be of the order of from 5 to 10 centimeters per year. Rising hot plumes of mantle material seem first to dome up the crust, then to rupture it with radial cracks. Two of the cracks generally continue to widen. Elsewhere, lithosphere must be consumed by subduction into the mantle at a rate equal to spreading. Plate subduction and partial melting due to heat beneath ocean trenches seems to account nicely for the volcanic arcs. Mountain ranges result either from upheaval of arcs by the force of subduction or from collision of two continents being rafted along on their lithosphere conveyor belts. Because ancient mountain belts are thought to represent old convergent plate margins, they provide important clues (together with paleomagnetic and other geologic evidence) for reconstructing ancient plate patterns. Plate tectonics seems to have dominated the earth for at least 2.5 billion years. We expect it to continue until the earth cools so much that the lithosphere will become too thick for plates to move. The moon reached this stage about 3.5 billion years ago.

Justice Oliver Wendell Holmes many years ago cautioned that "No generalization is worth a damn, including this one," and past history shows that "Here today, gone tomorrow" typifies many scientific ideas. Is the plate tectonic theory different? Somehow more permanent? While dissent has been unusually scarce, it is healthy to note the chief criticisms. All critics rightly emphasize the rather sparse geologic sampling of ocean floors as compared with continents, and that most of the theory is based upon a few perhaps nonrepresentative oceanic areas. A. A. and H. A. Meyerhoff, the most persistent dissenters, have emphasized that oceanic magnetic anomalies: (1) seem to be lacking from about 20 percent of ocean ridge areas; (2) are not everywhere parallel to ridge crests; and (3) are not everywhere symmetrical with ridges. They charge that assumptions have been added to assumptions until mere speculation became accepted fact. They suggest, instead, that ocean ridges are at least 2 billion years old, and that they are due to compression of the sea floor resulting from contraction of the earth à la Dana.

We can be certain that even the plate tectonic theory will change, but whether it will be relegated to the intellectual ash heap no one can yet say. In subsequent chapters we shall try to test the theory against the earth's historical record, and especially to use it to help make reconstructions of the earth's surface in the past.

Key Concepts

active leading edge
andesitic magmas
aulacogen
axis of spreading
convergent plate margin
coral reefs:
    atoll
    barrier
    fringing
divergent plate margin
East African rift system
lithosphere plates
magnetic anomalies
mantle plumes
ocean ridges
ocean trenches
partial melting
passive trailing edge
photic zone
plate collisions
plate sutures
plate tectonics
polarity reversals
reversal time scale
sea-floor spreading
seamounts
subduction
transcurrent or lateral fault
transform fault
triple junction

Readings

Burk, C., and Drake, C., 1974, The geology of continental margins: Heidelberg, Springer-Verlag.

Bird, J., and Goodman, K., 1976, Global tectonics: New York, McGraw-Hill. (Paperback)

Burke, K., and Dewey, J. F., 1973, Plume-generated triple junctions: key indicators in applying plate tectonics to old rocks: Journal of Geology, v. 81, pp. 406–433.

Cox, A. (ed.), 1973, Plate tectonics and geomagnetic reversals: San Francisco, W. H. Freeman.

Dewey, J. F., and Bird, J. M., 1970, Mountain belts

and the new global tectonics: Journal of Geophysical Research, v. 75, pp. 2625–2647.

Dietz, R. S., 1963, Collapsing continental rises: Journal of Geology, v. 71, pp. 314–333.

Hallam, A., 1973, A revolution in the earth sciences: Oxford, Oxford.

Heezen, B. C., 1960, The rift in the ocean floor: Scientific American, v. 203, pp. 98–110.

——, 1972, The face of the deep: Oxford, Oxford. (Nearly 600 underwater photos)

Heirtzler, J. R., 1968, Sea floor spreading: Scientific American, v. 219, December, pp. 60–70.

——, 1975, Where the earth turns inside out: Project Famous—man's first voyages down to the Mid-Atlantic ridge: National Geographic, v. 147, pp. 586–615. (Excellent sea floor photographs)

Hess, H. H., 1962, History of ocean basins, in Petrologic studies: a volume in honor of A. F. Buddington: Geological Society of America.

Isacks, B., Oliver, J., and Sykes, L. R., 1968, Seismology and the new global tectonics: Journal of Geophysical Research, v. 73, pp. 5855–5899.

Menard, H. W., 1964, Marine geology of the Pacific: New York, McGraw-Hill.

Mitchell, A. H., and Bell, J. D., 1973, Island-arc evolution and related mineral deposits: Journal of Geology, v. 81, pp. 381–403.

——, and Reading, H. G., 1969, Continental margins, geosynclines, and ocean floor spreading: Journal of Geology, v. 77, pp. 629–646.

Orowan, E., 1969, The origin of the ocean ridges: Scientific American, v. 221, November, pp. 103–119.

Seyfert, C. K., and Sirkin, L. A., 1973, Earth history and plate tectonics: an introduction to earth history: New York, Harper & Row.

Sullivan, W., 1974, Continents in motion: New York, McGraw-Hill.

Takeuchi, H., Uyeda, S., and Kanarmori, H., 1967, Debate about the earth; approach to geophysics through analysis of continental drift: San Francisco, Freeman, Cooper.

Vine, F. J., 1966, Spreading of the ocean floor: new evidence: Science, v. 154, pp. 1405–1415.

——, and Matthews, D. H., 1963, Magnetic anomalies over ocean ridges: Nature, v. 199, p. 947.

Wilson, J. T., 1965, A new class of faults and their bearing on continental drift: Nature, v. 207, p. 343.

Wyllie, P. J., 1971, The dynamic earth: New York, Wiley.

8

PREPALEOZOIC HISTORY
An Introduction to the Origin of Continental Crust

Mente et Malleo
By thought and dint of hammering
Is the good work done whereof I sing,
And a jollier crowd you'll rarely find,
Than the men who chip at earth's old rind,
And often wear a patched behind,
By thought and dint of hammering.

Andrew C. Lawson,
formerly of
the Geological Survey of Canada
and University of California

FIGURE 8.1
Early working conditions on the Canadian Shield. Geologic party in large bark canoe on Lake Mistassini, Quebec, 1885. (*Courtesy Geological Survey of Canada.*)

Prepaleozoic time, or the Precambrian as it is commonly termed, included about 80 percent of total earth history, that is, from nearly 5 billion years to about 700 or 800 million years ago. Yet for the first 1.0 to 1.5 billion years or so, there is no decipherable geologic record, and it now seems unlikely that much will be found. Apparently the crust had not developed sufficiently to become permanent until about 3.8 billion years ago. In Chap. 7 we outlined the probable earliest history and some of the consequences of such a history. Beginning in this chapter, we shall concentrate attention upon North America

as a tentative model of continental development. First, we shall examine in some detail the preserved geologic record for the interval from about 3.5 billion to 0.7 billion years ago, i.e., the Prepaleozoic. Although we find it logically more sound to refer to the first three-quarters of earth history either as Prepaleozoic or Prephanerozoic time, you should realize that to most geologists it is simply "Precambrian."

The economic importance of Prepaleozoic rocks is almost inestimable. Thousands of years ago Indians mined copper in Michigan and traded it widely over the continent. Since the industrial revolution, their inheritors have derived untold more wealth from these old rocks. The major source of iron ore is from peculiarly "*banded iron formations*," which are unique to the Prepaleozoic. Other major metallic sources of gold, silver, copper, nickel, chromium, and uranium, to list but a few, also are enclosed in

FIGURE 8.2
Banded high-grade metamorphic rocks exposed by glaciation along Sondre Stromfjord, southwestern Greenland. Some of the oldest dated rocks in the world (3.8 b.y.) occur near here. It is a question of long standing whether or not banding in such rocks represents relict stratification of original sediments. (*Photo by R. H. Dott, Jr.*)

these ancient rocks. Perhaps half of the world's metallic mineral resources come from Prepaleozoic deposits. This seemingly unlimited reservoir is, however, finite, and our increasing demands upon it are cause for alarm. Therefore search for more goes on.

At the outset, we must acknowledge that the Prepaleozoic record is far more obscure than that for subsequent time. To be sure, many Prepaleozoic rocks are severely deformed, metamorphosed, and deeply eroded (Fig. 8.2), but others are almost as youthful appearing as Cenozoic ones. Overwhelmingly the most important single characteristic of the older record is its lack of index fossils. As we saw in Chap. 2, the first use of index fossils for correlation and mapping was the major breakthrough that allowed the construction of a valid geologic time scale. But in Prepaleozoic rocks, so far we have been denied this important tool.

Considering the handicaps, it is remarkable what the pioneers of Prepaleozoic geology accomplished in unraveling a chronology that had to be based solely upon physical stratigraphic criteria of relative age. It was chiefly those pioneers who perfected the use of such criteria to a high order. More recently, isotopic dating has revolutionized the study of the Prepaleozoic even more than of younger eras. In general, it has confirmed much of the basic chronology established before by field geologic methods alone.

DEVELOPMENT OF A PREPALEOZOIC CHRONOLOGY

SEDGWICK IN WALES

You may recall from Chap. 4 that Adam Sedgwick first suggested that the eras, largest divisions of the geologic time scale, be named for stages of life development. He also recognized clearly the relations of older, unfossiliferous rocks to fossiliferous early Paleozoic ones in Wales. One of the first names proposed for Prepaleozoic time was Azoic, meaning "without life"; subsequently, Eozoic and Archeozoic ("ancient life"), and Cryptozoic were suggested when presumed fossils were found. The most

amusing term was Agnotozoic, proposed by a Wisconsin geologist who doubted that alleged fossils were truly organic.

Sedgwick recognized that Prepaleozoic rocks in Wales are somewhat more deformed and metamorphosed than overlying Paleozoic ones, from which they are separated by an unconformity. In many areas of the world, however, this is not so, and the two are virtually identical in appearance except for near-absence of fossils in the older. Because upper Prepaleozoic strata are not different from those of the lower Paleozoic, designation of a boundary is somewhat arbitrary where no unconformity is present, as is the case in many of the earth's ancient mobile belts. In general the lowest stratigraphic appearance of Cambrian index fossils has defined that boundary, but no human being was around to paint a stripe on the rocks for us; so, in continuous or conformable sequences of strata, what assurance is there that Cambrian index fossils appear at a position synchronous with their lowest position in Sedgwick's Welsh sequence? What if environmental or ecologic factors were unfavorable for the organisms in certain areas at the precise moment when the curtain raised on Act 1, the Cambrian Period, in Wales? Moreover, in strongly deformed or metamorphosed strata, precise location of this boundary is almost hopeless. Some day a criterion other than fossils may be found for defining this time boundary; even today it is selected by isotopic dating in some areas.

THE CANADIAN SHIELD

Though Prepaleozoic rocks were first recognized formally in Britain, the important development of a Prepaleozoic chronology occurred largely in the Great Lakes region of North America. Later it was worked out in Scandinavia as well.

The Prepaleozoic rocks on all continents are most widely exposed in the stable continental cratons. The region of more or less uninterrupted Prepaleozoic exposures in North America occupies primarily the eastern two-thirds of Canada, the United States margins of Lake Superior, and most of Greenland (Fig. 8.3). It is called the Canadian Shield. Younger strata also

SHIELDS AND OTHER MAJOR AREAS OF EXPOSED PREPALEOZOIC ROCKS

FIGURE 8.3
Shields and other major exposures of Prepaleozoic rocks of the world; Prepaleozoic rocks extend beneath all white areas of the continents, but not into the ocean basins. *(Bartholomew's Nordic Projection; with permission of John Bartholomew and Son, Edinburgh.)*

covered most shields in the past; therefore, shields are largely accidents of erosion where the stripping off of later deposits has exposed the Prepaleozoic basement (Fig. 8.3). The tectonic term craton is more useful because it defines the overall relative structural stability of a large portion of the earth's crust through a long time interval regardless of what aged rocks are exposed there today.

THE GREAT LAKES REGION

The first pioneer to probe the geologic secrets of the Canadian Shield was Sir William Logan, who in 1842 established the Geological Survey of Canada. Working conditions demanded heroic efforts, for much of the country was barely penetrable. In bush areas, transport was largely by canoe or on foot (Fig. 8.1); occasionally, boats had to be fashioned on the spot. Logan's successor as Director of the Survey even suffered the heartbreaking indignity of having a faithful, but hungry, horse eat an entire field

notebook at the end of a summer season.

Logan felt that granitic gneisses near Ottawa were among the oldest rocks of the continent, although they actually cannot be dated satisfactorily by field evidence alone. North of Lake Huron, Logan and his successors found and

FIGURE 8.4
Ellipsoidal or "pillow" structures in Early Prepaleozoic greenstones 15 kilometers west of Marquette, Michigan. The layers of metamorphosed lavas now are vertical, but pimplelike protrusions of the left sides of several ellipses indicate the original bottom direction. Spherical or cylindrical lava masses congeal quickly, when quenched by water. While cooling, ellipsoids are formed by flattening, but protrusions extend downward into spaces between underlying pillows. (Rotate page to compare appearance in different orientations.) *(Photo by R. H. Dott, Jr.)*

Northwest

③ KNIFE LAKE SEDIMENTS

Southeast

⑥KEEWEENAWAN SYSTEM

iron
Duluth gabbro (1.0 b.y.)

Gunflint Fm.

②SAGANAGA GRANITE

①KEEWATIN VOLCANICS

④VERMILION (ALGOMAN) GRANITE (2.5 b.y.)

③KNIFE LAKE

⑤ANIMIKIE STRATA

FIGURE 8.5

Diagrammatic cross section, north shore of Lake Superior to northern Michigan. Stratigraphic relationships studied by A. C. Lawson and early United States geologists are shown, as well as isotopic dates. Numbers indicate relative ages (1 = oldest). Prepaleozoic formations have been richly endowed with memorable—if unpronounceable—Indian names.

FIGURE 8.6

Contorted Early Prepaleozoic Soudan Iron Formation near Armstrong Lake, 15 kilometers east of Soudan, Minnesota. Note the characteristic prominent alternating bands of chert (light) and magnetite (dark). It was called the Soudan Iron Formation for the Soudan Mine, which was named (though misspelled) for the Sudan in Africa, the hottest place that some wag miner could think of after his first frigid Minnesota winter. *(Courtesy of Carl E. Dutton.)*

named several divisions of sedimentary, metamorphic, and igneous rocks. The oldest rocks there stand in vertical positions, making it difficult to determine their original superpositional sequence. A number of sedimentary, volcanic, and structural features allow distinction locally of original top and bottom in such sequences, and shield geologists perfected their application (Fig. 8.4).

In 1882 a young man named Andrew C. Lawson (author of the poem at the beginning of this chapter) joined the Geological Survey of Canada, and established a chronology for the Ontario-Minnesota border area. He proved conclusively that the oldest recognizable rocks there are metamorphosed andesitic and basaltic volcanic ones (named Keewatin; Fig 8.5). The most primitive crust was not granitic as had been assumed since pre-Wernerian times!

As mapping progressed in Minnesota, younger rock divisions were recognized and named (Fig. 8.5). The work was stimulated by discovery of iron ore near Marquette, Michigan in 1844, and the start of iron mining in northeastern Minnesota in 1884. The Minnesota ore-bearing rock (Fig. 8.6) occurs within a sequence of metamorphosed lavas called "greenstone." Outcrops are very discontinuous in the swampy, forested country around Lake Superisuperposition in the rocks; among these were graded bedding (Fig. 8.7), cross stratification (Fig. 8.8), as well as pillow structure in volcanic rocks (Fig. 8.4). Studies in this region were carried out largely by an army of United States Geological Survey workers under the guidance of C. R. Van Hise, later to become president of the University of Wisconsin. Copper, which has been mined in Michigan longer than iron, was an additional incentive for their work. It occurs in a thick succession of gently northward-dipping basalts and red-colored clastic sediments in the Keweenaw Peninsula of Michigan. A similar succession occurs on the north shore of Superior, where it dips south, thus forming a broad syncline beneath the lake (Fig. 8.5). Simple structure and distinctive rock types made correlation of these Keweenawan rocks across the lake simple. Where structure is much more complex, however, correlation and historical interpretation are far more difficult.

GREAT LAKES CORRELATIONS

From stratigraphy alone, four major divisions of sedimentary and volcanic rocks were recog-

FIGURE 8.7
Graded bedding (upward decrease of size in speckled layer at top) in an Early Prepaleozoic graywacke composed of volcanic detritus, Lake Vermilion Formation, Tower, Minnesota. The grains settled from a fluid in order of decreasing mass. Where strata are now vertical or overturned, the gradation of size indicates the direction of the original top. (*Courtesy R. Ojakangas, University of Minnesota, Duluth.*)

or; therefore, tracing of peculiar rock types, which serve as marker datums, facilitated mapping. Much of the iron ore is strongly magnetic and first was located because of the erratic behavior of compasses; for the same reason, iron-bearing strata were easily traced beneath concealed areas.

Strata containing iron ores south of Lake Superior soon were correlated with those of Minnesota, although in the former area they have been much more severely disturbed. Several features were important in determining

FIGURE 8.8
Cross stratification in the Baraboo Quartzite of Wisconsin (about 1.6 to 1.7 b.y. old). This structure typifies sands transported by vigorous currents; it represents the internal laminae of migrating dunes (see Fig. 8.15). Pen points in direction of current flow (right to left). Cross laminations tend to be sharply cut off at their tops and tangential at their bottoms. (*Photo by R. H. Dott, Jr.*)

FIGURE 8.9
Giants Range Granite (light) intruded into Early Prepaleozoic Knife Lake sediments (dark), north of the Mesabi Iron Range, Minnesota. The cross-cutting dikes clearly date the granite as *relatively* younger than the sediments (isotopic dating indicates that this granite is about 2.5 b.y. old). *(Courtesy Carl E. Dutton.)*

nized. They were known to be separated by unconformities that reflected at least four major mountain-building episodes called *orogenies* during which extensive granites were formed (Fig. 8.9). The orogenies punctuate the rock record in such a way as to provide several natural divisions (Table 8.1). But there were some thorny problems of correlation that could be resolved only through isotopic dating.

Isotopic dating has helped immensely to anchor age relations all around the Great Lakes. The numbers in Table 8.1 indicate the most important dates, and make possible a workable standard Prepaleozoic chronology for North America.

CORRELATION BEYOND THE GREAT LAKES

Relative degrees of deformation and of metamorphism commonly have been used to argue that certain rocks were very old or very young. Great caution is required in exercising such criteria because many misinterpretations are on record. To wit, recall the neptunian error in regarding all metamorphosed rocks as Primitive when, in fact, some of them turned out to be Mesozoic and even Cenozoic. The assumption that severe disturbance is necessarily related to great antiquity is too simple, for geographic or structural location may be more important. Cenozoic rocks in the Swiss Alps are far more deformed than are the late Prepaleozoic ones around Lake Superior. And Logan's "old" gneisses near Ottawa proved to be younger than the mildly folded Late Prepaleozoic strata on the shores of Lake Superior. More significant than age is whether a given group of rocks was deposited within a mobile belt or a craton.

Before the advent of isotopic dating, it became popular to think of the Prepaleozoic rocks as falling broadly into two groups, an older, intensely metamorphosed one (Archean) and a younger, less-disturbed one (Proterozoic). Far beyond the Great Lakes region, rocks were identified with one or the other of these solely on the basis of metamorphism and structure.

TABLE 8.1
Standard chronology of Prepaleozoic events for the Canadian Shield

Classification for Great Lakes states			b.y.	Classification for Canada	
Phanerozoic	Paleozoic Era		0.7	Hadrynian Era	
Prepaleozoic or Precambrian — Proterozoic	Late Era	Grenville orogeny		Helekian Era	Grenville orogeny
		Keweenawan System	1.2		Keweenawan System
		Unnamed orogeny			Elsonian orogeny
		Unnamed system	1.8		Unnamed
	Middle Era	Penokean orogeny		Aphebian Era	Hudsonian orogeny
		Animikean and Huronian Systems	2.5		Animikean and Huronian Systems
Archeozoic or Archean	Early Era	Algoman orogeny		Archaean Eon	Kenoran orogeny
		Timiskamian System	3.0		(Diverse local sequences)
		Saganagan orogeny			
		Keewatian	3.8+		

Many errors were committed that could be corrected only through isotopic dating, which provides the only valid basis for long-distance time correlations of Prepaleozoic rocks.

With the increase in isotopic dating, it has become clear that the continent contains several distinct isotopic date provinces (Fig. 8.10), that is, large regions within which a certain range of dates predominates. Such provinces are by no means simple, and they typically yield a considerable total range of ages, but statistically each has a discrete cluster of dates falling within about a 0.3 billion-year interval. It is important

to remember that most of the dates upon which Fig. 8.10 is based are dates of orogenic events; thus each province is a mosaic of related mobile belts. The rocks found in these old belts are composed of volcanic materials, sediments, and

FIGURE 8.10
Isotopic-date provinces of North America, showing regions within which respective ranges of plutonic and metamorphic dates cluster on a statistical basis. Structural trend or grain within each province also is shown. Several provinces overlap considerably; others are markedly discordant. Dates designate major periods of orogeny.

FRANKLIN

0.25-0.35

0.3-0.4

2.3

1.1-1.6

BELTIAN
0.9-1.2

SLAVE
2.6

GREENLAND
2.3-3.8

C O R D I L L E R A N

CHURCHILL
1.6-1.9

1.2-1.5

ALGOMAN-SAGANAGAN
2.3-3.3

G R E N V I L L E

0.3-0.0

1.6-1.9

PENOKEAN

0.4-0.8

1.8

0.8-1.0

1.2

CENTRAL
1.2-1.5

A P P A L A C H I A N

0.53

ISOTOPIC

AGE

PROVINCES

(billions of years)

0   1000 Miles

0   1000 Kilometers

large masses of granitic and high-grade metamorphic rocks, all of which attest to long-continued mobility. Many of the rocks have been metamorphosed more than once, thus their "isotopic clocks" may have been reset several times. Dates obtained from such rocks generally record only the last readjustment of isotopes during an episode of heating. Rarely the rocks retain two or more discordant dates, which reveal a complex history of multiple orogenies.

Most of the oldest isotopic dates (2.5 to 3.8 b.y.) tend to occur in the center of the continent from western Ontario southwest to Montana and Wyoming. But at least two other nuclei of similar antiquity also occur, one in the far northwest corner of the shield, and the other in southern Greenland. Surrounding each nucleus are more or less concentric mobile belt provinces of younger ages (Fig. 8.10).

As isotopic dating of mobile belts has supplied more data, it has become clear that the dates from all continents tend to cluster in a roughly similar way. The groups seem to represent crudely synchronous major granite-forming and metamorphic episodes; such peaks may eventually provide a basis for a worldwide standard Prepaleozoic time scale. Individual mobile belts seem to have existed for about 0.8 to 1.2 billion years, during which several periods of thick (geosynclinal) sedimentation, deformation, metamorphism, and granite formation occurred. The individual tectonic events, with periods of from 0.2 to 0.4 billion years, have been blended together to make up each of the Prepaleozoic age provinces.

INTERPRETATIONS OF CRUSTAL DEVELOPMENT FROM IGNEOUS AND METAMORPHIC ROCKS

THE IMPORTANCE OF GRANITE

Recall from Chap. 6 that, in light of the long erosional history of continents and of isostasy, some rejuvenation or addition of continental crust through time seems inescapable. Partly from these considerations, but also from the

igneous rock types of the early Prepaleozoic, together with the arrangements of old mobile belts, there has grown the *theory of continental accretion*. Geologists in the early twentieth century realized that rocks rich in silica, aluminum, and potassium, notably granites, seemed to be unique to continents. But field evidence suggested that granitic continental crust was not original, and so must have increased in volume through time. Granitic continental material, being of slightly lower average density than the remainder of the crust and mantle, stands topographically higher than the ocean basins in accord with isostasy (see Chap. 6).

Apparently the original crust was thin and composed largely of basalt. By later redistribution of lighter elements through weathering, erosion, and igneous activity, it was postulated long ago that some of the original crust was converted to granite to form embryonic continents. A century ago it was suggested that granite might form at depth in the crust simply by recrystallization—extreme metamorphism—of certain layers with geosynclinal sedimentary-volcanic sequences to produce new granitic material without involving any liquid magma (see Table 8.2). Through time, more and more *granitization* of volcanic and sedimentary rocks presumably had occurred, and the continents had grown thicker and larger as a result. Granitization was an extremely controversial idea during the early half of this century, but it has been discarded in favor of melting mechanisms rather than transformation in the solid state.

It is now thought that granites have formed from magmas produced by *partial melting* within the lower crust. Some seem to have arisen by partial melting of predominantly sedimentary rocks (S-type), whereas a second group seems to have formed by partial melting of predominantly igneous rocks (I-type). These types show subtle but consistent differences of mineralogy and texture. When melting occurs, the liquid state is less dense than adjacent material, so it rises. This is especially true if the magma is preferentially enriched in lighter elements (e.g., potassium, aluminum, sodium) as is required by the partial-melting hypothesis. If the magma then crystallizes slowly below the crust's sur-

TABLE 8.2
Comparison of chemical compositions of granitic rocks* with sedimentary and andesitic ones†

Rock type	Composition, %										
	SiO_2	TiO_2	Al_2O_3	Fe_2O_3	FeO	MnO	MgO	CaO	Na_2O	K_2O	H_2O
Granite	70.5	0.4	14.1	0.9	2.4	0.06	0.6	1.6	3.6	5.4	0.5
Quartz diorite	63.2	0.6	17.7	1.8	3.2	0.1	1.9	4.8	4.2	1.9	0.6
Average Middle Pre-paleozoic sediments	65.2	——	14.1	1.7	2.9	——	2.3	3.1	2.8	2.6	——
Average Post-Pre-paleozoic sediments	58.8	——	13.6	3.5	2.1	——	2.7	6.0	1.2	2.9	——
Andesitic lava	60.1	0.5	17.8	2.0	3.4	——	3.5	6.3	4.2	1.3	0.3
Basaltic lava	48.4	2.7	13.2	2.4	9.1	0.1	9.7	10.3	2.4	0.6	——

*True granites are much less common than diorites, which have less quartz and potassium feldspar. Therefore, the broader term *granitic rocks* is used to include the whole assemblage of coarse, crystalline-textured, light-colored, silica-rich igneous rocks.
†After Engel, 1963; Turner and Verhoogen, 1960.

face, coarse granitic rocks result, but if it reaches the surface, a fine-grained volcanic rock results.

Obviously if only older continental material has been partially melted to form a granite (or its extrusive cousin, rhyolite), then no increase in volume of continental crust would be possible, only a recycling. But rocks such as andesite are derived by partial melting along subduction zones and erupted in volcanic arcs (see p. 146). Such arc rocks then can be accreted to continents tectonically and eventually reconstituted to form new granitic material, because the composition of andesite is close to that of the most abundant granitic rocks, the diorites (Table 8.2). This is the way that we envision the accretion of continents today.

The chemistry of granitic rocks as a whole shows important trends through time. For example, younger granites are richer in potassium and radioactive daughter isotopes (such as strontium 87). This seems to reflect increasing chemical differentiation of the earth as a whole (as discussed in Chap. 5) and to provide evidence of a chemical maturing of continental crust through time.

SIGNIFICANCE OF ISOTOPIC DATE PATTERNS

The patterns of *isotopic date provinces* appear to have an important bearing upon the hypothesis

of continental accretion. The more or less concentric, younger-outward zones suggest that the continent may have accreted from a few old nuclei by the successive formation and ultimate consolidation of mobile belts around them (Fig. 8.11). First, the nuclei seem to have been knitted together into a single craton, which then grew still bigger through addition of newer mobile belts. It is thought that after long histories of orogenies, new granitic material was formed in each successive mobile belt, thus stabilizing and

FIGURE 8.11
Accretion of granitic continental crust through time. *A. Early Prepaleozoic (Archean) time:* Many small continental nuclei with most of earth's surface comprised of volcanic mobile zones (lines); continents *A, B, C* have already grown by fusion of smaller nuclei. *B. Later Prepalezoic (Proterozoic) time:* Much larger, more-stable continental cratons and narrower mobile belts.

A. 3.5 b.y. ago   B. 2.0 b.y. ago

transforming the behavior of each. Deep erosion accompanying isostatic uplift reduced the mountains to low plains, exposing the bowels of the old mobile belts that we see today. Shields, then, are mosaics of complexly interlaced ancient mobile belts now stabilized within cratons—all deeply eroded. It is estimated that the original depth of formation of some of the great granitic batholiths was as much as 20 to 30 kilometers.

The hypothesis of accretion presents an extremely appealing and relatively simple generalization that seems to unify many different types of geologic data and to offer a logical explanation of crustal development through enlargement and thickening of the continents. The whole mobile-belt concept, igneous and metamorphic petrology, and overall geochemical differentiation of the earth all seem to be explained in a related manner. But nature rarely is simple! Note that some orogenic zones of very different isotopic age are almost wholly superimposed or overprinted upon one another rather than being side-by-side. The Algoman orogeny around Lake Superior seems to have been so overprinted upon the much older Saganagan mobile belt (Fig. 8.5). In other cases, younger belts cut off older ones at high angles rather than being neatly concentric. For example, ghosts of reworked Middle Prepaleozoic trends can be discerned within the younger Grenville belt, where the rocks have been almost completely reconstituted (Fig. 8.10).

INTERPRETATION OF CRUSTAL DEVELOPMENT FROM SEDIMENTS—SOME BASICS

BIAS OF THE RECORD

Further insight into crustal history can be gleaned from Prepaleozoic sediments. The Lake Superior region contains very thick, deformed sedimentary and volcanic accumulations, which, for the most part, represent fillings of ancient mobile belts. They are typical for the shield areas. Deep downfolding in the mobile belts preserved sediments from subsequent erosion. Conversely, easily eroded cratonic sequences

are poorly represented in the older Prepaleozoic record. To further confound us, many Prepaleozoic mobile belts suffered so much metamorphism that little can be said about their original sediments. Nonetheless, in the Great Lakes region it is clear that at least by Middle Prepaleozoic time, there were sizable stable cratonic and mobile belt areas. In the following sections, we shall illustrate some important principles of interpretation of sediments, but keep in mind that the same principles apply equally to younger rocks.

TERRIGENOUS VERSUS NONTERRIGENOUS CLASTIC SEDIMENTS

The composition of any *clastic or fragmental sedimentary rock* reflects, more than anything else, the sources from which it was derived. Of course, climatic conditions may modify composition through weathering, as may chemical changes after deposition, but for the present we shall ignore these complications. It is necessary to distinguish *terrigenous clastic sediments* chiefly of silicate minerals such as quartz, derived from erosion of older rocks in land areas, from *nonterrigenous material* formed within an aqueous depositional environment. The latter includes chemically precipitated sediments, such as the evaporites (salt and gypsum), and carbonate rocks composed of fossil skeletal debris or precipitated calcareous particles. In this and the next chapter we shall consider primarily the terrigenous fragmental sediments; the nonterrigenous types are discussed in Chap. 13.

TEXTURAL MATURITY

Clastic textures reflect primarily the rates and intensities of physical sedimentary processes. Maximum size reflects the power of transporting agents such as running water or wind. Wind normally moves only sand and silt, whereas moving water can carry gravel as well. Mudflows and glaciers carry immense blocks for long distances because of their greater density and viscosity. In a general way, size tends to decrease with time and distance of transport. The reason is twofold, including a tendency for decrease of carrying power (or competence)

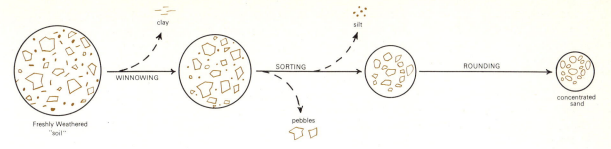

TEXTURAL EVOLUTION OF SAND

FIGURE 8.12
Idealized development of textural maturity of sand through extended abrasion and separation or sorting of different-sized grains.

with distance for most agents of transport, and also reduction of particle size by abrasion. Degree of *rounding* of sharp corners of fragments also is related to intensity and duration of abrasion as well as to toughness of the materials themselves. The range of sizes in a given clastic sediment, generally described as *size sorting*, reflects primarily the total time of transport and constancy of physical energy of transportive agents. A sediment subjected to long and constant agitation (e.g., beach sand) tends to be well sorted because there is maximum opportunity for the early dropping out of large particles and for the removal or *winnowing* away of fine materials. The latter are deposited ultimately in less-agitated environments. In this way, separation of different-sized materials occurs, with a final result being deposition of gravel near the source, well-sorted sand in another place, and well-sorted particles of fine silt and clay in still a third place. Generally, clastic sediments become finer as they are moved farther from their source.

From the above considerations, we can formulate a very useful generalization about the ideal textural evolution of terrigenous clastic sediments that will help us to interpret the history of any sample. Obviously with greater and greater abrasion and winnowing by currents or waves, size of particles will be reduced, sorting of sizes will improve, and rounding will increase as shown in Fig. 8.12. This idealized clastic evolution can be summarized as the degree of *textural maturity*. Usually ill-sorted, dark sandstones are loosely termed *graywackes*

(from an old German mining term, "wacken" for waste or barren); the sand grains in a graywacke may be of any composition. Bouldery and cobbly muds, as in glacial till or mudflows, represent the extreme of poor sorting. Such textural immaturity may result from very rapid deposition of sediments or from transport by an unusually viscous medium such as mud or ice, with much greater resistance to flow than water or air. In such media, winnowing of particles according to relative mass is ineffective (Fig. 8.13).

COMPOSITIONAL MATURITY

The *mineral composition* of a clastic sediment also will change as its particles are subjected to repeated physical crushing and chemical destruction of the less-stable minerals. Rock fragments tend to be ground down rapidly to their separate mineral grains, and dark (mafic) miner-

FIGURE 8.13
Comparison of relative sorting of sands by different processes (as measured by the statistic *standard deviation* in units of a special *phi* size scale). Several processes, e.g., surf and wind, produce markedly overlapping sorting characteristics. Additional criteria are necessary to distinguish completely the origin of most ancient sandstones; nonetheless, sorting readily eliminates some processes from consideration.

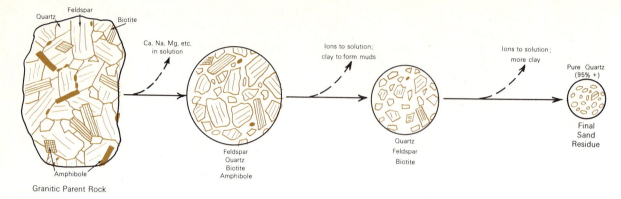

COMPOSITIONAL EVOLUTION OF SAND

FIGURE 8.14

Typical changes through time of the mineral composition of sand that was derived from erosion of a granitic source. Less-stable minerals are broken down both physically and chemically.

als, such as pyroxenes and amphiboles, suffer rapid chemical breakdown. This leaves a dominantly sand-sized (0.125 to 2.0 millimeters) ultimate residue of the more resistant material: quartz, chert, feldspar, and some mica, as well as very rare, but durable, accessory grains of minerals with high specific gravity (e.g., zircon, garnet, and magnetite). Of these, quartz is overwhelmingly the most abundant. Feldspar, though more abundant in parent igneous rocks, is of intermediate durability and so runs second place to quartz in sediments. The others, though more durable than feldspar, are simply far less abundant in source materials. Chert (tough, fine-grained sedimentary rock composed of silica) is the most durable material that originates in sedimentary environments and is common in many conglomerates and sandstones.[1]

An ideal evolution of *compositional maturity* exists (Fig. 8.14) with progressive loss of the less stable and concentration of more stable constituents as the vicissitudes of physical and chemical decay take their toll. Obviously the ultimate product of such an evolution is the concentration of a residue of very pure quartz- or chert-bearing sand and gravel. But there also is a close interrelation between destruction of the less-stable coarse particles and the increase of fine residue products of that destruction. For example, the most abundant sedimentary rock is shale, which is composed largely of fine clay particles derived from weathering of feldspars, and the most abundant minerals in the crust. Within shales there also is a spectrum of compositional maturity such that the clay species present reflect relative thoroughness of chemical decay of parent minerals.

Broad grouping into clans of sands and gravels according to composition is very useful as a simplification of knowledge. The clans recognized in this book are the *quartz-chert* (or *siliceous*) *sandstones*, the *feldspathic (or arkose) clan*, and the *rock-fragment (or lithic) clan*. These categories provide convenient summary nicknames, but their limits are statistical in nature. As with any classification imposed by man upon nature, there is a degree of unavoidable arbitrariness in defining categories.

STRATIFICATION

The nature of stratification also provides clues about processes of deposition. More than a century ago it was noted by the British genius

---

[1] Scientific revolutions resulting from invention of the telescope in the sixteenth century and compound microscope in the seventeenth century are well known. Of equal impact to geology was the invention in 1829 by a Scot named William Nicol of a device made from crystals of calcite for polarizing light. Nicol found that if minerals and rocks are cut and ground down to very thin, translucent slices, their compositions and textures can be studied microscopically if controlled polarized light is passed through the slices (see Fig. 8.18).

FIGURE 8.15
Origin of cross stratification and ripple marks by migration of dune forms produced by vigorous bottom (tractional) currents. Grains roll and bounce over the dune crests, coming to rest on the lee faces. Successive laminae, inclined in a downcurrent direction, form as the lee face migrates; each lamina is a buried fossil lee face. Cross stratification not only reveals the original top and bottom of deformed strata but also ancient current directions. (*A* indicates amplitude or thickness of cross stratification.)

H. Clifton Sorby that *ripple marks* and inclined or *cross stratification* in sand or fine gravel indicate moderately strong current action that rolled and bounced particles along by traction. At certain velocities, turbulence develops in such a way that a rippled or corrugated sand surface is

FIGURE 8.16
Experimental turbidity current of more-dense, muddy, turbulent water flowing from right to left down a sloping laboratory flume approximately 50 centimeters deep beneath less-dense clear water. Such currents are driven solely by gravity because of the density contrast between water masses. (*Courtesy G. V. Middleton and R. G. Walker, McMaster University, Ontario.*)

more stable than a perfectly smooth one. Because internally the inclined laminae reflect the lee faces of ripples or dune forms, they dip in the downcurrent direction (Fig. 8.15). *Tractional currents* are driven by gravity, as in rivers, or by the wind and tides, as in lakes and shallow seas. Very fine, horizontally laminated sediments, on the other hand, commonly suggest vertical settling of particles that had been carried up in suspension in the transporting fluid and not in contact with the bottom. This implies a minimum of direct current agitation of the bottom, therefore a relatively still environment. We may speak of cross stratification as representing a turbulent condition (i.e., agitated more or less constantly), and of delicate, parallel lamination in fine muds as indicating a nonagitated condition. In many sediments we actually find evidence of alternations from dominantly *traction transport* by waves or currents to dominantly *suspension transport*.

Graded bedding (Fig. 8.7) represents episodic introduction of abnormally coarse material into a usually still environment generally characterized by delicately laminated fine materials. Such spasmodic deposition occurs principally where infrequent density currents flow beneath less-dense water and carry coarser debris to normally tranquil areas. *Turbidity currents* are the most important of such agents. They derive their driving energy from the presence of fine sediments thrown into suspension by earthquake shocks or by severe storm activity. Sediment is kept in suspension by turbulence as the slightly more-dense, muddy water mass flows downslope beneath less-dense, clear water (Fig. 8.16).

Such a current maintains its kinetic energy and continues to flow until the turbid water encounters a topographic depression or until it becomes diluted by mixing with clear water. In Lake Mead, where natural turbidity currents have been studied most, muddy Colorado River flood water (density 1.003) sinks beneath clear lake water (density 1.001) and flows 100 kilometers along the bottom to Hoover Dam, where the currents are halted. Very fine sand and silt with graded bedding are deposited from them along the lake floor. Such currents were not anticipated when Hoover Dam was built, and they are filling the reservoir faster than expected.

About 1930 a British geologist, E. B. Bailey, noted that conspicuously cross-stratified sediments tend to occur apart from those with much graded stratification. He interpreted this dichotomy to reflect fundamentally different sedimentary environments and depositional processes. He also extended the interpretation of the environmental differences to a still more fundamental tectonic distinction. Cross stratification was deemed restricted to shallow, agitated water characteristic of seas over relatively stable cratonic regions, whereas graded bedding was considered characteristic of the most structurally mobile regions where presumed deep, usually less-agitated water prevailed, but into which coarser material occasionally was dumped. As a first approximation, this is a useful generalization.

INTERPRETATION OF PREPALEOZOIC SEDIMENTARY AND VOLCANIC ROCKS

EARLY PREPALEOZOIC OR ARCHEAN ROCKS

No record is known for the first 700 to 1,000 million years of earth history. Because of extreme heat, the entire surface must have been volcanic; it also suffered intense bombardment by large meteorites. As a result, the earliest crust must have consisted of ever-changing small chunks that had no chance of survival. As heat began to dissipate and large meteorite impacts ceased about 3.9 billion years ago, local crustal blocks began to survive. The oldest known rocks (2.5 to 3.8 b.y.) comprise two different assemblages, both of which have been more or less altered. Best known are *greenstone belts,* which are made up of mildly metamorphosed volcanic rocks and associated sediments. The second type includes high-grade metamorphic *gneisses* with very complex histories.

While their origins are warmly debated, the Archean gneiss assemblages include plutonic (granitic), volcanic, and sedimentary parent rocks. Granitic pebbles in some Archean conglomerates also attest to presence of some early granitic crust. Selective concentration of lighter elements, such as potassium, sodium, calcium, and aluminum, to form continental-type crust clearly had begun, but the chemistry of the granitic components differs in subtle ways from younger granites. Such things as isotope ratios and rare-earth-element distributions suggest derivation of the old granites from more chemically primitive mantle sources than for younger granites. We must remember that there was nearly three times as much heat generation through radioactive decay 3.5 billion years ago as now (see Fig. 5.9), thus widespread high-temperature metamorphism is not surprising. Embryonic continental crust must have been thin and unstable; partial melting of sedimentary and volcanic rocks seems to have been the rule. Recognizable sediments associated with the gneissic complexes include schists containing graphite, quartzites, banded iron formations (Fig. 8.6), and minor marbles. The schists probably were deposited as shales, but whether the graphite reflects organic or inorganic carbon is unknown. Rare cross-bedded quartzites suggest deposition in agitated (shallow?) water under temporarily stable tectonic conditions.

The greenstone belts, of which Keewatin rocks in Canada are examples (Table 8.1), were dominantly volcanic in character ("greenstone" derives from dark, green-colored minerals produced by metamorphism of basaltic igneous rocks). Basalts are most prominent, but andesitic and rhyolitic rocks also are present, typically in the upper part of a given sequence. Many of the basaltic greenstones are richer in magnesium than are younger basalts, which is interpreted as evidence of melting of mantle material at higher temperatures than now. Again, this is consistent with much greater radioactive heating. Pillow structures (Fig. 8.4) indicate eruption of basalts underwater, but abundant ash deposits suggest that some of the other volcanics were erupted from vents above water. Unusually large percentages of chromium and nickel in South African sediments older than 3.0 billion years suggest large, exposed sources of ultramafic rocks in the early crust. Applying the concepts of sandstone maturity, we find that the clastic sediments of the greenstone belts are very immature graywackes. Conglomerates and sandstones composed chiefly of poorly rounded and poorly sorted volcanic fragments and feldspar predominate. Clearly, volcanic material was being eroded and deposited very rapidly after only slight weathering and abrasion. In the words of E. B. Bailey, such sediments look "poured in or dumped." Prevalence of graded bedding (Fig. 8.7) suggests deposition in relatively quiet (deep?) water between volcanic vents.

By analogy with younger times, a more or less oceanic setting is implied for the greenstones; however, associated andesitic and rhyolitic rocks suggest at least local evolution of arc conditions. What were the relationships between greenstone belts and gneissic regions? One view is that they differ only in degree of metamorphism and reconstitution. Other explanations, however, assume that the gneisses are remnants of incipient continents, while the greenstone belts are remnants of oceans and volcanic arcs formed between continents.

Were plate tectonic processes as we recognize them in the younger geologic record operating in Archean time? Many geologists think not, and argue further that the Archean was unique. But the more we learn of Archean rocks, the more they seem to resemble younger ones after all; differences are more of degree. There seem to have been some lithospheric plates, but apparently they were smaller, thinner, and less rigid then so that relatively little material has survived in a recognizable form. A plausible scenario is that heat still was so great that volcanism occurred nearly everywhere. Mobile belts as such were not well defined—indeed, most of the earth's surface probably was mobile (Fig. 8.11A). Crust formed, was remelted, reformed, and remelted again like scum on a boiling pot or crust in a volcanic crater (Fig. 8.17). Finally, in late Archean time (about 2.5 to 2.6 b.y. ago), a major period of granite formation occurred more or less worldwide (Algoman

FIGURE 8.17
What the unstable early Archean surface of the earth probably looked like. Miniature plates and compressional zones on the chilled surface of basalt in Keanaki Crater, Hawaii, following a 1974 eruption. The confused pattern resulted from constant turmoil due to convection below and the formation and crumpling of a surface crust (compare Fig. 7.19). Area of view is approximately 100 meters wide. (By R. H. Dott, Jr.)

FIGURE 8.18

Photographs taken through a polarizing microscope of two contrasting Middle Prepaleozoic sandstones. *Left:* Graywacke sandstone composed of a mixture of quartz, feldspar, and diverse rock fragment grains surrounded by a considerable matrix of dark clay; note angularity and poor sorting of grains (Gowganda Formation, White Lake, Ontario). *Right:* Pure quartz sandstone composed almost solely of well-rounded quartz grains (note lack of dark matrix material); this is both compositionally and texturally mature (Ajibik Quartzite, near Marquette, Michigan). (Larger grains in each are about 1.0 millimeter in diameter.) *(By R. H. Dott, Jr.)*

or Kenoran orogeny of the Canadian Shield; Table 8.1 and Fig. 8.9). Apparently much of the present volume of continental crust was formed then by the aggregation, enlargement, and thickening of many minicontinents (Fig. 8.11B).

MIDDLE PREPALEOZOIC SEDIMENTS

Middle Prepaleozoic sediments show some striking contrasts with most of the older ones. Ill-sorted graywackes still formed, but light-colored, well-sorted pure quartz sandstones also became very abundant. The latter have much less fine matrix material, and, though most have been recrystallized to metamorphic quartzites, there is still evidence that they were texturally as well as compositionally mature. Middle Prepaleozoic sandstones thus are of two principal types, differing chiefly in texture: (1) well-sorted quartz sandstones and (2) quartz-rich graywackes (Fig. 8.18). Feldspar is sparingly present in the graywackes, and other constituents, such as volcanic rock fragments so abundant in older graywackes, are less common in these. The pure quartzites almost invariably show cross strata and ripple marks (Fig. 8.8). Conglomerates, where associated, generally contain well-rounded, durable pebbles of quartz or chert. Interstratified with the quartzites are thick sequences of mudstone or slate with zones of dark graywackes and prominent, banded iron formations consisting of alternating iron and chert bands (Fig. 8.6). Associated with some pure quartzites are limestones, many of

FIGURE 8.19
Part of a large algal stromatolite "reef" structure from Middle Prepaleozoic carbonate strata in the Northwest Territories, Canada. *(Photograph courtesy Paul Hoffman, Geological Survey of Canada.)*

which contain wavy laminated structures (stromatolites) presumed to have been formed by marine, bottom-dwelling primitive (filamentous) algae (Fig. 8.19). All of these rock types are prominent south and east of Lake Superior (Animikean and Huronian, Table 8.1).

What conditions could produce the first appearance of significant limestones associated with mature sandstones? A low (or very distant) stable land as a source of quartz and deposition in shallow, strongly agitated water are indicated. The wide extent of distinctive formations suggests, by analogy with younger strata, a marine origin. Algae fixed on the sea bottom cannot grow in water deeper than about 150 meters due to absorption by water of the sunlight needed for photosynthesis. Finally, other evidence of strong agitation over wide areas also argues for a very shallow sea susceptible to regular stirring by winds. For example, statistical analysis of measurements of cross stratification and ripple-mark orientations in Middle Prepaleozoic quartzites suggests persistent currents flowing dominantly from the north or northwest toward the south and southeast over most of the northern Great Lakes region. This sequence was deposited on a continental shelf.

Abundance of quartz sand indicates long and profound weathering of very large volumes of granitic or rhyolitic rocks. The volume of pure quartz sandstone provides a rough minimum index to the volume of material that must have been eroded to provide this concentrate. As an example, to produce the total volume of quartz residue represented by only one of several prominent quartzite-bearing formations north of Lake Huron would have required the complete weathering and erosion of at least 10,000 cubic kilometers (2,500 cubic miles) of granitic rock that contained a volumetric maximum of 25 percent quartz (Fig. 8.20). To account for all the incalculable quartz sand in the exposed part of the Lake Superior region, this figure must be multiplied at least a hundredfold. The staggering volume of Middle Prepaleozoic quartz sand found both in well-sorted sandstones (now quartzites) and in ill-sorted ones (graywackes) implies several cycles of weathering, erosion, and concentration within a few hundred million years immediately following the Algoman orogeny. In each successive cycle, there would have been removal of more and more unstable miner-

FIGURE 8.20
Relative volume of parent granitic or rhyolitic rock with a maximum of 25 percent quartz and ultimate residue represented by pure (mature) quartz sandstones. Many cycles of weathering and erosion are implied by such purification.

Northwest
(Ontario)   (Lake Superior)   Southeast
(Lake Michigan)

(ALGOMAN)

EARLY PREPALEOZOIC

(SAGANAGAN)

ANIMIKEAN

A. During Animikean Deposition

(ALGOMAN)

(SAGANAGAN)

(PENOKEAN)

B. Following Penokean Orogeny

FIGURE 8.21

Simplified cross sections summarizing the history of the Penokean mobile belt of the Lake Superior region. Penokean mountain building about 1.8 b.y. ago apparently resulted from subduction just south (to right) of these sections.

al grains and a gradual distillation of quartz. It is inescapable that the Middle Prepaleozoic continent already was large and contained a great deal of granitic rock. Much of it was relatively stable, and it had been deeply eroded from about 2.0 to 2.5 billion years ago. The sedimentary evidence, therefore, supports the concept of rapid early growth of continental crust (see Fig. 8.11).

All Middle Prepaleozoic strata in the Great Lakes area become thicker from north to south; volcanic rocks are present south of the lakes, and the degree of deformation and metamorphism increases markedly in the same direction (Fig. 8.21). Thus there is ample evidence of the existence of an east-west trending Middle Prepaleozoic (Penokean) mobile belt, which devel-

oped after deposition of the earlier shelf sequence. A stable craton still lay directly north of that belt, on which relatively thin and less-disturbed sandstones, shales, some limestone, and iron sediments were deposited. Clastic debris was derived almost wholly from the weathering and eroding of that craton. The mobile belt subsided profoundly, and, therefore, could receive very thick accumulations of sediments. Apparently it was the site of thrusting of a great plate of crust and mantle from the south (e.g., Fig. 7.18) beneath the craton to the north. Presumably a trench-arc system lay across Wisconsin and Michigan. But some Proterozoic mobile zones seem to have formed completely within cratons apparently as rifted troughs that were later compressed.

LATE PREPALEOZOIC ROCKS

The Penokean belt was upheaved, in its turn, into a vast mountain range, and, like the older mountains, was reduced by erosion. Late Pre-

Paleozoic and younger

Keweenawan sediments

Keweenawan mafic igneous rocks

Older Prepaleozoic

Bouguer Gravity Anomaly
contours in milligals

(+) Positive anomaly

(−) Negative anomaly

0   100   200   300 mi.

0   200   400 km.

FIGURE 8.22

The midcontinent gravity anomaly, one of the largest such features known. Large positive anomaly is associated with Keweenawan basalt and gabbro outcrops at Lake Superior. Extension of similar dense rocks southward beneath Paleozoic strata is indicated by extension of similar anomaly. Keweenawan lavas probably covered a still larger region originally. (*Adapted from E. Thiel, 1956, Bulletin of the Geological Society of America, v. 67, pp. 1079–1100; by permission of Geological Society of America.*)

scattered deep drilling indicate that a belt of *Keweenawan mafic, high-density rocks* extends southwest from Lake Superior to northeast Kansas, probably the world's largest such mass (Fig. 8.22). They also underlie central Michigan in a band extending southeast from Lake Superior. Magmas were erupted along these trends oblique to all older structures. The Keweenawan rocks are only mildly deformed, unlike those of the old mobile belts; therefore, they represent a hitherto unprecedented condition in the central North American crust. Basalts are rock types more characteristic of oceanic areas, so are anomalous within cratons. Profound rifting or fissuring of the continental crust occurred, allowing mafic magmas to well up to the surface from below the crust (Fig. 8.23). The lavas spread out on the surface of what was otherwise a stable, passive craton. Better-known late Cenozoic rifting and volcanism of similar scale are known in East Africa and in the Pacific Northwest of the United States (see Fig. 7.8). The lavas in all three cases rose through fissures and flooded immense areas, burying everything in sight and gradually leveling the landscape; hence they are called flood basalts. Collectively they represent a very important second-order class of tectonic features of the earth's crust. *Flood basalts* differ from andesitic volcanic rocks typical of mobile belts in that: (1) they are almost exclusively basaltic, (2) with few exceptions they occur in cratonic areas, and (3) they are only

paleozoic strata deposited over the old mountains are nearly flat-lying, and would seem to represent stabilization of the Lake Superior region at last. But sedimentation gave way to extensive outpourings of Keweenawan basaltic lavas accompanied by the intrusion of vast diabase sills and the Duluth gabbro mass (1.15 b.y. ago; Fig. 8.5). Following this unusual igneous episode, sedimentation resumed, producing widespread bright red-colored sandstones and shales. These strata have been considered river and lake deposits, though there is no proof that at least some were not marine.

Subsurface data from gravity surveys and

FIGURE 8.23

Extensional rifting and outpouring of lavas along a broad upwarp in cratonic crust as envisioned for the Prepaleozoic Keweenawan of north-central United States and the Cenozoic East African rifts. (*After H. Cloos, 1939, Geologische Rundschau, v. 30, p. 401; by permission of Geologische Rundschau.*)

mildly deformed. Clearly they require a structural explanation different from the mobile belts, namely profound extensional cracking or rifting of the crust; perhaps they are sites of rupturing apart of huge crustal plates as mentioned in Chap. 7. The East African rift valleys (especially the Red Sea; Fig. 7.8) are sites of incipient continental drifting, but in North America, splitting of the continent did not continue for some reason. Meanwhile, along the southeastern side of the continent, unusually severe deformation called the Grenville orogeny was taking place (Table 8.1). This event is now interpreted to be the result of a collision with another continent, which produced unusual heating and granitization of older crust across a belt 700 to 800 kilometers wide (Fig. 8.10). Apparently the Grenville orogeny and Keweenawan rifting were two related American manifestations of a global lithosphere plate reorganization.

SEDIMENTARY EVIDENCE OF THE NATURE OF THE PREPALEOZOIC ATMOSPHERE AND OCEAN

Many Prepaleozoic shales and graywacke sandstones are very dark due to presence of abundant unoxidized carbon and sulfide materials. Also, pebbles of pyrite (FeS_2) and uranium minerals that would be unstable in the presence of free oxygen are known from rocks older than 2 billion years. Iron carbonate ($FeCO_3$—siderite), very rare in younger sediments, is more common in Prepaleozoic ones. If free oxygen had been abundant, elements like iron, manganese, uranium, sulfur, copper, zinc, and vanadium all would tend to be present in their most highly oxidized forms, which is not the case. These facts are in accord with an original oxygen-poor or anaerobic atmosphere as outlined in Chap. 5.

Beginning in Late Proterozoic time, iron and silica tended to be chemically separated; the great affinity of iron for oxygen appears to have been largely responsible. Only a small amount of oxidized iron produces red color in sandstone, shale, and soil, and so the appearance of many red sediments should provide another indication of the onset of an oxidizing (aerobic)

atmosphere. Red shales are known locally in rocks as old as 2 billion years, but extensive red strata, or "red beds," first appeared in Late Prepaleozoic time. Besides the Upper Keweenawan red beds already discussed (Fig. 8.5), prominent examples occur in Glacier National Park, Montana, and in Scandinavia. Red sediments continued to form in all younger periods. Significant also is the presence of oxidized sulfur occurring as gypsum ($CaSO_4$) in Late Proterozoic strata in Australia. There can be no doubt that free atmospheric oxygen was abundant by about 1.5 billion years ago, and probably was becoming significant nearly a billion years earlier (see Fig. 5.12).

The banded iron deposits (Fig. 8.6) are known chiefly in Middle and Early Prepaleozoic rocks, which points to some unique chemical (or biochemical) condition. They have long been a great puzzle not only because of their confinement to Precambrian time but also because iron and silica show very different chemical behaviors today. Iron tends to be transported in acid and precipitated in alkaline conditions, whereas silica does just the opposite. Because the peculiar association of iron with silica largely predates widespread red beds, it would appear that it was related chiefly to oxygen-poor conditions, although there may have been some unknown biochemical influence, too. With CO_2 relatively much more important, sea water would have been slightly less alkaline than now and could have carried more calcium in solution. Silica apparently reached saturation in much of the Prepaleozoic sea. Because the entire early ocean and atmosphere were anaerobic (oxygen-free), iron could have been abundant in solution throughout the seas as ferrous ion (Fe^{+2}). Depending upon local chemical environments, it combined with sulfur (to produce FeS_2, pyrite), carbonate (to produce $FeCO_3$, siderite), or silica (to form various iron silicate minerals). Locally, however, in shallow agitated environments there was sufficient free oxygen to form iron oxides, even though the atmosphere was not yet oxygen-rich. It has been suggested that algal colonies may have created local oxygen-rich conditions. Once free oxygen became relatively abundant, practically all iron was deposited in oxide form. Iron was fickle, for it ceased its love affair with silica when the attractive newcomer, free oxygen, appeared. It is biologically surprising that free oxygen accumulated so slowly once photosynthesis began possibly 2.9 billion years ago. Probably the reason is because there were so much iron and carbon present that they could "soak up" all available oxygen to make iron oxide minerals and CO_2 respectively for perhaps a billion years. We estimate that atmospheric oxygen reached about half of its present abundance by 1.5 billion years ago (see Fig. 5.12).

EVIDENCE OF PREPALEOZOIC CLIMATE

Evidence about Prepaleozoic climates is scant. For the bulk of the record, we have discovered evidence to neither confirm nor deny an assumption of overall climatic conditions similar to later geologic time. Intuitively one suspects

FIGURE 8.24

Salt crystal casts in Proterozoic red shale from Belcher Islands, Hudson Bay, Canada (at least 1.8 b.y. old). More than 30 cube-shaped casts are recognizable (larger ones are marked). Together with associated sulfate crystal molds, these constitute one of the oldest known evaporite deposits; they are contained in some of the oldest red beds. *(Courtesy R. T. Bell, Geological Survey of Canada; see Canadian Journal of Earth Sciences, v. 11, pp. 722–728; used by permission.)*

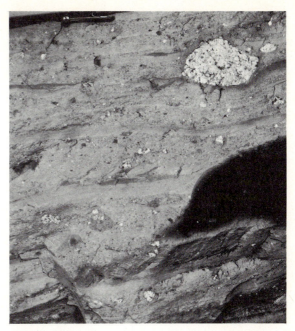

FIGURE 8.25

Mudstone with scattered pebbles and sand grains dropped from above (Gowganda Formation, near Blind River, Ontario). Widespread nature and association of this type of deposit with tills suggests dropping of stones from drifting icebergs (see Fig. 18.5). (By R. H. Dott, Jr.)

that many changes probably occurred. But, though we have little knowledge of the typical Prepaleozoic climate, we do have some evidence of climatic extremes. Evaporite deposits of Proterozoic ages are known on several continents (Fig. 8.24); these attest to dry (probably), hot conditions. Evidence of refrigerated periods is better known. Unfortunately, independent paleomagnetic evidence of latitudes during such events is unreliable.

At the base of the upper Huronian sediments in Ontario (Table 8.1), a peculiar, widespread assemblage of rocks called the Gowganda Formation occurs. The most distinctive rock type is a massive, almost completely unstratified and unsorted jumble of large boulders, pebbles, sand, and fine clay surrounded by a finer matrix of dark material (Fig. 8.18-left). Another striking type of deposit in the formation is delicately laminated mudstone that resembles Pleistocene glacial lake clays containing laminae interpreted as seasonal layers; some of these contain scattered pebbles (Fig. 8.25). A widespread unconformity marks the base of the Gowganda Formation. At three localities where the sub-Gowganda surface itself is exposed, fine parallel scratches strongly resembling striations made by glaciers are visible.

The peculiarities of the Gowganda Formation argue for glacial tills, river outwash deposits, and muds; some of the deposits doubtless accumulated in the sea. Today many glaciers extend into the sea along the southern Alaska, Greenland, and Antarctic coasts. Drifting icebergs carry all manner of frozen-in rock debris offshore to be deposited helter-skelter over large areas of the sea floor as they melt (see Fig. 18.5). The pebbles dropped into laminated Gowganda mudstones probably had a similar origin. A major episode of continental glaciation apparently occurred in Canada about 2.2 billion years ago. A large ice sheet spread out from the craton into the edge of the sea much as Pleistocene continental ice sheets spread across the North Atlantic continental shelf from New England and eastern Canada. Recognition of ancient glaciation is significant in showing that the well-known Pleistocene refrigeration was not unique in history. Other probable Prepaleozoic glacial deposits exist in the world, but they were not all synchronous.

LIFE RECORD BEFORE PALEOZOIC TIME

One curious fact has emerged from the exploration for fossils, that of the profound difference between the richly fossiliferous Paleozoic rocks and that of older units. While many Prepaleozoic rocks are either igneous or intensely metamorphosed, there are many sedimentary rocks that could retain evidence of life. Yet, almost none bear such evidence. Many of the barren sedimentary rocks of Late Prepaleozoic age are lithologically identical with Paleozoic strata replete with fossils. Unfortunately, what is known about life in the Prepaleozoic is not enough to build a satisfactory picture of paleoecology, paleozoogeography, or of the all-important origin of the animal phyla. However, there has been an intensive effort to find fossil and chemical evidence of life in the Prepaleozoic, and some patterns are beginning to emerge. These are discussed in Chap. 9.

SUMMARY

In spite of severe handicaps, a workable Prepaleozoic standard chronology was developed in the Great Lakes region. Isotopic dating allowed refinement of the chronology and provides the only valid basis for correlation beyond that region. Worldwide mapping of isotopic dates shows broad provinces with statistically clustered dates reflecting major periods of metamorphism and granite formation associated with mountain building Apparently these reflect major global convection and plate reorganizations. The Prepaleozoic basement of all continents represents a mosaic of ancient mobile belts delineated both by structural patterns and isotopic dates. The isotopic age provinces show a crude concentric pattern of younger-outward belts, which has been taken as evidence that new continental crust has been added in mobile belts at the margins of a growing craton by partial melting of volcanic and sedimentary rocks.

As noted in Chap. 5, timing of continental accumulation is of major interest. The Prepaleozoic record suggests that volcanism was universal over the surface prior to about 3.0 billion years ago. Before that, some granitic material was differentiated from the mantle, but significant volumes did not survive until 2.6 billion years ago. Great radioactive heat generation must have kept Archean crust thin and unstable. Apparently lithosphere plates formed, but they were small and had a very high mortality rate. The end of the Archean (2.5 billion years ago), however, was a great threshold for continent formation, for between 50 and 75 percent of modern continental crust had formed by then. Continent formation was not a uniform, continuous process, but has been spasmodic. Once formed, continental crust is virtually indestructible because of isostasy, which causes it to resist subduction. It may be, of course, greatly modified by compressional deformation, metamorphism, and rifting.

The composition of Prepaleozoic and younger igneous and metamorphic rocks attests to the overall cooling and chemical differentiation of the earth since early Archean time. For example, there are trends of increasing lighter elements (K, Al, Ca, etc.) and daughter isotopes, as well as more lower-temperature metamorphic rocks through time. Sedimentary rocks also give testimony to the chemical evolution of the earth. Like igneous ones, they show increasing potassium through time. And the oldest North American sediments are both texturally and compositionally immature and reflect great volcanic activity. But existence of large areas of granitic crust that had formed by 2.5 billion years ago is reflected in large volumes of Middle Prepaleozoic quartz sands. Features formed by currents provide a basis for reconstructing paleocurrent patterns, which help to delineate rather large, thoroughly weathered cratonic sources for the quartz. The Middle Prepaleozoic mobile belts, to a great extent, worked over old crust rather than always producing new crustal accretions, which is consistent with plate-tectonic theory. Indeed, the Prepaleozoic record shows clearly that shield areas are laced with old mobile belts and thus apparently are mosaics of old lithosphere plate fragments. Exactly when the plate-tectonic mechanism first began to operate is much debated, but certainly by no later than 2.5 billion years ago, clear cratons and mobile belts existed.

Prepaleozoic rocks are unique in some respects, though in most ways they are identical with younger ones. Fossils are far less abundant (and of different types) than in younger rocks. Early Prepaleozoic strata contain minerals that are very unstable in an aerobic environment, suggesting an oxygen-free atmosphere until about 1.5 to 2.0 billion years ago. Widespread Proterozoic sediments colored bright red by oxides of iron and rare sulfate ($CaSO_4$) evaporite deposits attest to an oxygen-rich atmosphere by at least 1.5 billion years ago. Banded chert and iron deposits are among the unique and most important trademarks of Early and Middle Prepaleozoic strata. Their depositional history, together with that of carbonate rocks, was the result of special chemical conditions of the sea and atmosphere; oxygen content probably was the chief factor, though biochemical agents also may have been important. There are other historical trends among sedimentary rocks. Dolomite exceeded limestone in Prepaleozoic time, suggesting that sea water was then richer in magnesium. Carbonate rocks constitute less than 1 percent of Archean and only about 5 percent of preserved Proterozoic sediments; they constitute about 25 percent of all younger ones. Similarly, evaporites make up less than 1 percent of Prepaleozoic and about 5 percent of Paleozoic and younger sediments. But were these rocks actually formed less frequently in the dim past, or have they simply been selectively recycled by chemical weathering? Durable quartz sand certainly has not been destroyed, but rather has been cycled countless times by erosion and redeposition—each time becoming more mature.

Key Concepts

Archean (Archeozoic)
banded iron formations
clastic (fragmental) sediments:
   terrigenous vs. nonterrigenous
continental accretion
cross stratification
evaporite deposits
flood (fissure) basalts
graded bedding
granitization
greenstone belts
high-grade gneisses
isotopic date provinces
orogeny
partial melting
Proterozoic
red beds
ripple marks
rounding of grains
sandstone clans:
   impure (graywacke)
   pure (quartzite)
sandstone maturity
size sorting
stromatolites
transport of grains:
   suspension
   traction
winnowing

Readings

Barghoorn, E., and Tyler, S. A., 1965, Microorganisms from the Gunflint chert: Science, v. 147, pp. 563–577.

Cloud, P. E., 1972, A working model of the primitive earth: American Journal of Science, v. 272, pp. 537–548.

Condie, K., 1970, Plate tectonics and crustal evolution: New York, Pergamon.

Moorbath, S., 1977, The oldest rocks and the growth of continents: Scientific American, v. 238, no. 3, pp. 92–104.

Patterson, C., 1964, Characteristics of lead isotope evolution on a continental scale in the earth, in Craig, H., et al., Isotopic and cosmic chemistry: Amsterdam, North Holland Press.

Ponnamperuma, C., 1977, Chemical evolution of the Early Precambrian: New York, Academic Press.

Price, R. A., and Doeglas, R. J. W., eds., 1972, Variations in tectonic styles in Canada: Geological Association of Canada Special Paper No. 11.

Rankama, K., 1963-1968, The Precambrian (4 vols.): New York, Interscience.

Sims, P. K., and Morey, G. B. (eds.), 1972, Geology of Minnesota: A Centennial Volume, Minnesota Geological Survey.

Society of Economic Geologists, 1973, Precambrian iron-formations of the world (a symposium): Economic Geology, v. 68, pp. 913–1179.

Symposium on Geochronology of Precambrian stratified rocks: Canadian Journal of Earth Sciences, v. 5, no. 3, June 1968, pp. 555–772.

Tarling, D. H. (ed.), 1978, Evolution of the earth's crust: New York, Academic Press.

Walter, M. R., 1977, Interpreting stromatolites: American Scientist, v. 65, pp. 563–571.

Windley, B. F., 1977, The evolving continents: New York, John Wiley.

9

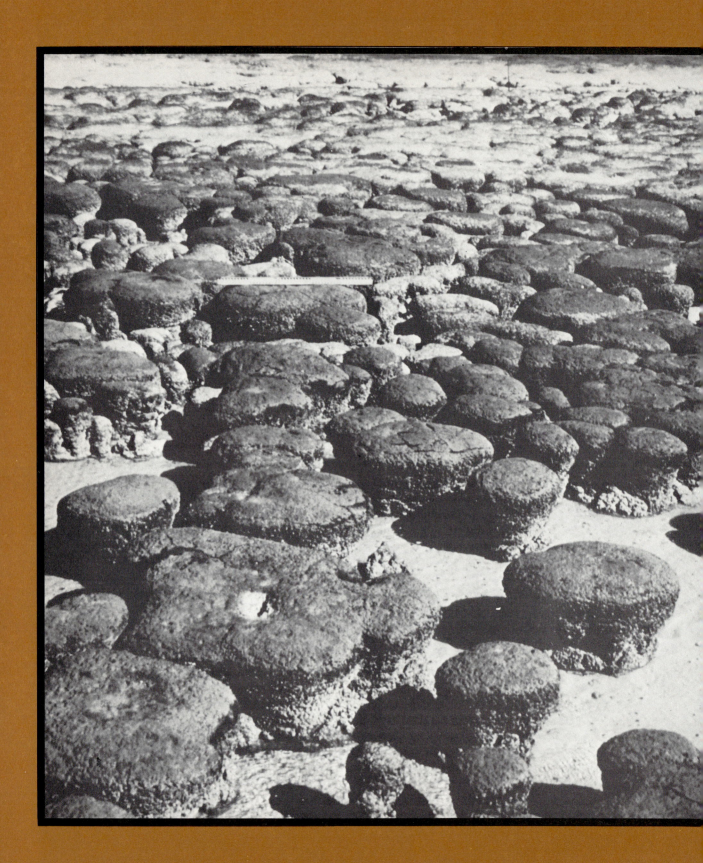

9

EARLY LIFE
AND
ITS PATTERNS

Now in vast shoals beneath the
brineless tide, on earth's firm crust
testaceous tribes reside; age after age
expands the peopled plain. The
tenants perish, but their cells remain;
whence coral walls and sparry hills
ascend from pole to pole, and round
the line extend.

Erasmus Darwin, "Temple of Nature,"

1803

9

When the astronauts returned from their voyage to the moon, they were asked what their most thrilling experience was. They were unanimous in saying that viewing the earth from space, and seeing how astonishingly beautiful and alive it is was the most thrilling event. All the fly-bys of the solar system by explorer satellites show the other planets to be lifeless, dull landscapes—only the earth is a living entity. Organisms, principally through the process of photosynthesis, have created this one and only beautiful planet as we know it today from a dull, lifeless body over 3 billion years ago (as shown in Chap. 5). We have seen how photosynthesis gradually converted the primitive atmosphere to an oxygen-rich one with life becoming more diverse and occupying the countless ecologic niches of today.

Before discussing the relationship of organisms to their environment, the study of ecology and paleoecology, we shall now look at the fossil record from the beginning to the time that modern types of ecosystems formed during the Ordovician.

THE LIFE RECORD BEFORE THE CAMBRIAN

As mentioned in the last chapter, the strata representing 3 billion years of the Prepaleozoic are almost devoid of fossils. There are only about 40 fossil localities known, in spite of an intensive effort to locate more during the last 20 years. Even the relatively rich Eocambrian (or Ediacaran) fauna of Australia (Table 9.1) would be described as very poorly fossiliferous by Cambrian and younger standards.

THE EARLY PREPALEOZOIC RECORD

The earliest known sedimentary rocks are found in Northwestern Australia (the Warrawoona Group at the "North Pole") and dated at 3.5 billion years. The rock types plus fossils indicate that the primitive atmosphere (whatever it was) had changed into one containing a dominance of CO_2. This is based, in part, on the presence of limestones by 3.5 billion years ago. These ancient deposits contain bacteria and photosynthesizing blue-green algae (see Table 9.1). Chemical evidence supports the fact that photosynthesis was taking place, because a number of complex organic compounds associated with that process have been found with these occurrences. In fact, most sediments in the Prepaleozoic have yielded chemical "fossils" associated with plant metabolism. The evidence strongly suggests that during this time a primitive type of photosynthesis was in operation in the CO_2 dominated atmosphere.

The Warrawoona is particularly significant, since it contains the earliest known reeflike *stromatolites* (finely laminated, usually hemispherical algal masses; see Fig. 9.2), which are quite large. Stromatolites gradually became more common during the Prepaleozoic and were havens or communities containing other plants, and probably in Late Prepaleozoic time, animals as well. Certain bacteria also construct stromatolites. If the earliest examples were bacterial, then their presence would not provide evidence of early oxygen generation as is generally supposed.

THE MIDDLE PREPALEOZOIC RECORD

Widely separated rocks dated between 2.5 and 1.7 billion years ago have yielded a modestly rich and surprisingly diverse marine flora, indicating increasing evolutionary activity. The discovery in 1954 by S. A. Tyler (described by Barghoorn and Schopf) of the first-known diverse fossil occurrence in the Prepaleozoic (Gunflint Formation, dated at 1.8 to 1.6 billion years ago) was largely responsible for a renewed interest in Prepaleozoic fossil studies.

The Gunflint fossils were found in black

TABLE 9.1

Evidences of organisms or related events contained in Prepaleozoic rocks

Partial list of strata containing Prepaleozoic fossils or other evidence of life compiled to show the best-substantiated records of organisms. The stratigraphic arrangement of the formations and systems is only approximate. (Not to scale)

Million years ago	Eras		Systems, formations (or rock units), and locations	Organism, other evidence of life, or events
570	Paleozoic	Cambrian	Lower Cambrian (worldwide)	Coelenterates, protozoans, poriferans, molluscans, worms, echinoderms, trilobites, trilobitomorphs, archaeocyathids, brachiopods. Stromatolites less common
		Eocambrian	Reed Dolomite (California)	*Wyattia* (Mollusca)
			Ediacaran strata (S. Australia)	25 species including medusae (coelenterates); *Rangea, Arborea,* and *Charnia* (frond-type octocorals); *Dickinsonia, Spriggina* (annelid worm); *Praecambridium* (a probable molluscan); and *Tribrachidium* (an echinoderm of the edrioasteroid type)
700±			Various localities (worldwide)	Evidence of glaciation dated between about 800-650 m.y. ago
900	Prepaleozoic	Late	Bitter Springs Formation (N. Australia)	Eucaryote cells, evidence that some reproduced sexually; fungi, filamentous blue-green algae, green algae, bacteria
1,100			Belt Supergroup (Montana)	Stromatolites, eucaryote cells, protozoa?
1,300			Beck Springs Dolomite (California)	Stromatolites, possible eucaryote cells (oldest reported)
1,700			Belcher Group (Hudson Bay)	Colonial bacteria, algal tubules, double-walled spheroids
		Middle	Gunflint Formation (Ontario)	12 species of procaryote blue-green algae, some filamentous stromatolitic bioherms, chemical synthesizing bacteria
1,900			(Worldwide)	Stromatolites increasingly common, atmosphere with oxygen component
2,000			Witwatersrand (S. Africa)	Stromatolites, blue-green algae
2,200			Banded Iron Formation (Australia)	Microspheres, bacteria
2,400		Early	Soudan Formation (Minnesota)	Blue-green algae, bacteria
2,700			Bulawayo Formation (Rhodesia)	Massive biohermal stromatolites formed by filamentous photosynthetic blue-green algae found in limestones
3,100			Fig Tree Formation (S. Africa)	Algal-like spheroids, bacteria
3,300 / 3,500 / 3,800			Pilbara (Northwest Australia) (Warrawoona)	Oldest sedimentary rocks Oldest stromatolites CO_2 probably a component of atmosphere

cherts in a banded iron formation on the north shore of Lake Superior. The microflora occurs between stromatolite layers (see Fig. 9.2) and consists of over 12 species of mostly blue-green algae, but representing newly evolved groups (see Fig. 9.3). Various bacteria and organisms not yet assigned to any floral family are also present. Similar assemblages of about this age

FIGURE 9.2
A microscopic photograph of accretionary layering of stromatolites, from the gunflint chert of Lake Superior, Canada. *(Photograph by S. A. Tyler.)*

have been discovered in South Africa, Greenland, Hudson Bay, and Australia, and they strongly suggest that by 2 billion years ago there was free oxygen in the atmosphere. This is strengthened by the fact that red beds began to appear 1.5 billion years ago.

It is important to note that all known plants discussed so far were *procaryotic cells* (although a few specimens from the Gunflint might be eucaryotic). Procaryotic cells today include those of very primitive organisms such as bacteria and the blue-green algae and are characterized by the lack of an organized nucleus and certain *organelles* (tiny units within cells that digest food and have other functions to keep the cell alive and capable of reproducing). Reproduction is accomplished exclusively by a simple cell division. All other cells, from single-celled

plants to cells in humans are *eucaryotic*, with an organized nucleus and organelles. The most important advance in the development of the eucaryotic cell is that the nucleus contains paired strands of genetic material (containing DNA) which make sexual reproduction possible. Sexual reproduction is one of two ways (the other being mutation) to generate new variation, and it greatly speeded up the evolutionary process leading to the origin of animals. Before sexual reproduction developed, mutations leading to better adaptation or new species were slow to appear because of the very low rate of production of variants in asexual reproduction (simple cell division). Just to emphasize the importance of sex: in an asexually reproducing population, the number of variants resulting from mutations is small, while in a sexually reproducing population, because of genetic recombination, a large number of variants is produced, resulting in more genetic information being passed on rapidly.

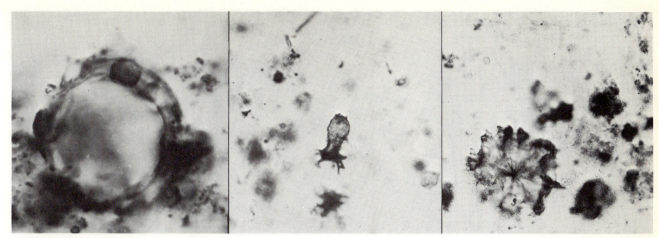

FIGURE 9.3
Greatly enlarged microscopic photographs of primitive
algae associated with stromatolites in the Gunflint cherts
of Canada. *(Photographs by S. A. Tyler.)*

THE LATE PREPALEOZOIC RECORD

The fossil record between 1.3 billion years ago
and the Paleozoic boundary of 700 million years
ago is somewhat better documented than the
earlier record, with over 20 localities scattered
over the world. It reveals very important evidence. In the Beck Springs Dolomite (1.3 billion
years old) of eastern California, there has been
found what may be the earliest occurrence of
green algae which are eucaryotic cells. There is
much dispute about when the eucaryotes first
appeared. They have been reported as early as 2
billion years, but the problem is that any dark
blob found in a fossilized cell or microsphere
could be, and has been, identified as a nucleus.
A very rich microflora was found in the Bitter
Springs Formation in the North Territory of
Australia (about 900 million years old) and
indicates that much evolution had transpired
since Gunflint time. There are 56 species of
blue-green, green, and red algae, fungi and
other plants, and spores, which may indicate
that some plants were reproducing sexually.

The Late Prepaleozoic was the acme of stromatolite development, for they are abundant
and widespread. They began to decline in the
Cambrian, and by mid-Ordovician they had
become a spotty and minor part of the flora.

Stromatolites live in a variety of ecologic niches,
but mostly are found in intertidal or shallow-
water environments; the giant living stromatolites of South Australia are intertidal (see Fig.
9.1).

Probably the most unfortunate gap of the
entire fossil record is the 200-million-year span
between Bitter Springs time and the beginning
of the Paleozoic Era. For in that interval, the first
multicellular plants and animals had evolved.
The multicellular green algae preceded animals
and are found in Norway and Australia in rocks
dated at about 700 million years. Some of the
algae were evidently planktonic (floating), while
others were attached to shallow-marine bottoms.

TRACE FOSSILS

The earliest evidence of animals reflects their
behavior. These are burrows of worms, as well
as trails and tracks of various worms, and, later,
of arthropods; collectively referred to as *trace
fossils* (indirect evidence of life; see Fig. 2.12).
The earliest reported are from the Grand Canyon series (about 1 billion years old) and vertical
burrows from the Buckingham Sandstone of
Australia (790 million years old). These (and the
younger Eocambrian or Ediacaran fauna) indicate that a variety of ecologic niches had developed by at least 700 million years ago.

These so-called trace fossils were formed by

FIGURE 9.4
A few of the complex Eocambrian metazoan fossils from the Ediacaran Hills, South Australia. *Left: Dickinsonia,* a possible worm. *Center: Spriggina,* also thought to be a worm. *Right: Parvancorina,* an animal of unknown affinity (see also Fig. 9.12 left). *(Photographs by G. R. Adlington.)*

different worm groups and are relatively common in sediments representing fairly quiet water throughout most of the fossil record. Studies have shown that there are three major trace-fossil zones representing from shallow- to deep-water bottoms. In the Prepaleozoic and Cambrian, only shallow-water types have been found. By Ordovician, deeper-water forms had evolved. From the Silurian to the end of the Cretaceous, there was a stable number of genera of trace fossils, most of which (20) had formed in shallow water. In the Cenozoic, the number of genera had jumped to 60 and remained so until today.

THE EOCAMBRIAN RECORD

Plant communities that must have existed prior to the beginning of the Paleozoic probably provided havens for the earliest animals, which were single-celled planktonic forms. However, our first glimpse of the animal record shows that much evolution had occurred, because quite complex and diverse animals with differentiated tissues existed over 600 million years ago. In the space of 200 million years (between the time of the Bitter Springs eucaryotic floral occurrence and the Ediacaran fauna of 700 to 680 million years; see page 185) for which we have no record, the eucaryotes had to clump together into colonies, become consumers of food rather than manufacturers, develop specialized cells in the colonies for locomotion, reproduction, food gathering, etc. This was followed by an evolutionary stage involving the formation of a body sac along with differentiated tissues and organs. All of this evolution was necessary before we encounter the Ediacaran fauna found in the Pound Quartzite in South Australia. Since that discovery, the same fauna has been found in scattered localities such as Newfoundland and England. Most of the fauna is composed of jellyfish (coelenterate medusae) of several types, which were probably floaters. The most interesting forms are attached corals (octocorals), several types of worms, possible molluscs, and echinoderms (see Fig. 9.4). This Eocambrian fauna provides us with the first evidence of an animal community from shallow-marine waters. The association, however, strongly indicates that it is an unusual or very specialized occurrence, like that of the Cambrian Burgess Shale (see page 194), so that there was probably more diversity and with more precursors not preserved or as yet undiscovered. One most-important observation is that none of these forms possessed a mineralized skeleton.

CAMBRIAN LIFE

SUDDEN ACQUISITION OF SKELETONS

The observation that, at the base of the Cambrian, skeletons suddenly appear in large numbers dominated by trilobites and brachiopods is based upon collecting originally done on or near the cratons. Subsequent and recent collecting now shows that there was a longer interval for the introduction of more groups of skeletalized animals, an interval that perhaps lasted as long as 100 million years—hardly a sudden event (see Fig. 9.5). The newly recognized portion of the lowest Cambrian represents a part of what was considered by some to be Late Prepaleozoic rocks and partly an interval found in Siberia called the Tommotian and now considered to represent the first 10 million years of the 30-million-year-long Early Cambrian. In these rocks the fauna is not as abundant as those in the higher Cambrian, but the first skeletalized groups are found (see Fig. 9.5). They include cephalopods and a few so-called experimental molluscs and echinoderms, along with sponges, archaeocyathids, worms, and coelenterates (the latter two groups had appeared earlier in the Ediacaran fauna).

Still, there are great thicknesses of sedimentary rock representing the long Late Prepaleozoic, which are barren of fossils. The problem is that we must account for a very large, diverse fauna that evolved within about 200 million years. In 1910, Wolcott tried to get around this by constructing the Lipalian interval, which he said was a major stratigraphic gap for which we have no fossil record at the base of the Cambrian. Many believe that some physical or chemical

threshold was passed, and that evolution was greatly speeded up, culminating in the sudden diversity we see.

The evidence that exists today shows that there was a gradual addition of newly acquired skeletons, and that this was due to a biological cause, that of developing the ability to secrete calcium. The development of mineralized skeletons was not confined to a single group, but cut across phylum lines, being found in seven phyla at about the same time in the Lower Cambrian. Not all phyla that were to develop skeletons evolved them at the same time; for example, the vertebrate skeleton is first found in the Early Ordovician of Russia (near Leningrad).

At least three factors may have been important in the formation of a skeleton, whether it is an outer (exo-) or inner (endo-) skeleton: (1) It would afford protection from ultraviolet rays (in the case of an exoskeleton), allowing animals to move into shallower habitats and also to be less vulnerable to predators (if any were around at that time). (2) Forms attached to the bottom could increase in size if they had a supporting skeleton. (3) Mobile bottom dwellers would have a base for the development of stronger muscles, hence increasing locomotor efficiency (both exo- and endoskeletons).

Two interesting points can be made in viewing the Cambrian fauna. First, the earliest faunas are not very diverse in terms of kinds of organisms, but, as new faunas appeared, diversity increased until by Ordovician time there was great faunal variety. The second point is that mineralized skeletons gradually appeared in the phyla during the course of the first two Paleozoic periods. The oldest cephalopod skeletons are almost 60 million years younger than the snails and brachiopods. Bryozoans and corals also took that long to develop skeletons after the calcareous (but unrelated) archaeocyathids appeared. Yet, the earliest Cambrian record still is rich in terms of numbers of organisms.

THE TRILOBITES

The fossil record of the Cambrian represents shallow-marine environments with few exceptions. Of this record, the greatest majority of

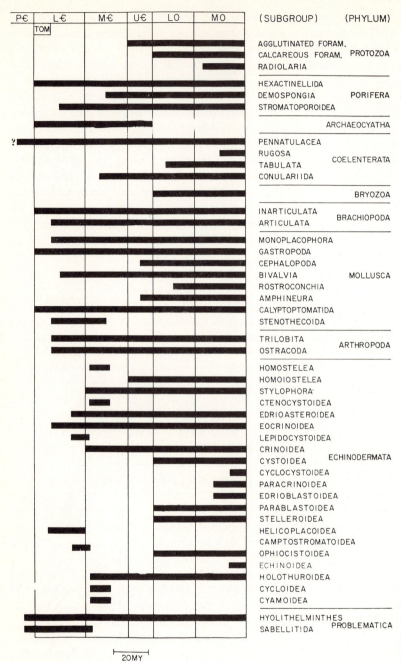

FIGURE 9.5
The ranges of all the earliest animal groups that developed skeletons. *(From S. Stanley, 1976, American Journal of Science, v. 276, p. 63.)*

organisms both in kinds and numbers were the trilobites, some 600 genera of which are known from the Cambrian. These strange animals became highly diversified and widespread during the course of the period and their many evolutionary divergences have permitted us to use the trilobites as time indicators. The early trilobites are thought to have been scavengers or mud-eaters living on the bottom, and most of them probably crawled rather than swam. As a result, they were more sensitive to the type of bottom sediments and tend to be facies fossils.

FIGURE 9.6

Cambrian faunal provinces on both sides of the North Atlantic Ocean, 2,000 kilometers apart. The "European" fauna characterizes graptolite-bearing shales, while the "American" occurs chiefly in shelly carbonate rocks. Long ago it was believed that land barriers separated and isolated the two faunas, but they are found mixed in the same strata in western Newfoundland and elsewhere. Close similarities of faunas and sedimentary facies on both sides of the North Atlantic evidence a former closer link between the opposing continents.

This is reflected in the very interesting distribution of trilobites in the Lower and Middle Cambrian rocks of most of North America and northern Scotland in contrast to those in the maritime provinces of Canada and western Europe (Figs. 9.6 and 9.7), and their close link with sedimentary facies.

There are two evolutionary trends that developed in many groups of trilobites and probably represent grades of evolution; that is, a similar evolutionary level reached by several independently evolving lineages. One was the gradual fusion of posterior segments to form a rigid paddlelike structure called the pygidium (see Fig. 9.8). This usually was accompanied by the reduction in spines along the margins. No one can account for the progressive trend to develop

FIGURE 9.7

Lower Cambrian trilobites. *Left: Paedeumias* is found in North America except for the eastern Appalachian belt (i.e., "American" province of Fig. 9.6). *(Photograph courtesy U. S. National Museum.) Right: Holmia* is a common genus of the "European" faunal province. *(Illustration from* Treatise on invertebrate paleontology; *courtesy Geological Society of America and University of Kansas Press.)*

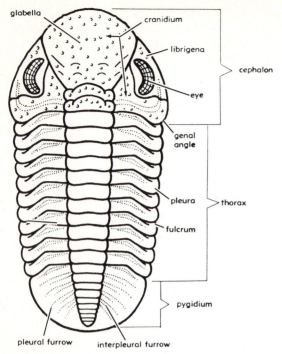

FIGURE 9.8
The morphology of trilobites. *(Courtesy the University of Kansas and Geological Society of America, from the* Treatise on invertebrate paleontology, *part O, 1959.)*

OTHER ORGANISMS OF THE CAMBRIAN

The trilobites make up more than half of the Cambrian fauna and show by far the most progression in terms of evolution. The rest of the fauna shows a curious mixture of evolutionary "experiments" (organisms that have a unique appearance, did not give rise to any other group, and were short-lived), along with quite conservative groups that did give rise to the vast faunas to come in post-Cambrian times.

The earliest strata of the Cambrian contain *archaeocyathids*, a sedentary bottom-dwelling group of solitary, vase-shaped forms with a double skeletal wall (see Fig. 9.9). In fact, they predate the trilobites. Their worldwide distribution when plotted on a map suggests that they were tropical, equatorial, much as the coral reefs of today (see Figs. WM. 8 and 11.2). They were quite widespread, forming undersea gardens that usually are devoid of other fossils. The genera are short-ranged, thus are used for correlation purposes; they became extinct at the end of the Cambrian.

The brachiopods, which are the next most

a posterior structure as large and as well formed as the head section or cephalon. Almost certainly the cephalon went through a similar period of fusion when trilobites became more active and mobile. It is a general rule among all animal groups that, with locomotion, the body becomes elongated and sense organs become highly developed and concentrated at the anterior end, while waste is discharged from the posterior.

Another very important trend was for the eyes to migrate away from the central bulb of the head (cephalon). In Early and Middle Cambrian time, they were connected by an eye ridge, but later became separated. In some groups the eyes gradually became smaller and disappeared altogether. Well-preserved specimens show clearly that trilobites had highly developed compound eyes similar to those of the crustaceans and other arthropods.

FIGURE 9.9
An archaeocyathid from the Lower Cambrian of Australia. *(Photograph by G. R. Adlington.)*

FIGURE 9.10

Lingula; an inarticulate brachiopod, from the Devonian of Eastern New York. *(Photograph by G. R. Adlington.)*

common group, are dominated by chitinophosphatic inarticulate forms (see Fig. 9.10). The shells are made up of a complex calcium phosphate molecule with much chitin (a plastic organic material) incorporated within the shell layers. These are the earliest brachiopods and were very conservative, showing negligible evolution after the Cambrian. They are considered, therefore, to have been unspecialized and primitive. For example, the modern *Lingula* is scarcely distinguishable from its Paleozoic ancestor. Their longevity may be due to their life in stable ecological niches of the intertidal zone or on the hard substrate of rocks or other shells. A few genera of calcareous inarticulates and articulates made up the rest of the brachiopods. While constituting only 30 percent of the Cambrian fauna, they were destined to dominate the faunas of the post-Cambrian.

Along with the archaeocyathids are some of the most fascinating of the Cambrian animals,

the primitive molluscs and mollusc-like forms, which constitute no more than 5 percent of all known groups. In the Early Cambrian, some cap-shaped shells appeared that are so similar to modern-day limpets in shape that they almost certainly lived in a similar rocky shore environment. These monoplacophorans (Fig. 9.11) are partly segmented molluscs and have the same body orientation as the chitons, Polyplacophora, which are a conservative group found living today relatively unchanged since the Paleozoic. The monoplacophorans are important because they gave rise to at least two groups of snails that dominated the later Paleozoic, which in turn gave rise to all later snails. By Middle Cambrian time, the clams made their appearance, but they were to remain a rare element until late in Ordovician time. Other molluscs were cone-shaped or limpetlike but had different internal structures; none of these survived the Cambrian.

Although the echinoderms are a very minor part of the entire Cambrian faunal realm, they show many interesting characters indicating experimentation on a level similar to the molluscs. The first echinoderm known is *Tribrach-*

FIGURE 9.11

Primitive mollusc of the Cambrian Period. *Helcionella* is a simple, cap-shaped monoplacophoran. *(Photograph by G. R. Adlington.)*

FIGURE 9.12

Left: Tribrachidium, earliest known echinoderm (Eocambrian). *(Photograph by G. R. Adlington.) Right:* Edrioasteroids from the Upper Ordovician, Cincinnati, Ohio, for comparison. *(Courtesy American Museum of Natural History.)*

idium from the Eocambrian fauna (Fig. 9.12, left). It appears to be related to the edrioasteroids of the later periods (see Fig. 9.12, right); yet nothing like it is known in the Cambrian. In the earliest Cambrian rocks, *Helicoplacus* (Fig. 9.13) has been found, which has spiral rows of polygonal plates and apparently was attached to the substrate. Stalked echinoderms appeared in the Middle Cambrian and are noteworthy in that they had but a few large plates. The echinoderms probably gave rise to the chordates no later than Ordovician time. This conclusion is based primarily on the fact that the larval stages of both are quite similar, but this hypothesis is only one of several concerning the origin of the chordates.

Finally, we must mention the Middle Cambrian Burgess Shale fauna from Field, British Columbia. There in 1910, C. D. Walcott made one of the most exciting fossil finds of the century (Fig. 9.14). The fauna consists of soft-bodied animals mostly belonging to various worms and arthropods. Almost all of these are known only from this one locality, but, due to the unusual preservation, they give rare insight into the soft anatomy of extinct groups. It enables us to have confidence in the correct classification of trilobites and other arthropods because arthropods are primarily classified on the basis

of appendages, which are rarely preserved. Present also are some worms belonging to the same families that are found in the seas today but which have no fossil record other than this. Plants and other types of animals present include sponges and jellyfish, all showing beautifully detailed anatomy. Such finds as this give us a rare glimpse of the past and suggest that perhaps the usual skeletal fossils represent a small percentage of the total fauna. Abundance of animal burrow structures in Cambrian strata over much of the continent attests to a rich soft-bodied element in the fauna everywhere (Fig. 9.15).

FIGURE 9.13

Helicoplacus, an "experimental" echinoderm from the Lower Cambrian of eastern California. *(Photograph by J. Wyatt Durham.)*

FIGURE 9.14
Examples of soft-bodied animals and arthropods from the Middle Cambrian Burgess Shale at Field, British Columbia. Note the exquisite detail of the arthropod appendages. *(Photographs courtesy U.S. National Museum.)*

FIGURE 9.15
Skolithos, an animal burrow tube (vertical dark features), in Upper Cambrian cross-stratified glauconitic sandstone in Wisconsin. *(By R. H. Dott, Jr.)*

FIGURE 9.16
Ordovician diorama based on an Upper Ordovician fossil site near Cincinnati, Ohio. Note large squidlike nautiloid cephalopod in the foreground. *(Photograph courtesy American Museum of Natural History.)*

ORDOVICIAN LIFE

CONTRASTS WITH THE CAMBRIAN FAUNA

Just as the Cambrian had its unique fauna, which is easily recognized by the predominance of trilobites, so, too, Ordovician organisms are distinct. When one first encounters Ordovician fossils, one is struck by the abrupt change in faunal composition as compared to latest Cambrian faunas (Fig. 9.16). A part of this difference may be related to changes of environment reflected by greater abundance of limestone, dolomite, and shale in the Ordovician sequence.

The Ordovician was marked by periods of continental submergence climaxed by a Late Ordovician inundation that was the most widespread ever recorded. This great period of submergence was accompanied by mild and uniform climate and a very extensive distribution of bottom-dwelling species. There were many new shallow-marine ecological niches available so it is not surprising to find a much larger number of species than in the Cambrian. By the end of the Ordovician Period, all major groups of animals capable of being preserved as fossils had appeared.

In examining collections from the two early systems, we are struck by several important differences. The trilobites are not as numerous in Ordovician strata as brachiopods and bryozoans. The evolution of the trilobites continued as vigorously as before, but along more restricted lines (there are about 1,000 species distributed through only 120 genera). The chief trends of evolution stressed changes of the eye and development of the central bulb and other gross features of the head section, which, to some researchers, reflects a shift to more active habits (Fig. 9.17).

Probably the most striking change in faunal composition involved the brachiopods. The phosphatic shelled inarticulate groups (Fig. 9.10) of the Cambrian became diminished in

FIGURE 9.17
Ordovician trilobites. *Left: Isotelus. Middle: Flexicalymene. Right: Cryptolithus. (Courtesy U.S. National Museum.)*

numbers, probably due to ecologic replacement by the calcareous, articulate forms of the Ordovician. Throughout the Ordovician, there was an extensive period of evolutionary expansion involving ovoid forms belonging to several major groups (Fig. 9.18).

FIGURE 9.18
Strophomenid and ovoid orthid brachiopods from the Trenton Limestone of New York. *(Courtesy U.S. National Museum.)*

The molluscs, represented primarily by the snails, underwent considerable evolution with the innovation of important new groups. But like the trilobites, they were masked by other, more common forms and did not become a substantial part of the fauna again until the Mesozoic Era. However, one very important new group, the nautiloid type of cephalopod, appeared and went through a rapid period of evolutionary expansion during the early and middle parts of Ordovician time when some ten orders arose (they actually are first known as rare forms in the Middle Cambrian). These Ordovician fossils are very common, so apparently the creatures had little competition from other organisms. This is suggested in part by the huge size some of them attained (exceeding 6 meters long; Fig. 9.16); cephalopods were the first large animals to inhabit the seas.

DEBUTS

We must now look at some of the new phyla that appeared during the period. From the Early Ordovician on, the calcareous, colonial bryozoans were to become constant companions of the brachiopods. Together they dominated most subsequent Paleozoic faunas. Not only do they occur together, but they seem to have been distantly related because they share several anatomical characters.

We have seen that the floating jellyfish type of

FIGURE 9.19
Streptelasma, a primitive, solitary rugose coral from the Upper Ordovician of Richmond, Indiana. *(Courtesy American Museum of Natural History.)*

coelenterate is found in rocks as old as the Middle Prepaleozoic. Its relative, the rugose coral, first appeared in Ordovician strata as a solitary form (Fig. 9.19). Colonial rugose corals, which also appeared during Ordovician time,

FIGURE 9.20
Hexagonaria, a colonial rugose coral from the Devonian of Michigan. *(Courtesy U.S. National Museum.)*

developed by gradual elongation and narrowing of the individuals, presumably in response to crowding. When individuals touched, they became compressed and then formed prismatic cross sections, which allowed the tightest possible packing (Fig. 9.20). Colonial dwelling has the advantage of mutual protection and benefit.

THE GRAPTOLITES

Perhaps the most important of the new groups were the graptolites. These, as the Bryozoa, were microscopic individuals grouped together in colonies (Fig. 9.21). They differed in possessing a chitinous rather than calcareous skeleton. A single group appeared in the Late Cambrian, but they radiated rapidly in earlier Ordovician time, spreading all over the world so that many species are shared by widely separated regions. Their very wide distribution is believed to indicate that they were planktonic or floating colonies (Fig. 9.21). Although they are known from many facies, the vast majority of graptolites are found in black shales, particularly in geosynclinal areas, for example on the eastern and western margins of North America. Because they evolved rapidly and became distributed instantly (geologically speaking) all over the world, and then became extinct equally abruptly, they are classic examples of ideal index fossils for stratigraphic zonation.

EARLIEST VERTEBRATE FOSSILS

The earliest known vertebrates are represented by dermal plates from the Upper Cambrian Deadwood Formation of northeastern Wyoming. These plates, which formed the outer armor of the most primitive fish group (Agnatha, or Heterostraca) are found disarticulated in a number of Lower and Middle Ordovician localities, but no *complete* skeleton has been found thus far and nothing therefore is known of their anatomy. The plates are very similar to those of well-preserved middle Paleozoic jawless fish. There are marine invertebrates (trilobites) in associated strata indicating that the fish were marine. There has been a controversy regarding the marine or freshwater origin of the fishes, but this has been resolved by

FIGURE 9.21
Tetragraptus, a typical planktonic graptolite colony from the Levis Shale of Quebec. Because they were planktonic the graptolites were very widespread; also because they underwent rapid evolution the species was short-lived making the graptolites excellent index fossils. (*Photograph courtesy U.S. National Museum.*)

finding such an association as in the Deadwood Formation. Recently, a very primitive protochordate (a small fishlike swimming form) was identified from the Burgess Shale. It has been hypothesized that this primitive group was the ancestor of the fish but, unfortunately, the morphology of this Burgess Shale creature does not help us understand much about the origin of the early fish.

MODERN ECOSYSTEMS ESTABLISHED

From looking at the fossil record discussed above, it is obvious that the Ordovician was vastly wealthier in animals than any previous time. Most of the marine ecosystems we find today, such as warm lagoons and coral reefs, were established by then. (An ecosystem is all of the organisms found in a particular physical environment and also their interactions.) There has been a kind of steady state in the types of marine ecosystems since the Ordovician. An Ordovician reef with a myriad of ecologic niches is not much different from a modern reef in its structure, but of course the *types* of organism are totally different. (An ecologic niche is a habitat occupied by a species especially adapted to that habitat.) Terrestrial ecosystems, like the tropical rain forest, took longer to evolve presumably because the atmosphere was not rich enough in oxygen during the Ordovician to support many types of land organisms, nor had their type evolved.

Raup and others believe that there probably was no more diversity in terms of numbers and kinds of marine organisms since the Ordovician. However, there is an apparent gradual increase in diversity up to the present time which they believe to be due to the fact that younger rocks are more accessible, hence are sampled more completely, a type of systematic bias of the fossil record (see p. 222).

LIFE PATTERNS

There are several ways of looking at life patterns (including evolution), a gradualistic view and a punctualistic view. In a *gradualistic view* of the fossil record, there was a gradual increase in atmospheric oxygen in the Prepaleozoic; of thickening of the ozone layer; of increasing salinity of sea water; the development of eucaryotes from the procaryotes by gradually complicating the number of organelles; the development of sex; the development of multicellular colonies; the formation of different types of tissues and the origin of multicellular plants (metaphytes) and animals (metazoa) all in a slow, even progression.

The evolution of ecosystems began in the Prepaleozoic with simple plant communities containing a few species and progressing to balanced communities of plant *producers* (plants manufacturing animal food) and animal *consumers*. Much of this gradualistic picture is "hidden" in the Prepaleozoic, but even the more complex communities throughout geologic time remain obscured since much of them are not preservable (most plants and nearly all of the microscopic animals).

In the *punctualistic view*, sudden events occurred owing to some key development, followed by a period of stability. A threshold was reached, such as a physical event, which caused the rapid expansion of diversity. For example,

there may have been a point at which the amount of calcium ions available in sea water made it possible for organisms to utilize calcium for the secretion of skeletons. Skeletons serve as body supports and muscle platforms. The latter allowed organisms to become larger in size. Eventually this resulted, in the vertebrates, in a body which was supported enough to leave the buoyant sea for a life on land, one of the most profound thresholds in evolution. After such thresholds have caused rapid expansion into new ecozones, there is a period of time when a steady state, in terms of number and kinds of species found, is maintained, during which time little evolution appears to occur. The nature of the fossil record suggests that this punctuated type of development happened during the development of the ecosystems and during the initial marine-animal diversity expansion period in the Eocambrian through the Early Ordovician. It was not until the Mesozoic that most major terrestrial ecosystems had evolved, even though some had formed in the late Paleozoic, such as the tropical rain forest. Again, the actual composition of the ecosystems has continuously changed.

COMMUNITY STRUCTURE

Faunas are the animal portion of the ecosystem and are composed of species held together in communities by a given set of environmental influences. Each community has its unique set of characteristics, as in the case of a swamp or coral reef. To survive in the community, an individual must gather food, assimilate food, protect itself, and reproduce itself.

The prime producers of food, as we saw in Chap. 5, are the microscopic plant plankton called *autotrophs*. All animal consumers are called *heterotrophs* and are the next and smaller level of energy on an energy pyramid. The direct feeders of plant food are the herbivores, which range in size from *Amoeba* to the elephants. Herbivores are not very efficient in recovering the energy from plant food so that most of it does not reach the next energy level of the pyramid, which is that of the carnivores (meat eaters) that live on herbivores (Fig. 9.22). Carnivores are larger than their prey, hence the

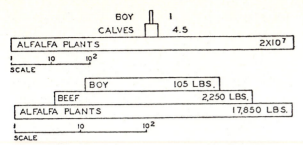

FIGURE 9.22

Two types of food pyramids based on 10 acres of alfalfa and one year of energy exchange. The upper pyramid shows the amount of alfalfa plants (the producer) needed to feed 4.5 calves (consumers-herbivores) needed to feed one boy (a consumer-carnivore) in one year. The lower pyramid involves the same amount of energy transfer but stated as weights at each energy level. (*From E. Odum, 1959,* The fundamentals of ecology: *Philadelphia, W. B. Saunders.*)

smaller amount of available energy on this level is shared by fewer individuals (the predator-prey ratio). There usually are three to five levels of energy, but the sequence from producers through one or two levels of carnivores can be complex depending on the nature of the community. This sequence is called a *food web* or *chain* (Fig. 9.23). Some of the lost energy in the community is accounted for by food that is not consumed. This material is eventually broken down by bacteria (decomposers) and transformed back into basic molecules used by the producers to manufacture new food. The methods of gathering and processing food by consumers are tremendously varied. We will exam-

ine some of the possibilities by looking at marine invertebrates because they dominate the fossil record.

MARINE ECOZONES

The bottom of the seas from the shore (littoral zone) down to the abyssal plain is composed of unconsolidated sediments with progressively finer sediments going in deeper waters (Fig. 9.24). Organisms that live within the bottom sediments (such as worms) are called *infaunal organisms,* and usually feed on small organic particles in the sediment; they are referred to as *deposit feeders.* Those organisms living on the surface of the bottom are *epifaunal,* and if attached to the bottom or on some object such as a clam shell they are called *sessile benthonic* (bottom) dwellers. If they can move about, they are *vagrant benthonic* forms. Worms and other infaunal types can also move through the sediments and are *burrowers.* Many benthonic organisms feed on particles floating in the water and these are called *suspension* or *filter feeders.* Organisms that live either permanently or temporarily above the bottom are classed as *planktonic* if they are passive floaters, such as the tiny food-producing algae, and if they actively swim are *nektonic,* such as the great white shark. Most of the planktonic forms live within the upper 100 or so meters of the ocean surface because plants require light to undergo photosynthesis. With food concentrated in this *photic zone,* most of the herbivores and carnivores are found here where the food chain begins. That is the reason that we find the greatest diversity of life and the largest number of individuals in shallow, warm

FIGURE 9.23

Two food chains in idealized ecosystems. (*From* Michigan Natural Resources, *September-October, 1972, by Ozz Warbach. Used with permission.*)

FIGURE 9.24
An idealized diagram of the edge of a continent showing the various ecozones in the oceans.

marine water, which constitute the bulk of the fossil record.

Needless to say, each one of these types of organisms has characteristics to make it viable in each of these ecozones. Swimming organisms tend to be streamlined to reduce water resistance, hence reducing the amount of energy required to move. Some organisms, such as *Nautilus*, can pump out their chambers to adjust their buoyancy because they must move vertically through hundreds of meters of water every day in search of food. In fact, most evolution involved such adaptations of organisms to their environment.

For an organism to survive and reproduce, it must perform such functions as feeding and moving about.

There are limitations to how much adjustment to the environment is possible, particularly during the life of an individual. This limitation is imposed by the genetic nature, which inherently dictates how large an organism can grow (see Chap. 10). The other limitation is that imposed by the environment. There are many environmental factors that exert an influence on organisms, such as temperature, sunlight (both are affected by depth of water), oxygen, carbon dioxide, salinity, water turbulence, the composition and nature of the sediment substrate, water and air circulation, and many others. It is important to remember that all factors, whether physical or biological, operate together. These will be discussed under the life sections to follow.

RELATIONS OF FOSSILS TO EARLY PALEOZOIC FACIES

Studies of early Paleozoic fossils and their relations to containing sediments have led to many important discoveries about continental histories. In Wales and western England, where such studies began, it was observed long ago that certain types of fossils tended to be restricted to particular sedimentary facies (Fig. 9.6). A shelly limestone facies is characterized by brachiopod and other shells, whereas a graptolitic shale facies is characterized by the fossils of the presumed floating organisms known as graptolites (see Fig. 9.21). Certain peculiar trilobites also occur in the latter facies (Fig. 9.17). Association of particular organisms with certain sediments reflects environmental or ecologic conditions. The shelly fossils commonly show breakage, abrasion, and some sorting as to size of shells, and so must have accumulated in relatively agitated water. But the delicate graptolites apparently were preserved best when deposited in relatively still water. Moreover, the dark color of the typical graptolitic rocks suggests that the environment was poorly oxygenated so that disseminated organic matter was not thoroughly oxidized or decayed. Scavengers probably were excluded by the paucity of oxygen, also favoring the preservation of the delicate graptolites.

Wales lies within the ancient Caledonian mobile belt, and most of England is on the edge of a northern European craton. Some of the earliest studies of sediments in mobile belts were conducted in Wales, and it was noted that the *graptolitic shale facies* tends to be confined to such belts, whereas the *shelly limestone facies* is most characteristic of the craton. In addition, the dark, poorly sorted graywackes showing graded bedding are characteristic of the graptolitic facies. On the other hand, better-sorted sandstones containing prominent cross stratification are rarely associated with graptolites, but instead are found intimately interstratified with shelly limestones. Both of the latter, then, seem to characterize strongly agitated environments.

Fossil-rich carbonate rocks reflect special conditions, namely a lack of terrigenous (silicate) detritus, which would suffocate many organisms, and favorable conditions of agitation as are found today especially in shallow tropical seas where invertebrate populations flourish. Countless carbonate-secreting organisms existed wherever extensive bodies of carbonate rock were formed, for skeletal material comprises a sizable proportion of such rocks. Apparently organisms play the dominant role in extracting calcium carbonate from sea water.

Some fundamental generalizations about relations of faunas and sediments are possible, as was verified in eastern North America by the early finding of graptolitic faunas identical with those of western Britain and in dark mudstone and graywacke rocks (Fig. 9.6). The observation that Cambro-Ordovician faunas are similar on both sides of the Atlantic tends to support hypotheses of continental separation outlined in Chap. 6.

SUMMARY

Earliest indications of life in some of the oldest known rocks include structures formed by marine algae and bacteria and traces of organic compounds known to be formed only by life processes. Through most of the Prepaleozoic succession, the preserved record of life shows no dramatic change in kind, but it shows an increase in abundance. After photosynthesis had proceeded for about 1.5 billion years, free oxygen began to accumulate in sea water and the atmosphere. Only then could efficient animal respiration become possible. No animal fossils are known in Late Prepaleozoic strata; first unquestioned animal fossils appear in Eocambrian rocks. When animal fossils did appear, they were of complex and diverse forms that reflected a great deal of evolution. It is still not clear why animals acquired the habit of secreting skeletons so "suddenly" in Cambrian time.

Fossils are rare in Eocambrian rocks, but as more are being found, it is realistic to redefine the beginning of the Paleozoic Era to include the Eocambrian. It was during Eocambrian time that a few primitive invertebrate skeletons appeared, but Cambrian strata contain the first truly abundant skeletons. Trilobites dominated the Cambrian fauna, but they show a grouping influenced by bottom environments. The same faunal groups and facies occur in America and Europe, suggesting a past closer connection between the two continents. Cambrian trilobites show several prominent structural evolutionary trends, notably reduction of the number of skeletal segments through time. Brachiopods, archaeocyathids, molluscs, and echinoderms make up most of the remainder of the preserved fauna, but there also is evidence that many soft-bodied, rarely preserved creatures existed; in life these may well have exceeded in number the better-preserved animal skeletons.

Ordovician faunas differ considerably from Cambrian ones. The medial Ordovician transgression brought an especially distinctive fauna, for the new epeiric sea teemed with brachiopods and bryozoans. Cephalopods, gastropods, diverse echinoderms and corals also were prominent. Vertebrate animals, represented by jawless, armored fish, appeared for the first time. Of greatest importance for dating and correlating Ordovician strata, however, were the rapidly evolving graptolites, whose tiny colonial fragments were widely dispersed over the ocean floor. The evolution of ecosystems can be viewed as a gradually unfolding process or one of punctuated events followed by stable periods. These events represent a rapid invasion of new ecozones after some threshold had been reached.

Organisms living in the oceans can be classed as benthonic organisms if they live on the bottom, or either planktonic or nektonic if they live above the bottom. Bottom dwellers are epifaunal if they live on the bottom or infaunal if they live within the bottom sediments. Most marine organisms live in the photic zone where plants receive enough sunlight to undergo photosynthesis. This is the base of the marine food chain. All plants are producers of food and all animals are consumers that convert food into energy in food chains. There is a delicate balance between the inherent genetic limitations of the individual and the factors of the environment.

Key Concepts

archaeocyathids
benthonic
blue-green algae
Burgess Shale
Cambrian fish
carnivore
consumers
ecologic niche
ecosystem
ecozone
Ediacaran fauna
epifauna
eucaryotic cells
food chains
gradualistic
graptolites
Gunflint Chert
herbivore
infauna
nektonic
organelles
paleoecology
photic zone
planktonic
procaryotic cells
producers
punctualistic
skeletons
stromatolites
trace fossils
trilobites

Readings

Cavalier-Smith, T., 1975, The origin of nuclei and of eucaryotic cells: Nature, v. 256, pp. 463–468.

Raup, D. M., 1976, Species diversity in the Phanerozoic: a tabulation: Paleobiology, v. 2, pp. 279–288.

——, Stanley, S. M., 1978, Principles of paleontology (2d ed.): San Francisco, W. H. Freeman and Co.

Rollins, H. B., Carothers, M., and Donahue, J., 1979, Transgression, regression, and fossil community succession: Lethaia, v. 12, pp. 89–104.

Schopf, J. W., 1977, Biostratigraphic usefulness of stromatolitic Precambrian microbiotas: Precambrian Research, v. 5, pp. 143–173.

——, 1978. The evolution of the earliest cells: Scientific American, v. 239, no. 3, pp. 111–138.

Tappan, H., 1974, Molecular oxygen and evolution, *in* Hayaishi, O. (ed.), Topics in molecular oxygen research: New York, North Holland, pp. 81–135 (Chap. 3).

Valentine, J. W., 1973, Evolutionary paleobiology of the marine biosphere: Englewood Cliffs, N. J., Prentice-Hall.

10

10

ORGANIC EVOLUTION
A Storm of Controversy and a Triumph of Human Intellect

A deer with a neck that was longer by half
Than the rest of his family's—try not to laugh—
By stretching and stretching became a giraffe
Which nobody can deny.

Lord Charles Neaves (1868)

(By permission of M. J. Sirks and Conway Zirkle
The evolution of biology. © 1964, The Ronald
Press, New York.)

FIGURE 10.1
Restoration of the Pleistocene Irish Elk. *(Courtesy American Museum of Natural History.)*

10

Since its beginnings, probably few theories have stirred so much controversy as evolution. Briefly stated, *organic evolution* is the cumulative change of organisms through time and is usually irreversible. By and large, the controversy over evolution is not so intense today as it was during the Victorian epoch, characterized in large part by religious conservatism and agrarian isolation. Into this austerity, the *Origin of Species* by Charles Darwin and the hammerlike logic and colorful persuasiveness of England's most popular lecturer and biologist, Thomas Huxley ("the Bishop eater"), shook the Victorians to their bootstraps.

As late as 1925 here in the United States, arguments about evolution blazed into national headlines during the famous Scopes "Monkey Trial" at Dayton, Tennessee. This trial was held to test a law prohibiting the teaching of evolution in the state. It was not until 1967 that Tennessee finally abandoned the "Monkey law," and in the fall of 1968 the U.S. Supreme Court declared all such laws unconstitutional. While legal aspects of teaching evolution have been clarified, there are still many people who are reluctant to learn about the subject because of religious precepts. In 1969 California adopted a requirement that the biblical account of the origin of plants, animals, and humans be given equal treatment with evolution in public school biology courses.

Failure to separate science and technology has, in recent years, caused a reaction of some people against science as the reason behind all of society's ills. It must be realized, however, in view of the trend toward urbanization and the problems inherent in a rapidly expanding population, that science is one human endeavor that can help solve today's problems.

The general concept of evolutionary change stands as one of the few really great generalizations of knowledge. Evolution is a concept that was almost inevitable, because it is a unifying principle that explains the distribution and diversity of life through its over 3-billion-year record. The theory of evolution was one of the last of such principles to develop—a full 200 years after Newton discovered the basic laws of mechanics. Yet, ideas about evolution, along with some of the questions that are fundamental to it, had been discussed as far back as classic Greek times. For example, the Greeks had a problem trying to explain why humans share so many features with animals such as the same shape and function of the heart, lungs, brains—or, for that matter, why organs of all animals from snails to humans, say, hearts (circulation pumps), are constructed of different tissues but function the same way.

SOME BACKGROUND AND A LOOK AT THE FUNDAMENTALS

Why is it that it took so long for man to realize that organisms have changed? Part of the answer lies in the fact that plants and animals evolved so slowly that humans could not see them change. Early thoughts on evolution were formed by explaining why there were so many different kinds of organisms. Here are some ideas that we feel might be some ingredients—thus a common denominator—in the various theories of evolution and that can provide us with a basis for asking some questions relevant to evolutionary concepts.

The appearance of a new character (or *mutation*) is one way we can recognize a new group on any given *classification* (taxonomic) level, for example, the sudden appearance of a new tooth cusp in a mammal or the enlargement of a preexisting feature such as the lengthening of a giraffe's neck or an elephant's trunk. It has been observed that not all new characters or modified characters constitute a new species, but if they are persistent and become widespread, it is with these that we can recognize differences among species and identify new species. From this oversimplification we can proceed to the "How" and "Why" questions, which serve to separate the various theories of evolution. How is the role of environment related to the origin of new species? How does inheritance affect evolutionary change? How and why do new characters come to be? And after new characters arise, how do they result in a new species?

We saw in Chap. 2 that the Greeks noted fossil shells far from the sea, indicating, they thought, that the waters of the earth were receding. Hence, for a long time the idea of environmental change was known. We shall come back to the all-important idea of such a change and see that it is closely linked to evolution and constitutes another basic ingredient of all such theories. The question that perhaps should have been asked is: "What effect does this change have on organisms?" But this question could not have been asked in Greek times because awareness of cause and effect relationships was not as acute as in later times. Furthermore, early people assumed that the fossil shells were simply dead modern shells. Recognition of possible extinct organisms generally did not occur in the modern sense until the 1600s. The question of the diversity of life was scarcely raised until relatively recently because of the belief that all life was placed on earth for the benefit or detriment of human beings.

One of the fundamental questions leading to a theory of evolution is: "Why are there so many kinds of organisms?" Before the 1700s, natural historians were preoccupied with other questions, such as the origin of life. All this was partly changed by the early classifiers because they brought before all a proof of the great variety of life. The giant among the natural historians who undertook encyclopedic description was Carolus Linnaeus, who published a series of works describing the fauna and flora of Sweden, and, finally, all known species of the world. By 1758, when his work was standard-ized, some 8,000 species of organisms had been defined. Imagine his amazement if he could realize that almost a million species now have been described and that no end is in sight!

BUFFON'S EVOLUTION

A leading intellect of eighteenth-century France was Georges de Buffon. He was the first influential man to convince some of his fellow scientists and philosophers that evolution must have occurred. He defined the species for the first time, noting that it exists as a separate entity and demonstrating that no species could interbreed with another. This is a fundamental concept in biology, and has taken on an even greater importance since the development of genetics (the science of heredity). He was the first to understand that the environment caused variation in the offspring, but he interpreted this to be important only on the local population level.

His writings are full of contradictions, but gradually through his life he moved toward the conclusion that all forms of life must have evolved from a fewer number of forms. His one most important contribution is that all animals are closely related to the environment in which they live and that the environment somehow molds and changes the organisms. He noted that there are also changes due to inheritance of characters from both parents. Thus he had established two of the absolutely fundamental attributes of evolution; he had asked the right questions at the right time in history. He put these two concepts together, saying that modifications of organisms are due to inheritance of characters and changes caused by environment; these in turn produced a natural descent that resulted in life as we see it today. In reading his statement we are not convinced that we should classify this as a theory of evolution, for he did not explain *how* new characters could arise, or *how* the environment could cause modifications *that could be inherited*. As we have noted, no theory of evolution should be called such unless it can explain how environment and inheritance interact and how new characters arise and change. As we said in Chap. 3, it was sometimes hazardous to suggest theories outside the climate of the times, and Buffon allowed for the special creation of species in his *Theory of the Earth*, even though he frequently implanted the idea that the Scriptures could be wrong.

Thus the conditions for a theory of evolution acceptable to the scientific body were established by Buffon by the middle of the eighteenth century as a reasonable, albeit vague, concept to explain life as we see it. The majority of the scientific community, including Charles Darwin, were strong dissenters and put their finger on the most serious problem to evolutionists, that of *time*. How could all of life become so diversified in such a short period of time since the Creation? Experiments involving breeding of domesticated plants and animals failed to produce a new species. The geologists realized that they needed more time also, and Buffon, as we have seen in Chap. 3, suggested that the earth was 75,000 years old. The problem of the span of time available for evolution was not settled until much later.

ERASMUS DARWIN: DID HE BEGIN IT ALL?

Buffon strongly influenced one of the important intellectuals of Great Britain, Erasmus Darwin, the grandfather of Charles. As an animal experimenter he saw that there are many changes going on during ontogeny (the life history of an individual): a caterpillar changes into a butterfly and a tadpole into a frog. He noted that in breeding, changes can be recognized from generation to generation, although the changes are very small. He realized also that there is a strong relationship between environment and heredity.

He went on to say that there are many forces which control evolutionary change. He maintained that lust (sexual selection) and the various weapons and ornaments and protective devices (deer antlers, etc.) attendant on sexual selection were important dynamic forces of evolution. The paraphrase of this in Charles Darwin's terms could be survival of the fittest. Another point he emphasized was that species

became adapted to their environment primarily by the method of gathering food. For example, the development of the elephant's trunk was to aid in tearing grass and moving it up to the mouth. The third element is security: protective coloration, wings, legs, shells, spines, etc. The results of these forces were *new characters which were acquired to fit the organism to its environment and these were passed on to its descendants*.

But, again, his statement, like Buffon's, lacks our criteria for a full-blown theory of evolution: How do the characters arise? He does tell us why they might have developed and that, if the characters make organisms better fit, they are passed on to the offspring. We believe that these ideas strongly influenced the early thinking of Charles. Natural selection is laced throughout his grandfather's *Zoonomia*, and Erasmus Darwin got many intellectuals in Britain interested in evolution and provided them with the beginnings of some method of approach to the problem.

LAMARCK: A STEP BACKWARD?

Lamarck's thinking represents the culmination of a long series of philosophers who attempted to explain all of the universe in a single concept. Lamarck's theory of evolution is in a sense one of the most misunderstood of all such concepts. If he were alive today and listened to the average zoologist state his theory, he would be puzzled. He summarized his philosophy in a book published in 1809. The thread going through it is: The fundamental aspect of nature is change. For example, minerals do not have a fixed composition but only stages in a continuous flux of disintegration and recombination. He believed that species as natural entities are creations of humans. Note how different this idea of species is from Buffon's fixed species. Life, Lamarck said, is a great stream of gradual complication. He believed that this system was replenished each day by spontaneous generation from inorganic material which filled the system up from the bottom, the simplest forms today being the most recently formed. The most complicated animals are the oldest groups. Even more amazing was his belief that no animal group ever becomes extinct. If some group (or better, stage) should be destroyed locally, it will be quickly replaced by those forms coming up from below just like a hole fills up with water. One facet of his theory, which has come to be thought of as the whole concept, and in fact is the only part considered important, is called the *inheritance of acquired characteristics*. Lamarck perceived that the effects of the environment somehow could be inherited by offspring. Buffon and Erasmus Darwin saw this, but neither could provide a mechanism to explain it. Lamarck found such a mechanism, and it is the most outstanding contribution he made, although few believed him.

THE LAMARCKIAN THEORY OF EVOLUTION AS WE KNOW IT

Lamarck believed that the environment does not directly cause animals to be different as Buffon and Erasmus Darwin believed, but changes in the environment will lead to changes in the needs of organisms. Changes in the needs will mean that the animals will be forced to change their habits. If these changes become constant, that is, if the environment continues to change, animals will acquire new habits, which will last as long as the needs that gave rise to them. New habits will give rise to new characteristics and thus a new species would arise. This was one of the few mechanisms suggested in the nineteenth century for the creation of new characters. It is a subtle concept in a way because it requires a mysterious inner force in the organism to develop a new feature, but it gets around arguments about the direct effect of the environment. For example, experiments have shown that a reduction of diet causes stunting in some animals, but such a decrease will not directly cause an animal to develop a new *genetic* line of descendants that are smaller.

One classic example Lamarck used to explain his theory is how the giraffe got its long neck. This somewhat ridiculous example was another reason why he was seldom taken seriously; other examples appear even more ludicrous.

The giraffe, so theorized Lamarck, started out as a grazing animal and, for whatever cause, there came to be a shortage of ground cover. Giraffes turned to eating leaves on trees, and as more and more turned to this source of food the supply of leaves lessened. The giraffes were forced to stretch their necks continuously to reach higher leaves. Thus, the long neck, and *voilà*, a new species. This is more or less a complete example of a theory of evolution and seems to answer all the questions we need to satisfy. The effects of a changing environment, a mechanism for developing a new character (hence a new species), and a way whereby these changes are passed on the offspring—a far more satisfying concept than Buffon's rather ill-defined direct environment approach. Lamarck also threw in a few "laws" to help amplify his idea. The first is built into other theories as well, namely, that everything gained by the phylogeny (the stream or history of a group of organisms) through the influence of nature is preserved by heredity, provided that the modifications are common to both parents. The second idea states that the more an organ is used (a stretched neck, for example), the stronger it becomes and the larger it will be in proportion to the duration of its use.

Most students of evolution do not subscribe to Lamarck's theory today for one important reason; there apparently is no experimental evidence to support Lamarck's ideas. There have been some classic experiments designed to test his ideas. In the famous Weimar experiment, over 200 generations of white mice systematically had their tails removed. Yet there was no trend for a shortening of tails in any descendants, which would be expected if Lamarck's thesis regarding use and disuse were correct. It can be argued, perhaps, that mutilation experiments hardly fall into the category of a valid test.

There are some other basic questions to be asked regarding his theory. How did the giraffe neck develop in the first place? How can new habits alter parents in exactly the same way so that offspring can receive the same new habits? It is important to realize that during the time of Lamarck, most people were of the opinion that species were fixed, immutable, created and destroyed at the whim of God. Those who did think about the problem of evolving organisms objected on the very good ground, as we have seen, that there was not enough time since the Creation for all species to appear; even heretics who believed with Buffon that the earth might be 75,000 years old could not envision the enormous diversity of life to be created in such a span of time. Lamarck got around this one very neatly; he said that as organic order got more complex, evolution accelerated.

Lamarck met opposition from his former collaborator, Baron Georges Cuvier, the most influential naturalist of France. Cuvier, working in the Paris Basin, was developing his theory of catastrophism, which fitted much better with the Bible, and seemed to account better for the diversity of life as documented by a real fossil record as proof. Where was Lamarck's proof in the fossil record? Nineteenth-century paleontologists, and indeed twentieth-century paleontologists as well, have not found any supporting evidence.

There is just one more point that can be raised: the characters that Lamarck chose to emphasize were functional or mechanical features such as legs, lungs, or muscles. That is, some organ that was used or disused got either larger, smaller, or more efficient. What about such things as mimicry and protective coloration, which are nonmechanical? How could they first appear?

CHARLES DARWIN AND NATURAL SELECTION

Lamarck produced his theory in 1809. The year 1809 is certainly an important one because it was on February 12, 1809 that two men were born who had a profound influence on the nineteenth century: Abraham Lincoln, emancipator of slaves, and Charles Darwin, whom we might call the emancipator of our primitive ideas on the development of life. Darwin attended the University of Edinburgh as a medical student, at which time he studied geology and became thoroughly grounded in all of natural

history. He later became dissatisfied with the course of his life, and went to Christ College at Cambridge University as a divinity student, acquiring a great interest in botany and the collecting of insects. The single most important event that was responsible for his gaining a notion of evolution, according to him, was the fortunate trip that he made in the years 1831 to 1835 aboard the *H.M.S. Beagle,* one of the early worldwide scientific cruises. This trip represents a milestone of scientific investigation; it was the first to study the bewildering variety and strange adaptation of plants and animals of the tropics and the Southern Hemisphere.

Darwin observed the fantastic variety of species that are found in the world, the tremendous numbers of individuals of species, and the amount of competition among populations for food. From his observation, he deduced that living organisms are enormously fertile; yet curiously, the number in a given population appears to be fairly constant. Populations are stable in size, he said, because there is a tremendous struggle for life, because of rivals or prey-predator relationships, and the inorganic environment itself. The result is that some organisms seemed to be better fitted for life than others. Notice that this is an idea that Erasmus Darwin had. Charles was profoundly influenced by this idea and also from reading Thomas Malthus's *An Essay on the Principles of Population,* which was written in 1798. Malthus said that populations increase geometrically while food increases arithmetically, and there is a continuous struggle for food. Thomas Malthus, of course, applied this to human population, saying that when the population increases have reached the saturation point for food, then famine occurs and the population will be reduced.

In the struggle for life, not all offspring survive; in fact, there is a high rate of infant mortality, and only those that are best fit for life survive. Darwin observed that the reason that some organisms are not fit for life is that they have inherited characteristics that prevent them from surviving in their environment. It appears that nature selects the animals best fit for the environment and thus "molds" the animals.

Through time, populations become better fitted to their environments as weaker members die before they are old enough to reproduce. Also, the longer an individual survives, the more it can reproduce to add more "fitted" offspring to the population.

He recognized, as did Lamarck, that there are environmental changes, and when the environment changes, this necessitates continual character modifications in populations. Darwin saw that as the environment changes nature will select those individuals that are best fitted for the particular changes occurring; that *only the fittest will survive*. But this is an oversimplification of his statement, and Darwin recognized that there are many different possibilities. For example, there may be no observable environmental change, and yet evolution seems to go on; populations are continually and *gradually*

changing toward being better adapted or progressively adapted to fit an environment. *Evolution is likely to move in the direction of greater adaptation*.

How did he come to these conclusions? To go back to the voyage of the *Beagle*, one of the stops made in their westward journey was in the Galápagos Islands (see Fig. 10.2). There Darwin recognized 13 species of finches. These finches were like none other in the world; they obviously were closely related to, but distinct from, those in South America. Interestingly, although finches are unable to fly great distances, their ancestors somehow had managed to get to the islands, which are 500 miles away from the Ecuadorian coast. Another interesting point is that only those species are found on the islands (other than oceanic birds), and they are living in environments unknown elsewhere to the family

FIGURE 10.2
Darwin's finches from the Galápagos Islands; note the differences in shapes and sizes of beaks reflecting differences in feeding habits. *(By permission of Biological Sciences Curriculum Study, High school biology, BSCS green version, 1968; Rand McNally.)*

of finches. Darwin was greatly puzzled by this, but gradually came to the view that they had evolved from perhaps a single species that had accidentally arrived from South America. Because there was no competition from other birds, they became adapted to bizarre habitats not otherwise occupied.

DARWINISM TACKLES THE GIRAFFE

Let's see how Darwin would have explained the long neck of Lamarck's giraffe. First of all, Darwin would say, there is in all populations a certain range of *variability*, which was inherited from ancestors. All populations show this; there is variation in coat color and pattern, length of neck, size of horns, etc. Even though there was variation in the length of neck among primitive giraffes, whether a neck was short or long may not have been important in the beginning when there was plenty of grass to eat. If their basic food should become so scarce, however, the protogiraffe would have to change its diet and there would be competition for food. Those forms having the shorter necks would have less food than those with longer necks, for the latter could eat tree leaves. Shorter-necked forms would die off because they would be less competitive. Thus there was a progressive adaptation for longer necks.

HOW GOOD IS DARWIN'S THEORY?

Is this a good theory of evolution? It satisfies all but one of our requirements for such a concept. It tells us how and why evolution occurs much more clearly and convincingly than Lamarck's ideas and even more importantly can be demonstrated experimentally (and has been so demonstrated many times over). But remember that Lamarck was the first one to come up with a plausible idea of how new characters arise. As changes occur in the environment, there is a change in the needs of organisms, resulting in a "developed" new character.

But Darwin had great difficulty in explaining how a new character arose. His theory works well after the new characters appear on the scene. Another flaw was his belief in a type of inheritance that came from breeding experiments, which implied that parental characters are blended in the offspring. This was a problem because the offspring should have characters that are *better* fit than the parents, not subdued by blending.

The biological critics pointed out that there was not a scrap of paleontological evidence to favor his theory. In fact, it was at this time that Cuvier clearly pointed out that the record showed catastrophism to be more likely (see Chap. 2). However, Lyell, Darwin's geological mentor, did produce from Cuvier's own work, and that of others, one piece of evidence to help out. He had subdivided the Tertiary rocks of western Europe on the basis of the percentage of modern species in each epoch. This implies that species were systematically changing in time. Neither Darwin nor Lyell, however, appeared to have used this as an example of evolution. Further, continental biologists were caught up with the idea of a central plan in nature, taking their clues from recent advances in embryology. They could see no "program" in Darwin's theory (how could an eye be developed under such a random theory?). This will be an important point in our story later on.

In light of the other theories, very little of Darwin's concept appears truly unique. After all, the concept of natural selection was known well before the *Origin of Species*. The Malthusian concept of the stable population was hardly new. Why then is Darwin singled out as a genius? Primarily because he was the first to put many loose, ill-defined ideas together into a working theory capable of being demonstrated experimentally. He so overwhelmingly documented his ideas that he conquered the reader with facts. Few biologists could deny the logic of his reasoning and proofs. Many of the public did react; they were either enchanted and elated by the *Origin of Species*, or were incensed because it clearly was at odds with Genesis and was "red in tooth and claw." Natural selection by itself appears quite innocent. Mother Nature does seem to select those that are best fit for the rigors of life. But put in a Victorian context as a seeming denial of the guiding hand of the Divinity, it appeared a very perturbing idea. T.

H. Huxley, a brilliant biologist and a sparkling orator, was largely responsible for placing Darwin in a convincing light before the public. He led many fascinating debates with the clergy and was known as the bishop-eater for his talent as a debator. When one bishop asked him which of his parent's ancestors was an ape, he said that he would rather have an ape for an ancestor than a bishop.

A TEST OF NATURAL SELECTION

The theory of *natural selection* should be given Darwin's name, and it has stood the test of time, although modified as we shall see. In fact, the concept of natural selection was independently conceived by A. R. Wallace during the 1840s while he was studying the floras and faunas of Brazil and Southeast Asia. Darwin and Wallace made the first announcement of the new theory in a joint paper offered before the Linnaean Society of London in 1858.

Natural selection is the simplest theory that satisfies most observations (see Chap. 3). We mentioned that protective coloration could not be explained by Lamarckianism. It was from this phenomenon that one of the most interesting experimental pieces of evidence was derived to prove Darwin's theory. In the midlands of England there is a species of moth called the pepper moth. Normally it is dull yellow in color with dark specks. It commonly lives on the bark of the plane tree (our sycamore). It is a favorite food for several species of birds. In Birmingham, as the industrial revolution progressed, the buildings and woods on the downwind, east side of Birmingham became blackened with coal soot. A very shrewd scientific worker in central England observed that pepper moth populations on the east side of the city, while showing some variation, were primarily dark in color. On the west side of Birmingham, toward the prevailing west wind and away from the soot-blackened trees and buildings, the pepper moth populations tended to be light-colored, reflecting the lighter, soot-free color of the bark on trees. He collected some of the light-colored pepper moths from the western part of the area and transferred them to the trees on the east

side of Birmingham and reversed the procedure for the dark-colored ones. Within a very short period of time, all the light-colored moths on the east side of Birmingham had been quickly gobbled up by birds, and, of course, the dark ones on the trees west of Birmingham likewise disappeared, indicating a very strong predator selection—a neat demonstration of the Darwinian theory of natural selection. Incidently, Darwin did not use the term "evolution" defined in the modern sense until 1871 when he used it in *The Descent of Man*.

THE FOSSIL RECORD AND EVOLUTION

The world in the mid-nineteenth century had essentially no evidence whatever to support the theory of natural selection from the fossil record, the only place where evolution really could be proved. The Reverend Sedgwick in 1850 reviewed the fossil record and thought he saw a general progression of forms. He saw that cephalopods were dominant in the early fossil record and that these were displaced by the more advanced fish, which, in turn, were replaced by reptiles, and then finally the mammals seemed to dominate the record. Such major changes were the basis, recall, for the Paleozoic, Mesozoic, and Cenozoic era subdivisions of geologic time just suggested by him. Even so, Sedgwick was not convinced that this progression was accomplished by transformation, but rather by the Creator adding these important groups to the total fauna through time. Charles Lyell at the same time said that progressive development of organic life from simplest to the most complicated has but slender foundation in fact. Darwin often stated that geology assuredly does not reveal any such finely graduated organic chain because the record is too imperfect, and recognized that this is perhaps the most obvious serious objection that could be urged against the theory. It was not until 1869 that Waagen and Karpinsky provided us with the first documented evolutionary trend in the fossil record, that of the ammonoids (see Fig. 13.4). This was followed the next year by Marsh's study of the evolution of the horse, which has become a classic example of organic evolution evidenced in the fossil record.

Lyell had illustrated a good case for evolution of species but looked at his evidence in reverse. He had subdivided the marine "Tertiary" (Cenozoic) into four divisions based on the decreasing percentages of modern species in older Tertiary strata. In reality species now living were gradually evolving through the Cenozoic, replacing older species, a process probably reflecting climatic changes.

THE RELATIONSHIP OF PHYLOGENY TO ONTOGENY

The year 1869 was an eventful one of evolution; not only did the first paleontological documentation of evolution become known, but a very important new concept appeared and was to have a profound influence on thinking for over 50 years, principally among paleontologists. The author of this new and exciting idea was Ernst Haeckel, the leading proponent of Darwinism in Germany. A shrewd and observant embryologist working with vertebrates, he noticed that there was a direct relationship between the development of the embryo and the history of the group to which it belonged. In attempting to understand embryological development, he saw that the mammalian embryo seemed to go through stages that reflected its ancestral history. He stated that at an early stage it developed "gill slits" and a tail similar to a fish. In fact, the embryo could not be distinguished from other vertebrate embryos during similar early stages. He said the *ontogeny*, or the life history of the individual, is a short history of *phylogeny*, which is the history of the race. This was counter to observations of others who said that only the young stages of the embryo resemble early stages of ancestors, later stages departing more and more from ancestral resemblance.

Using this "law" as a base, Haeckel went further and said that evolution proceeded by adding stages at the end of the individual life. This was a logical conclusion showing the relationship between the progression he saw in the embryo and the one he saw in the phyloge-

ny. Even though Haeckel believed in natural selection, he got caught in a trap. How could new characters be added at the *end* of life and be passed on to offspring after the breeding period had ended?

Much research has been done in embryology since Haeckel's day, and we now know that there are all too many exceptions to this simple analogy, and that ontogeny does not reflect accurately the course of evolution. For example, we know that teeth developed before the tongue in the evolution of the vertebrates; yet in the embryo the tongue appears first. We see now that phylogeny is the result of a series of ontogenies and not its cause (see Fig. 10.8).

STRAIGHT-LINE EVOLUTION

The next episode in our story came from paleontology. During the 1870–1880 period there was a frenzy of collecting and describing of fossils. Many complex phylogenies of the ammonoids were discovered that appeared to demonstrate Haeckel's idea. The very rich vertebrate fossil sites in the "Great American Desert" were being exploited by two colorful paleontologists—Cope and Marsh (see Chap. 16)—who tried to beat each other at describing the largest number of fossils. Thus began the golden age of paleontology in the United States. Out of this interest, intensive collecting resulted in great systematic collections of fossils. Many evolutionary sequences involving vertebrates were documented.

The picture that emerged was unexpected and formed one of the greatest puzzles of biology. Paleontologists found that many groups of organisms typically began as simple, unspecialized forms, gradually becoming more specialized and usually larger (Fig. 10.3). For example, the earliest horses had five toes like most mammals do, but because they specialized in running, they lost all but one toe. Furthermore, the horses increased in overall size, the most common evolutionary trend of all. Finally, at the end of an evolutional span, just before extinction, organisms became bizarre, frequently grotesque. Horned, crested, and duck-billed dinosaurs, for example, all highly specialized forms, appeared just before dinosaurs all be-

SUMMIT OF LOWER OLIGOCENE

Brontotherium platyceras

BASE OF LOWER OLIGOCENE

Brontotherium leidyi

UPPER EOCENE

Manteoceras manteoceras

Dolichorhinus hyognathus

MIDDLE EOCENE

Mesatirhinus petersoni

Palaeosyops leidyi

LOWER EOCENE

Lambdotherium popoagicum

Eotitanops princeps

Eotitanops gregoryi

FIGURE 10.3

H. F. Osborne's evolutionary scheme of the titanotheres of the Cenozoic, showing massive horns and large size achieved just before extinction. By today's standards this is not an acceptable evolutionary trend because we are unsure of the relationships among many of the genera. (*Courtesy American Museum of Natural History.*)

came extinct (see Fig. 16.34). The conclusion by some was that organisms seemed driven along the course of evolution to their goal—extinction. Antlers, for example, supposedly caused the Irish Elk (Fig. 10.1) to become extinct in the Pleistocene because the antlers became too large to be usable; organisms appeared to be trapped by their nonadaptive characters.

Today we feel that these concepts are oversimplifications and mostly descriptive. Yes,

there are evolutionary trends, but specialized forms can occur anywhere along the line, bizarre forms usually do not develop at all, and certainly extinction is not a goal. Even more importantly, characters do not become overspecialized, causing major groups to become extinct. The Irish Elk became extinct about 11,000 years ago probably due to a climate change which altered the flora from a subarctic tundra to grassy open country. Other large mammals of the Northern Hemisphere also became extinct at about this time. S. J. Gould has suggested that the huge antlers were male display symbols and were not used for combat because the tines are pointed backward. Still it is important to remember that the very orderly progression of development and the tendency to become specialized near the end of a period of adaptive radiation (a period when a group may evolve rapidly and may develop a number of different lines of descent) was unexpected by Darwin or his followers.

A STRIKING DISCOVERY

The year 1869 was important for another historical reason. It was that year that an Austrian monk by the name of Gregor Mendel published in an obscure journal the results of many years of observation and breeding of sweet peas. What Mendel had discovered was the system by which characters are transmitted from adult to offspring. He had, in short, discovered what Darwin only suspected in his later years, namely, the mechanism of heritable variation. This was one of the most important single discoveries after Darwin's work because it could have led the way to the very origin of new characters. Tragically, Mendel's work was lost and was not rediscovered for another 40 years, one of the great setbacks of science. Almost simultaneously in both Europe and North America, Mendel's theory of inheritance was discovered, after workers had independently arrived at the same conclusions. By 1910, when the field of *genetics* developed, we had learned how new characters arose and how they were inherited.

A striking discovery was made—new characters (or mutations) arise by chance, completely at random; thus, no purpose and no adaptation. For a period of over 70 years, studies made on fruit flies and other organisms demonstrated this over and over again. Here then were two opposing ideas—paleontologists believing that evolution is severely directed in straight lines leading to extinction, and the geneticists' principle that new characters arise by chance and at random. How could the complex dinosaurs be swept along on their "neat" course of evolution by waiting for a random new feature to come along? Surely even 4.5 billion years would not be enough time for its evolution. This was emphasized even more by experimental results showing that random mutations were either so small as to be insignificant (they could produce no new species) or too large so that offspring were not able to survive; besides, geneticists showed that almost all mutations were harmful. Furthermore, the mutations arise almost completely *independent* of the environment with the exception of those induced by natural high-energy radiation. The effect of radiation on genes is the reason nuclear reactor meltdowns are so feared. This effect was the telling blow to Lamarckianism, which would require the gene to mutate as a *result* of environmental forces.

THE HOME STRETCH

Such was the state of affairs until the 1930s, when a group of biologists in Europe and the United States began to see that these two ideas (genetic randomness of mutations and straight-line evolution) were not so far apart as they seemed at first. The different views were brought together to form a new *synthetic theory of evolution*. Essentially, Darwin's basic ideas appear to be correct. The real clue to the solution of many problems of hereditary variation came from genetics. We now know that the fundamental hereditary unit is the gene—a unit of DNA (see Chap. 7). Natural selection takes these hereditary products and preserves those that make the organism more adapted to its environment. It is also known that a gene may control the development of more than one character. Thus, the gene causing overall size increase in the Irish Elk may also have con-

trolled the size of the antler. The mere fact that the elk existed is proof enough that it was adapted to its environment.

THE OLD QUESTION—HOW DO NEW CHARACTERS ARISE AND SPREAD?

To return to the old question of how new characters arise, we can say that they may arise by chance and at random, but how? We know that the chromosome under unknown circumstances will be arranged during cell division so that it may cross over itself or form in such a way that the chemical code for the development of features in an embryo will be altered; thus, new variability (a new character) is introduced. These mutations have been induced in the laboratory by the use of x-rays and chemical-mechanical means. Variation was difficult for Darwin and other pregenetic evolutionists to understand in terms of how it is spread through the species. We know that half of the genes of the offspring come from each of the parents. Thus there is a "recombination" of hereditary material during each generation—one way of assuring the maintenance of variability in the population. The rate of mutation seems to be rather constant within a given group (ranging from 1 to 8 mutants per 100,000 genes in the human species). This, combined with the fact that natural selection weeds out all nonadaptive characters and preserves adaptive ones, would not seem sufficient to make any headway down the path to evolution. Yet, this is almost certainly the case.

HOW NATURAL SELECTION WORKS

Natural selection is a very creative and gradual process. We like the analogy that biologist-paleontologist G. G. Simpson used some years ago to explain it: first, the only part of a species we are interested in is the actively breeding population. Why? Because it is the only part that is involved in the exchange of *genes*. This is called the *gene pool*. Suppose that we have a large box filled with copies of all letters of the alphabet in great abundance and let each letter represent a mutation. Now we are allowed to

A: A population poorly adapted for the "ideal" 7-foot height. The broad distribution of sizes indicates low selective pressure, but the population can survive rapid changes in environment. **B:** A population very well adapted for the "ideal" 14-foot neck length. Narrow clustering around the 14-foot size reflects high selective pressure. But this population would become extinct if there were a rapid change in environment, requiring, say, a 20-foot neck length or a 4-foot neck, because no individuals have a 20-foot or a 4-foot neck length.

FIGURE 10.4

draw seven letters out at a time, at random. We select seven because we want ultimately to construct the word "giraffe"—equivalent in nature to the well-adapted real thing. Obviously, it might take us a lifetime to draw out—at random—the right combination. But supposing we can throw back into the box any *g's, r's, i's, a's, e's,* and *f's.* Even further, if we draw out the seven letters *kljnffe,"* we could paper-clip "*ffe*" together since it is the right code for getting our final adaptive result. All the time we can throw away the letters of the alphabet which do not lead toward our goal. It is obvious that we are

considerably increasing the chance of getting "*g-i-r-a-f-f-e*" than when we first dipped our hand into the box (our "gene pool"). Thus even a tiny "useful" mutation will quickly be preserved in the gene pool, while a more obvious and disadvantageous mutation will be eliminated as Darwin so clearly illustrated.

The increased length of the giraffe's neck was a process involving a constantly changing gene frequency for long neck (Fig. 10.4). Now if the environmental pressure (natural selection) is acting on a population, it will eliminate the unfit and carefully preserve the fit, those with longer necks. Evolution could thus keep up with a changing environment; understanding natural selection in this new light explains how random mutation, which is a creative process, produces characters that can be molded by the environment (Fig. 10.5).

ISOLATION

New species can evolve from another cause, which may not be involved in changes—that of *isolation.* Supposing that we look at a wide-ranging species such as the mountain lion. It ranges, in high, semiarid regions, from southern Mexico to Canada. The extreme member populations are isolated from each other, and thus their gene arrangements could be quite different. Supposing that two adjacent popula-

FIGURE 10.5
A series of frequency distributions showing a succession of giraffe populations with gradually increasing neck length, keeping pace with a constant environmental change.

tions, which have more in common, should become separated due to, say, a physical barrier such as a large river developing. If an advantageous mutation for a black coat color appears in one of them, it will spread eventually through that population *but not to the adjacent population* due to the barrier. Thus, a new, genetically isolated species would probably arise. If the mutation were a large one, and further isolation broke up the original population into smaller units with their own genetic "drift," the result might be considered a new genus.

The formation of new species by this process is increasingly being accepted as the most likely pathway by which evolution has occurred. In addition, the effects of isolation in a larger sense, that is involving whole biotas, are responsible for the differences we see between oceanic islands and continents.

The final chapter of the theory of evolution has not yet been written. There are many unknowns, many problems to be solved, and many questions yet to be asked. We do not know if mutations of a large nature such as the sudden appearance of a lung complex or wing results in the formation of a new order or class. Most of us feel that new classes and orders result from the accumulation of many small mutations. The reason for this puzzle is that, unexpectedly, most major groups appear suddenly in the fossil record, with few examples of transitions from ancestors, and with many groups already highly advanced when they first appear (as did the trilobites in Early Cambrian time).

THE FOSSIL RECORD

In fact, there are very few examples of continuous, gradually evolving lineages in the record such as that shown in Fig. 10.5. Eldredge and Gould believe that sudden appearances of new species can be explained by the likely possibility that new characters are more commonly introduced in small, isolated populations at the geographic periphery of the main species range. These populations become isolated and evolve into new species. The chances of these new

forms being preserved in the fossil record are small, but if successful, they could rapidly spread and become a common new species in the succeeding stratum. Hence the chances of finding a succession of species with gradually changing characters would be the exception.

Biologists and paleontologists are, by and large, convinced that this model of speciation helps explain the common observation that species abruptly appear in the fossil record and change very little in their characteristics and certainly not in any given direction nor usually in any given trend throughout their span in the strata. It also explains, on a higher level, the sudden appearance of faunas as well. In addition it helps explain how two or more species of the same genus can exist together by the development of barriers which first isolate populations until they become fully reproductively separated. Their requirements have then shifted so that they might have, for example, different food needs. When the barriers disappear, for example when a new isthmus (Panama) is reestablished connecting two land masses (North and South America) that were previously joined, the new species would not be competing for the same food. There are many types of barriers and many forms of competition as we have seen when we considered paleoecology in Chapter 9.

There are many examples, however, of so-called missing links which tell us that small-step mutation (micromutation) is an important factor in explaining the diversity of life that we observe. For example, the oldest feathered creature, *Archaeopteryx* (Fig. 16.40), has characters of both the birds and the dinosaurs, from which it probably evolved. Many examples in the fossil record of gradual changes in a lineage through time cannot be proved, because two or three stages in the sequence may be found together at the same stratigraphic level, or an earlier stage may be found later, out of sequence. Many of these problems are due to collecting or lack of preservation of critical forms.

In the 100 years or so since Huxley and others described the first known evolutionary trends,

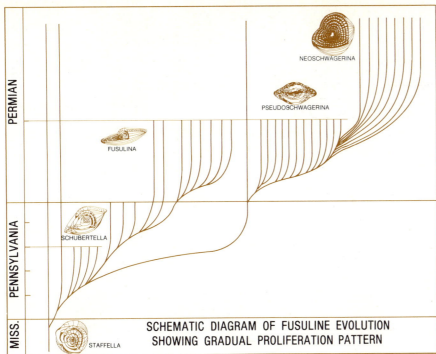

PERMIAN

PENNSYLVANIA

MISS.

NEOSCHWAGERINA

PSEUDOSCHWAGERINA

FUSULINA

SCHUBERTELLA

STAFFELLA

SCHEMATIC DIAGRAM OF FUSULINE EVOLUTION
SHOWING GRADUAL PROLIFERATION PATTERN

FIGURE 10.6
A schematic diagram of late Paleozoic fusulinid evolution, showing a relatively uniform increase in the numbers of genera.

very few new trends have been described that have stood rigorous testing. One example that has is *Myalina*, a clam from the Mississippian and Pennsylvanian (Fig. 10.8). *Myalina pliopetina* and *Myalina copei* are logical extensions of this evolutionary trend into the Permian; yet they are found in strata of equivalent ages in the Lower Permian. This does not necessarily mean that *Myalina pliopetina* did not serve as the ancestor of *Myalina copei* (it survived to live with its descendant), but we cannot prove it stratigraphically.

One final observation regarding the fossil record and evolution is that there is an apparent increase in diversity of life from its first appearance onward to the present when there is more known diversity than in all the rest of the fossil record combined. The fairly gradual increase would be consistent with progressive evolution producing more and more groups occupying more and more ecospace. David Raup calculated the sediment volume and the surface areas for each geologic system for the world. He found a surprising agreement among the

amount of faunal diversity for each system, as related to the average thickness of the rocks, and the area of exposed rocks available for collecting fossils. Thus, progressively younger rocks provide more fossils than older rocks and hence more apparent diversity. Raup concluded that there probably was a steady number of species per unit of time throughout the fossil record. This is borne out by the observation that fossil records are so fragmentary that we do not have a very good idea of what the diversification was really like, except under extremely rare circumstances such as the unusual Cambrian Burgess Shale fauna and flora (see page 194).

ADAPTIVE RADIATION

We shall see in Chaps. 11 to 17 that periodically in nearly all plant and animal groups, within a relatively short period of time, a large number of new species and genera will appear. In some cases, such as that of fusulinid protozoa (Fig. 10.6), there was a more or less gradual increase in numbers of genera. In other cases, the major diversification occurs suddenly (Fig. 10.7) and may gradually taper off throughout the rest of their geologic range, as in the case of the trilobites (Fig. 10.9). We are unsure of the reasons for these bursts of evolution, but the idea has been advanced that they are periods of *adaptive radiation*.

At the beginning of this event, a genus may move into a brand-new or "empty" ecological situation. As evolution proceeds and new species develop, the species occupy small environmental or ecological areas within the broader zone (Fig. 10.10), much like the Darwin finches did in the Galápagos Islands (Fig. 10.2). This process can be on any scale. In the case of the ammonoid cephalopods (see Fig. 16.31) in the Mesozoic, the whole subclass was involved in a radiation lasting most of the era. Whatever the cause, these bursts are very useful as stratigraphic tools because the species and genera tend to be short-ranged.

Throughout geologic time, adaptive zones have been vacated by episodes of extinction. The replacement organisms may belong to closely related groups as was the case during the Mesozoic extinctions of ammonites (see p. 436). Unrelated groups may invade an empty adaptive zone if they are broadly adapted for similar ecologic niches (an ecologic niche is the position that an organism occupies in a habitat). For example, following the extinction of the dinosaurs in the Late Cretaceous, the Cenozoic mammals radiated into many of the terrestrial habitats which may have been generally occupied by the dinosaurs. This process has been called *ecologic replacement*.

It is possible that the Cenozoic swimming mammals such as whales and porpoises could have replaced some of the Mesozoic swimming reptiles such as the ichthyosaurs in this way. Marine mammals and freshwater sharks are examples of evolutionary divergence in which groups evolve away from the central broad adaptive zone such as the terrestrial habitat of the mammals. Bats also provide good examples of *divergence*.

FIGURE 10.7

The invasion of land by amphibians represents an example of rapid expansion or *adaptive radiation* into a previously unoccupied environment. (*After A. S. Romer,* Vertebrate paleontology, *1966; by permission of The University of Chicago Press.*)

PHYLOGENY AND CLASSIFICATION

The course of evolution in a given group is manifested in its phylogeny, the history of the entire group. Figures 10.6 and 10.7 are examples of the phylogeny of the fusulinid protozoa and amphibians respectively. Both examples show the current interpretations of specialists regarding the derivation and subsequent history of these groups, and are subject to change as new fossils are found or as new ways of studying the characters are developed. Most specialists attempt to construct *classifications* of organisms to reflect phylogeny. Hence, most classifications are based on evolution.

Organisms most often are classified on the basis that "like things belong together"; implicit in this statement is the concept that they look alike because they are related. The species is the smallest formal unit, and individuals within it have more shared characters than higher, more inclusive classification groups *(taxonomic units)*. Species are grouped into genera, genera into families, families into orders, orders into classes, and classes into phyla. If two species share more features than other species of a genus, then we can assume that they very recently evolved from a common ancestral species. The genus usually shares some, but not all,

FIGURE 10.8

The evolutionary development of a series of clam species in the Upper Paleozoic. On the right is an example of recapitulation (ontogenetic column). Note that the changes in shape during the ontogeny of *Myalina copei* do not exactly reflect the actual phylogeny. *(From N. D. Newell, Late Paleozoic pelecypods; Mytilacea. State Geol. Surv. of Kansas, v. 10, pt. 2, 1942, with permission of the Kansas Geological Survey and N. D. Newell.)*

of the features of its contained species. The same is true for families, orders, etc. These features tend to be conservative, that is, established characters are less likely to be modified by selection.

On the whole this concept works well, but like most other ideas formed in the natural sciences there are annoying exceptions. One problem in comparing and analyzing relationships between organisms is that of convergence (see Fig. 10.12). In this case, unrelated organisms evolved characters quite different from their ancestors, but similar (almost never identical) to each other. This usually results from

living in similar environments. For example, the ichthyosaurs (see Fig. 16.41), which were marine swimming reptiles, are similar to sharks in outward appearance, but close inspection of the majority of characters of the entire organism leaves no doubt about their separate positions in the classification. Another example is the "wing" of bats, birds, and insects constructed of different skeletal materials, yet functioning in the same way.

The immense diversity of life in the past is intimately involved, as we have seen, with changing environments through climatic and other physical changes, the opening of "new" ecological zones (niches), or by geographic expansion or dispersal of species. For example, one important factor which controls diversity is temperature. Figure 10.11 shows that the num-

FIGURE 10.9

The trilobites underwent their most rapid diversification during the Cambrian, then gradually lost genera until they became extinct at the end of the Permian Period.

Number of trilobite genera per period

FIGURE 10.10

When a new group enters a new environment, *adaptive radiation* occurs. This process frequently involves the origin of many new species resulting from "breaking down" the broader environmental zone (the lowest and largest bar above). Successively small bars represent smaller and smaller ecological zones being entered by new groups. (*After G. G. Simpson*, The major features of evolution, *1953; by permission of the Columbia University Press.*)

FIGURE 10.11

A map showing decreasing diversity of snails toward the higher latitudes along the eastern coast of North America. (*From A. G. Fischer*, Latitudinal variations in organic diversity; Evolution, *v. 14, no. 1, 1960, with permission of A. G. Fischer and the Society for the Study of Evolution.*)

ber of species of snails decreases northward from warmer waters to cold. Warm climates support more species, but there are fewer individuals of each species. For example, in a 100 by 100-meter plot in the Amazon jungle, 60 species and 564 specimens were found. Twenty-two species were represented by a single specimen each, and only one species had as many as 33 individuals.

EVOLUTION—FACT OR THEORY?

Evolution, we believe, is the best hypothesis currently available to explain the diversity of life that we see, the faunal and floral changes that have been recorded in the geologic strata, and the ecologic changes observed for so long. For us, it is important for an understanding of earth history to question *why* these changes occurred and *how* they came to be. Our goal is to seek

physical and biological causes for these changes and to integrate these causes with ancient patterns of earth history so that we can understand diversity in a meaningful way. Darwin's theory has been remarkably successful in achieving this.

The goal of science is to work toward an understanding of the physical universe. We agree with Carl Sagan that science is not simply a body of knowledge or a group of facts, but rather a method used to determine what is fact and what is not, insofar as the present state of knowledge permits. We do this by means of observation, by questioning the meaning of those observations, and by constructing hypotheses which are then tested and accepted or rejected based upon further observations. Scientific theories are open-ended and help us to search for new knowledge. For example, Darwin himself said: "I am actually weary of telling people that I do not pretend to adduce direct evidence of one species changing into another,

FIGURE 10.12

Fossil natural casts of two genera of snails belonging to two different families, illustrating evolutionary convergence. Even though these two are almost identical and are found at the same place, critical features identify them as members of separate families. They illustrate the difficulty

sometimes encountered in proving the correct identity of index fossils. Other species of each genus do not look very much alike. *Left: Worthenia. Right: Glabrocingulum (Ananias). (Courtesy Niles Eldredge; photograph by G. R. Adlington.)*

but that I believe that this view in the main is correct, because so many phenomena can be thus grouped together and explained."

Science is simply trying to explain the workings and not the purpose of nature. We do not have all the answers—or even all the questions—regarding evolution so that we are inclined, at present, to feel it may not be important to argue whether or not evolution is a fact or a theory.

Key Concepts

adaptation
adaptive radiation
chromosomes
classification
DNA
divergence (evolutionary)
ecologic replacement
gene
gene pool
genetics
inheritance of acquired characteristics
isolation
mutation
natural selection
ontogeny
organic evolution

phylogeny
specialized organism
speciation
synthetic theory of evolution
taxonomy
variability

Readings

Cracraft, J., and Eldredge, N. (eds.), 1979: Phylogenetic analysis and paleontology: New York, Columbia University Press.

Darwin, C., 1859, On the origin of species by means of natural selection, or the preservation of favoured races in the struggle for life: London, J. Murray.

Eldredge, N., and Gould, S. J., 1972, Punctuated equilibria: an alternative to phyletic gradualism, Models in paleobiology, J. T. M. Schopf, (ed.): San Francisco, Freeman Cooper, pp. 82–115.

Glass, B., (ed.), 1959, Forerunners of Darwin: 1745–1859: Baltimore, Johns Hopkins.

Gould, S. J., 1977, Ontogeny and phylogeny: Cambridge, Harvard University Press.

Nelson, G., and Platnick, N., 1979, Cladistics and vicariance: New York, Columbia University Press.

Raup, D., and Gould, S. J., 1974, Stochastic simulation and evolution of morphology: Systematic Zoology, v. 23, pp. 305–322.

Savage, J. M., 1969, Evolution (2d ed.): New York, Holt.

11

HYPOTHETICAL MAP
OF THE
CURRENTS
OF THE
SILURIAN INTERIOR SEA
by
T. C. CHAMBERLIN
1881.

LEGEND

~~~~~ *Approximate outline of land.*
Shoals or Shallow Sea.
→ *Oceanic Currents.*
∿∿ *Lead and zinc deposits in Up*
Cambrian rocks.
++ *Ditto in Silurian rocks.*
∘∘ *Ditto in Pod. Silurian rocks.*
The arrows trial outlines and area
nic currents had nearby chan
get before the deposit of these.

11

EARLIEST PALEOZOIC HISTORY

A Study of the Interpretation Of Epeiric Seas

In the mud of the Cambrian main
Did our earliest ancestors dive;
From a shapeless, albuminous grain
We mortals are being derived.

Grant Allen,

The Ballade of Evolution

FIGURE 11.1
Hypothetical paleoge-ographic map of the United States for early Paleozoic time, one of the first paleogeo-graphic maps ever published. In 1881 the name Silurian often included all of what now comprises Cam-brian, Ordovician, and Silurian. This map re-ally refers to the Late Cambrian. Thin lines and arrows indicate supposed ocean cur-rents (compare Fig. 11.9). *(After T. C. Chamberlin,* Geology of Wisconsin, *v. IV, p. 530.)*

11

The beginning of the Paleozoic Era produced a great punctuation mark in the record of earth history because of the appearance in the stratigraphic record of the remains of animals. The Eocambrian fauna described in Chap. 9 was already moderately diverse and included some rather complex forms. It was followed by the sudden appearance of hard, preservable skeletons at the beginning of Cambrian time. It is the appearance of fossil animals, then, that defines the beginning of the Paleozoic Era.

In North America, as in other continents, many areas contain essentially unfossiliferous strata that underlie and are closely related to Cambrian ones. Because such strata contain so few fossils, their true geologic age is somewhat uncertain. In only a few cases have isotopic dates been available to resolve the dilemma. What dates exist suggest that these poorly fossiliferous strata spanned the interval from 570 to 700 million years ago. In many mobile belts, such strata are found conformable below others with Lower Cambrian index fossils, and a major angular unconformity is found beneath the essentially barren rocks; in eastern North America, the "basement" below these is Grenville-aged (800 to 1,000 m.y.) metamorphic complex (see Fig. 8.10). The barren strata clearly are more closely related to Cambrian sediments above than to deformed Precambrian rocks below. Therefore, it has become customary to refer to such unfossiliferous strata as Eocambrian, meaning the "dawn of the Cambrian" (also called Infracambrian or Vendian). At present, the term Eocambrian is an informal one, but we feel that the advent of the Paleozoic Era should be redefined formally to include these strata, especially now that fossils are turning up within them.

In this chapter we shall examine the early Paleozoic history of North America from 450 to 700 million years ago, with emphasis upon the structure of the craton and conditions in shallow epeiric seas that finally covered it during Late Cambrian time. The strata deposited during this long period of transgression of the craton have been called the *Sauk Sequence* for a county in Wisconsin (see Fig. 3.21). In order to interpret the lower Paleozoic rocks, it is necessary to introduce several important principles and types of diagrams and maps which, though shown here, apply equally to all periods; these same principles will be used in subsequent chapters. Comparisons with foreign regions will be noted as well, in keeping with our goal to test continually the working hypothesis of crustal evolution outlined in Chaps. 5 through 8.

EOCAMBRIAN DEPOSITS

EOCAMBRIAN GLACIATION

The most remarkable trademark of Eocambrian rocks is the presence of peculiar unsorted boulder-bearing deposits slightly below fossiliferous Cambrian strata. Since they were first described in northern Norway in 1891, they have been found on practically all continents (Fig. 11.2). Peculiar textures, wide distribution, and local scratched surfaces beneath the conglomerates led most geologists to regard them as tills (Fig. 11.3), representing a major episode of continental glaciation between 600 and 700 million years ago. Continental ice sheets spreading onto shallow marine shelves deposited till, which was more or less reworked by gravity sliding and currents, while icebergs dispersed and dropped boulders into normal marine sediments beyond glacier margins (see Fig. 8.25).

Eocambrian tills occur over an exceptionally broad latitudinal range, both in terms of present latitudes and of apparent Eocambrian paleolatitudes (Fig. 11.2). The postulated position of the paleoequator is based upon rock magnetism (explained in Chap. 6) and upon the distribu-

Pre Eocambrian tillites?

Apparent paleomagnetic Equator

FIGURE 11.2

World distribution of known Eocambrian till-like boulder deposits (circles) compared with paleomagnetically determined equator. For comparison, brown band shows distribution of Lower Cambrian archaeocyathid fossils thought to have been tropical. Known older Prepaleozoic tills also are shown (triangles). (*Modified from Harland and Rudwick, The great infra-Cambrian ice age; Copyright © 1964 by Scientific American, Inc. All rights reserved; Bartholomew's Nordic Projection used with permission.*)

tion of presumed tropical Lower Cambrian fossils called archaeocyathids (see Chap. 9). The very broad latitudinal distribution of the glacial deposits has been rationalized by assuming such a complete chilling of climate that glaciers could form at practically all latitudes, or at least that icebergs could carry debris into very low

FIGURE 11.3

Eocambrian (Bigganjargga) till and underlying striated rock surface in Varangerfjord, Finmark, northern Norway. Note unsorted texture of the conglomerate. Hammer handles indicate orientations of two sets of striations. (*Courtesy H. Reading.*)

FIGURE 11.4

Possible reconstruction of Siberian and North American continental nuclei as a supercontinent (left) that split between 1,200 and 1,500 million years ago (right). Note apparent match of structural trends (left) and formation of failed rifts (*aulacogens*) perpendicular to rifted margins of the new continents (right). (*After J. W. Sears and R. A. Price, 1978, Geology, v. 6, pp. 267–270; by permission of the Geological Society of America.*)

latitudes. But the Eocambrian represents more than 100 million years, therefore it is probable that the many tillite localities shown in Fig. 11.2 differ considerably in age. Perhaps drifting of continents during Eocambrian time brought different regions into high latitudes favorable for glaciation at different times. If true, then there is no necessity to try to imagine glaciation near the equator. But it has also been suggested that the earth's orbit became exceptionally elliptical, which would have caused much greater temperature contrasts between winter and summer. It is argued that such conditions might cause glaciation in mid-latitudes.

STRUCTURAL FRAMEWORK

Around the periphery of the continent, basaltic rocks interstratified with Eocambrian sediments suggest that deep fissures penetrated the crust there to allow molten mantle material to rise. It has been suggested that such fissures developed when large, Late Prepaleozoic lithosphere plates were rifted and dismembered to form embryonic North America and other continents (Fig. 11.4). As such separation (divergence) of plates proceeded, thick continental shelf sedi-

ments accumulated along the new, structurally passive trailing margins of the continents (Fig. 11.5). Apparently the Eocambrian was a time of profound worldwide reorganization of lithosphere plate patterns.

On both the western and eastern margins of North America, the distribution of late Prepaleozoic and Eocambrian strata indicates the presence of several narrow fault-bounded troughs that extended back into the craton (Fig. 11.4). Such troughs (named *aulacogens* by Russian geologists) are thought to be characteristic of newly rifted continental margins like the present East African rift valleys (see Fig. 7.8).

EOCAMBRIAN SEDIMENT DISTRIBUTION

Eocambrian strata are well developed in many mobile belts, but are nearly absent from most cratons. It is clear that the North American craton was a vast land area during latest Prepaleozoic and Eocambrian times. But the margins of the continent received thick Eocambrian sediments with some local volcanic rocks interstratified with them. The sediments represent shallow-marine continental shelf as well as nonmarine deposits.

Figure 11.6 shows an example from the southwestern United States of the relationship of Eocambrian and Cambrian strata to the craton margin. The diagram also illustrates how the geologist constructs cross-sectional diagrams to help interpret thickness, sedimentary facies, and age patterns. In this case we see at once that

FIGURE 11.5

Diagrammatic restoration of Eocambrian and Lower Cambrian strata from California to Utah, suggesting that western North America was a passive continental trailing edge. Basalts suggest possible rifting of North America away from some other continent by spreading. Typical of trailing edges is the accumulation of thick prisms of continental shelf sediments with mature quartz sandstones and carbonate rocks. (*After J. B. Stewart, 1972,* Bulletin of the Geological Society of America, *v. 83, pp. 1345–1360; by permission of the Society.*)

the thickest strata and most complete age sequence are represented beyond the craton in California and Nevada. Relative thickness of shallow-marine strata provides a direct guide to relative subsidence of the crust. In that region, Eocambrian strata are conformable beneath Cambrian ones, and there, subsidence amounted to thousands of meters. Farther east the Eocambrian and Lower Cambrian are missing entirely, however, and only a few hundred meters of Middle and Upper Cambrian strata rest unconformably upon the Prepaleozoic basement. This break · extends across the entire craton, and is one of the most profound unconformities in the entire stratigraphic record.

All around the continent we find these same patterns. First, thick Eocambrian and then younger and younger Cambrian and even Early Ordovician marine strata rest upon the Prepaleozoic basement progressively inward toward the center of the craton. Clearly the sea surrounded the continent at first and gradually spread over the craton. This entire transgression spanned about 300 million years—equal to almost the entire remainder of the Paleozoic Era after Cambrian time!

The pattern of facies in Figs. 11.6 and 11.8 shows that the clastic sediments were derived from the craton, because the sandy facies always occurred on the inner (craton) side of the shale and carbonate facies. Because the cratonic surface was weathered and eroded for nearly half a billion years before Eocambrian sedimentation began, a huge volume of clastic material must have become available. Undoubtedly it was deposited, eroded, and redeposited countless times during this interval by rivers, wind, and glacial ice. In many cycles of erosion and sedimentation, the compositional maturity of sands and gravels and the separation of clays by winnowing increases (see Fig. 8.12). This was the case, because pure, well-sorted and rounded quartz sand is prominent in both Eocambrian and Cambrian strata.

FIGURE 11.6
Cross sections showing Eocambrian and Cambrian strata in the southwestern United States. Columns (above) portray factually the character of preserved rocks exposed at several localities. The *restored cross section* (below) was constructed from the columns and other information in between. Subsequent deformation and erosion are ignored. It is an interpretive graph in which the vertical axis is stratigraphic thickness, and the horizontal axis is distance between data points. The geologist infers thickness and lithologic facies variations between separate columns of the upper diagram. Such cross sections are helpful in portraying regional relationships among strata in a vertical plane and are invaluable companions to facies maps, which show relationships in the horizontal plane. The vertical scale is exaggerated to show detail. Note the transgressive facies, and how time lines overlap progressively eastward onto the basal unconformity. *Olenellus* and *Paradoxides* are trilobite index fossils. (See Fig. 11.7 for symbols.)

THE CAMBRIAN CRATON

TRANSGRESSION

Flooding of the craton continued throughout Cambrian time, probably as a result of a worldwide rise of sea level (see Fig. 3.21). By the end of the period, probably two-thirds of the low surface of the continent was doused beneath a shallow *epeiric sea* (meaning "a sea over a continent"). The average apparent rate of transgression was about 18 kilometers (10 miles) per million years (or 0.05 feet per year), hardly a tidal wave. Actually the transgression was spasmodic, being interrupted by several partial

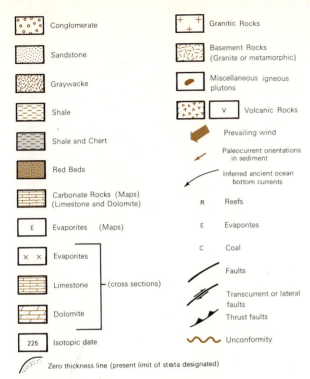

Conglomerate	Granitic Rocks
Sandstone	Basement Rocks (Granite or metamorphic)
Graywacke	Miscellaneous igneous plutons
Shale	Volcanic Rocks
Shale and Chert	Prevailing wind
Red Beds	Paleocurrent orientations in sediment
Carbonate Rocks (Maps) (Limestone and Dolomite)	Inferred ancient ocean bottom currents
	R Reefs
E Evaporites (Maps)	E Evaporites
× × Evaporites	C Coal
Limestone (cross sections)	Faults
Dolomite	Transcurrent or lateral faults
	Thrust faults
225 Isotopic date	Unconformity

Zero thickness line (present limit of strata designated)

FIGURE 11.7

Explanation of symbols used on facies maps, paleogeographic and paleogeologic maps, and most cross sections in remainder of book. Repeated in endpapers. (*Principal sources of data: Alberta Society of Petroleum Geologists, 1964, Atlas of facies maps of western Canada; Clark and Stearn, 1960, Geological evolution of North America; Douglas et al., 1963, Petroleum possibilities of the Canadian Arctic; Eardley, 1961, Structural geology of North America; Kay, 1951, North American geosynclines; Levorsen, 1960, Paleogeologic maps; Martin, 1959, American Association of Petroleum Geologists Bulletin; Donald E. Owen, unpublished maps; Raasch et al., 1960, Geology of the Arctic; Sloss et al., 1960, Lithofacies maps; Rocky Mountain Association of Geologists, 1972, Geologic atlas of the Rocky Mountain region.*)

retreats of the shoreline. Figure 11.9 is a paleogeographic restoration of the continent's appearance in mid-Late Cambrian time. It was constructed from the evidence contained in the *facies map* (Fig. 11.8). Considerable interpretation is necessary in preparing any *paleogeographic map*, especially where the strata have been entirely eroded from a large region in later time (as in northeastern Canada).

STRUCTURAL MODIFICATIONS OF CRATONS

As we pointed out in Chap. 6, cratons are only relatively more stable than other regions. Moreover, the cratonic surface in Cambrian time, though low in relief, was not perfectly flat. Cratons have undergone structural changes, largely of a mild (*epeirogenic*) sort, and these changes are reflected in the stratigraphic record.

The largest single feature of the Cambrian craton is the so-called transcontinental arch extending from Lake Superior southwest toward Arizona (Fig. 11.8). Along the sides of this trend, Cambrian sandstones lap unconformably against the Prepaleozoic basement, but along its axis, especially in the subsurface beneath the Plains states, Cambrian strata are entirely missing and younger rocks rest upon the Prepaleozoic there. Figure 11.10 shows the present distribution and thickness variations of preserved Cambrian sediments. Are these patterns entirely original ones, or do they reflect post-Cambrian erosion? Finally, what is the age and history of this great arch? Figure 11.8 reveals some clues not apparent from thickness maps alone. We see that dominantly quartz sandy facies border much of the arch, suggesting that it may have been a high feature supplying some of that sand during Cambrian time; perhaps it was an old Prepaleozoic drainage divide present as a range of low hills converted to islands when the Late Cambrian epeiric sea flooded the craton. But in some areas, boundaries between dominantly sand and the more shaly and carbonate-rich facies are perpendicular to the trend of the arch, suggesting that it did not influence Cambrian sedimentation much there, but was raised and eroded later. Presence of several major post-Cambrian unconformities over the transcontinental arch indicates a long history of periodic flooding alternating with erosion. Although it influenced sedimentation somewhat in the Cambrian, much warping of the feature occurred later.

Broad, gentle warpings of the cratonic crust, apparently in response to small changes of isostatic equilibrium due to density changes in

FIGURE 11.8

Upper Cambrian sedimentary *facies map* (note symbols in Fig. 11.7). Nowhere does the zero
thickness line (heavy brown) represent a Cambrian shoreline; erosional nature of this boundary is
proven, especially at the southeastern side of the transcontinental arch and in west-central
Canada, where the zero line intersects facies boundaries nearly at right angles. Clearly the facies
once extended beyond the zero line. Note southerly orientation of paleocurrent arrows.

LATE CAMBRIAN

PALEOGEOGRAPHY

(about 480 m.y. ago)

(Interpretation I)

FIGURE 11.9

Our interpretation of Late Cambrian paleogeography. Facies maps provide factual evidence for constructing *paleogeographic maps,* which inevitably are very interpretive. The latter may show land, sea, equator, poles, currents, winds, climatic zones, etc. Wherever we must restore geography beyond the zero limit of facies maps, there is much chance of error. For example, the single large low area shown in northeastern Canada may have had many small islands. (Compare Fig. 11.1.)

FIGURE 11.10

Cambrian *thickness map.* Contours connect points of equal thickness of total Cambrian strata. Marked contrasts of thickness are revealed (especially at the margins of the craton), but their causes must be interpreted carefully. For example, subsequent erosion has removed Cambrian strata from large areas. (See Fig. 11.7 for sources.)

the underlying lithosphere, not only have raised areas called *arches*, but have depressed areas called *basins* (Fig. 11.11). Faults also are present in cratons, and a few show great displacement. Sedimentation on cratons has been profoundly influenced by all such structural modifications. In addition, basins and arches have greatly influenced the accumulation of oil and natural gas, salt and gypsum, chemically pure limestone, and other economically important resources associated with sedimentary rocks. Therefore, the structure and history of cratons are of more than passing interest to humanity.

Compared with basins, arches typically contain thinner stratigraphic sequences that are interrupted by many unconformities. This implies that basins subsided relatively more rapidly and could therefore receive greater and more continuous sedimentary accumulations than arches. The arches intermittently were shoal areas experiencing little or no sedimentation or were islands undergoing active erosion.

INTERPRETATION OF THE HISTORIES OF ARCHES AND BASINS

In Fig. 11.11 note that thinning of strata over the arch is due both to originally thinner accumulations there and to erosion at unconformity surfaces. Interpretation of cratonic history rests heavily upon knowledge of the ages of warpings of arches and basins; therefore, it is crucial to make correct interpretations of the causes and timing of the thick and thin accumulations. In Fig. 11.12 we show an identical situation of thinner and thicker strata, but note that a different historical explanation is required in this case, for these structural features were

FIGURE 11.12

Differential warping *after* sedimentation to produce arches and basins. Note similarity of gross thickness patterns to those of Fig. 11.11; yet here the contrast resulted solely from erosional truncation *after* warping occurred. Histories of the two cases were very different, though results are superficially similar.

formed entirely *after* deposition of the rocks. In the first case (Fig. 11.11), thinning over the arch was caused by lesser *subsidence* and smaller accumulation rates as well as by periodic erosion. But Fig. 11.12 shows thinning caused solely by subsequent erosion or *truncation* (cut off by an unconformity). The "basin" in this case is purely secondary and is the result solely of downfolding and truncation after deposition. If one is exploring for petroleum entrapments, for example, this sort of analysis is of utmost importance because potential structural traps must have formed before petroleum began to migrate through permeable strata if any was to accumulate.

How do we acquire and synthesize the information for studying broad and subtle cratonic structures? Much of the evidence for arches lies in surface outcrops, but only the edges of basins are exposed. Therefore, subsurface information is required to define basins and some buried arches. Deep drilling for water and petroleum during the twentieth century has literally revolutionized the study of regional stratigraphy by providing the third dimension to our observations. Where deep drill-hole data are lacking, some insight into subsurface relationships can be gained from various geophysical devices. Magnetic and gravity surveys reveal information about special types and thicknesses of buried rocks. But seismology provides the greatest insight into thickness, structure, and, in some cases, even gross lithology of buried strata.

FIGURE 11.11

Contrasts of thickness and unconformity between a cratonic arch and basin being differentially warped sufficiently to influence sedimentation *throughout* their histories.

CAMBRIAN SEDIMENTS

PURE QUARTZ SANDSTONES

Let us next interpret the Cambrian sediments deposited in the epeiric sea. That the Upper Cambrian sediments of Fig. 11.8 are marine deposits is proved by the presence of fossil groups such as brachiopods and cephalopods known only to have lived in sea water. The green mineral glauconite (a potassium- and iron-bearing micalike silicate), which is known to form only in sea water, also is widespread in Cambrian strata.

Upper Cambrian sandstones, the dominant cratonic sediment, rank among the most mature in the world. They are unrivaled for perfection of *rounding* and *sorting* of grains, and may contain 99 percent quartz with traces of other very stable minerals (e.g., garnet, zircon, and tourmaline) (Fig. 11.13; also see Figs. 8.12 and 8.14). All of the clastic minerals point to ultimate metamorphic and granitic sources, but the grains had a long and complex sedimentary history prior to final deposition in the Late Cambrian epeiric sea.

The variation of sand-grain sizes is useful for interpreting their origin. By assuming that observable, present processes are keys to ancient ones, grain size characteristics of modern sands formed in known environments are useful for comparison with the textural properties of ancient examples formed by unknown processes (see Fig. 8.13). Sorting of the Cambrian sands suggests wind, surf, and vigorous marine

FIGURE 11.13

Microscopic photographs of typical Upper Cambrian sandstones, southern Wisconsin. *Left:* Quartz grains averaging 0.25 millimeter diameter. Rounding is so exceptional for larger grains as to suggest an abrasive history equivalent in distance to rolling by water around the earth 30 or 40 times. *Right:* Concentrated fraction of well-rounded mineral grains of relatively high specific gravity (2.8 to 3.5) from a Cambrian sandstone like that at left. Minerals present average 0.15 millimeter and consist of very durable zircon, tourmaline, and garnet. Such minerals constitute less than 1 percent by weight of typical Cambrian sandstones. (*Left view by R. H. Dott, Jr.; Right view courtesy J. A. Andrew.*)

FIGURE 11.14
Upper Cambrian (Galesville) sandstone along the Dells of the Wisconsin River, probably deposited here by wind. Lowest exposed stratum displays the simplest geometric type of cross stratification in which inclined laminae are almost perfectly planar, and surfaces truncating their tops are horizontal (see Fig. 8.15). *(By R. H. Dott, Jr.)*

currents as the most probable agents. Structures such as *ripple marks* and *cross stratification* (Fig. 11.14) prove that they were transported by vigorous currents of air or water. Judging from relative average thickness of Cambrian strata in the craton versus the mobile belts (a ratio of 1 to 10), the rate of sedimentation in the craton must have been slow. Very slow addition of new sediments to a strongly agitated environment would be ideal for producing the observed texture and stratification of these rocks.

Rounding reflects the physical durability of grains and the intensity and duration of abrasion. Experimental studies with transport of sand in laboratory troughs by water and in wind tunnels suggest that wind is more than 100 times as effective a rounding agent as running water. This is because impact between grains is cushioned slightly due to the greater viscosity of water. Cambrian strata are very uniform and widespread, and the presence of scattered marine fossils indicates that most of the sediments were last deposited in the sea. But the unusually high rounding values suggest a long history of wind transport. How can these facts be reconciled? It is significant that the early Paleozoic pure quartz sands show higher overall grain rounding than do modern sands that have been analyzed. The present does *not* provide a complete key to the past, for we cannot find good

examples today of all things found in the ancient record. No definite fossil record of land plants exists in rocks older than Devonian. Without a significant plant cover to hold soil in place, wind would have played a much greater role in erosion and transport of sediment than today. Therefore, the sand grains must have been acted upon by wind for long periods on the Late Prepaleozoic and Eocambrian continental surface during the nearly half a billion years before their redeposition in the Late Cambrian sea much as in the giant drifting sand fields on Mars. Confirmation is provided by the presence locally at the basal Cambrian unconformity of pebbles with peculiar flat sides identical with faces cut on pebbles today through blasting by windborne sand.

Some of the rounding and sorting of the Cambrian sand probably was also accomplished by strong tidal currents, which daily move grains back and forth across sand bars. In only 1 year's time, a grain might travel the equivalent of 2,000 kilometers within a relatively local area. In 1,000 years the grain would have rolled and bounced 1 million kilometers, equivalent to 50 trips around the earth! Even though it was being moved by water rather than wind, the constancy of strong agitation would take its toll by pulverizing less durable grains and rounding durable quartz at a geologically reasonable rate.

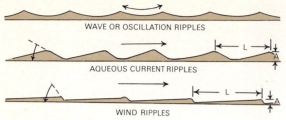

FIGURE 11.15
Profiles of ripple marks of different origins. Ideally, lee-face angle of inclination, wavelength *(L)*, and amplitude *(A)* vary for wind- versus water-formed current ripples. *(After E. H. Kindle, 1932, in* Treatise on sedimentation *(2d ed.), edited by W. H. Twenhofel; by permission of Williams & Wilkins Co.)*

Much sand produced by weathering through geologic time has not left the continents, but has been deposited and reeroded almost endlessly to become more and more purified. For many years, geologists thought that the most profound purification and concentration of the quartz sand occurred during Early Cambrian time, but as was shown in Chap. 8, much initial concentration actually occurred long before the Paleozoic Era began, as testified by large volumes of Middle and Late Prepaleozoic quartzites (see Fig. 8.18).

Where is all the clay that must have formed by decay of the immense volumes of igneous and metamorphic rocks indicated by the pure quartz sand concentrate? At least half of the original source rocks should have weathered to clay; yet shale is rare in Cambrian rocks of the craton. Apparently it was blown and swept from the craton to find its way into deeper, less-agitated zones of the sea beyond, for much shale does occur in Eocambrian and Cambrian strata on the outer edges of the craton beyond the Cambrian continental shelf edges (see Figs. 11.5 and 11.8).

RIPPLE MARKS AND CROSS STRATIFICATION

Cross stratification is one of the most prominent features of the Upper Cambrian sandstones of the craton from Ausable Chasm in northeastern New York to the Canadian Rockies (Fig. 11.14). Ripple marks also are prevalent. Variations in form between wave and current-formed ripple marks are well known, and the dimensional relations of current ripples differ slightly for the wind and water media due to viscosity differences (Fig. 11.15). It is said that certain cross-stratification types characterize wind dunes, others typify ordinary aqueous currents, while still others form only on beaches. Yet the form of cross strata actually varies greatly within any one modern environment. Sediment volume

Bottom flow
Two-dimensional wake
Three-dimensional wake

FIGURE 11.16
Relations of current turbulence patterns to different ripple or dune forms. Such forms may occur on any scale from amplitudes of a few millimeters to at least 100 meters and in either air or water. *(After J. R. L. Allen, 1966,* Sedimentology, *v. 6, pp. 153–190; by permission.)*

and coarseness, constancy of currents, and patterns of turbulence control the form of cross strata as well as whether the transporting medium is air or water.

Ripples and *dunes* represent different scales of roughening, be they wind or water-formed. Such phenomena have been studied experimentally for many years, beginning with water tank experiments in Germany in 1825. In those early experiments, brandy and mercury as well as water were used as fluids. Since then hundreds of artificial stream channels called flumes have been constructed to study erosion and transport of sediments by currents and waves (see Fig. 8.16). Similar studies have been made with air transport in wind tunnels. Cross laminae represent the buried lee (downcurrent) faces of ripples or dunes (see Fig. 8.15). If the ripple crests are straight, then the cross laminae are simple and planar; but experiments show that, due to turbulence, ripple or dune crests more commonly are irregular (Fig. 11.16). As individual dunes migrate, intervening depressions become filled with sand in laminae that dip toward the depression axes. These interdune depression fillings are what we see preserved in most cross-stratified sandstones.

It was long thought that large bodies of cross-stratified sandstone must represent wind-dune sands simply because such dunes were the only readily accessible features that geologists could compare with the rocks. Herein lies a fallacy, however, for as exploration of the sea floor progressed, large underwater ridges as much as 20 meters high and covered with dunes have been discovered on shallow, agitated sand bank areas. Examples are well known in the English Channel, on the Georges Bank and Nantucket Shoal off New England, and on the Grand Bahama Banks (Fig. 11.17). In these areas, very strong, turbulent currents develop due both to wind blowing over the shallow waters and to tidal fluctuations. The subaqueous dunes rival wind-formed ones in size, are

FIGURE 11.17
Air view of large submarine sand dunes (or sand waves) on the Bahama Banks, formed of oölites and other carbonate fragments by strong tidal currents at the margin of the banks in very clear water only 6 to 7 meters deep. (*Air photo courtesy R. H. Ginsburg.*)

similar in form, and they have essentially identical internal cross stratification. Given the apparently much larger and continuous area of very shallow sea during the Late Cambrian in North America, it is inescapable that subaqueous dunes characterized large areas of that sandy sea floor just as wind dunes must have covered much of the land.

CHANGE TO CARBONATE SEDIMENTS AND THEIR ENVIRONMENT OF DEPOSITION

As the burial of local islands and general submergence of most of the craton progressed, the supply of weathered quartz sand had all been redeposited by the end of the Cambrian Period, leaving little land exposed that could supply further *terrigenous clastic* material. It is as though a great sand faucet had been turned off after dripping steadily for 100 million years. Consequently, deposition changed to dominantly carbonate sedimentation over most of the craton. Carbonates had been forming already for millions of years on the cratonic margin (Fig. 11.8), and now, with final depletion of the quartz supply, similar deposition could spread over most of the shallow epeiric sea floor, which was otherwise favorable for it.

Much limestone proves to consist of macerated shell debris and may even display cross stratification, ripple marks, and other features typical of sandstones formed of terrigenous silicate detritus. Such rocks are called *clastic limestones*. They make up over half of all limestones and typify a "shelly facies" named long ago in Britain. Several conditions can be inferred from their characteristics. By analogy with living counterparts, most marine invertebrate fossil communities during life required agitated, well-oxygenated water for best growth, and the textures of the "shelly facies" confirm that the material accumulated in sea water with current activity sufficient to roll, abrade, and sort shells. Early Ordovician carbonate rocks imply a minimal influence of mud and sand from eroding lands and also a uniform, relatively shallow marine environment where calcareous-secreting algae and inverte-

brate animals could thrive. Today such creatures are most abundant and most diversified in the clear and well-lighted shallow, agitated, warm seas of the tropics and subtropics, but some carbonate sediments also form at higher latitudes.

IMPORTANCE OF OÖLITE

Fragmented fossils, scattered quartz grains, sporadic carbonate pebbles, and ripple marks in Late Cambrian–Early Ordovician carbonate rocks point to considerable agitation during their deposition. Also common in these rocks are layers of crowded spherical grains called *oölites*, consisting of fine, concentric carbonate laminae surrounding some nucleus such as a quartz or shell fragment (Fig. 11.18). It is clear that they are accretionary in nature, and where they have been observed forming today (as in Great Salt Lake and on the Grand Bahama

FIGURE 11.18
Microscopic photograph of modern oölites from the Bahama Banks. Note the central grain nucleus of each sphere and surrounding concentric layers (spheres average about 2 millimeters in diameter). Elongate grains are shell fragments also coated with calcium carbonate. *(Courtesy R. H. Ginsburg.)*

FIGURE 11.19
Lower Ordovician algal stromatolite showing characteristic wavy lamination (Shakopee Formation, Troy, Minnesota) (compare Fig. 8.19). Today marine filamentous ("blue-green") algae form identical laminated carbonate deposits by entrapment of calcium carbonate particles by a mucous coating on the algae. *(Courtesy R. A. Davis.)*

Banks), precipitation of carbonate occurs as they are rolled vigorously. Agitation of the water causes loss of carbon dioxide gas to the atmosphere, producing a saturated condition with respect to calcium ions. Calcium carbonate is therefore precipitated on all convenient surfaces such as sand grains. By inductive reasoning from the present to the past, we infer that ancient oölites also attest to strong agitation in shallow, calcium-rich waters; thus they provide an important paleoenvironmental indicator. The occurence of oölites in thin zones in Late Cambrian and Early Ordovician strata suggests local episodic changes either of degree of agitation or of chemistry in the epeiric sea.

DEPTH OF THE EPEIRIC SEA

The abundant evidence of agitation strongly suggests that the epeiric sea was shallow. Wavy, laminated, hemispherical *stromatolite* masses in the early Paleozoic carbonate rocks are thought to have been formed by primitive filamentous marine algae that lived on the sea floor (Fig. 11.19). Some of these stromatolites make up small reef mounds that are more abundant than were their counterparts in Prepaleozoic sediments. As was discussed in Chap. 8, algae cannot receive adequate sunlight for photosynthesis in water deeper than 100 to 150 meters because of abrupt absorption of most wavelengths of light by the water below the *photic zone*. Therefore they provide the best maximum-depth indicator for the Cambrian and Ordovician epeiric sea. Apparently it was not more than 150 meters deep over most of its

extent, and probably was much less in many places. Today simple algae that form laminated deposits like those of 500 million years ago commonly inhabit the intertidal zone near the edge of subtropical seas, so it is inferred that many Cambrian and Ordovician ones occupied a similar environment.

Many of the fine dolomitic sediments associated with algal structures show polygonal networks formed by shrinkage cracking due to drying of fine muds as happens today on exposed high-tide flats and in drying mud puddles (Fig. 11.20). *Polygonal cracks* so produced gave rise, through subsequent current-stirring, to *flat-pebble conglomerates* common in early Paleozoic carbonate rocks. All stages from mere cracking to transport and rounding of the chips exist (Fig. 11.21).

NORTHERN GULF OF MEXICO— ANALOGUE FOR EARLY PALEOZOIC EPEIRIC SEAS

The continental shelf of the northern Gulf of Mexico today provides some useful comparisons for environmental conditions in the early Paleozoic epeiric seas. The shelf area is less than 200 meters deep, has moderately strong currents over it (Fig. 11.22), and has varied sediment sources. The Mississippi River supplies the bulk of detritus. On the outer shelf, much clay occurs, but deposition there since the last rise of sea level has been nearly negligible. In Florida to the southeast, by contrast, the land is entirely Cenozoic limestone and the climate is humid subtropical. Therefore, essentially no

FIGURE 11.20
Desiccation shrinkage cracks in modern lime-mud flats of Florida Bay formed by drying of the flats during exposure at low tide. *(Courtesy R. N. Ginsburg; with permission of Society of Economic Paleontologists and Mineralogists.)*

terrigenous silicate clastic debris is supplied from it to the adjacent shelf. Environmental conditions off southern Florida favor carbonate sedimentation, and carbonate sands that contain considerable oölite are prominent on the shelf. At the south end of the peninsula, where nutrient-rich waters rise against the shelf edge, ideal growth conditions for organic reefs exist. Behind the Florida Keys reef zone, in the protected, less-than-3-meters-deep waters of Florida Bay to the north, very fine limy muds are forming.

While areas of respective facies differ, Gulf of Mexico sediments present good analogues for major early Paleozoic facies discussed for the craton. Quartz sands along the Alabama-Georgia coasts are texturally similar to those of the Upper Cambrian of the central craton, while

FIGURE 11.21
Flat-pebble conglomerate of fine dolomite chips formed by shrinkage cracking, then torn up by tidal currents or storm waves. Such conglomerates are especially common in Eocambrian, Cambrian, and Lower Ordovician strata (Upper Cambrian, southeastern Missouri).

Heterogeneous, incompletely
weathered debris from
varied bedrock and glacial till
under temperate climate

Debris from varied bedrock
but maturely weathered
(warm, humid climate)

Continental
Shelf Edge

BLAKE
PLATEAU
(no sedimentation)

Debris from varied bedrock
immaturely weathered
in arid climate

Deltaic

FLORIDA PENINSULA

Bank

BAHAMA ISLANDS

Bank

Sigsbee Deep

C U B A

YUCATAN

JAMAICA

C a r i b b e a n   S e a

Practically no debris
(carbonate rock and
humid, tropical climate)

| 0 | 200 | 400 km. |
| 0 | 100 | 200 | 300 mi. |

sand and silt

clay muds

carbonates (R = reefs)

Surface ocean currents

Relative sediment contributions

FIGURE 11.22

Recent sediments around southeastern United States for comparison with ancient epeiric sea deposits. Much of the sediment, particularly on the outer shelf, is relict from the last fall of sea level. On the average, sand on the outer shelf is moved by storm waves only about once every 5 to 10 years. Note the dominant surface-current movement parallel to the coasts. Note differences among the sources of clastic sediments in the north and the presence mostly of carbonate rocks in the south. (*Adapted from F. P. Shepard et al., 1960, Recent sediments of the northwest Gulf of Mexico; with permission of American Association of Petroleum Geologists.*)

FIGURE 11.23

Left: Matagorda barrier island, central Texas coast, immediately after Hurricane Carla, 1961. *Right:* The same coast a few weeks later. (Gulf of Mexico to left, Matagorda Bay to right in both views.) Hurricane-driven waves breached the island, spreading great volumes of sand both seaward and landward, but the breaches closed quickly, restoring equilibrium. *(Courtesy Shell Development Co.)*

carbonate sands and lime muds of the western Florida shelf are roughly similar to the Early Ordovician cratonic carbonate rocks. But, though the Ordovician rocks possess small algal reefs, ancient animal reefs like those of the present Keys area did not form on the craton until Silurian time. Likewise, nothing comparable to the Mississippi Delta sediments is known in Cambrian rocks, but ancient deltas are prominent in later Paleozoic strata.

On the outer Gulf shelf, clay-rich and slightly glauconitic sediments suggest analogies with finer silty and shaly glauconitic facies of Upper Cambrian strata. West of the Mississippi Delta, there are large sandy areas of "nondeposition" where new sediment is not being added and the sandy bottom is only influenced by major storms. Violent hurricanes profoundly influence sedimentation on the Gulf shelf and along the shoreline. Geologically speaking, hurricanes are common, and shorelines that "normally" are very stable are greatly modified within a brief span of time when a violent storm strikes (Fig. 11.23).

Do ancient sediments record largely average conditions or more violent, extreme conditions? As we have seen, vigorous current and wave agitation characterized much of the epeiric sea floor. Doubtless constant currents were present, but some sporadic and more violent processes such as storm waves also were important. For example, layers of conglomerate interstratified with normal quartz sandstone occur near former islands of Prepaleozoic rocks (Fig. 11.24), and flat-pebble conglomerates occur in carbonate rocks (Fig. 11.21). Each coarse layer probably records a single violent storm. It is likely that much, if not most, of the preserved epeiric sea record in such sediments was formed not so much by ordinary or average conditions, but by less-frequent, short-lived, violent events. Cambrian quartz sands may have been only slightly affected by normal currents, but would have suffered large-scale transport during brief, violent storms.

PALEOCLIMATOLOGY

HEAT BUDGET OF THE EARTH

Continuing our analysis of paleogeography, we next examine some basic principles for interpreting paleoclimate. The principles discussed here also apply to post-Cambrian time. The diagnosis of past climates is one of the most complex facets of earth study. Still, certain broad generalizations are warranted, and they

FIGURE 11.24

Thin layer of coarse conglomerate with well-rounded boulders between quartz sandstones. Boulders up to 1.5 meters in diameter were derived from nearby islands in the Cambrian epeiric sea that were composed of resistant Prepaleozoic quartzite. Tropical storm waves occasionally swept boulders away from sea cliffs to form many such local layers (Upper Cambrian near Baraboo, Wisconsin). *(By R. H. Dott, Jr.)*

serve at least as limits upon possible interpretations until such time as better guidelines appear. Climate represents a statistical average effect of the many meteorologic changes experienced by any given area, and those changes are related ultimately to atmospheric temperature and motion. Atmospheric motion is caused by differential heating and cooling and by the rotation of the earth. Landmasses, particularly if mountainous, obstruct and complicate atmospheric movements.

Average temperature of the earth depends upon many factors, including distance from the sun and tilt of the earth's rotational axis. Also important are the transparency of the atmosphere to both incoming and outgoing radiation as well as the area and height of landmasses. Under any configuration of the earth that we can conceive, there always would have been a significant temperature difference between the poles and equator. This results from the geometry of the curved surface of the earth and its atmosphere with respect to solar radiation (Fig. 11.25*A*). The *pole-equator temperature gradient* might have been increased or decreased by variable tilt of the axis, but it could not have been eliminated.

Sea water absorbs much solar radiation, becomes heated, and oceanic circulation distributes the heat over the earth. Land surfaces, however, retain negligible heat; most of the solar energy reaching them is radiated back to the sky (Fig. 11.26). Therefore, *the larger the total ocean surface, the warmer is the overall earth temperature*, and vice versa, which already was appreciated by Charles Lyell in 1830. Moreover, if the poles lay in open oceanic areas, then circulation of heated sea water from lower latitudes would tend to warm the polar areas. But, if poles lay within or near large landmasses, they would tend to be colder and the pole-equator temperature gradient would be accentuated (contrast Figs. WM. 1 and WM. 2).

CLIMATIC ZONES

The *Coriolis effect* resulting from the earth's spin causes air moving north or south to be deflected east or west by rotation; therefore its actual path describes an arc relative to the earth's surface. Near the equator the steady *trade winds* blow toward the equator because of the heating and rising of air there, but they are deflected westward (Fig. 11.25*B*). We assume that a trade

A. LATITUDINAL THERMAL GRADIENT

B. SURFACE WIND BELTS

FIGURE 11.25
Chief factors affecting atmospheric circulation. *A.* Solar radiation and the latitudinal thermal gradient of the earth's atmosphere due to varying angle of incidence of sun's rays (dark brown). *B:* Climatologists believe that the dominant zonal wind belts probably were present throughout geologic time (though variable in breadth and intensity). Spin of the earth causes deflection of winds from north-south paths; this deflection is known as the Coriolis effect.

FIGURE 11.26
Contrasting effects of solar radiation on land versus water, and the relative mixing of water layers. The shallow-water irradiated and agitated *photic zone* (above stable thermocline) is the habitat of algae and microscopic plankton, which form the base of the aquatic food chain.

winds belt existed on both sides of the equator throughout earth history. Midlatitude poleward-blowing *westerly wind* belts probably also existed throughout history, but they may have been subject to greater variability. In short, then, regardless of many unknown ancient climatic factors, almost certainly there has been some kind of latitudinal zonation of climate, but more distinct at some times than others.

With small, low continents, there would be little longitudinal (east-west) differentiation of climate, for *atmospheric circulation cells* could drift easily as the earth spun beneath the atmosphere. Such cells owe their existence to the combined effects of air pressure differences (in turn the result of temperature) and the Coriolis

FIGURE 11.27
Effect of a large continent on atmospheric circulation. Latitudinal zonal wind patterns are modified into major high- and low-pressure cells more or less locked into position by the land mass.

effect. But when large and mountainous continents existed, then the atmospheric circulation cells would become more fixed in place by these obstructions (Fig. 11.27). Year after year, seasonal high- and low-pressure cells would be fixed longitudinally as well as latitudinally as today. Under conditions of fairly large continents, we then could predict relatively drier areas on the west coasts of lands in subtropical latitudes and wetter areas on east coasts in such latitudes (see Fig. WM. 1). It would be dangerous to extend our ''predictions'' any further without a great deal of supporting evidence.

GEOLOGIC INDICATORS OF CLIMATE

Certain peculiar sediments and fossil organisms suggest some further constraints upon paleoclimatic interpretations. *Fossil plants* provide the best criteria for ancient land areas. Plant types can indicate hot versus cold, wet versus dry, and degree of seasonality of climate. Coal has formed mainly in warm to temperate humid swamps but some also could form in poorly drained, cold tundra regions. Specific plants present may help to establish paleotemperature conditions during formation of a particular coal. Large *organic reefs* form today only in the shallow, agitated warm seas found between approximately 30°N and 30°S latitudes. Assuming that reef-building organisms had similar ecologic requirements in the past, we can tentatively

infer ancient temperature and other conditions roughly analogous to those of present reef areas. Coupled with paleomagnetic evidence for past latitude, reefs can be of great paleogeographic value. Further discussion of this topic will be presented in Chap. 13. *Evaporite deposits* have climatic significance in that they indicate high evaporation potential, i.e., relative aridity. While this need not indicate excessively hot temperatures, the setting of most ancient evaporites is, in fact, consistent with warm temperatures. Proven ancient *glacial tills* and associated deposits (as already noted) provide evidence of unusual refrigeration of climate. Of these criteria, only a few apply to the Cambrian because no land plants existed yet, animals had not yet formed large reefs, and there was no glaciation. Evaporites are known in the subsurface in the eastern United States, however. In succeeding chapters, numerous examples of climatic interpretations will be cited and explored further as evidence arises, and we shall use all of these criteria to help infer ancient paleogeography.

SHALLOW-SEA CURRENTS

Having established the presence of a shallow epeiric sea characterized by persistent currents and periodically stirred by violent storms, we can now speculate about the relation of the craton to overall continental geography. By inductive reasoning from the present seas, we can say that circulation in a broad epeiric sea only a few hundred meters deep would be controlled primarily by winds blowing across the water surface and by tidal currents. Islands and shoals would, of course, deflect and modify circulation. The orientation of the bottom slope may have little influence in such a sea. Many currents in present shallow seas run parallel to the bottom contour (Fig. 11.22), and Cambrian epeiric sea currents roughly paralleled the transcontinental arch.

The Coriolis effect influences ocean water circulation as well as that of the atmosphere. It is known theoretically that water set in motion by wind does not move in exactly the same direction as the air. Due to the Coriolis effect, it is deflected, and, at successively greater depths, is deflected more and more. Obviously the relation may be very complex between surface currents and flow along the bottom, which affects sediments.

CAMBRIAN CLIMATIC ZONES

In shallow seas one might expect very erratic directions of currents to be indicated in the sediments, and indeed this tends to be so. Nonetheless, with reference to present directions, a south- to southwest-flowing average bottom-current direction is suggested throughout the central craton by cross-stratification orientations and the concentration of Cambrian shales at the southwest and west edges of the craton (Fig. 11.8). What sort of winds might produce this overall persistence of direction in the shallow sea? Trade winds, which today control the tropical surface circulation of the seas, could have produced the observed paleocurrent pattern had the craton then been within the low latitudes. Just such a position is, in fact, suggested independently by paleomagnetic studies (Fig 11.9) and evaporites in the eastern United States. All evidence points to North America having been a tropical continent rotated about 90° clockwise from its present orientation (see Figs 6.18 and WM. 8) in earliest Paleozoic time!

SUMMARY

Eocambrian time is designated as beginning about 700 million years ago after the Grenville orogeny. The periphery of the young continent was a broad continental shelf underlain by thick marine sediments. Eocambrian and Early Cambrian strata were confined to the periphery, but later Cambrian marine deposits spread across the craton. Fossils are rare in Eocambrian rocks, but as more are being found, it is appropriate to redefine the beginning of the Paleozoic Era to include the Eocambrian. It was during Eocambrian time that a few primitive invertebrate skeletons appeared, but Cambrian strata contain the first truly abundant and complex skeletons.

The continent was unusually stable. The craton, which has been thoroughly weathered and eroded for at least 200 million years, was gradually inundated by Cambrian transgression, forming a huge epeiric sea. This transgression probably resulted from a worldwide rise of sea level amounting to several hundred meters. Very pure quartz sand was redeposited widely on the sea floor as the shoreline shifted. Much of the sand had been moved ceaselessly over the landscape by wind before being reworked by the Late Cambrian inundation. Final deposition to form cross-stratified, pure quartz sandstones was accomplished largely by surf and shallow-marine currents, which apparently produced large areas of submarine dunes like those of the modern Georges Bank. Just before the advent of Ordovician time, carbonate sedimentation superseded that of quartz sand. Vast dolomite deposits characterized this accumulation in seas that could not have been more than 100- to 200-meters deep. Great tidal flats apparently existed at the margin of the sea. Current and storm-wave circulation dispersed sediments in the epeiric sea. In the central craton, prevailing transport by bottom currents was generally southerly and southwesterly, more or less parallel to the transcontinental arch.

Circulation in very shallow seas is mostly wind- and tide-controlled. Paleomagnetic evidence indicates that the Eocambrian-Cambrian equator lay across central North America in a nearly north-south direction (relative to present directions). This in turn suggests that the epeiric sea currents may have been driven by steady trade winds blowing in subequatorial zones that included most of the craton. Occasionally, however, storm winds affected sedimentation profoundly.

It has long been assumed that preserved sedimentary rocks record primarily normal or average conditions for past epochs. But in a region of very slow sedimentation, such as the Cambrian craton and outer continental shelf regions today, normal, day-to-day processes affect sediments very little. Only in the surf zone and on very shallow banks is sediment moved appreciably, except during rare, extremely powerful events such as hurricanes. Recognition of the importance of the so-called rare event requires an important shift of one's time perspective. Based upon ordinary human experience, what we may regard by probability as impossible becomes barely possible given millions of years of geologic time; what was regarded as possible becomes probable; and the probable, virtually certain. It follows that "catastrophic" storms judged rare by human standards must be regarded as common on the geologic time scale, and therefore may have been the principal agents to leave an imprint upon Cambrian sediments rather than more prosaic day-to-day processes.

"The history of any one part of the earth consists of long periods of boredom and short periods of terror." (D.V. Ager, 1973)

Key Concepts

arch versus basin
aulacogen
clastic limestone
climatic controls:
   atmospheric circulation cells
   Coriolis effect
   pole-equator temperature gradient
   ratio of land-to-sea surface
   trade winds
   westerly winds
cross stratification
epeiric sea
epeirogenic
flat-pebble conglomerate
oölites
paleoclimatic clues
   evaporites
   fossil land plants
   organic reefs
   tillites
photic zone
polygonal (mud) cracks
ripples versus dunes
Sauk Sequence
stratigraphic diagrams
   facies maps
   paleogeographic maps
   restored cross sections
   thickness maps
subsidence
terrigenous clastics
truncation (unconformity)
viscosity

Readings

Dott, R. H., Jr., 1974, Cambrian tropical storm waves in Wisconsin: Geology, v. 2, pp. 243–246.

Frakes, L. A., 1979, Climates throughout geologic time: New York, Elsevier.

Gretener, P. E., 1967, Significance of the rare event in geology: Bulletin of the American Association of Petroleum Geologists, v. 51, pp. 2197–2206.

Harland, W. B., and Rudwick, M. J. S., 1964, The great infra-Cambrian ice age: Scientific American, v. 225, pp. 28–36.

Hayes, M. O., 1967, Hurricanes as geologic agents: Bureau of Economic Geology, The University of Texas, Report of Inv. No. 61.

Holland, C. H. (ed.), 1971, Cambrian of the New World: London, Wiley-Interscience.

———— (ed.), 1974, Cambrian of the British Isles, Norden, and Spitsbergen: New York, Wiley.

Klein, G. deVries, 1970, Depositional and dispersal dynamics of intertidal sand bars: Journal of Sedimentary Petrology, v. 40, pp. 1095–1127.

Kuenen, Ph. H., 1960, Sand: Scientific American, April. v. 221.

Laporte, L., 1968, Ancient environments: Englewood Cliffs, Prentice-Hall. (Paperback)

McKee, E. D., 1945, Cambrian history of the Grand Canyon region: Washington, Carnegie Institution Publication 563.

Middlemiss, F. A., Rawson, P. F., and Newall, G. (eds.), 1971, Faunal provinces in space and time: Liverpool, Geological Journal Special Issue No. 4.

Raasch, G. O. (ed.), 1961, Geology of the Arctic: Toronto, Univ. of Toronto Press.

Rocky Mountain Association of Geologists, 1972, Geologic atlas of the Rocky Mountain region: Denver, Hirschfeld Press.

Schwarzbach, M., 1963, Climates of the past: London, Van Nostrand.

Shepard, F. P., Phleger, F. B., and Van Andel, T. H., 1960, Recent sediments of the northwest Gulf of Mexico: Tulsa, American Association of Petroleum Geologists.

Sloss, L. L., Dapples, E. C., and Krumbein, W. C., 1960, Lithofacies maps: an atlas of the United States and southern Canada: New York, Wiley.

Stearn, C. W., Carroll, R. L., and Clark, T. H., 1979, Geological evolution of North America (3rd ed.): New York, Wiley.

Woodford, A. O., 1965, Historical geology: San Francisco, Freeman.

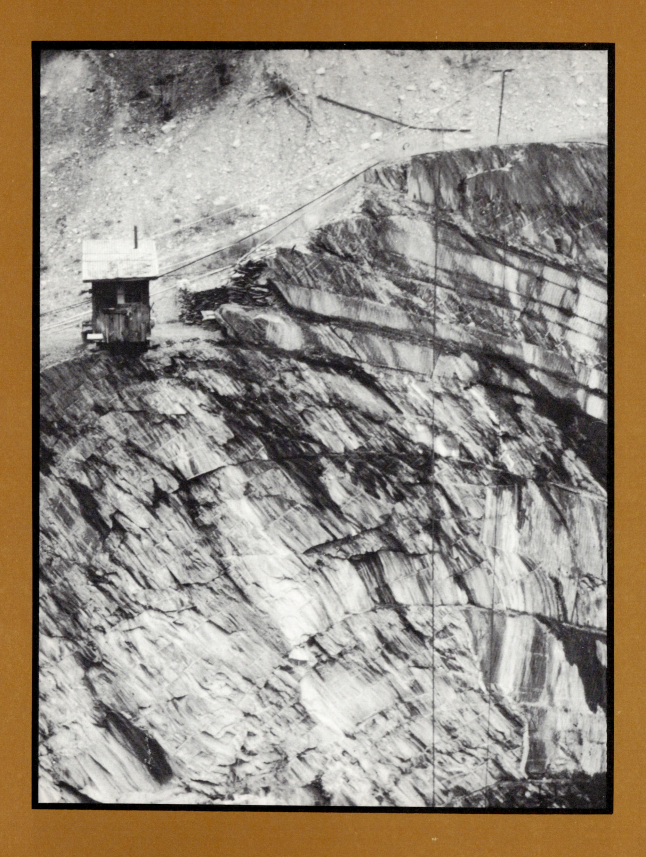

12

THE LATER ORDOVICIAN
Further Studies of Tectonics and the Paleogeography of Mobile Belts

We crack the rocks
and make them ring,
And many a heavy pack we sling;
We run our lines and tie them in,
We measure strata thick and thin,
And Sunday work is never sin,
By thought and dint of hammering.

Andrew C. Lawson

formerly of the Geological Survey of Canada and the
University of California

FIGURE 12.1
Anticline in black Ordovician slates (Martinsburg Formation), exposed in an an eastern Pennsylvania slate quarry. Slaty cleavage formed during the Taconian orogeny is inclined steeply downward to the right, parallel to axial plane of the fold but discordant with convex-upward stratification. Such structures typify mobile belts. (*Courtesy I. W. D. Dalziel.*)

12

In Chap. 11 we discussed briefly Early Ordovician cratonic sedimentation. Recall that diminution of the quartz-sand supply near the end of the Cambrian Period allowed carbonate deposition which continued into Ordovician time. In this chapter we consider the history from about 475 to 435 million years ago. Apparently a worldwide fall of sea level caused retreat of the sea completely from the craton for a few million years. Then transgression soon produced one of the largest epeiric seas ever. A cratonwide unconformity marking this important regressive-transgressive event defines the base of the widespread *Tippecanoe Sequence*, the second such major assemblage of Paleozoic strata on the craton (See Fig. 3.21).

The Ordovician record on the continental margins is clearer than that of the Cambrian. Volcanism was important both in the western (Cordilleran) and eastern (Appalachian) regions. In central Nevada, vast volumes of ellipsoidal lavas (Fig. 12.2) associated with stratified chert deposits occur within thick sequences of black mudstone (now slate) that contain many graptolites. Coarse clastic detritus is relatively rare, suggesting accumulation in still, probably fairly deep water far from land, as on a continental slope or deep ocean floor beyond the shelf edge.

Ordovician graptolitic slates and volcanic rocks also occur in the Appalachian belt, but are accompanied there by coarse clastic terrigenous material. Included are dark graywackes, local conglomerates, and red sandstones and shales, all reflecting elevation of a large land (or islands) in the mobile belt. Faunas in the Appalachian region continued to resemble strongly those of northwestern Europe and eastern Greenland. In Late Cambrian and Ordovician time, mountain-building disturbances and volcanism also affected the Caledonian mobile belt in Britain, underscoring its tectonic similarity with the restless Appalachian belt. The subduction of a now-vanished proto-Atlantic lithosphere plate beneath mobile belts on both sides of the Ordovician ocean basin is thought to have caused the orogenies and accompanying volcanic activity. In this chapter we examine especially the effects of orogenies and the evolution of mobile belts both because their records are clear in the Ordovician of the Appalachian region and also because that region was the cradle of the original geosynclinal concept.

FIGURE 12.2
Lower Paleozoic ellipsoidal lavas from Toquima Range, central Nevada. These lavas and associated chert (part of an *ophiolite sequence*) formed on the sea floor in what probably was deep water beyond the continental shelf margin. *(Courtesy M. Kay.)*

ORDOVICIAN OF THE CRATON

MEDIAL ORDOVICIAN REGRESSION AND THE ST. PETER SANDSTONE

After extensive dolomite deposition for several million years, a widespread unconformity was produced over virtually the entire craton. The sea retreated at least to the margins of the craton, and the interior was subjected to extensive erosion. Above the unconformity lies a very widespread pure quartz sandstone called St. Peter. Though the sea had retreated from the craton, the topography was very subdued everywhere. Maximum relief observed in erosional channels is about 100 meters. Remember that only a small relative vertical fall of sea level (or rise of the continental platform) was needed to cause exposure of a tremendous area of the former shallow sea floor. As soon as exposure occurred, weathering, solution, and erosion began to remove earlier Ordovician dolomites. In the center of the craton, all dolomite was removed to reexpose Upper Cambrian quartz sandstones (Fig. 12.3). These were reworked and the sand again was spread out (for perhaps the hundredth time since the Middle Prepaleozoic) from the cratonic core to form a veneer over much of the craton. The St. Peter is virtually identical with mature Cambrian sandstones (see Fig. 11.13). This enormously large sheet of sand totals more than 20,000 cubic kilometers!

Stream and wind dispersal distributed the sand, but it was finally redeposited mostly by the sea as transgression was renewed during the middle part of the Ordovician Period. Much of the preserved sandstone apparently represents shore and nearshore deposits formed at the advancing edge of the sea. Paleocurrent patterns were more complex than in the Late Cambrian sea, but geographic variations of grain size, and lateral facies relationships with other sediments of similar age, all point to derivation of the sand largely from the center of the craton with transport outward. Pure quartz sand drifted southwest into Oklahoma, and even as far west as central Idaho and Nevada. In the latter region, it is interstratified in an unusual association with black, graptolitic mudstones and lavas, which are the typical early Paleozoic rocks of the westernmost margin of the continent.

It can be shown that the sandstone is older near its outer limits than in the center of the craton (Fig. 12.4). It is a classic transgressive deposit whose deposition, as an educated guess, spanned about 5 million years. As noted in Chap. 11, much of the silicate sediment on outer shelf surfaces today is relict sand and mud originally deposited during Pleistocene lowerings of sea level and but slightly modified by currents and storm waves since. On carbonate-producing shelf areas, as off southern Florida, however, new carbonate material is being added more or less constantly as many generations of organisms contribute their skeletons, and as oölites are formed.

How can we reconcile the limited active clastic sedimentation on modern shelves with the presence of many ancient, very widespread quartz sandstones that appear to have formed in similar environments? Several unconformities present in the Upper Cambrian and Lower Ordovician successions of the craton reflect minor transgressions and regressions. Assuming that most of the ancient sand was deposited in proximity to the shore zone, then great lateral shifts of the shoreline must have gradually produced widespread transgressive and regressive sand deposits. Any sandstone formation so deposited must vary in age across the craton as shown in Fig. 12.4. On the shallowest bank areas, which probably were very large in the past, shifting of sand relict from earlier stages of transgression probably also occurred as on sand banks today.

STRATIGRAPHIC SEQUENCE DIVISIONS

In Chap. 3, we referred to regional rock units called *sequences*, which are assemblages or "packages" of many formations bounded by exceptionally widespread unconformities. The boundaries vary in age from place to place so that the sequences are not universal time divi-

PRE-MIDDLE ORDOVICIAN

(Pre-St. Peter) PALEOGEOLOGY

Present Edge of
Middle Ordovician
("St. Peter")

O   Lower
     Ordovician

€   Cambrian

Prepaleozoic

Rocks
unconformably
beneath
Middle Ordovician

1000 Miles

1000 Kilometers

FIGURE 12.4

Left: Diagram of mid-Ordovician strata in the southern craton showing northward decrease in age of transgressive quartz sandstone facies (St. Peter). *Right:* Integration of countless successive shoreline deposits formed during transgression to produce a widespread, tabular mass of sandstone. (Vertical scale is exaggerated.)

sions with strict age significance. Strata adjoining the bounding unconformities may vary in age several or even tens of millions of years. The unconformity beneath Cambrian rocks across North America and that beneath the St. Peter Sandstone together define the oldest sequence recognized, which has been named Sauk. The St. Peter Sandstone is the basal, transgressive deposit of the second (Tippecanoe) sequence. A total of six unconformity-bounded sequences are widely accepted for North America (Fig. 12.5).

The sequence concept applies chiefly to cratonic strata, but increasingly we see evidence that the changes reflected by the great unconformities (and accompanying transgressions and regressions) are probably related to lithosphere plate motions and so also to tectonic events in mobile belts at the plate margins. Cratons seem to have had a tendency to be more emergent when plate activity (spreading

and subduction) was modest, as in Eocambrian time, whereas they seem to have been widely submerged when plate activity was greater, as in the later Ordovician. It has been suggested that this may be due to more heat being retained beneath continents and causing isostatic rise in

FIGURE 12.5

Six major unconformity-bounded *sequences* designated by L. L. Sloss for the North American craton. Dark areas represent major unconformities, that is, space-time without any rock record present today. These major unconformities serve to subdivide the cratonic stratigraphic record into widespread "packages" of strata. Note that the more complete depositional records are at the craton margins. The St. Peter Sandstone lies at the base of the Tippecanoe Sequence. Compare Fig. 3.21. (*After Sloss, 1963,* Bulletin of the Geological Society of America, *v. 74, pp. 93–114.*)

FIGURE 12.3

Paleogeologic map of the medial Ordovician regional unconformity (approximately 470 to 475 m.y. old). Transgressive sandstones (St. Peter) were deposited upon this now-buried geology. The map is what one would see if all younger strata overlying the unconformity could be lifted off. Such a map helps reveal the history of large features such as the transcontinental arch (see Fig. 11.8) and provides clues to paleogeography. Note that Ordovician strata overlapped beyond Cambrian ones to rest widely upon Prepaleozoic basement in central Canada and on the transcontinental arch (proving the arch had formed *at least* this early). Evidence for preparing the map comes only from areas within the zero-thickness boundary of overlying strata. Subjective extrapolations must be made beyond that boundary where such strata are no longer present. (*After Levorsen, 1960,* Paleogeologic Maps.)

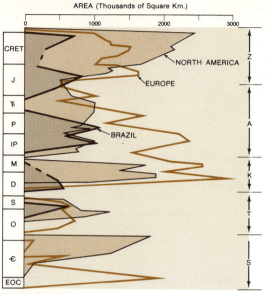

AREA (Thousands of Square Km.)

FIGURE 12.6

Comparison of stratigraphic records for three separate
cratons expressed as area of preservation. Overall similarity of curves suggests worldwide sea-level changes as a
major cause of transgressive and regressive events evidenced here. Of course local structural warping also affected the individual patterns, especially after Ordovician
time. North American *sequence* names are abbreviated at
right (see Fig. 12.5). (*Adapted from L. L. Sloss, 1976, Geology, v. 4, pp. 272–276 and P. C. Seares, et. al., 1978, Bulletin
of Geological Society of America.*)

the former case, and the opposite when plates
were more active. Another factor affecting sea
level may be the expansion and rise (or shrinkage and fall) of ocean ridges to raise (or drop)
sea level worldwide. But any plate tectonic
explanation of cratonic changes implies a worldwide effect. Can this be proved? Figure 12.6
suggests that, indeed, different cratons were
affected more or less simultaneously by sea-
level changes.

LATER ORDOVICIAN EPEIRIC SEA

During the regression of the sea discussed
above, mutations and natural selection combined to produce marked changes in the Ordovician shallow-marine fauna, which was temporarily restricted to a much-reduced shallow-
marine area on the borders of the continent.

Surely this shrinkage of environment produced
unusual selective pressures on the population,
tending to accelerate evolutionary diversification. Certainly, the rich later Ordovician fauna,
which invaded the craton as the epeiric sea
readvanced, was markedly different from the
early Ordovician one. It contained the first
vertebrate animals as well as the more diversified corals, bryozoa, stromatoporoids, cephalopods, and new groups of brachiopods discussed in Chap. 9.

Upper Ordovician limestones were somewhat
less extensively altered to dolomite than were
the lower ones, and so their faunas are better
preserved (origin of dolomite is discussed in
Chap. 13). Many are 80 or 90 percent shell
material, so constitute a shelly facies. The
Upper Ordovician strata of the Ohio-Indiana
region are among the most fossiliferous rocks in
the world, and they were the inspiration and
training ground for a "who's who" of outstanding American geologists. Upper Ordovician
fossil-rich carbonate strata also are the most
extensive marine deposits on the entire craton; a
cratonic basin near Hudson Bay contains a large
volume of them and smaller patches occur
elsewhere on the shield (see Fig. 12.7). The Late
Ordovician epeiric sea represented one of the
most complete floods experienced by any continent (12.8). A tremendous, uniform shallow sea
resulted, the like of which does not exist today.

SHALY DEPOSITS OF
LATER ORDOVICIAN TIME

In the Appalachian region, much of the later
Ordovician sequence is characterized by dark
mudstones with sporadic graywacke sandstones and chert layers. The top of the Ordovi-

FIGURE 12.7

Upper Ordovician sediment patterns for North America.
Widely scattered patches of sediments on the Canadian
Shield prove the great extent of the Late Ordovician sea.
Absence of Ordovician strata on several arches proves
subsequent further warping and erosion of them. Note
spread of red beds and marine shales westward from the
Appalachian region, forming a *clastic wedge* (see Fig. 11.7
or endpapers for symbols and sources).

UPPER ORDOVICIAN

SEDIMENTARY FACIES

(Cincinnatian)

(about 430 m.y. ago)

PEACE RIVER ARCH

CONTINENTAL ARCH

TRANS-

Is boulders

Edge of Craton

1000 Miles

1000 Kilometers

LATE

ORDOVICIAN

PALEOGEOGRAPHY

(about 430 m.y. ago)

cian sequence farther west also contains thin but persistent dark gray shale (Fig. 12.9). This shale is something of an oddity for the craton and must record some important change of terrigenous source areas. To determine its source, we must examine facies maps and cross sections (Figs. 12.7 and 12.10). The shale becomes thicker toward the east where practically the entire Ordovician sequence in the northern Appalachian region is composed of dark shales. Still farther east, important volcanic rocks occur in the Ordovician succession. Both lava and ash rocks are found, which suggest that important volcanic islands were formed here. In summary, the facies patterns point to the elevation of islands above sea level in the Appalachian belt, which shed sand and much fine mud that was carried westward. Most of this sediment was

deposited within the mobile belt itself, but as the volume increased, mud literally spilled out of the mobile belt onto the edge of the craton in New York. Near the end of the period, clay and volcanic ash were carried as far west as Iowa (Fig. 12.7). The distribution of the shale facies suggests southwesterly flowing currents; ripple marks and other features confirm this. By latest Ordovician time, the continent appeared as shown in Fig. 12.8.

VOLCANIC ASH—A CORRELATION TOOL

In the Appalachian Mountain region there is a major lateral facies gradation from dominantly *graptolitic shale facies* southwestward to a *shelly limestone facies* more characteristic of the craton (Fig. 12.7). Because of the differences of environment reflected by these facies, there is also a difference of fauna in each. In turn, the differences between their fossil assemblages make it difficult to correlate between the facies on the basis of index fossils; this illustrates the limitation of "facies fossils" discussed in Chap. 3.

Volcanic ash layers are found in the Ordovician strata of the Appalachian belt. They are

FIGURE 12.9
Ordovician black shales and thin white limestones (Utica Formation) along the New York Thruway, central New York. In this area, mud derived from the east spread beyond the Appalachian belt onto the craton. Differential flow under the influence of gravity on water-laden mud caused contortions.

FIGURE 12.10

Restored cross section showing relations of Ordovician facies to the Taconian land. Note the volcanic ash layers that transcend facies boundaries to provide useful time datums for correlation purposes. Note mid-Ordovician unconformity and Upper Ordovician clastic wedge in Pennsylvania. (*Adapted from M. Kay, 1951,* Geological Society of America Memoir 48.)

represented today by thin, clay-rich layers formed by the alteration of volcanic dust particles. Because each layer was erupted instantaneously in a geologic sense, individual ash strata provide time datum markers. Certain of the ash layers extend into both shaly and carbonate facies, and so they constitute unique datums for correlation between the two facies (Fig. 12.10). Also, those which have not been entirely altered can be used for isotopic dating to provide a numerical date for a sequence of strata much as isotopic dating of glauconite provides for Upper Cambrian strata of the craton.

PALEOWIND AND PALEOCLIMATE

Perhaps of even greater interest is the clue that volcanic ash distribution provides about the possible Ordovician wind pattern in eastern North America. This distribution suggests that the ash was derived from known volcanic centers of the northeastern Appalachian mobile belt (Fig 12.7). As in Cambrian time, apparently winds blew from northeast to southwest (in terms of present geographic coordinates; Fig. 12.8), which is the opposite of present winds. Marine currents would also influence ash distribution after it settled into the sea.

In spite of uncertainties, it is interesting to note the relation of the ash distribution to the possible position of the Ordovician equator with respect to North America as indicated by paleomagnetic data. In Fig. 12.8 you will note that the apparent equator extended from northeast to southwest—nearly bisecting the continent. This would place the Appalachian mobile belt within the trade wind zone and the ash distribution would be entirely consistent with such an orientation. All of the richly fossiliferous Ordovician carbonate deposits of the craton would fall within 40 degrees latitude of the equator, whereas evaporites and reefs in present Arctic Canada would lie within 20° of the equator. But if the equator then occupied its present position with respect to the continent, the carbonate rocks would span 70 degrees of latitude. As will be shown more fully in the next chapter, such sediments today form chiefly in low, warm latitudes, and so it would be less surprising to find them at 40°N than at 70°N.

ORDOVICIAN OROGENY IN THE APPALACHIAN MOBILE BELT

EVIDENCES OF INCREASING STRUCTURAL MOBILITY

Volcanic rocks of the mobile belts prove increasing structural mobility, and abundant dark muds deposited widely in the Appalachian belt and far out onto the craton also reflect unrest and uplift in the east. In the southern Appalachians, local red shales appeared in mid-Ordovician time, reflecting the elevation of land. In the northern Appalachian region, some spectacular, ill-sorted conglomerates with fragments from mixed Cambrian and Early Ordovician formations occur helter-skelter as local deposits within Ordovician black shale sequences; some of the blocks are many meters long (Fig. 12.11). These unusual deposits occur at several scattered localities. They represent submarine avalanches of debris derived from shallower water environments. The deposits became unstable and slid and rolled into the deeper, adjacent muddy environment. Such bouldery mudstones may reflect disturbances of the sea floor by earthquakes heralding much more profound unrest destined to produce major mountain-building episodes later in the period.

In the eastern United States and southeastern Canada, there is diverse evidence pointing to Ordovician mountain building. Ordovician black shale is succeeded eastward by red shale. The red bed facies coarsens eastward in Pennsylvania to sandstone and some conglomerate (Figs. 12.7 and 12.10). Such an assemblage is called a *clastic wedge* because of its shape in cross section (Fig. 12.10). Clastic wedges are associated with orogenies. The red deposits reflect a major change in depositional conditions. Because they lack fossils and show thorough oxidation of iron, it was suggested long ago that they were nonmarine deposits that bordered a large land somewhere to the east.

DIRECT EVIDENCE OF MOUNTAIN BUILDING

At a number of localities in the Appalachian region, important unconformities are visible within the Ordovician sequence and between Ordovician and Silurian or Devonian strata (Fig. 12.12). Extending from the St. Lawrence River valley south into eastern New York is a zone of major overthrust faulting that originated about the end of the Ordovician, or about the same time that the unconformities were developing. Farther east in New England and Maritime Canada, some isolated small granite masses are overlain by Silurian strata and were intruded into older, probable Ordovician ones. Isotopic dates confirm that some granites were formed there during Late Ordovician and Early Silurian orogenesis. Long, narrow ultramafic igneous rock masses were emplaced along the mobile belt (Fig. 12.12). You will recall (Chap. 5) that these unusual rocks represent mantle material faulted into the crust during severe deformation. Finally, Ordovician sandstones and conglomerates are heterogeneous in composition and show derivation from erosion of limestones, sandstones, shales, volcanic, and rare granitic rocks.

Evidence cited above points to a major interval of mountainous uplift, erosion, and intru-

FIGURE 12.11

Coarse conglomerate of Cambrian carbonate rock fragments interstratified with Ordovician shales, Trois Pistoles, St. Lawrence River, Quebec. These apparently originated by submarine sliding (see Fig. 12.19). *(Courtesy Jean Lajoie.)*

FIGURE 12.12

Summary of evidences of the Late Ordovician Taconian orogeny.

sion of some igneous plutons during the Ordovician Period. This event represents the first great Paleozoic mountain building in the Appalachian belt, and is called the *Taconian orogeny* for the Taconic Mountains of southeastern New York, where a major unconformity is exposed and where thrust faulting occurred during this upheaval.

PALEOGEOGRAPHY OF MOBILE BELTS AND THE ORIGIN OF GEOSYNCLINAL SEDIMENTS

EARLY IDEAS

Reconstructions of ancient geography of regions from their stratigraphic records is one of the highest goals of the earth historian. We have already demonstrated the reconstruction of cratonic geography for the Cambrian and Ordovician Periods. Now that we have examined some evidence from the Ordovician of the Appalachian region, discussion of paleogeography is appropriate for the mobile belts. The concepts developed in this section will apply as well to other cases.

The American J. D. Dana, one of the first to speculate about paleogeography of ancient mobile belts a century ago, believed that more or less permanent ridges of old Prepaleozoic igneous and metamorphic rocks lay at the borders of North America *throughout most of geologic time* (Fig. 12.13). This conclusion was based upon two principal observations: first, that many clastic sediments in the Appalachian geosyncline (such as the Upper Ordovician red beds) coarsen eastward; and second, to the east of those strata today lie predominantly granitic, gneissic, and schistose rocks considered a century ago to be entirely of Prepaleozoic age. The "Prepaleozoic-looking" rocks, which are particularly prominent in New England where Dana lived, naturally were assumed to be the source of the clastic sediments that had been derived from that direction.

Another American (C. D. Walcott) amplified Dana's concept in 1891 by recognizing the importance of two distinct sources of clastic geosynclinal material—the craton as well as a supposed Prepaleozoic borderland beyond the geosyncline on the other side (Fig. 12.13). He

FIGURE 12.13

The hypothesis of Prepaleozoic borderlands as the sources of most geosynclinal sediments. Since 1940, this model, which calls for large highlands *outside* geosynclines, has been replaced by a model of lands raised *within* geosynclinal belts (see Fig. 12.15).

suggested that the present Atlantic coastal plain once had been the site of a *borderland* from which much of the Paleozoic sediments of the adjacent, subsiding Appalachian geosyncline were derived. This concept of persistent borderlands of Prepaleozoic crystalline rocks, which is seen clearly in the old paleogeographic map of Fig. 11.1, was extended to all other North American geosynclines early in our century. Borderlands were assumed to have existed through millions of years along all margins of the continent, but when adjacent geosynclinal tracts ceased to subside, the borderlands disappeared beneath the marginal seas like mythical Atlantis. While important clastic sediments in mobile belts were derived from the craton, the derivation of clastic material from outside the craton (i.e., from the opposite direction) ultimately was of greater volumetric importance.

Inference that the marginal or extracratonic lands were comprised of Prepaleozoic complexes was a natural consequence of the long-held fallacy of age correlation of rocks solely on the basis of metamorphism and deformation. In 1930, Harvard University geologists discovered, in New Hampshire, Silurian and Early Devonian brachiopods in high-grade mica schists long assumed to be Prepaleozoic! So the originally designated borderland of "Prepaleozoic" rocks actually was in large part composed of much younger ones. But not only were the supposed borderlands composed mostly of Paleozoic and younger rocks, they could not have been permanent lands during Paleozoic and Mesozoic time, for the fossils found included forms known to have inhabited only marine environments. Intermittently, seas must have occupied almost the entire "borderland" area as well as the adjacent "geosyncline." Clearly, "borderlands" were neither permanent nor composed wholly of Prepaleozoic rocks. What, then, was the nature of the extracratonic lands?

MODERN CONCEPTS OF BORDERLANDS

At the same time that fossils were being discovered in the so-called borderland areas, a German, Hans Stille, and Americans Marshall Kay and A. J. Eardley were drawing the distinction between magmatic or *igneous-bearing* (eugeosynclinal) and nonmagmatic or *non-igneous-bearing* (miogeosynclinal) *subdivisions* of geosynclines. The important observation that ancient mobile belts contain a distinct zone of volcanic and intrusive igneous rocks of diverse ages became the key to a reinterpretation of borderlands.

It is important to realize that the original conception of a geosyncline by Hall, Dana, and Walcott included only the nonmagmatic (miogeosynclinal) portion of the expanded geosynclinal conception of Stille and Kay. *The original Appalachian example included only half of the mobile belt, and the less-disturbed half at that!* Synthesis and comparison of the characteristics of the entire Appalachian belt with other examples had led Kay and others to reject the old borderland concept in favor of a series of temporary islands periodically raised within the "magmatic" portion of the belts themselves. Details of stratigraphy in the latter zone show, besides abundant volcanic outpourings, many angular unconformities, coarse conglomerates, varying ages of igneous plutons, and metamorphism, as we have seen in our discussion of the Taconian orogeny (Fig. 12.12). In short, there is ample evidence of the persistence of great structural mobility throughout much of their histories. The rocks, moreover, attest to formation of both volcanic and nonvolcanic islands intermittently within the mobile belt itself. Some old mobile belts even extend seaward into still-active volcanic arc-trench systems (Fig. 12.14). Clastic detritus deposited in the so-called geosynclines has been derived from such islands, not from borderlands wholly composed of Prepaleozoic rocks and lying outside the belt. The rocks exposed to erosion in the islands were mostly slightly older geosynclinal deposits augmented periodically by additions of igneous extrusions and intrusions.

SOURCE FOR THE SEDIMENTS IN MOBILE BELTS

Having examined some of the stratigraphic record for Prepaleozoic and early Paleozoic mobile belts, we can now formulate generalizations about the sources of sediments found

FIGURE 12.14
Continuity of ancient mobile belt with a modern volcanic island arc-submarine trench system, which, together with stratigraphic evidence, has led to the paleogeographic comparison of many such ancient belts with modern arcs.

therein. Prior to onset of structural mobility in mid-Ordovician time, the Appalachian region was a broad, tranquil continental shelf receiving clastic sediments only from the *cratonic source* (Fig. 12.15). Mature quartz sand continued to enter from the craton when the region became a mobile belt in later Ordovician time, but it soon became subordinated to less mature sands from other sources (Fig. 12.16).

The *volcanic-source* type (Fig. 12.15) is exemplified by some of the Ordovician of the eastern Appalachian belt. As indicated earlier, during mid-Ordovician time, extensive volcanism occurred there, and volcanic islands were formed

as evidenced by volcanic-rich clastic sediments associated with lavas and by the fine volcanic ash blown far to the south and west. Apparently structural mobility had accelerated sufficiently so that volcanic accumulation exceeded subsidence and built islands similar to those of modern volcanic island arcs.

A *tectonic land* source (Fig. 12.15), which was emphasized particularly by Marshall Kay, is exemplified by Upper Ordovician and Lower Silurian strata of the Appalachian region. As shown above in Figs. 12.7 and 12.10, these sediments coarsen toward the east, and the uppermost Ordovician red beds apparently represent deposition on the western margin of such a large land. Microscopic study of Ordovician sandstones shows that they belong to feldspathic and lithic graywacke clans, for they contain quartz, feldspar, and fragments of older sedimentary and minor metamorphic rocks (Fig. 12.16); volcanic detritus is subordinate. Their composition and great volume indicate that a large land comprised chiefly of older, partly metamorphosed sedimentary and some igneous rocks was the principal source of these sediments. Such a land was raised structurally during the Taconian orogeny, and that tectonic land was the dominant source of Ordovician terrigenous clastic sediments in eastern North America. How large might it have been? Considering that Upper Ordovician clastic sediments of the western Appalachian belt comprise approximately 100,000 cubic kilometers and

FIGURE 12.15
Three major source types for sediments deposited within subsiding parts of mobile belts. Composition of ancient sediments indicates the influences of each type (or combinations) for different past times.

FIGURE 12.16
Microscopic photograph of a typical immature Ordovician (Normanskill) graywacke from eastern New York. Grains include quartz (white), feldspar (gray), and varied sedimentary rock fragments (dark) derived from erosion of earlier Paleozoic strata in the ancient Taconian tectonic land to the east. Note the very poor sorting and rounding (grains average about 2 millimeters in diameter).

assuming symmetrical deposition of sediment on both sides, erosion of a land approximately 1,000 kilometers long by 50 kilometers wide, and an average of 2 kilometers high, would be adequate. Of course the land was elevated as it was eroded, thus it need not have been as high at any one time as these dimensions suggest. But comparison with some of the modern island arc-trench systems shows that such dimensions are by no means excessive. Apparently we can account for the great volume of Ordovician geosynclinal sediments by envisioning a dynamic mobile belt that was elevated and eroded in one part, while the resulting sediment was deposited in another, subsiding part. In this manner, many mobile belts have been filled to form so-called geosynclines.

SEA LEVEL, SEDIMENTATION, AND SUBSIDENCE

In Chap. 6 we noted that American geologists assumed from the outset that all geosynclinal sediments were deposited either in shallow-marine water or on land just above sea level. This view was based upon the strata of the western nonvolcanic portion of the Appalachian Mountain belt, where many rocks contain algal stromatolites and other marine fossils, whereas others are clearly river-delta deposits. Meanwhile, European geologists were arguing that typical geosynclinal sediments were deep-marine in origin—perhaps even trench deposits. It is now clear that many different environments have existed in mobile belts. As shown in Fig. 12.17, the relation of sedimentation to sea level is the most critical factor.

Very thick sediments, whether beneath continental shelves, in mobile belts, or in cratonic

FIGURE 12.17
Models of different possible sedimentary environments within mobile belts. The difference is relative position of sedimentary and volcanic accumulations (black) with respect to sea level, which determines degree of agitation by waves and currents as reflected in sediment textures. Relative size and height of land and climate also influence the types of sediments, especially determining whether carbonate rocks will form.

WHOLLY SUBMARINE   CHIEFLY DEEP MARINE   CHIEFLY SHALLOW MARINE   CHIEFLY TERRESTRIAL   SEA LEVEL

ACCUMULATION OF THICK SHALLOW MARINE
AND ALLUVIAL PLAIN SEDIMENTS

FIGURE 12.18
Regional crustal subsidence produced by lateral as well as vertical accumulation of shallow marine and nonmarine alluvial plain sediments. Through sedimentation over a large area, the crust is warped down a limited amount; the locus of active accumulation constantly shifts, spreading the load over an increasingly wide region and producing further subsidence. This appears to be what happened along the eastern margin of the craton during Late Ordovician time and on the Gulf of Mexico shelf more recently.

basins, imply subsidence of the crust during deposition in order to provide room for them to accumulate vertically. Only if deposition began in deep water, as the Europeans believed, could considerable accumulation occur without any subsidence. If the sediments are all shallow-water types, then subsidence must just balance accumulation to maintain a constant shallow environment. This is what many early Americans believed happened in geosynclines. But what causes the subsidence?

James Hall of New York argued over a century ago that a large mass of sediments must depress the crust in accord with Archimedes' principle (see Chap. 6). We know that ice caps 3,000 meters thick have depressed the crust 1,000 meters. But Hall neglected the fact that, because

sediments like ice are less dense than the solid earth beneath, they have to pile up on the crust to a great thickness to cause only modest *crustal subsidence*. Unlike ice, however, sediments cannot accumulate much above sea level without being spread laterally by wind, rivers, or waves. Apparently shallow-marine sediments cause subsidence only when large volumes are spread over broad areas, for example on the continental shelf around the Mississippi Delta. Because sea level inhibits vertical accumulation, deposition must spread laterally, and over time it surely does depress the crust isostatically (Fig. 12.18). The early Paleozoic continental shelf of eastern North America apparently behaved in this fashion.

Where deposition begins in deep water, a great thickness can accumulate, and it will depress the crust significantly. But once sea level is reached, currents and surf will spread the sediments laterally. If the rate of sedimentation continues to exceed subsidence, then deltas will build outward, literally crowding the sea away to produce regression. Onshore, rivers will build a broad alluvial plain slightly above sea level. Thus we might expect to find vertical transitions from deep to shallow-marine and then to nonmarine sediments as "holes" were filled. In fact, exactly such sequences are common in clastic wedges in mobile belts; for example, the Ordovician of the Appalachian belt was capped by the red beds deposited at the height of the Taconian orogeny.

While *isostatic sediment loading* must cause some subsidence, it is not the primary factor in the profound subsidence evidenced by very thick sediments deposited in active mobile belts. Other possible causes include phase changes that thin the crust or lithosphere (see Chap. 6) and forcible downbuckling. A century ago, Dana, you may recall, invoked shrinkage of a cooling earth to cause downbuckling beneath geosynclines, but there is no evidence that the earth is contracting. Later, *thermal convection currents* in the mantle were thought to buckle down the overlying crust. Still more recently, *subduction* of one lithosphere plate beneath another (probably also driven by convection) has been postulated to explain most of the subsidence evidenced in mobile belts.

FIGURE 12.19
Restoration of eastern North America showing evolution from a passive to an active continental margin culminating in arc collision, which caused the Taconian orogeny. As the continental margin was downwarped in Late Ordovician time, carbonate sedimentation gave way to deeper-water *flysch* deposition by turbidity currents. Finally, Taconian upheaval resulted in westward spreading of nonmarine red beds *(Queenston molasse). (Adapted from J. F. Dewey and J. M. Bird, 1970, Bulletin of the Geological Society of America, v. 81, pp. 1031–1060 and unpublished work of W. F. Kidd.)*

ENVIRONMENTS OF THE APPALACHIAN REGION

EARLY PALEOZOIC SHELF

Let us now apply the preceding discussions to the Appalachian region. From Eocambrian to mid-Ordovician time, there was no mobile belt there. Instead, as we have noted, a broad, structurally passive continental shelf faced a proto-Atlantic Ocean basin, which was completely destroyed before the present Atlantic formed (Fig. 12.19). Sedimentation closely balanced subsidence so that thick, shallow-marine Eocambrian and Lower Cambrian sands followed by later Cambrian and Ordovician carbonates accumulated on the shelf, while fine

muds and occasional limestone slide blocks (Fig. 12.11) accumulated on the continental slope beyond. Because no high lands existed nearby and the entire shelf was located within the subtropics, carbonate deposition was favored. Organisms flourished to produce a shelly carbonate facies. With the beginning of subduction of lithosphere, the Appalachian mobile belt was born (Fig. 12.19). The shelf deposits eventually became deformed, together with the younger, truly geosynclinal strata as the mobile belt developed.

THE GRAPTOLITIC FACIES

With the onset of structural mobility, the region became engulfed in black muds and dark sands (graywackes). Because of the large organic content of the muds and few obvious current features in the sands, it was assumed long ago that the region had subsided in mid-Ordovician time to produce quiet, deep-water environments lacking oxygen and current agitation. Floating graptolite colonies drifted over the area, and their delicate remains sank to the bottom and were preserved in the muds.

Understanding of the origin of the important graptolitic shale facies was slow in coming. In the Alps, early Mesozoic dark shales and cherts that lie upon serpentine and basalt with pillow structure all together form what the Europeans have long called an *ophiolite suite* (Fig. 12.20; see also Fig. 12.2). The igneous members of the suite were interpreted as oceanic crust, and the overlying fine sediments as deep-marine deposits. The Alpine Sequence became more sandy in Cretaceous and Eocene times, and its alternating, very evenly stratified sandstones and shales (called *flysch* in Switzerland) resemble the graptolitic facies of the early Paleozoic. But even if the ophiolites are deep marine in origin, how could the overlying sands be so? A deep-water origin was proposed for the similar early Paleozoic graptolitic shale facies of Wales in 1930. The immature dark graywackes, which contain graded bedding so characteristic of that facies (Fig. 12.21), were interpreted as rapidly "poured into" a deep, tranquil zone of the sea

FIGURE 12.20
Idealized vertical sequence from oceanic igneous rocks and chert *(ophiolite sequence)* upward through graptolitic-graywacke *(flysch)* strata to red clastic *(molasse)* deposits. Such successions are commonly seen in orogenic belts and represent a progression from deep marine to nonmarine conditions, as long ago argued by Europeans.

floor. Since then, graded bedding has been widely recognized in similar sequences all over the world, including the Alps. Presence of ripple marks led many to believe that these sequences were shallow deposits, because ripples were thought to form only there, the deep seas being assumed static. With the recognition in 1950 of the importance of episodic turbidity currents (see Fig. 8.16) in transporting relatively coarse sediments into otherwise quiet environments and the discovery of ripples on the

FIGURE 12.21
Graded bedding with finer texture upward in Ordovician graywacke (Martinsburg Formation) near Middletown, New York. These sandstones, which are typical of *flysch*, are interpreted as turbidity current deposits. *(Courtesy Earle F. McBride.)*

deep-sea floor (Fig. 12.22), a reinterpretation of many sandy geosynclinal strata as relatively deep-water deposits became possible. The turbidity current revolution finally settled the controversy between the American and European views of geosynclinal sedimentation.

Returning to the Appalachian belt, we find Ordovician ophiolite sequences (most clearly in Newfoundland but also as ultramafic rocks farther south; Fig. 12.12) overlain by graptolite or *flysch facies* (Fig. 12.20). Basalts extruded onto the deep-ocean floor were buried first by cherts and black muds containing manganese concentrations. By mid-Ordovician time, however, immature sands showing graded bedding were introduced rapidly by turbidity currents and were apparently deposited on *deep-sea fans* (Fig. 12.23). Volcanic rocks continued to be erupted intermittently. Apparently volcanic and other islands were being raised nearby as a result of subduction of lithosphere (Fig. 12.19). Although relationships are not so clear farther south, the presence of identical rock types there suggests similar conditions. Throughout the mobile belt, facies patterns and paleocurrent criteria indicate that sands of the graptolitic facies were derived from the east, where the Taconian volcanic and

FIGURE 12.22
Deep-marine ripple marks on a fine sand bottom at a depth of 3,500 meters in the South Pacific Ocean near the Antarctic Circle. Prior to the development of submarine photography around 1950, it was assumed that the deep seas were static and currentless; therefore ripple marks could form only in shallow water. Compare Fig. 3.8. (*Official NSF photo,* USNS Eltanin, *Cruise 15, courtesy Smithsonian Oceanographic Sorting Center.*)

tectonic lands were rising. At the turn of the century, an American paleontologist noted that many graptolites in eastern New York had been oriented by bottom currents. More recently, current-sculptured *sole marks* found on the bottoms (soles) of sandstone strata where interstratified with shale (Fig. 12.24) have provided much additional paleocurrent data to help us interpret the paleogeography. As shown in Fig. 12.19 much of the sediment derived from rising lands was deposited in a small ocean basin between the continent and the island arc to the east. This basin was probably similar to the modern Sea of Timor, which has an oceanic crust, between the Indonesian island arc and mainland Australia (see Fig. 17.32). During the Taconian orogeny, the Ordovician small ocean basin was closed as the advancing island arc and the basin's sediments were crushed against the rigid continent.

The evenly stratified, dark Swiss Alpine

FIGURE 12.23
Diagrammatic portrayal of deep-sea fans west of California (view looking east toward continent; vertical scale exaggerated). Such fans are fed largely by turbidity currents, with submarine canyons that cut across continental shelves acting as conduits for movement of sediment to the fans. Most sediment transported by turbidity currents is deposited on such fans. Note the gradual burial of topographic features. (*From H. W. Menard, 1960,* Bulletin of the Geological Society of America, *v. 71, pp. 1271–1278; used by permission.*)

FIGURE 12.24

Flute structures on the bottom (sole) of an Ordovician graywacke (Martinsburg Formation), near Staunton, Virginia; the bulbous ends (right) point upcurrent. Turbulent bottom currents scoured elongate flutings in cohesive muds and deposited sand within them. Long after lithification, erosion exposed the structures on the now more-resistant sandstone. *(Courtesy Earle F. McBride.)*

(*flysch*) strata pass upward into nonmarine light-colored, irregularly stratified coarse conglomerate, cross-stratified sandstone, minor shale, and coal. These deposits (called *molasse* in Switzerland) reflect the final uplift of the Alps above sea level. Some of the *molasse facies* is red-colored, and is in every way analogous to Late Ordovician–Early Silurian red beds of the Appalachian belt (Figs. 12.10 and 12.20). They all constitute clastic wedges associated with orogenies.

RELATIONS BETWEEN EUROPE AND NORTH AMERICA

There is a long-recognized similarity of early Paleozoic faunas between eastern North America and northwestern Europe (Fig. 12.25), which has suggested some kind of connection. The tectonic history of the two sides of the North Atlantic also suggests a former physical relationship, for sequences of rock types and major orogenies are almost identical. The presence of an early Paleozoic mobile belt in Britain and Norway that has an almost perfect mirror image in east Greenland and Newfoundland is especially compelling. A few ardent advocates like Wegener invoked continental drift many years ago to explain these similarities, but their restorations had Europe and North America joined throughout the Paleozoic Era (see Fig. 6.1). A plate-tectonic interpretation, however, postulates that they were separated until Devonian time. During the Ordovician and Silurian Periods, these two continents and Africa were moving toward each other as the intervening proto-Atlantic plate was being consumed by subduction on both sides (Fig. 12.25). Such subduction is postulated in order to explain volcanic and orogenic activity on both continents. Such activity began earlier in the northern region between Greenland and Scandinavia (Early Ordovician) than in the Appalachian region, where it began in mid-Ordovician time and culminated with the Taconian orogeny around the end of the period. Taconian deformation is now interpreted as the result of the

LATE ORDOVICIAN

Lines of Mesozoic opening
of the Atlantic

E Q U A T O R

SUBDUCTION

NA

ORDOVICIAN

SUBDUCTION

PROTO-ATLANTIC

OCEAN

South Pole

O R D O V I C I A N

T I L L I T E S

G

Proto-Atlantic Plate Subduction

Lower Cambrian Paleocurrents

Motions of continent-bearing plates

FIGURE 12.25

Hypothetical Late Ordovician paleogeographic map showing the subduction of a proto-Atlantic Ocean plate as
North America (NA), Europe (E), and the southern supercontinent Gondwanaland (G) began their approach. Relative latitudes are revealed by paleomagnetic data, but
exact relative positions cannot otherwise be determined.
Note that continental glaciation occurred in northern Africa. All that remains of the proto-Atlantic today are mafic
oceanic igneous rocks and deep-sea sediments found
within the old mobile belts on either side of the present
Atlantic. *(Data from J. F. Dewey and J. M. Bird, 1970,
Bulletin of the Geological Society of America, v. 81, pp.
1031–1060; P. E. Schenk, 1971, Canadian Journal of Earth
Sciences, v. 8, pp. 1218–1251; K. Swett and E. Smit, 1973,
Bulletin of the Geological Society of America, v. 83, pp.
3223–3248; and S. Beuf, et al., 1971, Institute Francais du
Petrole, Special Pub. No. 18.)*

collision of an arc-trench system with eastern North America. Ophiolites, *flysch*, volcanics, and even pre-Ordovician continental slope deposits were crushed together and thrust onto the continent margin to make a *subduction complex* (Fig. 12.19).

Based upon the above concept, the map of Fig. 12.25 has restored Prepaleozoic and lower Paleozoic strata of northwestern Scotland, western Norway and Svalbard to their apparent original positions as part of the ancient continental shelf along eastern Greenland and western Newfoundland. The lower Paleozoic rock sequences and paleocurrent directions as well as the faunas already noted (Fig. 12.25) match perfectly. Normal cratonic quartz sandstones of northwesternmost Scotland with obvious derivation from the northwest—now a deep ocean—were especially hard to explain without drift. Opponents had to invoke a sunken Atlantis continent for a source, but there can be little doubt that cratonic North America was that source. Eastern Newfoundland and a portion of the mainland farther south apparently were then either part of Africa or Europe (Fig. 12.25). All these tectonic elements were crushed together by continental collision in Devonian time, as detailed in the next chapter. When the present Altantic basin began forming in Mesozoic time, irregular separation of the two continents left these fragments of one attached to the other.

SUMMARY

A major regression of the sea from the entire craton occurred in early mid-Ordovician time, due apparently to a fall of sea level. The subsequent transgression may have resulted from the beginning of subduction of the proto-Atlantic lithosphere plate. Strata just deposited in the previous 30 million years or so were laid bare to erosion. Thin Lower Ordovician dolomites were quickly stripped off the central craton to expose Upper Cambrian quartz sandstones. Sand was eroded and redeposited during retransgression of the sea in medial Ordovician time to form a very widespread sheetlike unit demonstrably older at the margins of the craton (where transgression commenced) than near the center. The quartz sand supply dwindled again as the sea advanced to flood the entire craton. In later Ordovician time, epeiric-sea fossiliferous carbonate deposits were laid down everywhere until the end of the period when dark muds began to encroach from the east.

Ordovician faunas differ considerably from Cambrian ones. The mid-Ordovician transgression brought an especially distinctive fauna, for the new epeiric sea teemed with brachiopods and bryozoans. Cephalopods, gastropods, diverse echinoderms, and corals also were prominent. Vertebrate animals, represented by jawless, armored fish, appeared on many continents for the first time. Of greatest importance for dating and correlating Ordovician strata, however, were the rapidly evolving graptolites, whose colonial fragments were widely dispersed over the ocean floor.

In the mobile belts, very thick Eocambrian, Cambrian, and Ordovician marine sedimentary and volcanic rocks accumulated. Most of the lavas were submarine, except in the Appalachian belt, where volcanic islands appeared. Near the end of Ordovician time, a large, mountainous tectonic land was raised there during the Taconian orogeny. Whereas terrigenous clastic sediments deposited in the mobile belts during earliest Paleozoic time had been derived largely from the craton, during Ordovician time in the Appalachian belt, more and more clastic material was derived from volcanic and tectonic lands raised within the belt itself. As presently interpreted, those lands were composed of only slightly older geosynclinal rocks and igneous masses rather than ancient Prepaleozoic granitic and metamorphic rocks lying outside the belt to the east. Mobile belts have been structurally disturbed more or less continuously throughout their histories. Uplift of one portion by mountain building produced land sources for sediments that were deposited in another, subsiding portion of the same belt. The weight of the thick sediments themselves depressed the crust somewhat, but more profound structural disturbances—chiefly viselike compression between two lithosphere plates—must have been the primary cause of subsidence that made possible the accumulation of thousands of meters of sedimentary and volcanic rocks.

During times of relative structural serenity, carbonate rocks were deposited in shallow parts of many mobile belts, but at other times, subsidence produced deeper water environments in which muds were deposited. Where sedimentation began on oceanic crust, sequences can be seen with ultramafic igneous rocks (presumably from the upper mantle) overlain by thick pillow basalt (presumably oceanic crust), bedded chert, and black shale. When sand was introduced intermittently by turbidity currents into relatively deep water, the very evenly stratified graptolitic (or *flysch*) facies resulted. During the most active time of uplift of new mountains, so much sediment was produced by erosion that a great terrigenous clastic wedge or prism of red strata (the *molasse* facies) accumulated so that the sea was literally crowded away and the shoreline migrated cratonward.

The patterns summarized here were forerunners of things to come. We shall see that the early Paleozoic tectonic and sedimentary framework, and the concepts developed in this and the preceding chapter for interpreting them, will be applicable to much of the later record as well. In subsequent chapters, therefore, we shall not dwell at length upon similar patterns, but rather shall emphasize new ones as they appear. Keep in mind that details of the stratigraphic record are not ends in themselves, but rather provide vehicles for illustrating how geologic history can be interpreted and how restorations are made. Our objective is not to bury the reader in a mass of murky detail, as it were, but rather to exhume from the strata some principles of historical reasoning and some broad insights into the evolution of the earth.

Key Concepts

crustal subsidence:
   isostatic sediment loading
   lithosphere subduction
   thermal convection currents

current sole marks
deep-sea fans
flysch versus *molasse* facies
graptolitic shale facies
mobile belt subdivisions:
   igneous rock-bearing (eugeosyncline)
   non-igneous rock-bearing (miogeosyncline)
ophiolite suite
paleogeologic maps
Prepaleozoic borderlands
St. Peter Sandstone
sediment sources:
   cratonic
   tectonic lands
   volcanic islands
shelly limestone facies
subduction complex
Taconian orogeny—effects:
   clastic wedge
   granitic plutons
   immature graywackes
   thrust faulting
   ultramafic rocks
   unconformities
Tippecanoe Sequence
volcanic ash:
   correlation between facies
   paleowind fallout patterns

Readings

Bird, J. M., and Dewey, J. F., 1970, Lithosphere plate-continental margin tectonics and the evolution of the Appalachian orogen: Bulletin of the Geological Society of America, v. 81, pp. 1031–1060.

Dapples, E. C., 1955, General lithofacies relationship of St. Peter Sandstone and Simpson Group: American Association of Petroleum Geologists, v. 39, pp. 444–467.

Dott, R. H., Jr., and Shaver R. H. (eds.), 1974, Modern and ancient geosynclinal sedimentation: Tulsa, Society of Economic Paleontologists and Mineralogists Special Publication No. 19.

Eardley, A. J., 1962, Structural geology of North America: New York, Harper & Row.

Kay, M., and Colbert, E. C., 1965, Stratigraphy and life history: New York, Wiley.

King, P. B., 1977, The evolution of North America (2d ed.): Princeton, Princeton.

Levorsen, A. I., 1960, Paleogeologic maps: San Francisco, Freeman.

McBride, E., 1962, Flysch and associated beds of the Martinsburg Formation (Ordovician), central Appalachians: Journal of Sedimentary Petrology, v. 32, pp. 39–91.

Price, R. A., and Douglas, R. J. N. (eds.), 1972, Variations in tectonic styles in Canada: Geological Association of Canada, Special Paper II.

Schenk, R. E., 1971, Southeastern Canada, northwestern Africa, and continental drift: Canadian Journal of Earth Sciences, v. 8, pp. 1218–1251.

Sloss, L. L., 1976, Areas and volumes of cratonic sediments, western North America and eastern Europe: Geology, v. 4, pp. 272–276.

Stewart, J. H., et al. (eds.), 1977, Paleozoic paleogeography of the western United States: Los Angeles, Pacific Coast Section, Society of Economic Paleontologists and Mineralogists.

Swett, K., and Smit, D., 1972, Paleogeography and depositional environments of the Cambro-Ordovician shallow-marine facies of the North Atlantic: Bulletin of the Geological Society of America, v. 83, pp. 3223–3248.

13

13

THE MIDDLE PALEOZOIC
Time of Reefs, Salt Deposits, and Forests

He who with pocket-hammer
smites the edge
Of luckless rock
or prominent stone, disguised
In weather-stains
or crusted o'er by Nature.
The substance classes by
some barbarous name,
And thinks himself enriched,
Wealthier, and doubtless
wiser than before.

William Wordsworth,
The Excursion (1814)

FIGURE 13.1
The Gilboa forest of Devonian age in the Catskill Mountains, eastern New York. *(Courtesy American Museum of Natural History.)*

13

The tectonic configuration established in Ordovician time also characterized the middle part of the Paleozoic Era, that is, from 435 to 350 million years ago. Continued subduction of the proto-Atlantic lithosphere plate resulted in volcanic activity and deformation on both sides of the old ocean, which shrank and finally was destroyed by the collision of northwestern Europe and northeastern North America in Devonian time. As the deep-ocean barrier disappeared, the faunas and floras of both continents became more and more alike. The edges of each colliding plate were intensely deformed and uplifted to form a single great mountain chain between two cratons. Granite batholiths formed within this great intercratonic mobile belt, and huge prisms of red clastic sediments were shed from the mountains onto both cratons. The cratons themselves also were affected by the continental collision, as indicated by extensive warping of arches and basins.

Apparently organisms first invaded the land in middle Paleozoic time, indicating that an atmospheric ozone shield from deadly ultraviolet radiation existed by now (if not earlier). The oldest proven land plants are latest Silurian, and coastal forests were commonplace by Late Devonian time. Appearance of land vegetation must have altered processes of erosion and deposition enormously (Fig. 13.2). The earliest definite air-breathing animals known were Early Devonian spiderlike creatures, whereas the oldest land vertebrates were Late Devonian amphibians. Epeiric seas were widespread over all mid-Paleozoic cratons. Unusually rapid evolution produced many new faces among both marine invertebrates and vertebrates. Especially important was the first development of large animal reefs, whose locations were chiefly determined by warping of cratonic basins. Major evaporite deposits also occurred within many of the basins rimmed by reefs.

Microorganisms produce the raw materials of petroleum after burial in sediments. Their organic compounds undergo chemical changes at temperatures of 100 to 200°C that lead to petroleum. If favorable permeable rocks overlain by impermeable ones are present at the right time and place, oil and gas can be trapped in significant quantities. Buried middle Paleozoic organic reefs are among the most prolific of such traps, but sandstones are also important. The famous Drake well drilled near an oil seep in 1859 at Titusville, Pennsylvania, tapped a petroleum-saturated Devonian sandstone. Exploration for elusive hidden traps requires sophisticated geologic and geophysical tools. As a by-product of practical oil-seeking, science has been rewarded with a wealth of otherwise inaccessible subsurface geologic data, which adds greatly to the accuracy of our historical analysis, especially for the deeply subsided basins. And such greater accuracy is itself essential if we are to find new petroleum reserves for this energy-hungry world of ours. Even with the best possible skill and luck, however, it is a losing battle, for petroleum—like all mineral resources—is a nonrenewable resource in terms of the human time scale. Our best estimates indicate that in less than 100 years from now the world's petroleum reserves will have been exhausted. Long before that, however (certainly by the year 2000), we must have other major energy sources to supplement petroleum. Coal and nuclear fuels with all their environmental hazards are most likely to be the first supplements, but they are nonrenewable, too, and will probably be exhausted by the year 3000.

MIDDLE PALEOZOIC LIFE

INVERTEBRATE ANIMALS

With the spread of the great Upper Ordovician sea, many new creatures appeared along with a

FIGURE 13.2
Martian landscape of sand dunes and rock fragments. Violent windstorms blow sand freely without the impediment of vegetation. Great importance of wind deposition on Mars today gives a fair impression of its tremendous importance on Earth before land plants appeared in middle Paleozoic time. As noted in Chap. 11, wind abrasion and sorting were keys to the development of very mature quartz sandstones. *(Photographed by* Viking Lander 1, *1976, whose leg appears in center; NASA photo P17430; courtesy Jet Propulsion Laboratory.)*

corresponding increase in diverse *ecological niches* (a niche is an organism's way of life in a habitat). Favorable conditions permitted vast populations to develop, and we have a rich heritage of these in exceptional fossil accumulations around Cincinnati, Ohio, just as rich, middle Paleozoic faunas occur farther west at the Falls of the Ohio. As discussed in Chap. 9, we believe that by the end of Ordovician time most of the different types of large-scale marine *ecologic zones* or biological communities had been established with, of course, different organisms occupying them as time and evolution proceeded to the present.

At first glance, the Silurian fauna has much of the aspect of later Ordovician life in the predominance of brachiopods and bryozoans, but corals and echinoderms increased in importance. The brachiopods of the middle Paleozoic changed, with elongated and deep-bodied forms replacing many of the dominant thin, ovoid Ordovician types (Fig. 13.3).

By Silurian time, two groups that had undergone rapid evolution before—the nautiloid cephalopods and graptolites—had virtually disappeared. A few important graptolite genera are useful in correlating Silurian strata. Other more primitive and long-ranged graptolites persisted until Early Mississippian time, but are very rare. Thus far we have seen that, since the beginning of the Cambrian Period, there was a rapid development and decline to near extinction of three groups, the trilobites, nautiloids, and graptolites, which were important elements in the early Paleozoic faunas and are so useful as index fossils. We do not know the cause of this pattern, but the most common hypothesis is that some predator group evolved and either

FIGURE 13.3
Mucrospirifer, a common Devonian spiriferid brachiopod, from Arkona, Ontario. *(Courtesy U.S. National Museum.)*

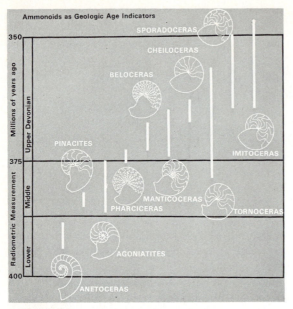

Ammonoids as Geologic Age Indicators

FIGURE 13.4
Evolution of Devonian ammonoid cephalopods, showing changes in the sutures (which separated the internal chambers) with time. In Paleozoic ammonoids, suture patterns are called *goniatite sutures. (Redrawn from a figure in* Bursts of evolution *by M. A. House in* The Advancement of Science, *1963, p. 501; by permission of British Association for the Advancement of Science.)*

ravaged the species themselves or ate their principal food source, thus disrupting the food chain. It is possible, for example, that trilobites fell prey to large, predatory nautiloids. On the other hand, before the height of nautiloid diversity, the trilobites were well on their way to extinction, and thus may not have been affected by the nautiloids. Despite the lack of diversity of the middle Paleozoic trilobites, they are somewhat common in certain facies as represented by the phacopids (see Fig. 9.8). These trilobites have amazingly complex eye lenses composed of two elements, much like that of certain microscopes.

By Devonian time (Fig. 13.4), coiled ammonoids had evolved from a nautiloid stock. They were to become the dominant invertebrate group of the Mesozoic Era. Even when they first appeared, they were distinctive enough to be useful for intercontinental correlation, particularly of Upper Devonian strata. It is felt by most

authorities that they were dominantly swimmers and floaters, which made them ideal index fossils as were the graptolites before. First, they could be geographically dispersed instantaneously (by geologic standards), and, secondly, they could cross sedimentary facies boundaries. They might flourish in well-circulated waters above a foul, stagnant mud bottom on which no benthonic organism could survive, and after death sink to the bottom, where chances of preservation would be increased in an oxygen-poor environment.

The most obvious change in the Silurian marine fauna was an increase in the number of colonial rugose and tabulate corals. Some of their colonies became so large as to form the first coral reefs in several parts of the world; Devonian reefs were even more widespread (Figs. 13.5 and WM.7). A companion to these coral-reef formers was the calcareous sponge Stromotoporoidea group, which originated in the Ordovician. It, too, formed Silurian and Devonian reefs and "gardens," and so was a common rock-former.

The trilobites still were present, and in some facies they are fairly abundant, but are represented by only a few groups. A distant arthropod cousin appeared in fair numbers during the Silurian—the sea scorpions (eurypterids; Fig. 13.6). These strange forms superficially resemble the air-breathing scorpions, but probably were not related. They were monsters of the sea during the Silurian, and some attained lengths of 3 meters before becoming extinct in late Paleozoic time.

DEVONIAN FAUNAL PROVINCES

A number of biological communities with interrelated species in a large region defines a *faunal province.* A faunal province always is characterized by having a number of *endemic* (native) *species,* (see page 495). The Devonian marine faunas of the world can be more clearly delineated into such provinces than those of any previous periods. For example, an eastern Americas' faunal province was subtropical, marine, and can be recognized today only in eastern North America and northern South America, in con-

FIGURE 13.5

A diorama based on a fossil site in the Middle Devonian of central New York; note the rugose coral colony in the left foreground. *(Photograph courtesy American Museum of Natural History.)*

FIGURE 13.6

A sea scorpion *(Eurypterus)* from the Silurian Bertie Water Lime at Litchfield, New York. *(Courtesy U.S. National Museum.)*

trast with the more widespread Old World Realm province of Fig. WM.7.

THE FISHES

We turn next to a group that adds immeasurably to our knowledge and is important historically, though not abundant, namely the fish (Fig. 13.7). In the Silurian, and more particularly the Devonian, many complete skeletons of the jawless and armored fish have been preserved. These primitive fish, the Agnatha, were heavily armored and had flattened anterior shields. They undoubtedly were poor swimmers and in all probability were filter feeders living on the bottom. The internal skeleton was never preserved, and so we suspect that they had a

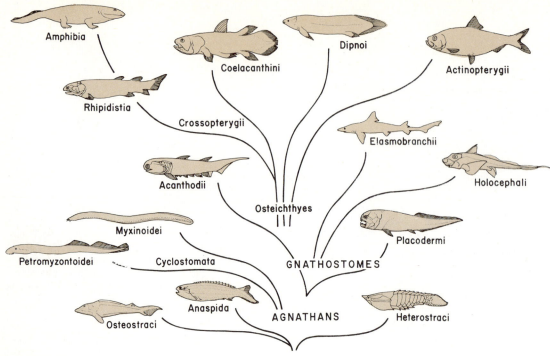

FIGURE 13.7
Evolution of the fishes. (*Drawing courtesy American Museum of Natural History.*)

cartilaginous skeleton like primitive fish of today.

Sometime during the Silurian Period, though perhaps earlier, the first vertebrate jaw developed. It was modified from several gill arches, which are anterior equivalents of ribs with the specialized function of supporting part of the respiratory system. Probably at first they were imbedded in muscle, but early in the evolution, the upper jaw became fused to the head shield, thus forming a rigid and far more efficient eating machine. Primitive jawed fish are called placoderms (Fig. 13.7), and, while some of them retained the ancestral flattened body, most became streamlined and armor was reduced. These changes presumably were in response to selective pressure for better swimming. Some of the primitive jawed fishes, such as *Dunkleosteus* (Fig. 13.8), became large and voracious predators. One other carnivorous group worthy of mention appeared. The sharks, which are a

primitive group that lost the heavy armor (though vestiges occur as dermal plates), developed highly efficient muscles for swimming. The first of the sharks probably were marine

FIGURE 13.8
Restored head shield of *Dunkleosteus*, a large placoderm, from the Devonian of northern Ohio. Note the awesome jaws of this predator. (*Courtesy American Museum of Natural History.*)

FIGURE 13.9
The lobe-finned coelacanth above and rhipidistian below, which likely gave rise to the first tetrapod amphibian. (*From* The encyclopedia of science and technology, *by permission, McGraw-Hill Book Co.)*

forms, but several important groups very early invaded fresh waters.

The development of a jaw in the fishes is yet another example of how a major new character enables a group to undergo adaptive radiation into an unoccupied adaptive zone. The fishes rapidly expanded after the Devonian and are some of the most common *pelagic* (swimming) organisms of today.

There has been a great deal of controversy regarding the original environment of the vertebrates. Professor Romer of Harvard University felt that the evidence favors a freshwater habitat because marine invertebrates are not found associated with the primitive fish fossils. But Dr. Denison of the Field Museum of Chicago, after reviewing newer evidence, has presented a strong case for the marine origin of fishes.

The Osteichthyes are the most successful of all fish because they developed bony skeletons and reduced the rigid armor to thin scales, thus greatly increasing their agility and speed. The development of paired fins, gill slits protected by bony opercula, and bony rays to strengthen the fins, all served that end. Of the two recognizable subgroups of the bony fishes, the ray-finned fish (Actinopterygii) has greatly exceeded all other kinds in numbers and diversity today. The other bony group (Choanichthyes), which includes the *lobe-finned fish* or Crossopterygii (see Figs. 13.7 and 13.9) and the lungfish (Dipnoi) developed internal nostrils that enabled them to respire with their mouths closed. It is because of the internal nostrils that some of the forms gradually began to breathe air, probably out of necessity when trapped in drying tidal or freshwater ponds. The Choanichthyes are represented today by the lungfish and the "living fossil," the coelacanth, considered extinct since Mesozoic time until 1938, when the first living specimen was caught off East Africa. Other important characteristics of this group are that the fins are lobed and muscular and have articulated rays that allowed some members to "walk" on the bottom. The amphibians are assumed to have derived from this group. Current thinking among specialists is that the Rhipidistia of the Upper Devonian are likely the ancestral group (see Fig. 13.7).

INVASION OF THE LAND

The very first amphibian made its appearance during the Late Devonian and is called a labyrinthodont (because of the labyrinthine infolding of tooth enamel). It was an awkward model, looking like something a committee put together. Its limbs were nothing more than jointed lobed fins; its head and tail were fishlike, too. Nonetheless, it did breathe air. Our record of them comes from some scraps in eastern Canada and several good specimens from northeastern Greenland. They would hardly be classified as common fossils, yet they are of enormous importance as the evolutionary stem of all air-breathing, vertebrate land animals. Further discussion of early land vertebrate development and the rise of land plants, which appeared at least as early as Silurian time, is deferred to the next chapter.

We shall see that a relatively few characters that were "preadapted" for life on land in both the plants and animals were strongly selected for terrestrial environments. Consequent elaboration of these characters, along with the development of additional functional anatomy, permitted a very fast invasion of the land.

FOSSILS AS CALENDARS

We have seen that isotopic dating gives us the best estimate of "absolute" time in terms of years. But the smaller units of time—the month and day—are so brief geologically that they cannot be resolved by any isotopic method known. As we have seen, fossils were used to establish the "relative" time scale, and now it appears that they also will prove useful for limited "absolute" chronologies. A very ingenious line of investigation initiated by Professor John Wells of Cornell University about 1960 helped to bridge the gap between years and the smaller units of time by the use of fossils.

Biologists have observed that modern corals deposit a single, very thin layer of lime once a day. It is possible, with some difficulty, to count these diurnal (day-night) growth lines and to determine how old the coral is in days. More important, seasonal fluctuations will cause the growth lines to change their spacing yearly so that annual increments can also be recognized, much as in growth rings of trees.

Out of curiosity, and because he is a paleontologist, Wells began looking for diurnal lines on

FIGURE 13.10
Devonian coral showing a wide annual band; the white lines bracket that band; the fine lines are presumed to be daily growth lines. *(Photograph by G. R. Adlington.)*

fossil corals. He found several Devonian and Pennsylvanian corals that do show both annual and daily growth patterns. But he was astonished to find that the Pennsylvanian forms had an average of 387 daily growth lines per year-cycle, and that the Devonian corals had about 400 growth lines (Fig. 13.10).

By making counts between annual marks, Professor Wells found an average of 360 growth lines per year on modern corals he examined. He then constructed a graph (Fig. 13.11) based on the latest isotopic dating of the periods back to the beginning of the Cambrian, which suggested a systematic decrease in the number of days per year through geologic time. Several years later, Pannella and his associates (1968), using many more organisms from Recent to Cambrian, have shown that while a systematic decrease is probably valid, their points on a graph formed an S-curve rather than a straight line. This is probably caused by the differing effects of tidal action accompanying the changes of continents, ocean basins, and epeiric seas. Studies using clams (such as the common East Coast edible clam *Mercenaria mercenaria*) have shown that the growth lines on and within the shells record different environmental events such as seasons or breeding periods, which are also rhythmic. They may record episodic events such as storms, unusual temperature fluctuations, attacks by predators, and the like. Much can be learned of past environments from such studies.

Meanwhile, geophysicists and astronomers estimated that there has been a deceleration (presumably due to tidal friction) of the earth's rotational velocity in recent centuries amounting to 2 seconds per 100,000 years. If one extrapolates this rate (which may or may not be

valid), the Cambrian day would have been 21 hours and the year 420 days long. By the same reasoning, the length of a Devonian year should have been 400 days long. Thus, the astronomic extrapolations and Wells's coral data for the Devonian are remarkably similar. More recent studies using both corals and brachiopods have shown that Wells may have undercounted the growth lines. Some estimates place the length of the year in Early Silurian time at 421 days. Much more sampling and control will have to be done before this method can be used with accuracy.

Algal stromatolites also look promising for giving information on lengths of ancient years. It has been found that modern stromatolites in Bermuda waters lay down a daily growth increment, and study of Cambrian algal stromatolites from Washington shows that there are from 400 to 420 second-order increments in a series of growth bands.

Much research is being done to explore this exciting new development in the use of fossils to understand earth history. For example, stromatolites similar to types that today form chiefly in the intertidal zone existed 2.7 billion years ago. They were several centimeters high, suggesting that the moon was present at least by then to influence tides. But it is interesting to note that the height of such algal colonies increased to a maximum of 6 meters in late Prepaleozoic time. If these stromatolites *were* restricted to the intertidal zone, their increase in height would suggest a gradual decrease in the distance of the moon from the earth as the cause of increased tides. Since Prepaleozoic time, however, seemingly the tides have decreased while the length of the day has increased, which changes presumably were due to movement of the moon away from the earth (conservation of angular momentum requires that the earth must accelerate if the moon approaches it, and vice versa). Perhaps we might be able to learn more about the history of the earth-moon system from fossils than from actual moon exploration (see the Wade reference, 1969).

Kahn and Pompea in 1978 studied the living nautiloid *Nautilus pompilius* and noted that the number of growth lines between each septum, which separates the chambers, averages about

FIGURE 13.11
Changing length of the day during Paleozoic time on the basis of coral growth bands (two curves show limits of uncertainty). If correct, the data suggest the moon was 3 to 3.5 percent nearer the earth and the equilibrium tide was nearly 10 percent greater in early Paleozoic time than now. (*Data from Wells, 1963*, Nature; *Scrutton, 1964*, Palaeontology.)

30, which they believe represents a lunar month. They examined a number of different nautiloids going back into the fossil record to the Ordovician and noted that the oldest nautiloid had about 10 growth lines between septa. They observed that there was a progressive increase in the number of days in a lunar month and made the assumption from this that the rate of the moon's recession from the earth has averaged about 94 centimeters per year. Astronomers had previously calculated a rate of 5.8 centimeters per year. Astronomic data as well as geophysical and geological observations strongly indicate that the earth-moon distance must have been much greater than Kahn and Pompea indicate (their calculation would place the moon at half its distance 400 million years ago, while such evidence as lunar rock dates indicates that the earth-moon spacing was closest 4 billion years ago).

THE SILURIAN CONTINENT

AFTERMATH OF THE TACONIAN OROGENY

In the Appalachian mobile belt, erosion of tectonic islands raised during the Late Ordovician Taconian disturbance was already advanced by Early Silurian time. Deposition in the western part of the belt continued more or less uninterrupted from the Ordovician, but an unconformity marks the base of Silurian strata farther east. Gradual eastward encroachment of the sea over the erosional surface is clearly documented. Sandy and gravelly facies gradually shifted eastward as transgression proceeded.

Early Silurian clastic sediments clearly were derived from the core of the mobile belt east of New York. This is shown by the coarsening of sandstones and increase of conglomerate eastward (Fig. 13.12) and by orientation of current-formed features. Silurian quartz sandstones are widespread in the Appalachian Mountains, making up roughly 100,000 cubic kilometers (25,000 cubic miles). This staggering volume of sand with scattered quartz pebbles represents weathering, winnowing, and concentration from a tremendous volume of source rocks in the Taconian Mountains (see Fig. 8.20).

Unusual Silurian iron-rich sedimentary deposits in the southern Appalachian Mountains provide ore for the important Birmingham steel industry; a similar ore used to be mined in Newfoundland, too. Apparently there was intense tropical weathering (perhaps of iron-rich mafic igneous rocks) in the Taconian mountains. Iron was introduced by rivers in unusual concentrations to a somewhat restricted, marginal-marine environment. The sediments forming there, including marine shells, became replaced and cemented with red hematite.

NEW MOUNTAIN BUILDING IN THE NORTH

In the eastern part of the mobile belt, Silurian volcanism was important in New England and Maritime Canada. Many of the eruptions were submarine, for lavas are found interstratified with richly fossiliferous limestones, sandstones, and shales. Only local, small islands existed.

In Newfoundland, northeastern Greenland, and northern Canada, there were new mountain-building disturbances during the Silurian Period. Possibly structural upheaval began first in the south during the latest Ordovician (the Taconian orogeny) and then spread progressively northward during the Silurian. Late Silurian terrigenous clastic sediments occur in the Arctic (Franklin) mobile belt, and especially in northern Greenland. Locally in the Arctic, an angular unconformity separates Ordovician and Silurian strata, and another, even more profound angular unconformity, marks the boundary between Late Silurian and Early Devonian ones.

GENERAL PALEOGEOGRAPHY OF THE CRATON

Silurian strata are scattered widely over the craton (Fig. 13.12). Their scattered distribution led to a long-held view that the Silurian epeiric sea was restricted in area and was studded with many low islands. Such an interpretation was strongly influenced by the hazardous assumption that present limits of marine strata closely approximate their original distribution. As we have seen, one must be careful in cratonic

sequences to evaluate the importance of unconformities that may account for a great deal of
erosion of formerly more extensive strata. Such
a discontinuity within the Devonian System is
known to occur over all of the craton!

If we examine the sedimentary facies for the
Middle Silurian, we see that most of the preserved rocks are carbonates and some shale that
must have formed under a minimum influence
of lands. Moreover, the deposits and their
fossils suggest a former shallow sea with uniform conditions over an immense region rival-
ling that of the Late Ordovician (Fig. 12.8).
Therefore, Fig. 13.13 shows our preferred interpretation of Middle Silurian paleogeography in
which it is assumed that marine Silurian strata
originally covered all of the craton, and were
eroded during Devonian and later times to
produce their present distribution.

ORGANIC REEFS

GENERAL CHARACTERISTICS

Although animals inhabit all portions of the
seas, the greatest number and diversity of
marine animals and plants live in warm, shallow waters within the zone of light penetration
and strong agitation (the photic zone; Figs.
11.26, 9.24). The floating microscopic *plankton*
forms the beginning of the long *marine food chain*
(see Fig. 9.23); microscopic plants begin the
chain, followed by microscopic animals, which
feed on the plants and each other. Bottom-
dwelling invertebrates in turn feed upon both
kinds of plankton and upon each other (see Fig.
9.24). Various swimming invertebrates, fishes,
and marine mammals in turn are predatory on
the other creatures. Because plants provide the
ultimate basis of all animal diets, the food chain
must begin in shallow, lighted water where
photosynthesis is possible. Furthermore, agitation tends to make shallow waters well oxygenated for animal respiration. But nutrients must
also be available (see Chap. 9).

Optimum growth conditions for reef-building
organisms exist where the upwelling of deep,
fertile waters occurs. *Upwelling* (rising) characterizes many areas of the sea, but is especially
important where currents impinge against sub-

merged slopes or escarpments and are deflected
upward, as along the northwest side of the
Florida Straits and the eastern, windward edge
of the Bahama Banks (Fig. 13.14). At such
locations in tropical and subtropical latitudes,
crowded and diverse marine animal and algal
communities develop, and the growth of skeletal organisms is so great that wave-resistant,
moundlike masses of calcium carbonate are
built up on the sea floor to sea level. Cementa-
tion of such carbonate deposits can occur within
but a few thousand years. These are called
organic reefs (Fig. 13.15). Reefs are the most
"urbanized" and therefore most ecologically
complex areas of the sea floor.

Reefs are geologically as well as biologically
important, for large masses of carbonate rocks
are built by reef-forming organisms. Many
ancient reefs exist and are of special interest
because of their significance both environmen-
tally and as petroleum traps. The main reef-core
rock commonly is characterized by a distinctive
fabric of interlocking skeletal material. Typically
it forms a massive, lenticular body surrounded
by clearly stratified sediments (Figs. 13.16 and
13.17).

Today corals and calcareous algae are the
most prominent reef builders, but in the past
many other organisms contributed as well.
These included algae alone (especially prior to
the Silurian Period) and stromatoporoids,
sponges, bryozoans, crinoids, brachiopods, and
certain molluscs. Many ancient limestone reefs
have been more or less converted to dolomite,
causing considerable modification of original
textures (origin of dolomite is discussed shortly). Organic reefs, both modern and ancient,

On the following pages:
FIGURE 13.12
Middle Silurian lithofacies; note large patches of carbonate rocks over the Canadian Shield region and widespread
organic reefs, R. Also note limestones in the outer volcanic zones of the mobile belts (see Fig. 11.7 or endpapers for
symbols and sources). Patterns on western margin of
continent probably were not yet in their present locations.

FIGURE 13.13
Middle Silurian paleogeography. Most of the continent
lay beneath an epeiric sea. Western margin of continent
cannot be restored with confidence due to structural com-
plexities.

MIDDLE SILURIAN

SEDIMENTARY FACIES

(Niagaran)

(about 410 m.y. ago)

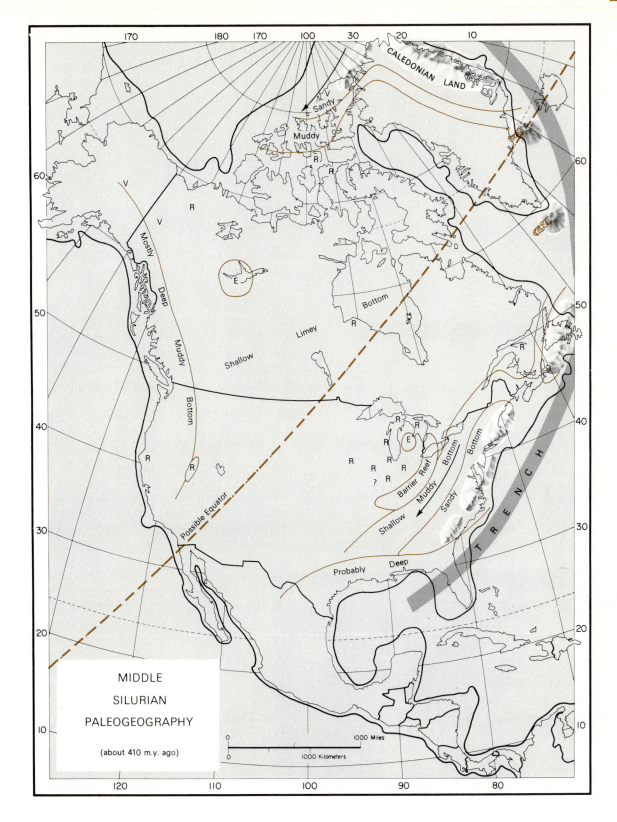

MIDDLE

SILURIAN

PALEOGEOGRAPHY

(about 410 m.y. ago)

FIGURE 13.14

Modern reefs and carbonate banks of the Bahama Islands–Florida Keys–Cuba region. The Bahama Islands and the keys are emerged dead Pleistocene reefs; living reefs lie seaward of them. Reefs have grown almost continually in this region since late Mesozoic time. Carbonate sands cover the bank (see Fig. 11.17). Contour line indicates 200 meters below sea level. (*After Newell and Rigby, 1957,* Society of Economic Paleontologists and Mineralogists Publication 5.)

vary greatly in size and form. Some occur as long, linear *barrier reefs* at the edges of shelf or bank areas. Examples include the Florida Keys and Bahaman reefs (Fig. 13.14), and the largest of all modern examples, the Great Barrier Reef of Australia (see Fig. 13.36), 1,700 kilometers long. Also important are the more circular *fringing reefs* developed around islands, and the Pacific *atolls* built on submerged prominences (see Chap. 7).

Reefs encompass many subenvironments characterized by differences in the organic community and sediments. The *windward side* is generally rather steep-faced and is constantly battered by waves; it is the zone both of most active growth and of destruction. Fragments of reef rock periodically are torn loose to slide down the reef front into deep water (Fig. 13.17). Thus an apron of coarse, poorly sorted, angular reef debris characterizes the *fore-reef facies*. The debris shows crude stratification inclined as much as 40 degrees away from the reef front. *Back-reef* facies consists chiefly of stratified clastic carbonate sand derived from the reef, though oölite also may be present (see Fig. 11.18) as well as evaporite layers.

SILURIAN AND DEVONIAN REEFS

One of the most interesting aspects of Silurian time is the great development of large organic reefs involving for the first time important

FIGURE 13.15

Underwater photo of a modern reef top near Key Largo, Florida. Living brain coral (center) and staghorn coral (lower) are surrounded by sand produced by wave erosion of the reef. The water is only a few meters deep, allowing photography under ordinary sunlight. (*Courtesy George Lynts.*)

FIGURE 13.16
Air view of inaccessible reeflike limestone lens (center) in Upper Devonian strata exposed in the Canadian Rocky Mountains near Mount Mac-Kenzie, Alberta. Massive light-colored rock extends as tongues into adjacent, stratified material, which dips away from the lens-shaped mass. Light-colored masses at right edge presumably are slide blocks derived from the reef. (*Courtesy W. S. MacKenzie,* Geological Survey of Canada.)

contributions from animal skeletons as well as from algae. Silurian examples extend from Tennessee to the Arctic and from Ohio to Alberta (Fig. 13.12), but the prevalence of reefs is by no means confined to North America. The original Silurian strata studied and named in eastern Wales contain important reef masses along the eastern side of what we call the Caledonian mobile belt. Similar reefs also are well known elsewhere in Europe, but Devonian reefs were the most widespread ever (see Fig. WM.7).

Small reefs had been formed by algae since Early Prepaleozoic time. Primitive filamentous algae, most characteristic today of the intertidal zone, have built laminated structures throughout history (see Figs. 8.19 and 11.19), but beginning in early Paleozoic time, more complex types became important reef contributors. Animals began building small reefs in Cambrian and Ordovician times, but why there was such a sudden burst of highly complex animal and algal reef building in Silurian time is not clear. Probably it was due to a combination of evolutionary changes and existence of shallow-marine conditions that gave some selective advantage to ''urbanized'' living.

Figure 13.17 shows the various facies typically associated with ancient organic reefs. These facies are delineated both by different lithology and fossil communities. Stromatoporoids and corals built most Silurian and Devonian wave-resistant cores, while crinoids, molluscs, and

brachiopods dominated the back-reef and fore-reef deposits. Because shoal reefs grow in the surf zone only a meter or so below sea level, the depth of surrounding water can be estimated accurately in some fossil examples by tracing a reef-front stratum down into basin deposits and noting the vertical difference of level. In Fig. 13.17 we see that the depth was 300 meters, which is reasonable for many of the Devonian barrier reefs of western Canada and for some Silurian examples in the Midwestern states. Presence of fossil algae also may provide a depth indicator, because photosynthesis is possible only down to depths of about 150 meters. Thus we see that ancient reefs are unusually rich in clues to past environments. Besides providing indications of depth and wind direc-

FIGURE 13.17
Diagram of typical ancient reef facies seen in many Silurian and Devonian examples. Back-reef includes clastic skeletal sand and oölite; fore-reef includes blocks derived from windward side of reef. Other reef shapes also exist, such as so-called pinnacle reefs that grade into basin facies in all directions.

A. CAMBRIAN THROUGH MID-SILURIAN   B. UPPER SILURIAN

FIGURE 13.18
Two thickness maps showing that the circular Michigan basin originated in Late Silurian time (contours in meters). The basin suddenly subsided as much then as it had throughout all of earlier Paleozoic time. *(Adapted from Cohee, 1948, U.S. Geological Survey Oil and Gas Chart 33; Alling and Briggs, 1961, Bulletin American Association of Petroleum Geologists, v. 45, pp. 515-547; by permission.)*

tion, their regional distribution is also a clue to ancient warm latitudes, as we shall see shortly.

ORIGIN OF MARINE EVAPORITE DEPOSITS

THE MICHIGAN BASIN

Other prominent Silurian sediments include evaporites found particularly in New York, Ohio, Michigan, and in western Canada (Fig. 13.12). These deposits are the basis for major plasterboard and chemical industries. Potassium, important as fertilizer, also is obtained from evaporites, for example, in Saskatchewan. Evaporites characterize chiefly Upper Silurian strata that formed immediately after the maximum reef development shown in Fig. 13.13. In Michigan, especially, subsidence accelerated in a circular area (Fig. 13.18). Subsidence commenced rather suddenly far from any rising arch or mountainous uplift. An increase in density or a decrease in thickness of the crust beneath Michigan must have upset local isostatic equilibrium. Probably there was some indirect relation to structural movements in the Appalachian mobile belt.

In the Michigan Basin, up to 1,500 meters of sediments were deposited, chiefly dolomite rock $[CaMg(CO_3)_2]$, but with as much as 750 meters of rock salt (NaCl) and anhydrite-bearing $(CaSO_4)$ strata. These represent sedi-

ments precipitated from sea water under conditions such that evaporation exceeded total replenishment of water by rainfall, river flow, and inflow from the open sea. Evaporite strata required concentration of brines from an immense total volume of sea water. If the Silurian sea were as saline as that of today, it would have required evaporation of the equivalent of a column of normal sea water nearly 1,000 kilometers deep (about 600 miles) to deposit 750 meters of evaporite strata! Certainly the water was never 1,000 kilometers deep over Michigan; rather it was apparently fairly shallow at all times. Therefore, we must postulate continual replenishment of water as evaporite sediments were precipitated over a subsiding basin floor.

RESTRICTED CIRCULATION

A presumably analogous situation is seen today in central Asia in the Kara-Bogaz Gol, an embayment of the Caspian Sea. The Gulf's waters are replenished more or less continuously across a shallow bar, but because the region is very arid, evaporation causes continuous precipitation of salts from very concentrated, dense waters at the bottom of the Gulf. These brines cannot escape to the Caspian Sea because of the restricting bar. As a result, a *dynamic equilibrium*[1] exists between precipitation of evaporite sediments at the bottom and replenishment and evaporation of water at the top. As long as this equilibrium persists, evaporite sedimentation continues. The Kara-Bogaz Gol provides a kind of model to help us understand evaporite deposition in ancient sedimentary basins (Fig. 13.19).

---

[1]Equilibrium commonly is thought of only in static terms, but in nature practically nothing is static. Development of the concept of equilibrium (or a *steady-state* condition) in dynamic systems was a major advance for geology. In an open system, there may be a constant flow of material and energy, as in an evaporite basin, but the system looks much the same from one time to the next. A beach also exemplifies dynamic equilibrium or steady state, for energy is constantly being expended in the system and sand is constantly in motion, yet the beach does not change in form significantly through long spans of time (except for severe storms, which temporarily upset the equilibrium). Most large rivers with steady discharges also illustrate dynamic equilibrium, at least over modest time spans.

FIGURE 13.19
Equilibrium evaporite basin model illustrating precipitation due to restricted circulation. Normal sea water flows continually into the restricted basin at a rate closely balanced by evaporation; this produces dense brine that sinks and precipitates different evaporite minerals according to concentration. Numbers indicate water density. (*After Briggs, 1957, by permission of Michigan Academy of Science.*)

Circulation of sea water within many ancient basins was restricted by a variety of causes so that *dense brines* sank and precipitation of salts occurred. Though aridity of climate is indicated, excessively hot temperatures are not required,

for dry winds could accomplish the evaporation effectively. But if we assume both moderately warm temperatures and dry winds, then evaporites can be explained readily wherever oceanic circulation in shallow seas was impaired. In Michigan, the fringing reef complexes that began developing around the basin in Middle Silurian time apparently caused restriction of circulation leading to evaporite precipitation (Fig. 13.20). Also, a slight lowering of sea level apparently occurred near the end of Silurian time (Figs. 3.21 and 12.6), which would have produced more islands and shoals around the basin margin, further restricting circulation. In New York, circulation was restricted not only by

FIGURE 13.20
Late Silurian paleogeography of the Michigan–New York–Ohio evaporite basins. Barrier reefs restricted marine circulation into the basins; evaporites occur in basin centers. (*After Alling and Briggs, 1961, Bulletin American Association of Petroleum Geologists, v. 45, pp. 515–547; by permission.*)

shoals on the west and south (Fig. 13.20), but also by land to the east. Red-colored fine clastic sediments are found intimately interstratified with the marine evaporites there.

Ordinary sea water today contains about 3.5 percent dissolved salts, of which sodium chloride (NaCl) is the most familiar and most abundant (Table 13.1). Theoretically, complete evaporation of sea water should produce sequential deposition of a series of evaporite minerals in reverse order of their relative solubilities in water. Laboratory experiments performed as early as 1849 showed the theoretical salt precipitation sequence to be expected. It is the rule to find incomplete sequences, however, indicating that a variety of events may disturb the evaporation process. For example, cessation of precipitation or solution of earlier salts results from seasonal temperature (and humidity) changes or from destruction of circulation barriers to allow dilution of brines by normal sea water. Furthermore, a lateral sequence of evaporite mineral facies is frequently encountered, as shown in Fig. 13.19, with a continuous flow of water undergoing constant evaporative precipitation of carbonate first, sulfates farther along, and chlorides at the farthest end of the flow.

SUPRATIDAL *SABHKA* EVAPORITES

A fruitful line of research in evaporite (and

FIGURE 13.21

A: Evaporites and dolomite formed due to infrequent washovers onto supratidal salt flats (called *sabhka* in Arabic) and some seepage from normal marine waters at right. *(Adapted from Illing et al., 1965, Society of Economic Paleontologists and Mineralogists Special Publication 13.)* B: Dolomization of limestone (CaCO₃) by seepage reflux of very saline brines derived from restricted lagoon (left) where evaporation increases brine density. Dense, magnesium-rich brines are "pumped" through limestone toward right; calcium is carried away and replaced by magnesium to produce dolomite [CaMg(CO₃)₂]. *(Adapted from Adams and Rhodes, 1960, Bulletin American Association of Petroleum Geologists, v. 44, pp. 1912–1920; by permission.)*

TABLE 13.1
Proportions of principal salts dissolved in normal sea water

Salts	Amount dissolved, %	Thickness if all were precipitated from 2,000 meters of water
NaCl (salt)	2.72	47.5
MgCl₂	0.38	5.8
MgSO₄	0.16	3.9
CaSO₄ (anhydrite)	0.12	2.3
K₂SO₄	0.08	
CaCO₃ (calcite)	0.01	0.6
MgBr₂	0.01	
	3.48%	60.1 meters

dolomite) formation indicates that *supratidal deposition* also is of major importance. For example, salt flats several kilometers wide border intertidal lagoons of the Persian Gulf. The flats, called *sabhkas* in Arabic, are but a few centimeters above average high tide and occasionally are flooded. Salt water fills the pores of sediments beneath the salt flats, but it is abnormally saline due to the high evaporative potential of arid climates. Gypsum and dolomite are precipitated within the salt-flat muds (Fig. 13.21). Under conditions of balanced subsidence and deposition, significant thicknesses of evaporite-bearing strata could accumulate, and lateral shifts of shoreline would cause transgressive or regressive migration of the evaporite facies.

POSTDEPOSITIONAL CHANGES

Evaporites undergo many changes after deposition. These involve chiefly reactions with ground water because of the great solubility of the evaporite minerals. The most common

TABLE 13.2
Relative abundance of four chief bases dissolved in river and sea waters

	River water, %	Sea water, %
Calcium	73	3
Magnesium	11	10
Sodium	9	84
Potassium	7	3
	100%	100%

change is simple solution of evaporite layers, causing subsidence or collapse of overlying and interstratified insoluble sediments into solution caverns. Collapse causes fragmentation of non-evaporite rocks, producing a jumbled breccia of angular fragments. Another typical change is the conversion of anhydrite to gypsum near the surface by *hydration*, that is, the combination of two molecules of water with each molecule of calcium sulfate. As a result of such changes, we only rarely see normal, original evaporite sequences in surface outcrops. Undisturbed, natural evaporite deposits are largely confined to the subsurface below the depth of penetration of fresh ground water. Indeed, many evaporite deposits were unknown until deep drilling penetrated hitherto unexplored basins.

ORIGIN OF DOLOMITE ROCKS

CALCIUM AND MAGNESIUM IN THE SEAS

Magnesium is about three times as abundant in sea water as calcium; yet calcium is added six or seven times faster by rivers (Table 13.2). At present rates of input, calcium abundance should be doubled in only 1 million years, and magnesium in 18 million years. Because the composition of sea water seems to have remained fairly constant, at least over later geologic time, both calcium and magnesium must be constantly removed from sea water so that some sort of balance of their abundances—a dynamic equilibrium—is maintained. The figures also show that calcium is less soluble than

magnesium, so is more readily extracted. The carbon dioxide content of sea water largely controls deposition both of calcium and magnesium, but carbon dioxide is in turn sensitive to temperature. With an increase of temperature, carbon dioxide escapes from water to the atmosphere. This causes dissolved calcium to become less soluble, and inorganic precipitation of calcium carbonate may ensue, especially in warm climates. The reaction is as follows, in which loss of CO_2 drives it to the right:

$$Ca^{2+} + 2HCO_3^- \rightleftharpoons CaCO_3 + H_2O + CO_2 \uparrow$$

A decrease of temperature accompanied by an increase of pressure, as in the deep seas, increases both calcium and magnesium solubilities. Therefore, many of the skeletons found more than 5,000 meters below sea level are composed of silica, which is less soluble there than calcium carbonate.

Given an abundant supply of calcium and magnesium from weathering of rocks on land, the quantity of carbonate sediments that can form in the shallow seas is a function of the carbon dioxide content of sea water, which, in turn, is controlled ultimately by the amount of that gas in the atmosphere. Sea water is more nearly saturated with respect to calcium carbonate than more soluble magnesium carbonate, even though magnesium is more abundant. It follows that special conditions are required to precipitate magnesium carbonate to form dolomite.

IMPORTANCE OF SALINITY

From the above discussions, it appears that some carbonate sediments could form by direct, inorganic precipitation from sea water through evaporation. Dolomite $[CaMg(CO_3)_2]$ is the most common evaporitic carbonate mineral. Experimental as well as observational evidence indicates that dolomite commonly forms in an alkaline medium such as sea water with slight excess salinity and slightly elevated temperatures. Under such conditions, magnesium now becomes insoluble and substitutes for some calcium atoms to form either impure calcite or dolomite.

Although a theoretical case can be made for direct precipitation of dolomite under evaporative conditions, there are immense volumes of dolomite rock that show no evidence of having so formed. Most Paleozoic dolomites formed initially as accumulations of calcareous skeletons, yet no known organism secretes the mineral dolomite. Therefore, much dolomite rock must have been converted from limestone after deposition by the following reaction:

$$2CaCO_3 + Mg^{2+} \rightarrow CaMg(CO_3)_2 + Ca^{2+}$$

Full discussion of this reaction is beyond the scope of our book, but one of the reasons that calcium now becomes more soluble than magnesium has to do with relative concentrations of the two elements; if magnesium becomes more than 8 to 10 times as abundant as calcium, dolomite will form.

Recrystallization of limestone to dolomite rock is called *dolomitization*, but this occurs in a variety of ways and at varying times after initial deposition. Magnesium must have been introduced, and, from previous discussions, percolation of high-salinity brines through limestone is one important mechanism. In recent years, extensive search of modern sediments (using x-ray identification techniques) has shown that modern dolomite, long thought nonexistent, is common in supratidal muds. Evaporative increase of salinity in pore waters, as in the example of supratidal evaporites cited above (Fig. 13.21), causes magnesium to be precipitated. The brines undergo continuous seepage reflux, the reality of which has been confirmed by experiments as well as field observations of modern sediments. This makes possible dolomitization of large volumes of limestone. Very recently, yet another mechanism has been suggested that involves repeated mixing of fresh with sea water, as in certain permeable layers beneath Florida. The chemical explanation of this postulated mechanism is not clear, but repeated changes of saltiness seemingly would affect relative solubility of magnesium versus calcium.

COMPARISONS OF MODERN AND ANCIENT CARBONATE SEDIMENTATION

We should next seek suitable environments in the modern seas that might match those of early and middle Paleozoic carbonate deposition. Remember, however, that there is little assurance that all ancient environments have modern counterparts (and vice versa). But it is reasonable—even essential at first—to assume that the more persistent ancient environments are at least approximated somewhere today. Already we have made comparisons of modern reefs with ancient ones, but equally important analogies can be drawn for nonreef carbonates as shown in Fig. 13.22 (also 11.22).

We look to the modern shallow seas for analogies with ancient epeiric seas for reasons outlined in Chap. 11. Broad, smooth continental shelves, together with the Baltic Sea, Black Sea, and Hudson Bay, represent small contemporary epeiric seas. From paleogeographic evidence, it is clear that total land area today is many times larger than during the first two-thirds of Pale-

FIGURE 13.22
Restoration of depositional environments for Lower Devonian Manlius Limestone, southeastern New York, based upon carbonate lithology, faunal content, and comparisons with modern carbonate environments of the Florida–Bahama region. Note supratidal dolomitization due to evaporation (left). (*After LaPorte, 1967*, Bulletin of the American Association of Petroleum Geologists, *v. 51, pp. 73–101; by permission.*)

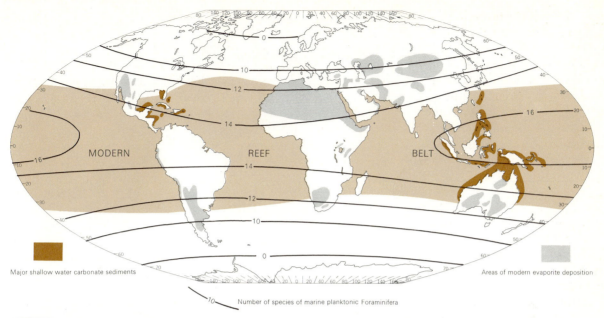

Major shallow water carbonate sediments

Areas of modern evaporite deposition

10 — Number of species of marine planktonic Foraminifera

FIGURE 13.23

Distribution of modern organic reefs, major shallow-marine carbonate deposition, and evaporites (chiefly nonmarine). (*Adapted from Lowman, 1949*, Geological Society of America Memoir 39; *Rodgers, 1957*, Society of Economic Paleontologists and Mineralogists Special Publication 5; Goode's world atlas, 1964.) Also shown are contours indicating latitudinal diversity (number of species) of modern marine planktonic Foraminifera as a function of temperature. Foraminifera provide an illustrative example, but most marine and nonmarine organisms show similar patterns of increasing diversity in warm, tropical latitudes. (*After F. G. Stehli*, Science, v. 142, 22 November 1963, pp. 1057–1059; copyright 1963 by American Association for the Advancement of Science.)

ozoic time (see World Maps, p. 369). Therefore, we can only hope for small modern areas analogous to past carbonate-forming epeiric seas. Figure 13.23 shows the chief areas where carbonate sediments are forming today. Disregarding the pelagic foraminiferal oozes of the deep-sea floor, which first appeared in the late Mesozoic, carbonates shown are shallow, tropical sea products. A minimum of contaminating terrigenous clastic detritus is the most important factor in precipitation either organically or inorganically of significant carbonate strata. Nonetheless, there is no doubt that warm, well-lighted water provides optimum condi-

tions. It taxes the imagination little to translate these conditions to the Paleozoic carbonate-producing epeiric seas. Lands were relatively small and low until late Paleozoic time, and shallow seas were large. Calcareous-secreting organisms thrived, and their skeletons contributed the bulk of carbonate material (see Fig. 2.11). The wide distribution of ancient carbonates with abundant and diverse invertebrate faunas long has been taken to indicate wider-than-present Paleozoic tropics.

Note that so far we have assumed relatively constant seawater chemistry and other ecologic requirements for most marine organisms since Eocambrian time. The assumption of near-constancy of ecologic requirements of respective types of organisms is based upon simple probability. Many fossil communities show the same associations of types of organisms and of sediments as do their modern counterparts. It is improbable that all members of very complex communities could have changed requirements to the same degree and at the same rate through time. Therefore, the finding of ancient communities with two or three dozen organisms showing mutual relations closely duplicating their modern counterparts argues for similar require-

DEVONIAN LATITUDINAL DATA

Reefs
E   Equator
G   Possible Glacial Tillites

Red Beds
Evaporites
Possible Poles   P ▲

Approximate maximum transgression

FIGURE 13.24
Presumed paleoclimatic and latitudinal indicators for Devonian time compared with paleomagnetically indicated
equator position (E) shown on modern geography for contrast. Coral reefs, evaporite deposits, and red-bed facies
all fall near and parallel to the paleomagnetically indicated
equator; Devonian tills in South America lie far from the
equator near the paleomagnetically indicated south pole
for South America (note north pole in Pacific). *(Adapted
from Schwarzbach, 1963; Irving, 1964; McElhinny, 1967,
UNESCO Symposium on Continental Drift, Montevideo;
compare Habicht, 1979, American Association of Petroleum
Geologists Studies in Geology No. 9.)*

ments through time. Sulfur isotope variations in
evaporites suggest some changes of seawater
composition, but probably they were small.

PALEOCLIMATE AND PALEOGEOGRAPHY

Abundance of iron oxides in certain Silurian and
Devonian sediments indicate a strongly oxidizing atmosphere. Important evaporite sediments
prove a relatively high evaporation potential
over large parts of North America, and so one is
tempted to conclude that the climate was warm.

Over Devonian land in eastern North America,
apparently there was moderate humidity as
suggested by forests that cloaked the region
(Fig. 13.1). Presence of great organic reef complexes and rich, diverse marine fossils in both
Silurian and Devonian marine strata suggest
warm, shallow, agitated seas by analogy with
the restriction of modern reefs to shallow tropical seas (Fig. 13.23). We also find today that the
greatest *diversity* of marine organisms occurs in
the warm subtropics and tropics (Fig. 13.23),
where optimum (though not exclusive) ecologic
conditions for many organisms exist (see also
Fig. 10.11).

Devonian land plants are similar the world
over, suggesting that climate was rather uniform. Wide distribution of richly fossiliferous
middle Paleozoic marine carbonate rocks, and
especially the great latitudinal spread of fossil
reefs (Fig. 13.24; Fig. WM.7) suggest subtropical
conditions for North America, Europe, Siberia,
and Australia. Based upon the above reasoning,
it has long been felt that the average climate of
the earth through time has been milder and
more homogeneous than it is today. The pre-

sent certainly is not a very good key to the past in terms of climate!

When compared with modern distributions, middle Paleozoic reef and carbonate rock patterns present a serious paleogeographic problem because they are found at 70°N latitude with respect to the present equator, 30° higher than reefs now grow! Three hypotheses can be presented to explain this seemingly anomalous latitudinal spread: (1) The past warm-water (subtropical) oceanic belt was so wide as to extend to about 70 or 80°N, thus bathing the entire North American continent in warm seas (Fig. 13.25A); (2) the subtropical belt, whether wider than today or not, was differently oriented with respect to North America and the present equator; i.e., the continent or the pole-equator system (and climatic zones) have somehow shifted relative to each other (Fig. 13.25B); or (3) reef-forming organisms have changed their ecologic adaptations slowly through time so as to become restricted gradually in latitudinal range.

The third possibility above seems least likely for reasons already discussed, and it is essentially untestable. Choice between the first and second hypotheses is not easy. An argument in favor of the second comes from paleomagnetic studies, which suggest that the middle Paleozoic magnetic field was oriented very differently than now with respect to North America. This relationship would bring the reefs within a subtropical belt parallel to that equatorial position. Comparing Fig. WM.7 and Fig. 13.24, we see that worldwide reef distribution and paleo-

magnetic data concur with other geologic evidence summarized before that points to a very different past arrangement of lithosphere plates.

DEVONIAN STRATA OF THE CRATON

REGRESSION AND TRANSGRESSION AGAIN

In Late Silurian and Early Devonian time, marine deposition became restricted to a few basins connected by narrow seaways and the marginal mobile belts. In the latter regions, marine Upper Silurian and Lower Devonian strata are perfectly conformable. Over most of the craton, however, regressive facies developed, and finally most of the craton emerged as a lowland for a few million years. Younger, transgressive Devonian strata rest unconformably upon a variety of older rocks, including even Prepaleozoic ones on several arches (Fig. 13.26). The Early Devonian unconformity is one of those rare cratonwide breaks like that beneath the Ordovician St. Peter Sandstone. It is the boundary between the second (Tippecanoe) and the third (Kaskaskia) cratonic sequences illustrated in Figs. 12.5 and 12.6. Pure quartz sandstone south of the eastern Great Lakes (Oriskany of Fig. 13.32) resembles the lower Paleozoic quartz sandstones in every respect, and, like them, it was deposited at the margin of a transgressive sea. It was derived from erosion of those same

FIGURE 13.25
Two alternate hypotheses to explain distribution of middle Paleozoic reefs and associated richly fossiliferous carbonate strata containing great diversity of species. A: Wider tropics (up to 70° latitude) then than now (darkest area) due to overall warmer average climate. B: Different relative orientation of the tropics with respect to North America.

PALEOGEOLOGY
BENEATH EARLY
DEVONIAN UNCONFORMITY

S Silurian

O Ordovician

€ Cambrian

Pp Prepaleozoic

Earliest Devonian?

TRANSCONTINENTAL ARCH

Devonian

Earliest

Present edge
of Devonian

0   1000 Miles

0   1000 Kilometers

FIGURE 13.26
Paleogeologic map showing rocks beneath the widespread Early Devonian unconformity. A great deal of warping affected the craton prior to Middle and Late Devonian transgression, but Prepaleozoic basement was exposed widely only in Canada (B represents basins). Note that youngest strata beneath a regional uncomformity tend to be less extensive than successively older ones; Lower Devonian strata are least widespread, Upper Silurian slightly more so, Middle Silurian and Ordovician ones, respectively, still more so. *(After Levorsen, 1960,* Paleogeologic maps, *and D.E. Owen, unpublished maps.)*

FIGURE 13.27
Effects of Devonian regression and transgression of cratonic arches and basins. Note differential erosion beneath the unconformity and differential deposition above. Subsurface information is mandatory to make such reconstructions in basin areas.

older sands, reflecting the repetitive nature of early cratonic history. Apparently the Ordovician and Silurian carbonate rocks previously deposited were dissolved by ground water, and any shale (of which there was very little) was weathered, with the resulting clay carried beyond the edges of the craton.

CRATONIC BASIN DEPOSITION

Warping of the North American crust was widespread in Early Devonian time, and practically all of the craton and even portions of the mobile belts were above sea level and being eroded to produce profound changes on the continent. Basins, arches, and domes became more sharply delineated than ever before. The net result of warping and regression was much differential erosion (Fig. 13.26). As transgression occurred near the middle of the Devonian Period, marine deposition resumed first in basins, which were most easily flooded, and gradually encroached upon arches and domes (Fig. 13.27). As a result, Upper Devonian marine strata are most widespread (Fig. 13.28).

The Michigan Basin continued to subside and more evaporites were deposited. The Williston or Elk Point Basin farther west also received important evaporites. An immense barrier-reef complex developed around its margin and extended far northwest in Canada (Fig. 13.29). Apparently it formed along a zone of deeper water upwelling against a shallow carbonate shelf. Devonian reefs long had been known in the Canadian Rocky Mountains (Fig. 13.16), and petroleum had been produced from some since 1920, but reefs beneath the plains were unknown until about 1947, when drilling encountered phenomenal petroleum reserves trapped therein. The discovery triggered one of the continent's greatest oil booms and provided a wealth of information about buried rocks previously unknown. Indeed, existence of the basin itself was hardly appreciated before that time.

On the following pages:
FIGURE 13.28
Upper Devonian sedimentary facies. Note importance of reefs and evaporites in western Canada, and isotopic dates for widespread granite rocks (see Fig. 11.7 or endpapers for symbols and sources).

FIGURE 13.29
Late Devonian paleogeography. Note the importance of marginal tectonic lands in the east and north. Wind direction in the east is inferred from volcanic ash distribution; in the west, from reef patterns. Note collision of Europe with northeastern North America and approach of Africa.

UPPER

DEVONIAN

SEDIMENTARY FACIES

(about 360 m.y. ago)

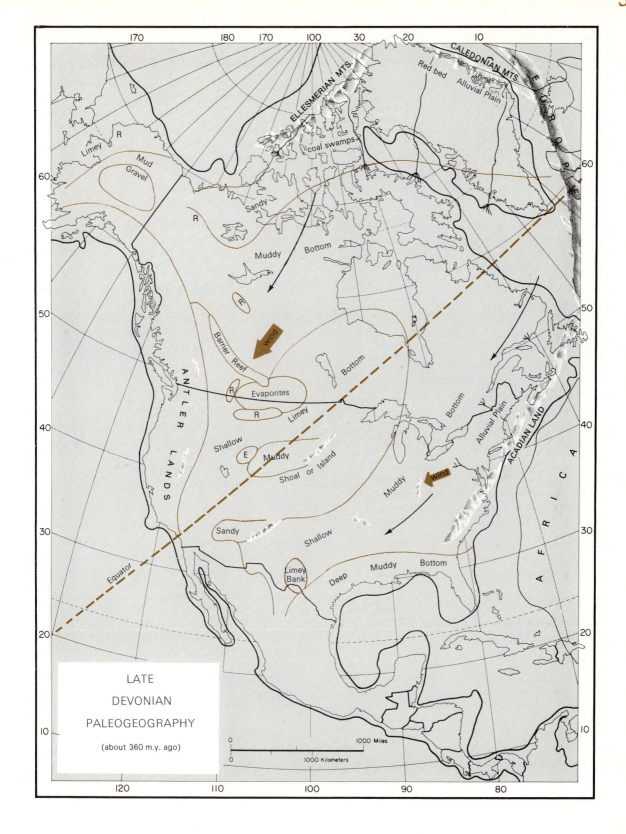

LATE

DEVONIAN

PALEOGEOGRAPHY

(about 360 m.y. ago)

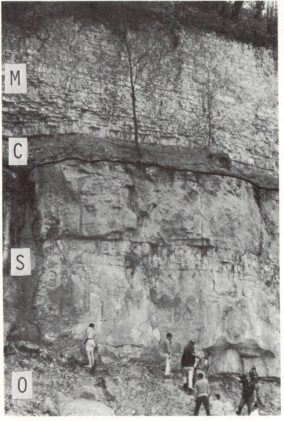

FIGURE 13.30

Outcrop on the northeast side of the Ozark Dome near Louisiana, Missouri, showing Ordovician, Silurian, and Mississippian strata. C is Mississippian Chattanooga Shale, which overlies a subtle but widespread regional uncomformity. Presence of the uncomformity here is apparent only from the absence of Devonian fossils. Compare Fig. 13.31. *(By R. H. Dott, Jr.)*

CRATONIC ARCHES

Because the stratigraphic record on arches is less complete than in basins, and because some of the unconformities there are very subtle (Fig. 13.30), it is more difficult to date their up-warpings. At least some Cambrian and Ordovician strata are present on most, but Silurian and Early Devonian rocks are absent from all the arches (Fig. 13.26). This situation naturally tempts us to guess that warping and erosion began in Silurian time, but there are other clues for us to consider, too. If shoals or islands were present in Silurian time, sandy or muddy facies would be expected around them; yet only Silurian carbonate rocks are found (Fig. 13.12). Moreover, the paleogeologic map of the mid-Devonian unconformity reveals that present distribution of Silurian rocks on the craton was controlled mainly by Early Devonian erosion. Therefore we conclude that many of the domes were not upwarped until Devonian time (for example Fig. 13.31). Because of the number of such features, we also conclude that this was a time of unusually severe cratonic deformation.

THE ACADIAN OROGENY IN THE APPALACHIAN BELT

EVIDENCE OF DEVONIAN OROGENY

James Hall's original observations leading ultimately to the concept of a geosyncline were based partly upon comparisons of Devonian

FIGURE 13.31

Section across a typical cratonic arch, the northern Ozark dome, Missouri, showing several unconformities as they appear today (vertical scale exaggerated). Uniform thickness of Cambrian and Ordovician strata across the arch indicates that upwarping occurred after their deposition. For reasons listed in text, initial warping is dated as Early Devonian. Later warping produced Mississippian and Pennsylvanian unconformities.

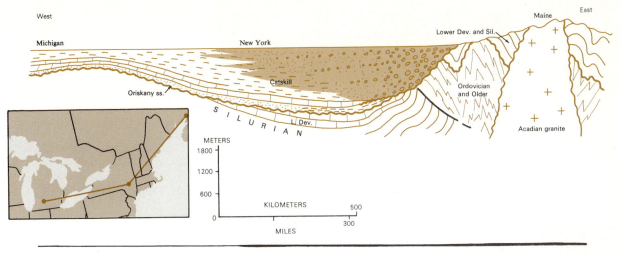

West   East

Michigan   New York   Maine

Lower Dev. and Sil.

Oriskany ss.

Catskill

Ordovician and Older

S I L U R I A N

L. Dev.

Acadian granite

METERS
1800
1200
600
0

KILOMETERS
500
300

MILES

Section without vertical exaggeration

FIGURE 13.32

Restored cross section of Devonian rocks in eastern United States showing effects of the Acadian orogeny and the upper Lower Devonian (pre-Oriskany) unconformity. Oriskany sandstone was derived from erosion of older sandstones in the central craton. Note superimposing of Acadian on older Taconian folding (upper right).

strata of the craton with those of the northern Appalachian Mountains. Whereas the entire Devonian sequence averages less than 300 meters thick in the craton, it is nearly 1,800 meters in the Appalachian region of eastern Pennsylvania and New York (Fig. 13.32). In New England and southeastern Canada, Devonian rocks include considerable lava and volcanic ash; and throughout most of the Appalachian belt, later Devonian strata include red sandstone, conglomerate, and shale (Figs. 13.28 and 13.33), a *molasse* facies named for the Catskill Mountains of southeastern New York.

Like the older red clastic succession of Late Ordovician age in the same region, Catskill sediments also coarsen toward the east, and so must reflect elevation and erosion of another prominent land there. We know that this land was mainly composed of but slightly older fossiliferous Paleozoic rocks. The Devonian tectonic land encompassed most of the area of the older eroded Taconian land (Fig. 13.29). Thus a major Devonian orogeny was superimposed upon the older one. In the maritime region of southeastern Canada and in northern New England, granites of Devonian age and unconformities related to this orogeny are well displayed. Therefore, this mountain-building episode has been named the *Acadian orogeny* for the old French colonial name for that region.

More granite and regional metamorphism (Fig. 13.34) developed during the Acadian than the Taconian event, and probably some thrust faults also were active. The Acadian orogeny, therefore, was a more severe disturbance of the

FIGURE 13.33

Microscopic photograph in polarized light of an Upper Devonian (Catskill) sandstone from southeastern New York. Immature sand composed largely of metamorphic rock fragments (e.g., quartzite and schist) was derived from erosion of the Acadian tectonic land in present New England. Larger grains are about 2 millimeters in diameter. *(By R. H. Dott, Jr.)*

FIGURE 13.34
Superimposed or overprinted metamorphic events reflected in discordant K–Ar isotopic dates. Stratigraphic relations show that Hudson Highlands gneisses are all Prepaleozoic; in the west, isotopic dates reflect their true age (850 to 730 m.y.). Fossiliferous Cambrian-Ordovician rocks to the north yield metamorphic dates reflecting the Taconian orogeny (435 to 400 m.y.); but some Prepaleozoic gneisses also yield Taconian dates due to remetamorphism. Finally *all* the rocks in the eastern half of the map, regardless of original age or type, yield overprinted Acadian dates (380 to 350 m.y.). (*After Long, 1962, p. 998, and Long and Kulp, 1962, pp. 980 and 983, Bulletin of the Geological Society of America, v. 73; by permission.*)

intruded by granitic plutons serve to date the orogeny rather closely. Also, Acadian granites have been dated extensively by isotopic methods (Fig. 13.28). In Nova Scotia and New England, many yield dates of from 360 to 330 million years ago. In Maryland a large mass known as the Baltimore Gneiss has been dated by the K–Ar method, using biotite mica, as from 350 to 300 million years (Early Mississippian). But zircon from the same rock yielded U–Pb dates of from 1,100 to 700 million years (Late Prepaleozoic)! These were among the first *discordant isotopic dates* discovered for a single rock (See Chap. 4). In recent years such discordances have proved to be more common than was supposed; indeed, they are now more or less expected in complex mobile belts that have suffered several superimposed or overprinted orogenies.

Discordant dates reflect two major events in the history of the Baltimore Gneiss. This rock apparently was formed first as a granite during the widespread Grenville orogeny (1,000 to 700 m.y.), which affected the entire southeastern margin of present North America and produced the complex metamorphic and igneous rock basement of the Paleozoic Appalachian mobile belt (see Fig. 8.10). The Maryland granite was again heated and deformed during the Acadian orogeny, at which time biotite was recrystallized so that it shows an isotopic age of 350 to 300 million years. Zircon is very resistant to temperature changes; therefore it has retained isotopic ratios of uranium and lead that reflect the date of original crystallization of the granite. Biotite is very sensitive to heating in excess of 200°C during metamorphism, and argon 40,

earth's crust and represents the greatest orogeny for the Appalachian mobile belt.

DATING THE OROGENY

At several scattered localities in the Acadian region, angular unconformities with Mississippian strata resting variously upon deformed Lower Devonian or older rocks that were being a gas, tends to leak from the mica crystal lattice. The result is resetting of the mica's isotopic clock; all the ^{40}Ar now present in the biotite of the Baltimore Gneiss has accumulated from decay of ^{40}K since Acadian metamorphism and cooling of the rock below 200°C.

Isotopic evidence, unconformities, and the thick Middle and Late Devonian Catskill clastic rocks all together indicate that Acadian mountain building occurred during Late Devonian and Early Mississippian time through an interval of perhaps 25 to 30 million years (360 to 330 m.y. ago), which is a significant span even to the geologist.

THE CATSKILL CLASTIC WEDGE

Conditions under which the Late Devonian clastic sediments accumulated in the eastern United States were almost identical to those under which the Late Ordovician sediments of the same region had formed. But the total volume and coarseness of the Catskill rocks exceed those of the Ordovician, and the pebbles and sand grains of the Catskill are chiefly composed of metamorphic and granitic rock fragments, feldspar, mica, and quartz (Fig. 13.33). The red color is due to the presence of a small percentage of iron oxide between the grains.

The Catskill sediments form a *clastic wedge* deposited as a vast alluvial coastal plain that sloped gently westward from the eroding Acadian mountains to the epeiric seashore. This plain was on the order of 300 to 500 kilometers wide at its maximum extent. At its eastern side, it was built of gravels and sands debouched by rivers from the foot of the mountains. Many river channel deposits are evident in the eastern Catskill facies. Along the western shoreline, however, finer sediments were deposited on river deltas and black marine muds and some thin sands spread beyond the deltas onto the craton (Fig. 13.28) as in Late Ordovician time (see Fig. 12.8). Volcanic ash also blew westward over the eastern craton.

In southeastern New York, the red sediments contain remains of a spectacular buried fossil forest (Fig. 13.1). Tree stumps about 30 centimeters in diameter are found buried in their rooted

positions, but some trees were rafted down rivers to the epeiric sea, there to be buried in black shales. Identical land plants occur in Devonian red beds in eastern Greenland, and coal is present in Arctic Canada. The oldest known vertebrate land animals also occur in the Greenland strata.

THE CHATTANOOGA SHALE ENIGMA

Black muds, which first began accumulating in the region west of the Catskill deltas in Middle Devonian time, spilled westward onto the craton in Late Devonian time (Figs. 13.28 and 13.32). By the end of the period, such muds were deposited all across the craton from New York to Nevada and north into Canada. This unprecedented expanse of mud continued to accumulate on the craton for perhaps 5 or 6 million years into Early Mississippian time, resulting in the Chattanooga Shale of Fig. 13.30. Erosion of numerous landmasses all around the continent (Fig. 13.29) gave rise to the vast quantity of clay. The shale contains much organic carbon, traces of uranium and phosphate minerals, and iron sulfide. Slow deposition in an oxygen-poor (anaerobic) environment uninhabitable by carbonate-secreting organisms is indicated. Yet, as we have seen, epeiric seas typically were agitated and oxygen-rich (aerobic), therefore the Chattanooga is a great puzzle.

Some authorities have postulated unusually deep, stagnant water for the deposition of the black shale, as is indicated for the older graptolitic facies in mobile belts. This idea may be supported by an apparent sea level about 200 meters above present as indicated by the worldwide sea-level fluctuation curve of Fig. 3.21 (or endpapers). But within only a few million years, deposition returned to normal, shallow-water carbonate conditions (late Early Mississippian). Another possible explanation is that excessive worldwide salinity developed due to preceding evaporative conditions. This could cause density stratification of the epeiric sea, resulting in a stagnant-water layer over the sea floor. Still another suggestion is that the epeiric sea was so clogged by floating marine vegetation (as in the present Sargasso region of the mid-Atlantic)

FIGURE 13.35
Approximate volume of Catskill clastic sediments deposited west of the source area (left) plus an assumed equal volume shed eastward together would require erosion of a mountainous mass about 1,000 x 200 x 12 kilometers high (or 7 kilometers higher than Mt. Everest). But erosion would proceed simultaneously with uplift; therefore, the mountains need never have been so high.

that mixing of bottom water was inhibited. Further discussion of black shales will be found in Chapter 16.

A MODERN ANALOGUE

The volume of preserved coarse Catskill sediments is on the order of 288,000 cubic kilometers, as contrasted with about 105,000 cubic kilometers of Ordovician ones. The Devonian clastic wedge must represent the erosion of at least half of a mountainous Acadian landmass with dimensions shown in Fig. 13.35. But assuming that uplift was more or less continuous over 20 to 30 million years, the land need never have been more than about 2 kilometers high.

For comparison it is instructive to note that a modern, analogous tectonic land exists in New Guinea, which is 2,000 kilometers long by 600 wide and averages nearly 2 kilometers in elevation (Fig. 13.36). New Guinea also is marginal to a craton, that of Australia, and an epeiric sea now floods the northern edge of that craton much like (although smaller than) the Devonian epeiric sea over the North American craton. Finally, sedimentation is also similar in that carbonate deposits, including reefs, are forming on much of the epeiric sea floor. Terrigenous clastic material is accumulating on advancing deltas along the south side of New Guinea, where large rivers drain the mountainous upland of the tectonic land and flow south across a broad, heavily forested alluvial plain. South-

ward (cratonward) regression of the sea appears to be happening today in southern New Guinea due to sedimentation just as it did in North America 350 million years ago.

COLLISION OF EUROPE AND AFRICA WITH NORTH AMERICA

THE CALEDONIAN MOBILE BELT

An early Paleozoic mobile belt, called the Caledonian, has long been recognized in Britain and Norway along the northwestern margin of the European craton. It was from studies of this belt that James Hutton formulated his revolutionary eighteenth-century views of the earth. The Caledonian belt is a twin of the northern Appalachian–East Greenland system, and it contains similar rocks, which reveal a similar history. Episodes of volcanism and deformation are recorded in the Caledonian belt at least as early as Cambrian time; more severe ones in the Ordovician correspond to the Taconian, whereas the culminating *Caledonian orogeny* (although slightly older—end of Silurian) corresponds closely to the Devonian-Acadian event of America. Even major cratonic events show some correlation between the two continents. The warping of basins and arches as well as major transgressions and regressions correspond to a remarkable degree (see Figs. 12.5 and 12.6),

FIGURE 13.36

Relations of modern tectonic lands of New Guinea and Indonesia to the Australian craton, a close analogue to Late Devonian conditions in North America. New Guinea is a complex tectonic and volcanic borderland with dimensions comparable to the Acadian land; Australia is about the size of the United States. Rivers carry clastic sediments southward from the mountains toward the craton where they are deposited in deltas at the margin of an epeiric sea, which, like its larger Devonian counterpart, has carbonate sedimentation and active reefs (dark brown areas) over much of its area (rotate page 90° clockwise for a closer comparison with Fig. 13.29). *(Analogy originally suggested to writers by T.S. Laudon.)*

suggesting that cratonic structure is also controlled by plate movements.

The first geologists to study eastern Greenland were Europeans, which was a happy circumstance, because what they found was so familiar that they might easily have forgotten they had left home. (It was in East Greenland that Alfred Wegener, father of the continental drift theory, lost his life in a blizzard in 1930). We noted in Chap. 12 the great similarity of lower Paleozoic strata and fossils there with those of northwestern Scotland (see Figs. 9.6 and 12.25). Devonian and Carboniferous sediments are also virtually identical with those of northwestern Europe as well as with those of the Appalachian belt, and species of fossil land plants and fishes in these rocks are the same on both continents.

STRUCTURAL SYMMETRY

Structurally the two margins of the ocean are mirror images. In East Greenland as in the Appalachian belt, thrust faulting carried thick mobile-belt rocks westward against the craton.

In Norway, thrusting was eastward against the European craton. In both cases intense metamorphism and large granitic batholiths developed in what is now the seaward side of those belts. Locally a definite bilateral symmetry still can be seen across the mobile belts today. In northwestern Scotland, Caledonian thrust faulting carried metamorphic rocks northwest over flat, lower Paleozoic carbonates, but eastward thrusting occurred nearby in Norway (Fig. 13.37). In Newfoundland, a central zone of

FIGURE 13.37

Simplified restored cross section of the Caledonian mobile belt, northwestern Scotland to southern Norway. Rocks of the mobile belt were thrust toward *both* margins; the belt apparently was symmetrical, being an *intercratonic* one formed between colliding Greenland and Europe.

FIGURE 13.38

Coarse, angular Devonian ("Old Red") conglomerate, Solund District, western Norway, one of several areas of thick, nonmarine sediments deposited in downfaulted basins following the Caledonian orogeny. Note heterogeneous composition representing varied older metamorphic rocks. *(Courtesy Tor H. Nilsen.)*

intensely deformed Ordovician and Silurian oceanic rocks lies between two zones of mildly deformed continent-margin rocks. From the above clues, as well as from paleomagnetic data, we can reconstruct the continents as they must have looked before and after a Late Devonian collision, which incidentally also resulted in the destruction of the proto-Atlantic Ocean shown in Fig. 12.25.

CATSKILL—OLD RED SANDSTONE *MOLASSE* FACIES

The Devonian red-bed deposits of eastern Greenland and northwestern Europe (see Fig. 3.23), like those of the eastern United States, reflect the rapid erosion of high mountains (Fig. 13.38). In Fig. 13.39 we see that the Old Red Sandstone of Europe was the mirror image of the American (Catskill) red beds. The sediments were transported by rivers in both directions away from a single great Caledonian-Acadian mountain range. Approximately equal volumes occur on both sides. The identity of land plant and fish fossils contained therein is readily explained as well.

INTERCRATONIC MOBILE BELTS

We now see more clearly that the *intercratonic* type of mobile belt noted in Chap. 6 results from

continent-continent collision. Moreover, it follows that many apparently marginal mobile belts like the Appalachian actually were intercratonic at some time and have been subsequently dismembered by continental drift. It is only since the development of plate-tectonic theory, however, that such belts could be adequately explained. While the similarities of northwestern Africa with eastern North America were not as striking as those of Europe, an apparent bilateral symmetry of the Appalachian belt in light of plate theory suggested collision

FIGURE 13.39

Hypothetical evolution of northern Appalachian–Caledonian mobile belts during closure of the postulated proto-Atlantic Ocean basin, which culminated in continental collision in Devonian time. Note that both continents had passive trailing edges with carbonate shelf zones facing the proto-Atlantic until mid-Ordovician time when subduction began. Note major change of subduction in eastern America after arc collision during Taconian orogeny (see Fig. 12.19). The old ocean basin completely disappeared later except for a mafic-rock suture zone (bottom). Note the symmetry of red Catskill and Old Red clastic wedges.

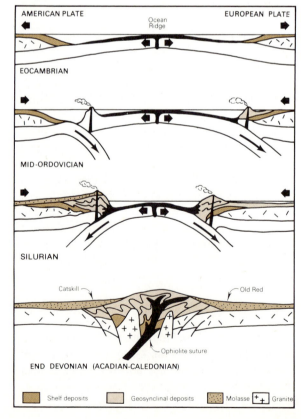

also with Africa. Comparative studies were initiated, and based upon the similarities found, we infer that much of the southeastern edge of North America was originally part of Africa, as shown in Fig. 12.25. Figure 13.39 summarizes the probable evolution of the Appalachian-Caledonian system in terms of lithosphere plate movements; the scheme shown also applies to other intercratonic belts in a general way. Of particular importance in unraveling these complex belts is the identification of *suture zones*, which are crushed oceanic rocks (typically the mafic ophiolites) that mark the old boundary between converging plates. Such a zone is generally all that remains of the ancient ocean basin that was consumed between two converging continents. Besides the suture, the central zone characteristically also has deformed volcanic arc rocks and granitic plutons, and shows high-temperature metamorphism. Bounding the central zone on both sides are intensely deformed sedimentary rocks thrust outward toward adjoining cratons (Fig. 13.39). Marginal mobile belts formed between converging oceanic and continental crust do not show such bilateral symmetry.

MOUNTAIN BUILDING IN THE ARCTIC AND CORDILLERA

In the Franklin mobile belt of Arctic Canada and in northern Alaska, Devonian conglomerates and sandstones (Fig. 13.28) reflect uplifts simultaneous with Caledonian-Acadian mountain building. The Franklin and Caledonian belts even merge in northeastern Greenland, but structural relationships with Alaska are unclear even though the facies and history were very similar. Indeed, it is not even clear that most of Alaska was attached to North America yet, and the same is true for many parts of the present western edge of Canada and the United States. We shall hear more about this problem in later chapters.

Granitic rocks of latest Devonian and earliest Mississippian ages are widely scattered in the Arctic (Fig. 13.28). Local ultramafic and volcanic rocks of early and middle Paleozoic ages present on the northernmost tip of Canada apparently reflect oceanic and island-arc rocks crushed

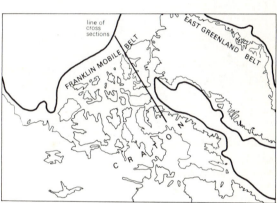

FIGURE 13.40
Sequential restored cross sections showing deformation in Arctic Canada produced by the Mississippian Ellesmerian orogeny. Note also distribution of reefs and evaporites. (Not to accurate scale.)

against the northern edge of the continent, causing the upheaval of mountains and thrust faulting of Paleozoic strata southward against the craton during the Late Devonian–Early Mississippian *Ellesmerian orogeny* (Figs. 13.40 and 13.41). This event probably resulted from collision either of an arc or of a continent such as Siberia with northern North America, but proof is lacking.

Disturbances also began to affect the Cordilleran region of western Canada and the United States for the first time. Important chert-bearing conglomerates were deposited in northern Alaska and northwestern Canada during Late Dev-

FIGURE 13.41
Devonian nonmarine strata in eroded anticlines and synclines of Franklin mobile belt, Arctic Canada (large plunging anticline at left, syncline in middle, anticline at right). Folding occurred in Mississippian Ellesmerian orogeny. *(Courtesy National Air Photo Library [Canada], Surveys and Mapping Branch, Department of Energy, Mines and Resources; Photo No. T419R-132.)*

onian and Early Mississippian times. In Nevada, at least 1,000 meters of Mississippian and Early Pennsylvanian sandstone and chert-rich conglomerate lie unconformably upon folded and thrust-faulted early Paleozoic oceanic sediments and pillow basalts (Fig. 13.42; also 12.2). Together with granites yielding Late Devonian ages in western Canada (Fig. 13.28), these rocks indicate an episode of mountain building called the *Antler orogeny*. Unlike the orogenies discussed before, however, deformation and metamorphism were mild, there was little volcanism, and no extensive nonmarine red-bed clastic wedge was formed. Here most of the clastic

sediments eroded from modest-sized tectonic lands were deposited below sea level by turbidity currents and at sea level on small deltas. The

FIGURE 13.42
Evidence of the Antler orogeny in northern Nevada. Folding and thrusting began in Late Devonian and continued through Mississippian time; note westward overlap of later Paleozoic conglomerate as lands were eroded (C and R denote index fossil zones). *(After R. H. Dott, Jr., 1964, Kansas Geological Survey Bulletin 169.)*

sediments are characterized by an abundance of chert and quartz fragments resulting from erosion of uplifted early Paleozoic deep-marine cherts and sandstones deposited beyond the western edge of the continent (see Fig. 11.5). What could have caused this peculiar Antler orogeny? There is no evidence of collision with another continent here, but in northwestern California there are Devonian rocks that may be the remnants of a volcanic arc that collided with western North America to cause the orogeny. Their location during Devonian time is not yet satisfactorily proven however.

SUMMARY

The middle Paleozoic was a time of unusual evolutionary diversification among practically all organisms—marine and land; animal and plant; invertebrate and vertebrate. Corals and crinoids became extremely important, and ammonoids appeared. Fishes enjoyed their most rapid evolution ever, culminating in exploration of land by a lobe-finned hopeful amphibian. Land plants, which appeared by the end of Silurian time, developed rapidly so that diverse forests with large trees cloaked the Devonian lowlands. Primitive land plants were dependent upon water to complete their reproduction, so probably could not inhabit drier uplands yet.

Middle Paleozoic time was one of continued extensive carbonate sedimentation in shallow epeiric seas on the craton, but with increasingly important influences of arches and basins. There was a great increase both of organic reefs and evaporite deposits. All of these were unusually widespread, but reefs had their all-time maximum development in Devonian time. Obstructions to marine circulation, especially through activities of reef-building organisms, repeatedly enhanced evaporative precipitation. Development of abnormally saline waters in turn enhanced the conversion of limestone to dolomite. Distribution of the richly fossiliferous carbonate rocks, organic reefs, and evaporites, together with paleomagnetic and paleowind evidence, indicate an equatorial setting for

North America during Silurian and Devonian time.

Important oscillations of relative land and sea levels occurred. The most profound unconformity that resulted is within the lower part of the Devonian succession, rather than at a system boundary (just as in the Ordovician). Extensive warping that affected almost the entire continent probably was related to the beginning of severe Acadian and Caledonian disturbances on the eastern side of the continent. In the northern Appalachian region and eastern Greenland, as well as in northwestern Europe, that middle Paleozoic mountain building produced large tectonic lands, erosion of which resulted in the deposition of immense volumes of coarse, red clastic sediments adjacent to them.

Geologic, paleontologic, and paleomagnetic evidence all point to a Devonian collision of northwestern Europe with eastern North America as the culmination of consumption of proto-Atlantic lithosphere by subduction beneath those continents. Plate-tectonic theory explains the mountain building in the Appalachian–East Greenland and the Caledonian mobile belts as the result of that subduction and collision to produce a bilaterally symmetrical intercratonic mobile belt. Orogenies affecting the Cordilleran and Arctic regions cannot yet be explained as fully in terms of plate theory, but it is likely that these resulted from collisions either of volcanic arcs and/or small continental blocks. In any case, the entire continent was now feeling the effects of lithosphere plate interactions.

Key Concepts

Catskill clastic wedge
Chattanooga Shale
discordant isotopic dates
diversity of organisms versus latitude
dolomitization
dynamic equilibrium or steady state
ecologic niche
ecologic zone (communities)
endemic species
evaporites:
    dense brines
    supratidal *sabhkas*

faunal province
hydration
intercratonic mobile belts
lobe-finned fish
marine food chain
Middle Paleozoic orogenies:
   Acadian (Appalachian region)
   Antler (western North America)
   Caledonian (Europe and Greenland)
   Ellesmerian (Arctic Canada)
Old Red Sandstone clastic wedge
organic reefs:
   barrier versus fringing versus atoll
   fore-reef versus back-reef facies
   reefs versus latitude
Oriskany Sandstone
pelagic
plankton
suture zones
upwelling

Readings

Alberta Society of Petroleum Geologists, 1964, Geological history of western Canada: Calgary. (An atlas of stratigraphic maps)

Alling, H. L., and Briggs, L. I., 1961, Stratigraphy of Upper Silurian Cayugan evaporites: Bulletin of the American Association of Petroleum Geologists, v. 45, pp. 515–547.

American Association of Petroleum Geologists, 1973, Arctic geology: Tulsa, American Association of Petroleum Geologists Memoir 19.

Bird, J. M., and Dewey, J. F., 1970, Lithosphere plate-continental margin tectonics and the evolution of the Appalachian orogen: Bulletin of the Geological Society of America, v. 81, pp. 1031–1060.

Boucot, A. J., 1974, Evolution and extinction rate controls: Developments in palaeontology and stratigraphy; v. 1, Amsterdam, Elsevier.

Haller, J., 1971, Geology of the East Greenland caledonides: New York, Wiley-Interscience.

House, M. R., 1979, The Devonian system: London, Palaeontological Association Special Papers, v. 23.

Kahn, P. G. K., and Pompea, S. M., 1978, Nautiloid growth rhythms and dynamic evolution of the earth-moon system: Nature, v. 275, pp. 606–611.

Kay, M., ed., 1969, North Atlantic geology and continental drift: American Association of Petroleum Geologists Memoir 12.

Ladd, H. S. (ed.), 1957, Treatise on marine ecology and paleoecology: Geological Society of America Memoir 67, v. 2, Chaps. 10, 11.

Lovell, J. P. B., 1977, The British Isles through geological time: London, Allen and Unwin.

Millot, J., 1955, The coelacanth: Scientific American, December.

Oliver, W. A., Jr., 1976, Biogeography of Devonian rugose corals: Journal of Paleontology, v. 50, pp. 365–373.

Oswald, D. H. (ed.), 1968, Proceedings of the international symposium on the Devonian System: Calgary, Alberta Society of Petroleum Geologists.

Pannella, G., MacClintock, C., and Thompson, M. N., 1968, Paleontological evidence of variations in length of synodic month since Late Cambrian: Science, v. 162, pp. 792–796.

Pray, L. C., and Murray, R. C., 1965, Dolomitization and limestone diagenesis: Society of Economic Paleontologists and Mineralogists Special Publication 13.

Scrutton, C. T., 1964, Periodicity in Devonian coral growth: Palaeontology, v. 7, pp. 552–558.

Sloss, L. L., 1972, Synchrony of Phanerozoic sedimentary-tectonic events of the North American craton and the Russian platform: Montreal, 24th International Geological Congress Proceedings, sec. 6, pp. 24–32.

Strand, T., and Kulling, O., 1971, Scandinavian caledonides: New York, Wiley-Interscience.

Wade, N., 1969, Three origins of the moon: Nature, v. 223, pp. 948–950.

Wells, J. W., 1963, Coral growth and geochronometry: Nature, v. 197, pp. 948–950.

Wilson, J. L., 1975, Carbonate facies in geologic history: New York, Springer-Verlag.

14

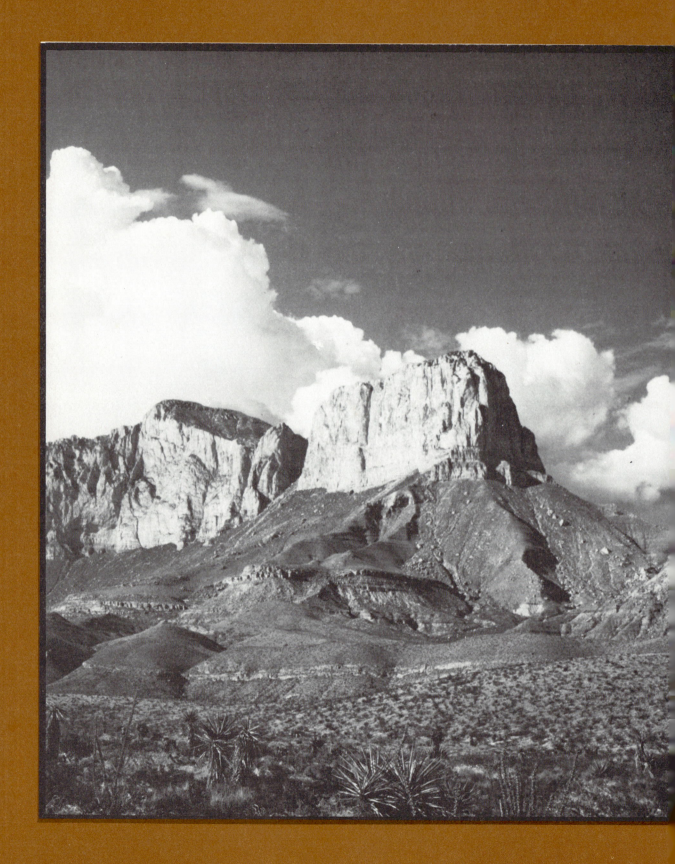

14

LATE PALEOZOIC HISTORY
A Tectonic Climax and Retreat of the Sea

Here about the beach I wandered,
nourishing a youth sublime
With the fairy tales of science,
and the long results of time.

Alfred Tennyson

FIGURE 14.1
Famous Permian reef complex of the Guadalupe Mountains, southwestern Texas. Vertical limestone face of El Capitan peak at right is approximately the shallow-water reef front. Sea level was just above the tops of the mountains, while strata below El Capitan in the foreground were deposited at a depth of almost 400 meters. *(Courtesy G. A. Crawford.)*

14

We have shown that during early and middle Paleozoic time, North America experienced repeated and widespread transgressions and regressions by epeiric seas. Except at times of considerable mountain building, carbonate rocks tended to form widely, even in the mobile belts. But during major orogenic episodes clastic wedges engulfed mobile belt areas and even influenced cratonic sedimentation as we saw for Ordovician and Devonian times.

Whereas the North American Paleozoic stratigraphic record had been dominated by marine conditions before, during Carboniferous and Permian times (350 to 230 m.y. ago), it was increasingly influenced by tectonic disturbances that raised much of the craton above sea level. Early Carboniferous or Mississippian strata represent a transition from middle Paleozoic marine to later nonmarine conditions. By the end of the era, practically the entire continent stood above sea level and was inhabited by cosmopolitan plants and animals. What had been a paradise for denizens of the sea was to become one for swamp forests, lurking reptiles, and giant insects. Wholesale extinctions of large groups of shallow marine invertebrates and explosive proliferation of land organisms accompanied the shrinkage of shallow seas.

General increase of structural unrest resulted in drastic changes in sedimentation in North America and Europe. Near the end of the Mississippian Period (roughly the Early Carboniferous of Europe), deposition of typical carbonate rocks almost ceased on the craton. Pennsylvanian (or Late Carboniferous) and Permian strata contain great volumes of terrigenous clastic sediments derived from peripheral mountains and also from large uplands within the craton itself. Pennsylvanian strata contain our largest coal deposits, which formed under special climatic and topographic conditions favoring luxuriant growth of swamp forests. Identical conditions and floras existed in Europe and North Africa as well. The coals occur in repetitive strata showing a conspicuous pattern of cyclic alternations of rock types that seems to be of worldwide significance. The Late Paleozoic, therefore, represents a time of revolutionary changes.

Uplift of the Caledonian and Franklin mobile belts ceased in late Paleozoic time. Erosion reduced the mountains there so that marine deposition resumed in Pennsylvanian and Permian times. Meanwhile, farther south, Africa and South America, which were fused together as part of the supercontinent Gondwanaland, completed their collision with North America and Europe. This last compression resulted in extensive thrust faulting and uplift in the southern Appalachian belt, and even more profound orogeny across Europe. Simultaneously, the collision of Siberia with eastern Europe formed the Ural Mountains, while several microcontinents collided and fused with southern Siberia to become China. Thus by the end of the Paleozoic Era, all the continents were more or less interconnected and total land area was very large. This situation greatly influenced the evolution of land life by facilitating dispersal of the organisms and providing a great diversity of ecologic niches.

MISSISSIPPIAN ROCKS

LAST WIDESPREAD CARBONATES

The Mississippian Period was characterized by the last widespread carbonate-producing epeiric sea in North America. Carbonate sediments spread over most of the craton and large portions of the mobile belts (Fig. 14.2). Limestones

FIGURE 14.2
Middle Mississippian paleography. Note large lands in eastern North America, Antler lands in Cordillera, Ellesmerian land in Arctic, and the approach of Africa from the southeast. Compare Figs. 13.29 and 14.8 to appreciate transitional nature of Mississippian period.

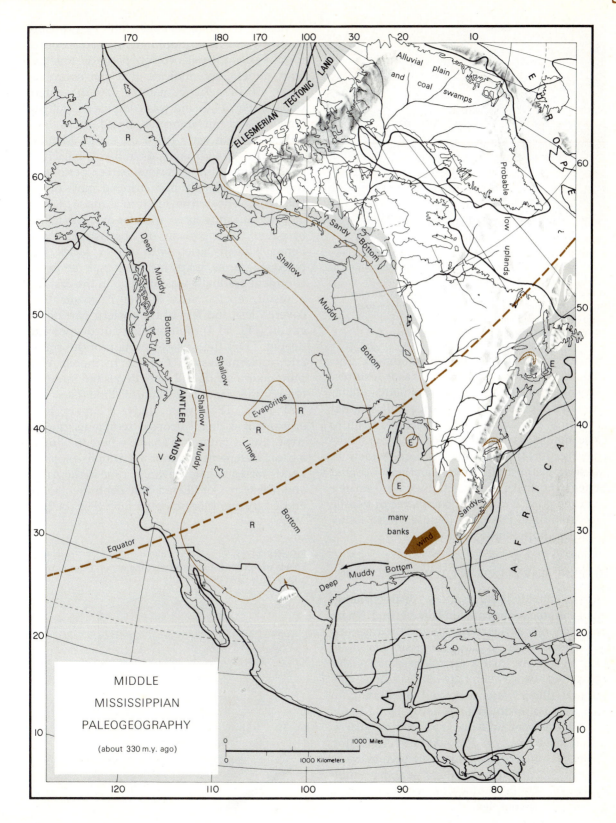

MIDDLE

MISSISSIPPIAN

PALEOGEOGRAPHY

(about 330 m.y. ago)

FIGURE 14.3
Cross stratification in pure crinoidal Mississippian Burlington Limestone near Hannibal, Missouri. This formation, with a total volume of about 300×10^{10} cubic meters, contains the skeletal remains of approximately 28×10^{16} individual crinoid animals. After death, the skeletons were disarticulated and the fragments dispersed like sand as shown here; it is a typical clastic limestone.

rich in crinoid fragments predominated (Fig. 14.3). They display conspicuous clastic textures. Breakage and sorting of fossil fragments, oölite, cross stratification, ripple marks, and scoured structures are common. These features point to close analogies with the modern Bahama Banks (see Fig. 13.14). Organic reefs occur widely in Mississippian strata from southern United States to northern Alaska, but they were smaller than the great barrier-reef complexes of middle Paleozoic time. Evaporite deposits were somewhat less important as well.

SEDIMENTARY AND TECTONIC CHANGES

During Late Mississippian time, distinct changes in cratonic sedimentation commenced. There was a change in the composition of terrigenous sediments. Prior to Late Mississippian time, only very pure quartz sandstones and shales composed overwhelmingly of a single clay mineral (illite) were deposited. Beginning in later Mississippian time, heterogeneous sands bearing feldspar and mica, as well as quartz, began to appear on the southeastern craton (eastern United States), and they became more widespread during subsequent periods. Simultaneously, the mineralogy of associated shales became more complex.

Most of the above changes indicate exposure by erosion of new, more heterogeneous source rocks. Facies and paleocurrent patterns, together with mineralogy, indicate that the source of much of the clastic material lay in eastern Canada (Fig. 14.2) and was composed chiefly of igneous and metamorphic shield rocks long buried beneath early Paleozoic strata. Material also was derived from lands in the Appalachian belt. Erosion during various earlier periods of regression gradually had stripped most of the Paleozoic veneer from eastern Canada, laying bare significantly large areas of old, crystalline basement. Comparison of Mississippian and Pennsylvanian facies with earlier facies and paleogeographic maps indicates that the entire continent gradually was tilted up in the east. As a result, epeiric seas of late Paleozoic time covered less and less of the eastern part of the craton (Fig. 14.4), and erosion cut deeper and deeper. Over the western portion of the craton and adjacent eastern Cordilleran mobile belt, pure quartz sands still were deposited intermittently throughout late Paleozoic time (Figs. 14.5 and 14.7). They were derived largely from erosion of earlier Paleozoic sediments on the west-central craton in Canada, where the Prepaleozoic basement was not yet widely exposed. Lands raised in Late Devonian time in the Arctic and western Cordilleran belt continued to be eroded during Mississippian time (Fig. 14.2).

FIGURE 14.4
Effects of uptilting and erosion of the craton beginning in middle Paleozoic time.

FIGURE 14.5
Large-scale cross stratification in Pennsylvanian-Permian Weber Sandstone, Yampa Canyon, Colorado. Most of this and related sandstones covering Utah and Wyoming represent wind deposits, but interstratified marine limestones indicate periodic widespread inundations by a shallow sea. It is as if a quarter of the Sahara Desert were flooded periodically. *(By R. H. Dott, Jr.)*

At the end of the Mississippian Period, a major regression drained the entire craton, and the earliest Pennsylvanian seas were confined to marginal mobile belts. As a result, a major unconformity occurs beneath Pennsylvanian strata over the craton (Fig. 14.6). A great deal of warping and faulting occurred before the sea returned to the craton in mid-Pennsylvanian time (Figs. 14.7 and 14.8).

LATE PALEOZOIC REPETITIVE SEDIMENTATION—AN ENIGMA

SEDIMENTARY CYCLES

Beginning in Late Mississippian and continuing through Early Permian times, the strata deposit-

ed over the craton and inner parts of the mobile belts displayed a striking *repetitive pattern*, which is present to varying degrees in late Paleozoic strata on other continents. Upper Mississippian deposits in the eastern states show clear repetitions of sandstone-shale-limestone sets repeated several times vertically. Illinois geologists have shown that the sandstones and shales represent in part deltaic deposits formed by river systems flowing from the southeastern Canadian region. In Pennsyl-

On the following pages

FIGURE 14.6
Paleogeology beneath widespread Early Pennsylvanian unconformity showing great warping and differential erosion of the craton. Old arches and basins were reactivated and new ones formed. Features in Canada are inferred because they lie beyond the present eroded (zero) edge of Pennsylvanian strata (compare zero edge in Fig. 14.7). *(From A. I. Levorsen, Paleogeologic maps, W. H. Freeman Co., Copyright© 1960; and unpublished maps by D. E. Owen.)*

FIGURE 14.7
Middle Pennsylvanian sedimentary facies; note eastern coal-bearing regions and western coarse clastic areas reflecting uplifted regions (see Fig. 11.7 for symbols).

PRE-PENNSYLVANIAN
PALEOGEOLOGIC MAP

Mu Upper Mississippian

M Mississippian

Lower and Middle Paleozoic

Prepaleozoic

Edge of
Pennsylvanian

M or Mu

1000 Miles

1000 Kilometers

MIDDLE PENNSYLVANIAN

SEDIMENTARY FACIES

(Des Moinesian)

(about 275 m.y. ago)

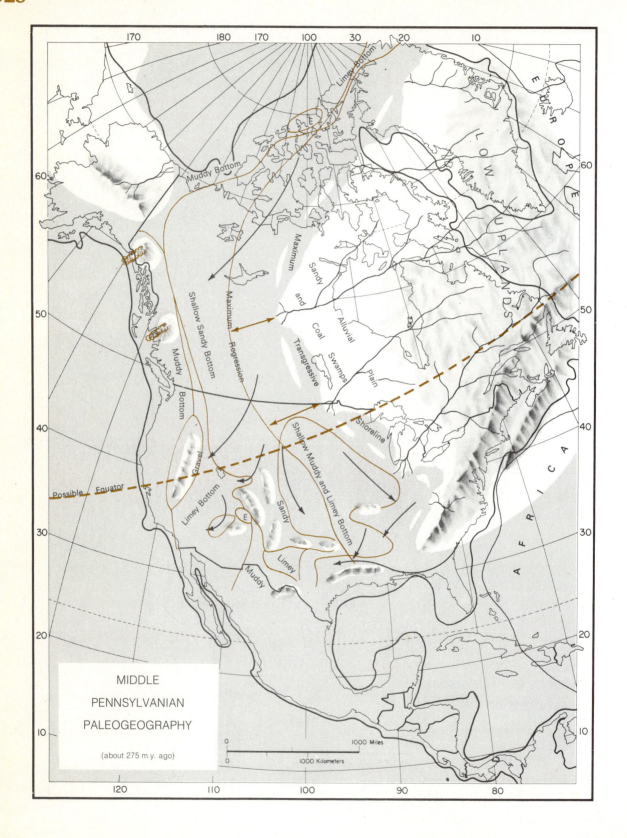

MIDDLE

PENNSYLVANIAN

PALEOGEOGRAPHY

(about 275 m.y. ago)

FIGURE 14.8

Middle Pennsylvanian paleogeography. Note enlargement of land area in eastern North America and many local lands in the southwest as compared with Mississippian time. Note approach of Africa from the southeast which caused upheaval of the Appalachian Mountains.

vanian time, influxes also came from the Appalachian region (Figs. 14.2 and 14.8).

Practically all Pennsylvanian strata on the continent show some kind of repetitive pattern, but the most striking occurs in coal-bearing sequences. At least 50 late Paleozoic cycles are known, many of which can be traced widely over the southern craton. A typical cycle (Fig. 14.9) commences at the base with cross-stratified sandstone and conglomerate resting unconformably upon older strata; channel structures, fossil logs, and relatively poor sorting of grains indicate these were formed by river

FIGURE 14.9

Idealized cross section showing lateral and vertical relations of one Pennsylvanian depositional cycle (cyclothem) reflecting major transgressions T and regressions R of the sea and shoreline.

processes. The middle of the cycle contains coal and plant-bearing shales, while the upper part generally contains brackish-water or marine fossiliferous shales and limestones. Many such cyclic sets of strata occur vertically stacked upon one another. Lateral variations within the sets are just as important as vertical alternations, for both record drastic environmental shifts through time.

In the central states, the upper marine rocks of the cycles are best developed, whereas the coal-bearing, nonmarine rock types predominate in the east. Meanwhile, the western craton was covered by a vast sand-dune desert (Fig. 14.5). Periodically it was flooded by a shallow sea that deposited thin but widespread marine limestones—again in a *cyclic pattern*.

Alternation of marine and nonmarine deposits points to many transgressions and regressions of the sea over wide areas. With more than 20 distinct cycles representing about 60 million years, however, it is apparent that these were geologically rapid oscillations, occurring at much shorter intervals than earlier Paleozoic

Section without vertical exaggeration

FIGURE 14.10

The Netherlands before diking began—model for ancient, low, swampy coasts. Note peculiar, nonprotruding delta of the Rhine River, which carries negligible sand; vigorous tidal and longshore currents due to winds sweep sand from the shallow (epeiric) North Sea floor northeastward to form long sand spits and barrier islands. Contours at lower right indicate elevation of land surface.

transgressions and regressions. Apparently they require a different explanation. Possible causes of these oscillations include: (1) *worldwide rise and fall of sea level* due either to advance and retreat of continental glaciers or to large-scale warping of the deep-sea floor; (2) *spasmodic tectonic up-and-down motions* of the entire continent; or (3) fluctuations of clastic sediment supplied to deltas in response to *cyclic climatic changes* affecting erosion in the uplands; as more sediment was supplied, deltas would advance, but transgression would occur when the supply decreased. The last cause, while an important process, is too localized to explain Pennsylvanian oscillations. The second (a kind of Saint Vitus's dance hypothesis) overtaxes credulity because so many short-term oscillations were involved. We prefer the first mechanism, with changes in known late Paleozoic glaciers on Gondwanaland as the ultimate cause.

Pennsylvanian strata provide perhaps the clearest examples of complex interactions of land-sea relationships. *A long-term rise of eastern North America* is evident (Fig. 14.4), but more localized differential warping of the craton also occurred on an intermediate time scale and accentuated many arches and basins. Finally, superimposed on both of those effects were the short-term transgressions and regressions presumably caused by sea-level oscillations. The unique rate of each process was superimposed upon that of the others to produce a very complex stratigraphic record.

COAL SWAMPS

Basal sandstones and shales of a Pennsylvanian cyclic set are interpreted as *river and delta deposits,* while the coals formed in vast coastal swamps containing junglelike vegetation. The *Dismal Swamp* of North Carolina (2,500 square kilometers), *Dutch lowlands* (15,000 square kilometers; Fig. 14.10), *Florida Everglades* (25,000 square kilometers), and comparable-sized south *Louisiana swamps* all provide useful modern comparisons.

The ancient *coal swamps* were dominated by large, scaly barked trees (lycopsids) that had evolved first in Late Devonian time. Fossil seeds show adaptations for dispersal by floating on water. Other plants also were present, and amphibians, primitive reptiles, air-breathing molluscs, and insects inhabited the soggy forests as well. Land floras and faunas of Europe and North Africa were identical. Presence of cold-blooded animals and the nature of the vegetation over the three continents indicate that the climate must have been very humid and probably warm. The Carboniferous trees lack growth rings, which together with the presence of large, thin-walled cells in the tree trunks, suggest a lack of distinct annual seasonal changes either of temperature or humidity, and that growth was rapid. This flora is considered to have been a tropical one.

As trees and shrubs died and fell to the swamp floor, much of their debris was protected from decay by rapid submergence and burial, thus excluding oxygen and preventing attack by all but *anaerobic bacteria*. With time, it was compacted by the weight of subsequent sediments to form peat. The typical ratio of thick-

FIGURE 14.11
Pennsylvanian coal above light-colored sandstones deposited in a channel that crossed an immense coastal swamp.
Note lenticular cross section of main channel deposit
(bottom); differential compaction of coal and shale overlying the sandstone accentuated the lens form of the channel. (Cagles Mill Dam, near Terre Haute, Indiana.)

nesses of uncompacted peat and coal is about 10
to 1. To form a sizable commercial coal seam
(Fig. 14.11) therefore required a fantastic quantity of vegetation (especially when we know that
much of it must have decayed completely).
Through gradual escape of more volatile hydrocarbon compounds from plant tissues, peat
changed to coal, which is more compact and
contains a higher percentage of carbon; increased *burial and compaction* over time produced
better grades of coal. Principal areas of commer-
cially important Pennsylvanian coal are the
Illinois basin, the central Appalachian Mountains, and the Maritime Provinces of Canada
(Fig. 14.7).

PALEOGEOGRAPHIC RECONSTRUCTION

Ultimately the sea flooded the low, coastal
swamps, and the marine upper phase of each
cyclic set was formed during such transgressions, thus burying the peat. As with sediments
previously discussed, the understanding of
Pennsylvanian strata has been facilitated greatly
by studies of modern sediments. Much of the

craton must have been exceedingly flat and very
near sea level so that the coal swamps then, like
the present southern Florida Everglades, Dutch
lowlands, and Louisiana swamps (Fig. 14.12),
were practically at sea level when the peat
originally accumulated. Therefore, only a small
rise of sea level or sinking of land could cause
very widespread inundation of former swamps.
Conversely, a very small fall of sea level would
cause an equally widespread regression, enlarg-
ment of land area, and a great expansion again
of rivers, deltas, and swamp forests. These are
exactly the conditions recorded in the cyclic
strata, namely many geologically rapid vertical
fluctuations of relative sea and land levels,
causing widespread *oscillatory transgressions and
regressions* over nearly one-fourth of the craton.
Between low mountainous uplands along the
Appalachian mobile belt and a persistent epeiric
sea over the western part of the craton, the
shoreline oscillated across a region 800 to 1,000
kilometers wide (Fig. 14.8). Overall sedimentary
facies, paleocurrent patterns, and mineral composition of sandstones indicate that large rivers
flowed west from the Appalachian region and
southwest from eastern Canada to debouch into
the epeiric sea along its eastern margin (Fig.
14.13*A*).

Much of the Pennsylvanian sandstone and
shale in the central United States represent
environments associated with deltas (Fig.

FIGURE 14.12

The Mississippi Delta and swampy alluvial plain—model for certain Pennsylvanian deposits. *Left:* modern protruding or "birdfoot" delta (A.D. 1500 to present) and swampy deltaic plain integrated from three older deltas, now partially submerged. Sand from delta feeds chenier plain and barrier islands farther west (see Fig. 11.22). *Right:* Effect of a 5-meter rise of sea level for comparison with the transgression phase of Pennsylvanian cycles (no part of map area is more than 10 meters above sea level today). *(Adapted from R. Leblanc and H. Bernard, 1954, in* The Quaternary of the United States; *and H. N. Fisk and J. McFarlan, 1955,* Journal of Sedimentary Petrology.)

FIGURE 14.13

Regressive and transgressive fluctuations of environments for a typical Pennsylvanian sedimentary cycle in the central United States (compare Fig. 14.12). *(Suggested by data from H. R. Wanless et al., 1963, Bulletin of the Geological Society of America.)*

14.12). As the shoreline oscillated, so did the deltas (Fig. 14.13). River-channel sandstones pass laterally into black shales and coals formed in swamps between channels. Channel sands grade westward into silty shales containing marine and brackish-water fossils. Frequent pulsations of deposition produced sedimentary features of great variety (Fig. 14.14).

TERMINAL PALEOZOIC EMERGENCE OF THE CONTINENT

PERMO-TRIASSIC GEOGRAPHY

The tendency of the eastern side of North America to tilt upward continued from late Paleozoic into Mesozoic time. Recall that changing facies and paleogeographic patterns indicated that normal marine deposits became progressively more restricted, and westward encroachment of terrigenous detritus increased many-fold (Fig. 14.15). These tendencies culminated in Permo-Triassic time, and so there was more land area, which was of greater average eleva-

FIGURE 14.14
Contorted Pennsylvanian sandstones west of Tulsa, Oklahoma. Water-saturated fine sands were deposited on a gently sloping delta fringe and were so unstable that they slipped and became contorted. Such deformation is common in rapidly deposited sand-shale sequences.

tion, than ever before during the Paleozoic Era (Fig. 14.16).

By Early Permian time, a sea comparable in size to the Black Sea occupied the south-central craton, but final stands of the Permian epeiric sea were in present southwestern Texas and the northern Rocky Mountains. In the former area, a great Late Permian *organic reef complex* flourished around the edge of a basin that was more than 300 meters deep (Figs. 14.1, 14.17), but as the sea retreated farther, the reefs died. Thick evaporite and red-bed deposits rapidly filled the basin and buried them. The evaporites include rare, valuable potassium minerals. The reef complex had swarmed with an amazing variety of life. Brachiopods, including bizarre forms which were adapted to corallike growth, were prominent, together with sponges, bryozoans, and advanced forms of algae; corals and crinoids were less important than in middle Paleozoic reefs.

At the west edge of the craton, *phosphate and chert* formed with black shales, probably as a result of *upwelling* of nutrient-rich cold waters at the edge of the continent, which is common today along western coasts. Meanwhile, evaporites and red beds accumulated widely in western United States and evaporites and car-

bonate rocks in northern Canada where mid-Paleozoic mountains had stood (Fig. 14.15). Undoubtedly evaporites and red beds also were deposited across central Canada, but erosion has removed them. This vast expanse of Permian evaporites records the slow drying up of epeiric seas as the land area of the earth enlarged to nearly an all-time maximum.

THE RED COLOR PROBLEM

Origin of the red color of many sedimentary rocks has been a subject of controversy for generations. Red beds are the trademark of Permian and Triassic strata on five continents; these two systems account for nearly half all red strata. The vivid color adds to the beauty of many landscapes, especially throughout western United States. Such deposits long have been considered to bear witness to unique paleoclimatic conditions. For example, an intriguing early speculation (with little evidence of support) was that prolific coal-swamp plants generated excess atmospheric oxygen, which then produced abundant red iron-oxide minerals in sediments. Paleogeographic distributions of many ancient red beds have long been taken to indicate formation at low latitudes in arid regions. (Indeed, circular reasoning often has been used to infer ancient low-latitude deserts from distribution of red beds.) But nature is rarely simple and there are important exceptions to the generalization, especially in Gondwanaland where red beds occur in proximity to tillites. So let us examine this ambiguous *red-bed problem* more closely.

On the following pages:
FIGURE 14.15
Middle Permian lithofacies and isotopic dates of granite plutons associated with volcanic rocks in west. Note the importance of evaporites and red beds in the western craton and phosphate and widespread volcanic rocks with associated limestones in the Cordilleran belt. (See Fig. 11.7 or endpapers for symbols and sources.)

FIGURE 14.16
Middle Permian paleogeography. Note the almost complete exclusion of the sea from the craton and eastern North America, and volcanic islands in the Cordillera.

Exact age uncertain
(Permian or Late Penn.)

red ss

PHOSPHATE-BEARING

240

250

Early Permian
(Dunkard)

260

250

Edge of Craton

250

225

255

MIDDLE PERMIAN

SEDIMENTARY FACIES

(Leonardian-Guadalupian)

(about 250 m.y. ago)

1000 Miles

1000 Kilometers

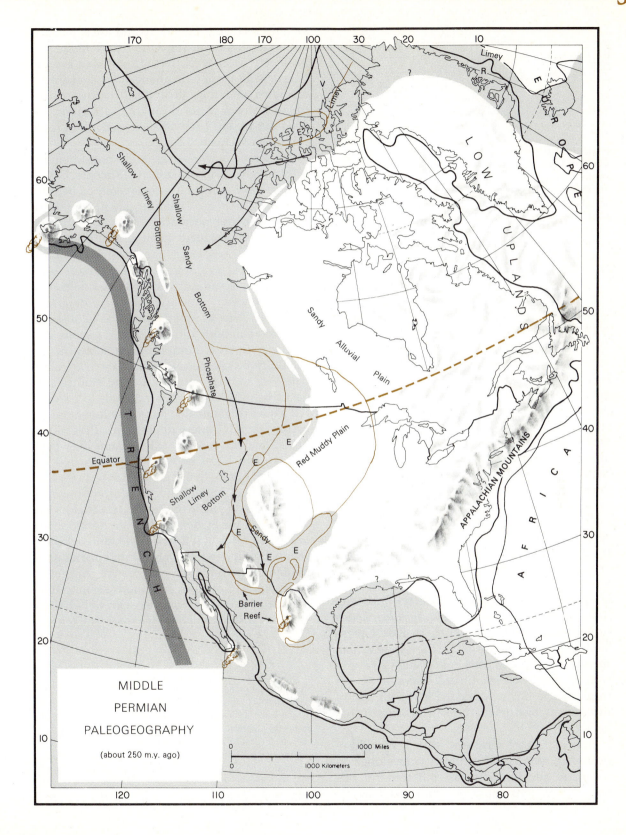

170 180 170 100 30 20 10

Limey

R

E
U
R
O
P
E

60

L
O
W

U
P
L
A
N
D
S

50

Shallow Limey Bottom

Shallow

Sandy

Bottom

Phosphate

T
R
E
N
C
H

Equator

Shallow Limey Bottom

Sandy

E

E

E

E

Barrier
Reef

Sandy

Alluvial

Plain

Red Muddy Plain

E

?

?

APPALACHIAN MOUNTAINS

A
F
R
I
C
A

V

Limey

E

60

50

40

30

20

10

MIDDLE

PERMIAN

PALEOGEOGRAPHY

(about 250 m.y. ago)

0   1000 Miles
0   1000 Kilometers

120 110 100 90 80

FIGURE 14.17
Complex Permian facies around the Capitan reef, south-western Texas. *(After P. B. King, 1948,* U.S. Geological Survey Professional Paper 215.) "Those of us who grew up with West Texas Permian geology . . . learned facies the hard way . . . as we looked into the interior of the range, we saw all of our fine units [formations] dissolve before our eyes, merging into a monotonous sequence of dolomite" (from left toward center in the figure). (See also Fig. 14.1.) *(P. B. King, 1949.)*

What can we learn from North American occurrences? Abundance of Permian evaporite deposits indicates high evaporation rates over the epeiric sea and adjacent lowlands. Pennsylvanian-Permian wind-dune sands covering nearly a million square kilometers of the western part of the craton (Fig. 14.5) have been cited as evidence of a desert, but dunes alone are no certain climatic proof, for coastal dunes are common today in humid as well as arid regions. But the huge area of these deposits does argue for a desert origin, and Permian fossil plants in Arizona confirm aridity because they have small, thick, hair-covered leaves and spines typical of modern dry-climate floras. Farther east, however, sporadic Permian and Triassic coals as well as aquatic plants, freshwater molluscs, fish, amphibians, and some aquatic reptile fossils indicate moderate humidity there. Late Triassic floras show *climatic zonation* from *mild-temperate* conditions in East Greenland through *wet tropics* in the Appalachian region, *tropical savannahs* (alternately dry and wet) in southwestern United States, and wet tropics in southern Mexico (Fig. 14.18). All of these areas also have Permian or Triassic red beds.

From paleontological evidence, it would appear that red-colored sediments developed in both arid and humid areas, but some of this seeming paradox could be rationalized by as-suming relatively humid upland areas blanketed with abundant vegetation and more arid lowlands. Moreover, trees grow today even in true deserts along permanent rivers, and animal carcasses and logs float long distances from uplands down rivers through arid regions and even out to sea. Therefore it is difficult to reach a simple conclusion.

What red color indicates, first and foremost, is thorough *oxidation of iron* in sediments. Only a trifling 1 to 6 percent of iron is required because it is such a potent coloring agent and is so unstable in the presence of free oxygen. The chemical environment favoring red color must be strongly oxidizing and slightly alkaline. Though most examples appear to have been nonmarine, some red beds interstratified with marine limestone and gypsum were deposited in the sea. And, as noted above, red color seems to develop in both arid and humid regions. *Red lateritic soils*, for example, form today in the humid tropics, especially where there are alternate wet and dry seasons. Recent studies suggest that oxidation of iron minerals to produce red oxides disseminated among sediment grains occurs largely after deposition. Progressive breakdown of iron in sand grains has been observed in late Cenozoic sediments that are now turning red. In finer, clay-rich sediments, it appears that brown, hydrated iron minerals "age" to red hematite by *dehydration* after burial. The bulk of the evidence points to development of red color through intense oxidation of iron in sediments under relatively warm conditions in nonmarine or marginal evaporative marine environments. Optimum conditions seem to have included a stable land surface with well-drained soils and distinct wet and dry seasons. Plenty of

FIGURE 14.18
Hypothetical paleoclimatic map for late Paleozoic and early Mesozoic time (land is brown). Accepting the paleomagnetically indicated equater position, we see that most of North America, Europe, and North Africa would have been in the Trade Winds belt, whereas today they lie chiefly in the Westerly Wind belt (see Fig. 11.25B). The restoration explains fossil and sediment evidence very well. The Tethys seaway (see Figs. 15.21, and WM.6) apparently supplied moisture to eastern North America.

PALEOCLIMATOLOGY
OF
PERMIAN AND TRIASSIC
TIME
(200-225 m.y. ago)

↙ Wind Patterns

time apparently could make up for either cool temperature or great aridity, which we assume would slow down but not stop the chemical reddening process. Large Permian and Triassic land areas seem to have met these conditions to an unusual degree.

Once established in sedimentary rocks, indestructibility of red color is legendary, as evidenced by dirt roads and river waters in regions of red beds, or in the dust that reddened skies during the "dust bowl" days of the 1930s. Few distant recipients appreciated that their skies and eyes were being reddened by 200-million-year-old dust blown hundreds of miles. Red color even survives cooking, as pioneer vertebrate paleontologist E. C. Case (University of Michigan) found on collecting trips into Oklahoma and Texas around 1900. Occasionally Case mixed biscuits for his field parties, but they were so brightly colored by the local water that only the chef was enthusiastic about his ferruginous creations.

PERMO-TRIASSIC PALEOCLIMATE

The position of the equator indicated by paleomagnetic evidence for late Paleozoic time is noteworthy in relation to distribution of various sediments, fossil types, and paleocurrents. Most significant is that coal-bearing deposits would lie at or within 20 or 25 latitudinal degrees of the apparent paleoequator (Figs. 6.12 and 14.8), thus in tropical to subtropical zones, where mild, humid, nonseasonal climatic conditions would be expected. Land apparently was low enough and small enough during Mississippian and Pennsylvanian time that rainfall was uniform over the continent. During Permian and Triassic time, position of the paleoequator seemingly was not greatly different, but both area and elevation of land had increased. This would cause greater differentiation of climate (Fig. 14.18); therefore, it is plausible to postulate relatively humid Permo-Triassic uplands, but drier lowlands. Much of North America would fall in the late Paleozoic tropics and thus in the Trade Winds belt. The western side of the

continent consequently would be leeward of the Appalachian Mountains, which could account for aridity there. Fossil evidence summarized above confirms these deductions and indicates a latitudinal zonation of climate.

The presence of land areas in Europe and Africa east of the old Appalachian Mountains would affect climate significantly. The *subtropical high-pressure cell* shown over Africa (Fig. 14.18) would be weaker and drier than over open ocean. Moisture, however, was available by evaporation from seas covering central Europe and northern Africa (see Fig. WM.6). Such moisture would have been carried westward to the windward side of the Appalachians, where humid-forest fossil plants have been found. The paleogeographic picture strongly suggests that the Appalachian region had a *monsoonal climate* (very markedly seasonal rainfall due to seasonal reversal of prevailing winds) like that of India and southeastern Asia today.

TECTONICS

CRATONIC DISTURBANCES

As noted above, the craton suffered more severe deformation in Pennsylvanian time than ever before. Many basins and arches were warped and eroded again, but the most intense disturbances occurred in the southern part of the craton (Fig. 14.19). Both the *Colorado* and *Oklahoma Mountains* were of the order of 1,000 meters high, and erosion rapidly removed Paleozoic strata to expose Prepaleozoic igneous and metamorphic rocks. Thick, coarse, red gravels, immature sandstones, and shales (Fig. 14.20) were deposited on alluvial fans and deltas between the mountain fronts and the nearby sea; they grade abruptly into marine strata. As shown in Fig. 14.19, the mountains must have been large fault blocks, but it is very unusual for such high mountains to be raised within cratons. Their ultimate cause is uncertain, but we believe that somehow they were by-products of the collision of Gondwanaland with southern North America.

Until late Paleozoic time, the southern margin of the continent was a passive stable shelf and continental slope much like the Nevada region during early Paleozoic time (see Fig. 11.5). Unlike the main Appalachian region farther northeast, tectonic disturbances did not affect the Marathon-Ouachita region (Fig. 14.7) until the Mississippian Period. Only thin, fine, deep-water shale and chert had accumulated there earlier. But now sand derived from the eroding mountains farther east spilled over the edge of the continent and was dispersed by turbidity currents (Figs. 14.21 and 14.22). Volcanic ash entered from the south, indicating that a volcanic island arc had formed, thus implying that subduction had begun. During the Pennsylvanian Period, the trough between the arc and the shelf to the north filled rapidly with clastic

APPALACHIAN OROGENY—FINAL COMPRESSION

sediments, and by Permian time, the entire Marathon-Ouachita sector was uplifted and deformed (Fig. 14.21), together with the Appalachian belt.

Farther northeast in the main Appalachian region, folding and thrust faulting of Pennsylvanian and older strata toward the craton also occurred (Fig. 14.23). While results of late Paleozoic deformation are the most obvious, they were superimposed upon effects of the earlier orogenies. Evidence for dating the final compression is as follows: (1) in southwestern Texas and south-central Oklahoma, folded Pennsylvanian strata are overlain unconformably by Early Permian ones; (2) in the central Appalachian Mountains, Pennsylvanian rocks were folded and faulted, together with older ones; (3) from Nova Scotia to Florida, nearly flat-lying Upper Triassic strata rest with unconformity upon folded Paleozoic rocks (Figs. 14.24 and 14.25); and (4) isotopic dating of scattered

FIGURE 14.19

Colorado Mountains elevated in Pennsylvanian time showing coarse, red facies deposited adjacent to them. Lower Paleozoic strata were so thin that erosion quickly exposed Prepaleozoic basement. Farther west, the Antler orogenic belt became buried as the Oquirrh basin subsided abruptly and received 7,000 meters of Pennsylvanian strata. These features and other cratonic basins shown underscore the new instability of the crust.

OKLAHOMA MTS.

N. MEX. ARIZ.

OKLA. KANS. UTAH COLO. NEV.

COLORADO MTS.

OQUIRRH BASIN

0   100 mi

0   200 km

meters
0
4000

CRATON   MOBILE BELT

Probable faults

COLORADO MTS.

OQUIRRH BASIN

UP. PENN.-L. PERMIAN

Penn.

Miss.

(ANTLER BELT)

other strata became schists or quartzites. In the western and eastern Appalachians (see Fig. 14.24), the older shales were converted to slate and some coals to anthracite (hard coal). Ultra-mafic (serpentine) rocks and basalts in the central zone represent sutures of old oceanic material crushed between colliding continents (see also Fig. 13.39).

Recall that the rocks of the eastern, high-grade metamorphic belt of New England and the Atlantic Coast states long were considered to be entirely Prepaleozoic because they ''looked old.'' Later they were shown to include much Paleozoic rock, and so are simply more intense-ly disturbed equivalents of the strata of the Ridge-and-Valley region to the west. As is typical of the deep root zones of mobile belts, it is difficult to differentiate the deformation and metamorphism produced by each successive orogeny because the effects have all been super-

granitic bodies formed during this final uphea-val indicates an average age between 200 and 250 million years. It follows that the final deformation of the mobile belt, which is called the *Appalachian orogeny*, occurred between Late Pennsylvanian and Late Triassic times, the exact timing varying somewhat with locality. Figure 14.26 summarizes the Paleozoic history of the Appalachian belt in plate tectonics terms.

Along the axis of the mobile belt, Paleozoic and Prepaleozoic rocks were most intensely deformed and metamorphosed by the various orogenies. Carbonate rocks became marble, and

FIGURE 14.20
Coarse conglomerate and sandstone of the Pennsylvanian and Permian Cutler Formation in western Colorado (larg-est boulders are about 3 meters long). These sediments were deposited by a series of mudflows during flash floods on alluvial fans on the arid western side of the Colorado Mountains (see Fig. 14.19). (*Courtesy A. Schultz and L. J. Suttner.*)

FIGURE 14.22
Steeply tilted Pennsylvanian sandstone and shale deposited by turbidity currents (Marathon Mountains, southwestern Texas). Identical sequences of rhythmically alternating sandstone and shale in Europe are called *flysch*. Each sandstone was deposited instantaneously, while the thinner shales represent far more time, having been deposited very slowly. *(Courtesy L. M. Cline.)*

FIGURE 14.21
Changing late Paleozoic geography of the Oklahoma region. The Ouachita region was a deep-water area starved of sediments until Late Mississippian time. Turbidity currents then introduced large volumes of sand from Appalachian region to the east, which gradually filled the deep troughs. By Late Pennsylvanian time, uplift produced lands in the region from which much red clastic sediment was shed. *(From published and unpublished work by L. M. Cline and associates.)*

APPALACHIAN TECTONIC DIVISIONS

- Large folds
- Steep faults
- Low angle thrust faults

- Late Triassic rocks in fault basins
- Lov. grade metamorphic rocks
- High grade metamorphic rocks
- Paleozoic granitic plutons

New England—Maritime region

Allegheny Plateau

Ridge and Valley

Blue Ridge

Piedmont

C R A T O N

C O A S T A L P L A I N

Wichita-Arbuckle

Ouachita region

Buried Triassic and unmetamorphosed, little deformed Lower Paleozoic rocks

Marathon region

| 0 | 400 | 800 km. |
| 0 | | 500 mi. |

FIGURE 14.23
Tectonic map of the Appalachian mobile-belt system. Note apparent zonation across the belt from obvious thrust faulting at the cratonic margin to a metamorphic and granite pluton belt with superimposed Triassic fault basins. (*Adapted from* Tectonic map of North America.)

FIGURE 14.24

Cross sections of the Appalachian belt (see Fig. 14.23 for location). Upper interpretation of 1950 implied a bilateral symmetry to the belt and showed thrust faults steepening downward. Lower interpretation of 1979 based upon deep seismic reflections shows no asymmetry and thrusts flattening downward to form one major fault zone at depth.

Most dramatic is evidence that metamorphic and granitic basement rocks are part of this immense thrust sheet. *(Top adapted from P. B. King, 1950,* Bulletin American Association of Petroleum Geologists. *Bottom adapted from F. A. Cook, et al., 1979,* Geology.*)*

FIGURE 14.25

Angular unconformity (arrow) between intensely folded Mississippian strata (Horton Group) below and Upper Triassic red-bed deposits (Wolfville Formation) at Rainy Cove, Nova Scotia. This unconformity reflects the late Paleozoic to Middle Triassic Appalachian orogeny. *(Courtesy G. de Vries Klein.)*

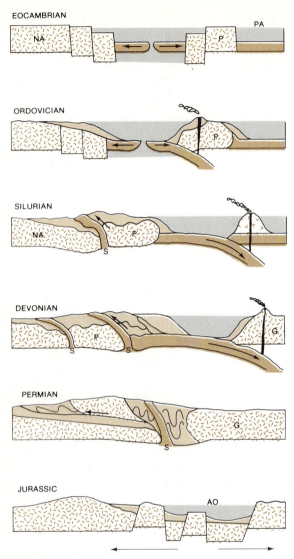

FIGURE 14.26
Hypothetical evolution of Appalachian mobile belt. *Eocambrian rifting* formed proto-Atlantic Ocean (PA) and Piedmont microcontinent (P). Ordovician subduction and closing of marginal ocean resulted in collision of microcontinent with North America (NA) *(Taconian orogeny)*. Subduction of proto-Atlantic plate caused Late Devonian arc collision *(Acadian orogeny)*. Continued subduction resulted in Permian collision of Gondwanaland with North America, which in turn caused large-scale overthrust faulting *(Appalachian orogeny)*. Early Mesozoic rifting initiated present Atlantic Ocean (AO) (S—collision suture zones). (Compare Figs. 12.19 and 13.39).

imposed. Granitic rocks are also hard to differentiate; some are Prepaleozoic, but some also formed during each episode of Paleozoic orogeny. Isotopic dating shows that the largest share of the granite and metamorphism was of Acadian age, however.

SIGNIFICANCE OF THRUST FAULTING

The Appalachian region was among the first mobile belts to be studied in detail, and the fold structures long ago were taken to indicate a shortening of the circumference of the earth. In Chap. 6 we noted that this idea was consistent with the long-held assumption of a hot origin for the earth followed by cooling and shrinkage. And early laboratory experiments using vise boxes to squeeze layers of sand and clay or wax produced amazingly faithful replicas of the natural folds (see Fig. 6.7). The recognition of thrust faults lent much weight to crustal shortening across mobile belts because lateral displacements on some thrusts could be shown to amount to several kilometers. But, as also noted in Chap. 6, there have been different ways of interpreting the thrust faults and folds. Some geologists suggested that they flatten downward and merely represent superficial sliding of strata over a more rigid basement (see Figs. 6.8 and 6.9). This is the so-called *thin-skinned hypothesis*. Most workers, however, have considered the basement to be involved as well in a more fundamental compression of the entire crust—the *thick-skinned hypothesis*. Maximum displacement on thrusts was envisioned in terms of tens of kilometers. With deeper drilling and modern high-resolution deep-seismic-reflection techniques, it is now possible to "see" to greater depths and to test these hypotheses. The lower section of Fig. 14.24 shows a result that indicates thin-skinned thrusting on a breathtaking scale. Apparently a huge sheet 6 to 15 kilometers thick, and including Prepaleozoic basement, has been displaced westward at least 260 kilometers. It is internally complex with many subsidiary thrusts that flatten downward. Most dramatic is the evidence that apparently none of the rocks at the surface along this section are rooted in place (contrast upper section). Rooted rocks all lie below the main thrust. Apparently these are simple structurally and may have some potential for petroleum entrapment.

How does the thrust faulting fit into the overall development of the Appalachian mountain belt? Figure 14.26 summarizes a hypothetical evolution of the belt in terms of plate tectonics. After initial rifting of some ancient supercontinent between 700 and 800 million years ago, eastern North America was a passive continental margin characterized by shelf carbonate banks (see Fig. 12.19) until mid-Ordovician time when compression began due to closure of a small *marginal ocean* between North America and either an arc or a *microcontinent* (isolated sliver with continental crust). This culminated in the *Taconian orogeny* due to collision (see Fig. 12.19). A second, apparently similar, collision resulted in the *Acadian orogeny* in the central and southern parts of the Appalachian belt while a continent-continent collision (with Europe) occurred farther north (see Fig. 13.39). Finally, near the end of the Paleozoic Era, Gondwanaland, which had been approaching for some time (see Figs. 12.25 and 13.29), collided dramatically with North America to produce the great thrust faulting characteristic of the *Appalachian orogeny*. Significant thrusting has been produced by all kinds of collisions (and even by subduction alone), but, clearly, continent-continent collisions take the prize for the most spectacular thrust faulting! Indeed, we are finding that structural mobility has been far greater than ever imagined before the plate tectonic theory was developed.

LATE PALEOZOIC MOUNTAIN BUILDING IN EURASIA

The end of the Paleozoic Era was a time of unusually widespread mountain building, the result largely of several continental collisions. Besides North America, central Europe was the site of great mountain building in its *Hercynian mobile belt*, an extension of the Appalachian belt (Fig. 14.27). The *Hercynian orogeny* was caused by the collision of Northern Africa with Europe. Similar-aged deformation around the southern perimeter of the *Siberian craton* may have resulted from the collisions of several small Chinese cratons (Fig. WM.12). Upheaval of the *Ural belt*

at this time is thought to have resulted from collision between lithosphere plates carrying Siberia and Europe. Late Paleozoic mountain building affected all other continents, too, but discussion of that is deferred to the next chapter, on Gondwanaland.

Both the Hercynian and Ural belts experienced earlier deformation, especially in Devonian time, following which Europe was widely covered by Early Carboniferous (Mississippian) carbonate rocks. But this was a lull before an orogenic storm, for the major Hercynian mountain building, characterized by widespread metamorphism and granitic plutonism, began in Late Carboniferous (Pennsylvanian) time. In northwest Africa, as in Europe, erosion of tectonic lands produced large quantities of sand and mud, which were spread out into the surrounding epeiric sea to form huge deltas and coal swamps like those of eastern North America at the same time. The resulting sediments show characteristics of repetitive or cyclic sedimentation, too. Most of Europe's mineral wealth of coal and ores was formed at this time. As in North America, the last epeiric sea deposition occurred in restricted basins of later Permian time. Evaporites and nonmarine red beds became more and more widespread, and by the beginning of the Triassic Period, all of Europe was land. Late Paleozoic plants, freshwater fish, amphibians, and reptiles, as well as the sediments, were almost identical with their counterparts in North America and northern Africa.

Siberia is much like North America in possessing a central craton completely encircled by mobile belts (Fig. 14.27). Marine Eocambrian strata are well represented in marginal mobile belts, and most of the craton was inundated by Cambrian and later epeiric seas. Great volcanism characterized an early Paleozoic mobile belt on the south, and Ordovician, Silurian, and Devonian systems all include important red beds there. By Carboniferous time, coal-bearing deposits became widespread both in mobile belts and upon the craton itself. Although it is generally believed that the Ural belt resulted from Late Paleozoic collision of Siberia with Europe, it is at present difficult to relate the other Paleozoic mobile belts of Asia to plate motions (but see Fig. WM.14).

FIGURE 14.27
Northern continental reconstruction shows supercontinent Laurasia (continents matched at − 900 meters). Eastward rotation of North America refits disjunctive mobile belts and shorelines, closes most of the Arctic Ocean. It also places eastern North American fossil assemblages closer to nearly identical European cousins, yet keeps equally similar western North American and Asiatic ones in proximity. Note rotation also reconciles Eurasian and North American paleomagnetic data for poles and equator. (*Adapted from many sources, especially Carey, 1959; Irving, 1964.*)

LATE PALEOZOIC ANIMAL LIFE

One of the most striking events in all geologic history occurred during a very short interval of time in the late Permian. For an as yet poorly understood reason, the balanced, fairly stable major shallow-marine ecosystem collapsed, resulting in mass extinction. Some groups such as the trilobites and the productid brachiopods were completely wiped out, while others, such as the molluscs, were devastated, but a few primitive or conservative families that were more adaptable to rapidly changing conditions survived. Most of the extinctions involved the marine floras and faunas, but land organisms were also affected. The differences between the diverse and prolific Middle Permian faunas and the Upper Permian faunas is striking and worldwide. The Upper Permian and Lower Triassic marine faunas, where found, consist of relatively few species, but large numbers of individuals, suggesting either very high- or low-saline waters. The causes of mass extinctions and the consequences are further discussed in Chaps. 15 and 19.

GENERAL PATTERNS

In many respects, life late in the Paleozoic Era was quite different from that of the earlier Paleozoic. At first, some groups did not change very much, but quickly the entire aspect of the

FIGURE 14.28
An ostracod *(Trachyleberis),* a calcareous bivalved crustacean from the Pliocene of Italy. *(Photograph courtesy P. A. Sandberg.)*

fauna became distinctive. New environments, such as vast coal swamps, provided a uniqueness not seen before nor fully duplicated after.

Slow, westward withdrawal of the epeiric sea was accompanied by spread of nonmarine sediments, which provide a more complete record of nonmarine environments than at any previous time. Thus we can gain much information on land organisms from, for example, the excellent record of Pennsylvanian plants associated with coal deposits.

INVERTEBRATE ANIMALS

Arthropods

After the Devonian Period, the trilobites became very rare, and few species survived until ultimate extinction in Late Permian time. The ostracods (see Fig. 14.28), while fairly abundant from Ordovician time onward, reached a degree of diversity permitting them to be used for stratigraphic zonation. The Pennsylvanian has been called, somewhat whimsically, the age of cockroachs by allusion to occurrences of insects in lake and lagoonal deposits of coal-bearing regions. The record is sufficiently well known to gain a picture of the first animals to invade the air. Although insects first appeared in the Silurian, nothing is known about them until the above occurrence. They must have undergone rapid evolution, since such diverse forms as dragonflies with wing spans of 30 inches and Texas-sized cockroaches 4 inches long are known by Pennsylvanian time.

Mollusca

The gastropods and pelecypods again are relatively minor parts of the fauna, although some primitive groups are abundant locally (Fig. 14.29). Freshwater clams, for example, are used in Europe for correlating nonmarine deposits. Most of these molluscs became extinct by the end of Permian time, but a few survivors gave rise to advanced Mesozoic molluscan faunas. We have seen that during Late Devonian time the ammonoids became widespread, and evolutionary expansion occurred rapidly in many groups. This trend continued and the ammonoids are of great service for upper Paleozoic correlation (Fig. 14.30).

Brachiopoda

The whole aspect of upper Paleozoic brachiopod assemblages is different from that found in Devonian faunas; however, differences are primarly in proportions. Spiriferid brachiopods are common (some being as large as 12 inches), but

FIGURE 14.29
Knightites, a bellerophontid gastropod from the Middle Pennsylvanian near Gunsight, Texas. This is a bilaterally symmetrical, coiled form; the group is the most primitive of the snails. *(Courtesy U.S. National Museum.)*

FIGURE 14.30
Left: A straight nautiloid cephalopod from the Permian of West Texas. *Right:* A goniatite cephalopod *(Imitoceras)* from the Mississippian of Indiana. *(Courtesy U.S. National Museum.)*

they were exceeded in numbers by productid brachiopods, which are rarer in older rocks. The productids invaded a host of environments which were not available to bivalved forms before because of remarkable adaptability provided by spines on their shells. Spines permit-

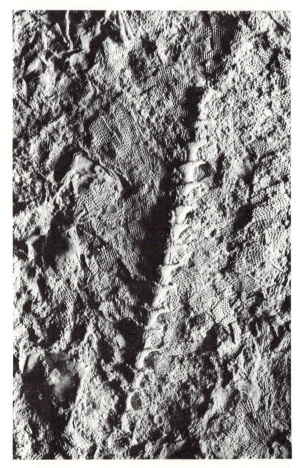

FIGURE 14.31
Archimedes, a unique, screw-shaped colony of fenestellid Bryozoa; a commonly encountered Mississippian fossil. *(Courtesy American Museum of Natural History.)*

ted them to either fasten to a hard surface or to serve as pilings in soft mud. They also served as a food-straining sieve and for protection. The productids continued to dominate until the end of the Permian, when they became extinct. Only four brachiopod groups (rhynchonellids, terebratulids, spiriferids, and ultraconservative inarticulates) survived into Mesozoic time. Thenceforth, brachiopods were a very minor part of the fauna.

Bryozoa

Up to Carboniferous time, the dominant group of Bryozoa was the massive or branching colonial trepostomes. During late Paleozoic time, the fenestellid Bryozoa overshadowed other groups. Fenestellids, which began in Ordovi-

cian time and gradually became important, tended to have delicate, lacy colonies. One bizarre genus (*Archimedes*), an index fossil of the Mississippian, had a corkscrew-shaped colony, which was simply a twisted cone (Fig. 14.31).

Echinodermata

We have seen that, in the Silurian and Devonian periods, organic reefs became increasingly important and widespread. In Mississippian time, this trend continued in many regions, but the composition of reef communities was much different from previous coral-hydrozoan-bryozoan-brachiopod reef assemblages. Commonly, the younger reefs and "gardens" possessed a crown of crinoids (Fig. 14.32), for massive reefoid limestones in many cases are composed almost entirely of crinoid stems. Reef paradises resulted in evolutionary expansion of crinoids (Fig. 14.33), and during the Mississippian they reached their climax, constituting the

FIGURE 14.32

Mississippian diorama. Most stalked forms are crinoids; note the screw-shaped colonies of *Archimedes,* a bryozoan, in left foreground. *(Courtesy American Museum of Natural History.)*

FIGURE 14.33
Camerate crinoid *Dichocrinus* from Lower Mississippian of eastern Iowa. Crinoids began during the Ordovician with two or three circles of skeletal plates on the head (below the arms). As evolution proceeded, more plates were added and there were many changes and elaborations involving the head and arms. *(Photograph courtesy U.S. National Museum.)*

largest portion of the fauna, if not in numbers of species certainly in bulk (Fig. 14.3). Although crinoids are living today, they were never again as diverse as during Mississippian time.

Corals

Some accumulations of colonial corals in Carboniferous rocks qualify as reefs, but in general, conditions did not favor coral-reef development. Colonial coral genera as well as solitary corals are fairly abundant. In Europe the corals, together with goniatites, are used to subdivide

FIGURE 14.34
Fusulinids *(Parafusulina)* from the Permian of West Texas. These are large forms shown at near true size. *(Courtesy U.S. National Museum.)*

stratigraphic units, but no dramatic evolutionary developments are discernible.

Protozoa

The protozoans, though the simplest organisms in terms of organic structures, have undergone an amazingly complex and varied evolution, suggesting that all are not the primitive *Amoeba*-like forms that we have come to think of from elementary biology classes. Protozoan fossils are first encountered in Lower Cambrian rocks (possibly also in the Prepaleozoic), but are rare below the Upper Devonian. It was not until Mississippian time that we see the first active diversification, when Foraminifera became common. The Foraminifera possess calcareous skeletons and delicate, threadlike pseudopods used for locomotion and food gathering.

During Pennsylvanian and Permian times, fusulinid Foraminifera (Fig. 14.34) underwent important evolutionary expansion and proliferated so that by Permian time many genera had evolved. The evolutionary pattern is unusually well documented (see Fig. 10.6). Long-ranged genera tend to be conservative, while short-ranged forms are likely to be highly specialized and have had lower survival potential. Many of the short-ranged genera are extremely useful as correlation tools because they are known throughout the world and are nearly everywhere in the same stratigraphic order. Permian fusulinids are found in two distinct faunal realms. One is best developed in central Asia and Japan, but also is recognized along the coast from Alaska south to California (discussed further in Chap. 16). This fauna can be recognized in a narrow belt from Spain eastward along the Mediterranean countries into the Middle East to northern India, thence southeast through Burma and Malaysia to the North Island of New Zealand. Then, curiously, another narrow belt can be traced from northern India into southeastern China up through the areas mentioned above. In addition to the fusulinids, most other invertebrate groups are represented by different genera, which are unique to this narrow belt called the *Tethys Faunal Realm* (see Chap. 15 and Figs. 15.21 and WM.5 for the significance of this distribution). The other realm includes all of North America east of the latter zone. Differences between the two are striking, particularly when comparing faunas as close together as California and Nevada. All fusulinids vanished by the end of the period.

EVOLUTION OF EARLY LAND LIFE

INVASION OF THE NEW HABITAT

Because the remains of countless plants gave rise to the great Pennsylvanian coals just discussed, it is appropriate to consider next the rise of land plants. Aquatic (marine?) plants are dominant elements in the Prepaleozoic record as compared to animals. The blue-green calcareous algae formed small reefs in the Prepaleozoic, and examples are also abundant throughout younger rocks. As far as we know, land surfaces were barren of any plant life until Late Silurian time, though primitive ground-hugging forms likely appeared earlier as noted in Chap. 9. The first known appearance of plants on land predates that of land vertebrate animals by about 40 million years. The atmospheric ozone layer discussed in Chap. 5 may not have developed fully until mid-Paleozoic time, which could explain this rather late appearance of land life.

Both plants and animals faced formidable problems in adapting to the land habitat. Mechanisms had to be developed: (1) to prevent desiccation, (2) to provide a strong supporting structure, and (3) to permit reproduction and facilitate dispersion of species in the new environment. Many obstacles were solved by plants through development of a vascular system, which assured all cells from top to bottom of the plant receiving moisture and nourishment; photosynthesis also could work more efficiently. As a part of the total system, more elaborate roots increased absorptive surface areas for acquisition of nutrients from soil and provided stability for large trees. The key to the system was the origin of long tracheid cells, which allow liquid to rise by capillary action. In addition, these cells provide woody supportive tissue to strengthen the main stem. Both the vascular system and the development of durable spores occurred in some water plants, thus they were preadapted for life on land. Spores are the simplest device for reproduction and dispersal because they are moved readily by air currents. Desiccation was alleviated by development of protective plant epidermal cells and analogous protective cells around seeds. Such protection, first accomplished by the primitive conifers, allowed growth and reproduction in drier habitats. Triradiate spores can be produced by algae and they are known in the Cambrian. The tetradiate spores and tracheidlike tubes produced by vascular plants have been reported from possibly Late Ordovician or Early Silurian rocks of Virginia and New York, but the first definite vascular plant remains were found in the Late Silurian of Wales. It is thought that the vascular plants developed in water, most probably in ponds, lagoons, and bogs, because they are found in near-shore marine and nonmarine rocks.

PALEOZOIC LAND PLANTS

By Early Devonian time, the first diversification of the primitive psilophytes occurred (Fig. 14.35), and very rapid evolution ensued so that, as mentioned in Chap. 13, by Late Devonian time lowland forests were in existence. The origin of the early psilophytes is unknown, but they probably were leafless vascular plants, which likely evolved from a filamentous green-algal stock.

FIGURE 14.35
An example of a Devonian primitive vascular plant *(Psilophyton)* from the Lower Devonian of Gaspé, Canada. *(Photograph courtesy F. M. Hueber and U.S. National Museum.)*

The famous Gilboa forest in the Upper Devonian red beds of the Catskill Mountains of New York (see Fig. 13.1) contains five groups of plants derived from the Psilophytales, which were eclipsed at the end of the Devonian by hardier descendants. Trees in this forest are estimated to have been over 10 meters high and inhabited a predominantly low-lying, river-margin environment. These new plant groups spread rapidly over a large portion of the Northern Hemisphere so that by Carboniferous time they were the chief contributors to swamp vegetation and peat. The reservoir of energy stored in the vast, buried coal swamps of Europe, Asia, and eastern North America was destined to power the industrial revolution. Paleozoic plants have contributed both directly and indirectly to the housing, production, and general well-being of Western civilization for over 200 years. They still constitute one of the greatest reservoirs of stored energy.

The true ferns (Filicineae) first appeared in Late Devonian time, and they were extremely abundant through the Carboniferous. Some were at least 20 meters in height, but, despite their prominence, they probably were not principal contributors to the accumulation of peat until the Late Pennsylvanian. For many years they were confused with the similar seed ferns (pteridosperms), which reproduced, as the name implies, by seeds rather than spores. The seed ferns were a dominant group in the Gilboa forest, some species reaching 10 to 13 meters in height. The seed ferns are the possible stock from which more advanced cycads and flowering plants (angiosperms) arose during the Mesozoic Era.

A visit to any coal strip mine in the eastern part of the United States likely will yield, within a few minutes, examples of the lycopsids and sphenopsids, the most important coal-forming groups of the Carboniferous. The lycopsids were relatively diverse in the late Paleozoic. They are represented today by the temperate-forest ground pine, a very small, bushy plant, pale by comparison with its towering ancestors, which grew over 30 meters high (Fig. 14.36). The lycopsid trees were strange-looking because they were without branches except at the very top of the trunk. The leaves were long and spikelike, similar to the individual palm leaf of today. As the trees grew, leaves were replaced from the top, and as old leaves dropped off,

FIGURE 14.36
Diorama of a Pennsylvanian coal forest; large trees with
scar patterns are lycopsids, single tree *(Calamites)* in the
right foreground with the horizontal grooves is a jointed
sphenopsid. *(Courtesy American Museum of Natural History.)*

scars were left in rows or spirals. Reproduction
was by spores contained in cones. The other
common plant group found in association with
coal is the Sphenopsida. They are characterized
by being jointed, and by having prominent
horizontal grooves (Fig. 14.36). In life, a circle of
small branches protruded from each joint, and
these bore a circlet of leaves and occasional
cones. They undoubtedly formed thickets not
unlike the biblical Nile River bulrushes (papy-
rus), and were probably ensconced in a similar
habitat *sans* Moses. They averaged about 5 to 6
meters tall. Like the lycopsids, they were never
very diversified, and are represented today by
the curious little *Equisetum* (or horsetail), a sort
of "living fossil" found along swamp margins
and railroad embankments.

NORTHERN VERSUS SOUTHERN FLORAS

The fossil record of Carboniferous plants is the
earliest to be complete enough to recognize
three floras in different climatic belts. The
tropical to subtropical flora was dominated by
the lycopsids until Late Pennsylvanian time
when the lycopsids were over-shadowed by the
ferns. This flora (Eurameric) is found in North
America, Europe, North Africa, and China. A
less widely distributed temperate flora is found
in eastern Siberia, consisting of seed ferns,
some lycopsids, and primitive conifers with
seasonal growth rings. By Permian time these
northern floras had changed to conifer-
dominant types, with the Chinese and
Eurameric separated into distinct units. The
Permian record is not so good as that of the
Carboniferous because the low-lying swampy
environments favorable for plant preservation
were replaced by terrain of higher elevation and

FIGURE 14.3
Typical fossil leaf of the seed fern *Glossopteris*, principal genus and namesake of the southern late Paleozoic flora. Note tonguelike shape, central rib, and fine venation. Lower Karroo strata, Waterburg Coal Field, Transvaal, South Africa. (*Courtesy E. P. Plumstead.*)

FIGURE 14.38
Present distribution of late Paleozoic land floras of the world. Note the separation of lycopsid and *Glossopteris* floras. Climatic zones presumably formed the barriers between the floras. (*Adapted from Just et al., 1952,* National Research Council Report on Paleobotany N22; *and Gothan and Weyland, 1954,* Lehrbuch der Palaeobötanik.)

drier climates. Primitive conifers were better adapted than other plants for drier conditions.

The southern Gondwana or *Glossopteris* flora, named after the dominant genus (a large seed fern; Fig. 14.37), is relatively undiversified, but is widespread (Fig. 14.38). It is associated with glacial deposits and contains tree trunks with prominent seasonal growth rings (see Fig.

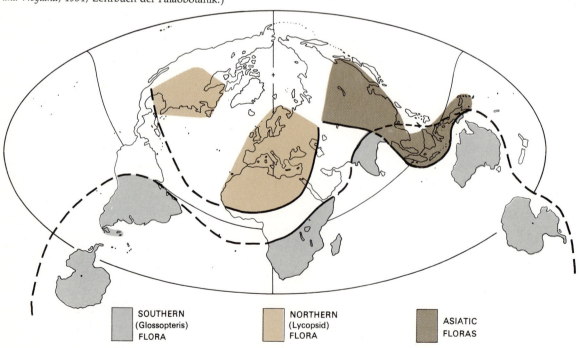

SOUTHERN
(Glossopteris)
FLORA

NORTHERN
(Lycopsid)
FLORA

ASIATIC
FLORAS

FIGURE 14.39
Seasonal growth rings in petrified wood in the *Glossopteris* floral province (Permian or Triassic), Senekal, Orange Free State, South Africa. Such rings could reflect hot-cold, wet-dry, or dark-light seasons, but probably resulted mainly from temperature changes. (Diameter of log is about 40 centimeters.)

14.39). Therefore, it is assumed that, unlike its northern contemporary, the southern flora lived under cooler, temperate climatic conditions. It, too, produced immense coal deposits. The *Glossopteris* flora is critical because of its bearing on continental drift and the evidence it provides of worldwide climatic zonation. We shall examine it further in Chap. 15.

VERTEBRATE ANIMALS

Amphibians

Water, besides being the elixir of life, has certain properties making it possible for organisms living in it to have an easier way. For example, because of its density and buoyancy, it helps support an animal's weight. This is critical to our understanding of the skeletal support essential for a relatively large animal to live in air—a factor all-important to the primitive amphibian. Many evolutionary modifications were involved in developing strong skeletons. In the amphibians, "experimental redesign" of the vertebral column provided both protection of the spinal cord and support for leg and back muscles. In several ways, the vertebral elements were interlocked to allow flexibility of the column for movement, but at the same time to achieve greater support than in fishes (Fig. 14.40).

The earliest amphibians (Late Devonian) were fishlike in appearance, but very quickly in some forms the body and head became flattened and limbs shortened. Eyes shifted to the top of the

FIGURE 14.40
Types of vertebral columns in early tetrapods modified for body weight support. Left-hand figure is a crossopterygian fish. The middle two figures are two types of interlocking amphibian vertebral columns, and the right-hand figure is that of a primitive reptile. (*Modified from Weller*, The course of evolution, *McGraw-Hill, 1969.*)

FIGURE 14.41
Seymouria; a lower Permian amphibian that developed many reptilian characteristics and has been used as an example of transitional "experimentation." *(Courtesy American Museum of Natural History.)*

head, suggesting that the creatures may have spent much time in very shallow water.

Development of adaptive skeletal mechanisms for land dwelling solved only part of the problem presented by the new habitat. The amphibian had inherited a fish-type egg, one which encased the embryo in a membrane that allowed oxygen to enter and wastes to pass out into surrounding water. This type of egg is simple and effective in a water medium, but amphibians had to remain near the water or at least return to it to lay eggs. The problem was solved by reptiles. The reptilian egg has an outer, leathery case or shell that allows oxygen to enter, and various membranes within the egg prevent liquids from escaping through drying in air. In addition, such eggs contain food for the embryo (the yolk) and a storage area for wastes. It is a completely self-contained unit.

If we are to discriminate between a fossil amphibian and a reptile, we need more information, for eggs are very rarely preserved. The skulls have many diagnostic features; most important is the presence in amphibia and fish of grooves called lateral lines that mark the presence of a sensory device in aquatic forms. The reptiles lack this system, seemingly because of selection against this (for them) useless device. It goes without saying that the oldest reptile fossils, which are encountered in Lower

Pennsylvanian strata at Joggins, Nova Scotia, are so close to their amphibian forebears that, lacking skulls, we would be hard pressed to be sure of their identity (Fig. 14.41).

Reptiles

The first reptiles (cotylosaurs) left a poor record, but there is no doubt that they experienced some Pennsylvanian diversification. As the seas regressed during Permian time, large, low lands appeared, and in Oklahoma, Texas, and New Mexico we find a rich and varied reptilian fauna first studied in detail by E. C. Case. Rivers meandered across broad flood plains dotted with lakes and swamps. A diverse fauna of bizarre reptiles together with amphibians and forms transitional between them (Fig. 14.41) inhabited these environments.

The cotylosaurs gave rise to all other reptiles, two groups of which were of particular note during the Permian Period. One group, the pelycosaurs (Fig. 14.42) was quite diversified, including several genera that had long, elaborately branched dorsal spines. Undoubtedly these spines had skin stretched between them. The most plausible function of this contraption was regulation of body-heat budget. Some workers believe that such weird adaptive structures mean that the climate was very hot. In any event, these pelycosaurs were so specialized for a narrow environmental niche that, when conditions changed at the close of Permian time, they became extinct. The pelycosaurs possessed teeth differentiated into incisors, canines, and

FIGURE 14.42
Edaphosaurus, a pelycosaur from the Permian of Texas.
(Courtesy Field Museum of Natural History.)

molar teeth. This contrasts with other reptiles, which had teeth all of the same or very similar shape, varying only in size. Most reptiles do not masticate food, but swallow it whole. No molars or other specialized teeth are needed.

The other group, which is in the same subclass (Synapsida) as the pelycosaurs, is the mammal-like reptiles (Therapsida). These, together with the pelycosaurs, came to dominate Permian lands, for by this time there was a sharp decline in amphibians. The mammal-like reptiles underwent very rapid expansion and there is much variation in morphology. For example, in contrast to the waddle of other reptiles and amphibians of late Paleozoic time, they must have enjoyed a faster, smoother gait because a small pelvis served as an abutment for legs that had rotated to be more parallel to the body. This characteristic and the nature of their teeth suggest that some of these forms were effective predators. The therapsids had a second-

ary palate, a mammalian feature; however, they had small braincases and other clearly reptilian characters. Mammal-like reptiles survived the crisis of terminal Paleozoic extinctions, and some time during Triassic or Jurassic time, gave rise to true mammals. Having spawned them, the therapsids—like the homing salmon—expired (See Fig. 15.18).

Not much can be said regarding the zoogeographic distribution of land vertebrates during the Late Permian and Triassic. Both the therapsids and other reptiles are found in widely separated localities. In addition, the therapsids are dominant in the Upper Permian (170 genera), in which the other reptiles are represented by 15 genera. The therapsids were physiologically adapted to warm humid conditions, which apparently were common in the Late Permian, whereas the other reptiles apparently were more suited to higher temperatures with seasonal droughts. The latter conditions are thought by some to have been more widespread by Late Triassic time when the other reptiles came into dominance.

SUMMARY

The close of the Paleozoic Era was originally designated on the basis of a major break in the development of marine life. Half of all invertebrate marine families and perhaps 80 percent of all species disappeared during this greatest crisis ever in the history of life. Conversely, land life evolved and expanded rapidly as new terrigenous habitats multiplied, not only in North America, but worldwide. Many explanations have been offered for the marine extinctions, including a recent suggestion that, because of unusually extensive Permian evaporite deposition, saltiness of the entire world ocean had changed too much for many organisms, or another that climate became too hot for many tropical forms to survive. Our favorite explanation is simply that the virtual disappearance of shallow epeiric-sea environments for a period much longer than any previous Paleozoic regression created too great a stress. Whereas earlier regressions typically lasted a few million years, this one endured 15 to 20 million years.

Remarkable correspondence of explosive evolution of Devonian land plants and land animals with the marked increase of land area by mountain building hardly can be sheer coincidence, and provides an outstanding example of mutual relations between the physical and organic evolution of the earth. Epeiric seas reached their maximum sizes in later Ordovician, Silurian, and Devonian times, but beginning in the Devonian in eastern North America and Europe, mountain building provided diverse new habitats for invasion by neophyte land organisms, which were quick to respond to fill the ecological vacuums. But why did they not invade the Late Ordovician Taconian land, which apparently was similar to the Devonian ones? It has been speculated that for a considerable time after the atmosphere became oxygen-rich, there was lethal ozone at ground level. Much oxygen presumably had to accumulate before the ozone layer rose to a higher level safe for advanced forms of life. Possibly this did not occur until Devonian time. Clearly the development of life was closely tied to physical history at this time. First land plants were tied closely to liquid water for their reproduction so could inhabit only the wet lowlands. By Permian time, drier uplands also became vegetated as primitive conifers (gymnosperms) evolved and solved the need for liquid water for reproduction.

Insects as well as amphibians had first invaded the land in middle Paleozoic time, and now they proliferated. By the Pennsylvanian Period, reptiles had appeared; they were much more adaptable, thanks especially to the development of the shelled egg, which (like the coniferous plants) freed the organism from water for its reproduction. Reptiles evolved rapidly, and in Permian time the mammal-like group appeared.

During late Paleozoic time, tectonic unrest increased in tempo all over the continent, including even parts of the formerly tranquil craton. Accompanying the tectonic revolution were marked changes of sedimentation. First, there was the introduction of much more heterogeneous clastic material, reflecting erosion of earlier Paleozoic strata to expose Prepaleozoic basement in Canada. Second, nonmarine and deltaic sediments became more widespread as the continent was tilted up and the epeiric sea was crowded farther westward, causing regression and spread of clastic sediments on a much greater scale than during previous orogenies. As the continent was tilted, black and gray-colored nonmarine and brackish-water Pennsylvanian sediments accumulated in oxygen-poor swampy environments near sea level in repetitive (cyclic) patterns, which indicate many oscillations of the shoreline amounting to several hundred kilometers. During Permian and Triassic time, however, as land became still more elevated, red sediments accumulated widely in oxygen-rich environments, apparently especially those with distinct wet and dry seasons.

The greatest tectonic activity was in the Appalachian and Hercynian mobile belts, where complex folding, thrust faulting, and local granite formation occurred as a result of collision of northwestern Gondwanaland with southern Europe and southeastern North America. Simultaneously, other lithosphere plate collisions were occurring, for late Paleozoic mountain building affected nearly every continent. Siberia collided with Europe, southeast Asia with Siberia, and some adjustments occurred also in South America, Africa, Australia, and Antarctica. By Permian time, the continents were grouped together in one huge mass called Pangea. Land area was extremely large at this time so that land life had little difficulty dispersing over the entire supercontinent. As a result, land vertebrates were not very diverse.

The possible position of the equator as suggested by paleomagnetic evidence during late Paleozoic time is noteworthy in relation to distribution of various sediments, fossil types, and paleocurrents. Most significant is that coal-bearing deposits would lie within 20 or 25 latitudinal degrees of the apparent paleoequator, thus in tropical to subtropical zones, where mild, humid, nonseasonal climatic conditions would be expected. Land apparently was low enough and small enough during Mississippian and Pennsylvanian time that rainfall was uniform. During Permian and Triassic time, position of the possible paleoequator seemingly was not greatly different, but both area and elevation of land had increased. This caused greater differentiation of climate; therefore, we postulate relatively humid Permo-Triassic uplands, but drier lowlands. Much of North America lay in the late Paleozoic Trade Winds belt. The Appalachian Mountains probably had a monsoonal climate, while the western side of the continent would be leeward of the mountains, which accounts for great aridity there. A distinct global latitudinal climatic zonation is clearly indicated by the distribution of the tropical lycopsid flora of North America and Europe, the southern temperate glossopterid flora of Gondwanaland, and a northern temperate Asiatic flora. Within North America, Permo-Triassic floras also show a zonation with latitude on a finer scale.

Key Concepts

Appalachian and Hercynian orogenies:
  collisional suture zones
  marginal ocean basins
  microcontinent
thin-skinned versus thick-skinned tectonics
anaerobic bacteria
burial and compaction of peat
coal swamps
Colorado and Oklahoma cratonic mountains
Glossopteris versus *Lycopsid* versus Asiatic floras
Marathon-Ouachita belt
mass extinction
Oquirrh basin
Permian phosphate, chert, and evaporites
Permian reef complex of Texas
Permo-Triassic climatic zonation:
  mild temperate
  monsoonal
  tropical savannah
  wet tropical
red-bed problem:
  dehydration
  lateritic soil
  oxidation
repetitive or cyclic sedimentation
river and delta deposits
Siberian and European cratons
Tethys Faunal Realm
tilting of craton
upwelling
Ural mobile belt

Readings

Alberta Society of Petroleum Geologists, 1964, Geological history of western Canada: Calgary. (An atlas of stratigraphic maps)

Anderton, R. A., et al., 1980, A dynamic stratigraphy of the British Isles: London, Allen and Unwin.

Andrews, H. N., Jr., 1961, Studies in paleobotany: New York, Wiley.

Beloussov, V. V., 1962, Basic problems in geotectonics: New York, McGraw-Hill. (English translation)

Cook, F. A., et al., 1979, Thin-skinned tectonics in the crystalline southern Appalachians: Geology, v. 7, pp. 563–567.

Douglas, R. J. W., et al., 1963, Geology and petroleum possibilities of northern Canada, in Proceedings Sixth World Petroleum Congress (sec. 1): Frankfurt, pp. 519–571.

Fisher, G., et al., 1970, Studies of Appalachian geology—central and southern: New York, Wiley-Interscience.

Gray, J., and Boucot, A. J., 1978, The advent of land plants: Geology, v. 6, pp. 489–492.

Hatcher, R. D., Jr., 1972, Developmental model for the southern Appalachians: Geological Society of America Bulletin, v. 83, pp. 2735–2760.

Klein, G. de Vries, (ed.), 1968, Late Paleozoic and Mesozoic continental sedimentation, northeastern North America, Boulder: Geological Society of America, Special Paper 106.

Logan, A., (ed.), 1974, The Permian and Triassic systems and their mutual boundary: Calgary, Canadian Society of Petroleum Geologists Memoir 2.

McKee, E. D., Oriel, S.S., et al., 1967, Paleotectonic investigations of the Permian system: U.S. Geological Survey Professional Paper 515.

Nalivkin, D. V., 1960, The geology of the U.S.S.R.: London, Pergamon.

Newell, N. D., et al., 1953, The Permian reef complex of the Guadalupe Mountains region, Texas and New Mexico: San Francisco, Freeman.

Roberts, R. J., et al., 1958, Paleozoic rocks of north-central Nevada: Bulletin of the American Association of Petroleum Geologists, v. 42, pp. 2813–2857.

Robinson, P., 1971, A problem of faunal replacement on Permo-Triassic continents: Palaeontology, v. 14, pp. 131–153.

Rodgers, J., 1970, The tectonics of the Appalachians: New York, Wiley-Interscience.

Stewart, J. H., et al., (eds.), 1977, Paleozoic paleogeography of the western United States: SEPM Pacific Coast Paleogeography 1.

United States Geological Survey, 1979, The Mississippian and Pennsylvanian (Carboniferous) systems in the United States: Professional Paper 1110.

Wanless, H. R., et al., 1963, Mapping sedimentary environments of Pennsylvanian cycles: Bulletin of the Geological Society of America, v. 74, pp. 437–486, also GSA Map MC-23.

Zen E., et al., 1968, Studies of Appalachian geology: northern and maritime: New York, Wiley-Interscience.

WORLD MAP
SUPPLEMENT

WORLD MAP
SUPPLEMENT

The face of places,
and their forms decay;
And that is solid earth,
that once was sea;
Seas, in their turn,
retreating from the shore,
Make solid land,
what ocean was before.

Ovid

Metamorphoses, XV

PURPOSE

The sixteen maps on the following pages provide several kinds of information on a global basis unlike most of the maps scattered through the remainder of the book. Each provides at a glance an overview either of paleogeographic or tectonic features of the whole world for some moment in time. The information on each map supplements discussions at many widely scattered points in the text, therefore these maps are located centrally for easy reference while one is reading any part of the text. Moreover, it is valuable to be able to compare easily and quickly features for several different ages, thus providing an additional reason for grouping these maps rather than dispersing them through the book. The reader is urged to refer frequently to the maps, and to facilitate this, a number of citations appear through the text with designations such as ''see Fig. WM.1'' (WM refers to ''World Map'').

WHAT IS SHOWN

There are three categories of maps in this section—climatic, paleogeographic, and tectonic. The first two maps summarize some *climatic data* for the present earth to be used for analogies with past climates. The six *paleogeographic maps* that follow show land, sea, mountains, inferred surface marine currents, tropical (including subtropical) versus temperate floras and faunas, continental glaciation, and major evaporites. Two maps show major *tectonic provinces* as they are seen today. The first (Fig. WM.9) shows isotopic date provinces for Prepaleozoic and earliest Paleozoic time. In other words, it shows in a broad way major subdivisions within the basement rocks of continents, and supplements especially the discussions of early continental evolution. The second map (Fig. WM.10) shows mobile belts and cratons for the past 700 million years, which provides a broad tectonic framework for all discussions of Phanerozoic tectonic history and for the following six maps. The last six are global *paleotectonic* or *lithosphere plate maps* for the same times as the six paleogeographic maps. These show inferred plate configurations with subduction, spreading, and transform margins indicated where possible. These

maps are intended to supplement all discussions of tectonic history, especially whenever plate interactions are treated.

Preparation of all of these maps has drawn upon countless sources and upon our own experiences over many years. It is impossible to cite all here, but the principal references for the reconstructions are listed below, and many other relevant references are cited throughout the book.

LIST OF MAPS

CLIMATIC MAPS:

WM. 1   Warmer Climate Features
WM. 2   Colder Climate Features

PALEOGEOGRAPHIC MAPS:

WM. 3   Mid-Cenozoic (Early Miocene)
WM. 4   Late Mesozoic (Cretaceous)
WM. 5   Jurassic
WM. 6   Late Paleozoic (Carboniferous)
WM. 7   Middle Paleozoic (Devonian)
WM. 8   Early Paleozoic (Cambrian)

TECTONIC PROVINCE MAPS:

WM. 9   Prepaleozoic and Earliest Paleozoic Isotopic Date Provinces
WM.10   World Mobile Belts of Past 700 Million Years

GLOBAL PALEOTECTONIC MAPS:

WM.11   Early Miocene
WM.12   Late Cretaceous
WM.13   Jurassic
WM.14   Late Paleozoic (Carboniferous)
WM.15   Middle Paleozoic (Devonian)
WM.16   Early Paleozoic (Cambrian)

ABBREVIATIONS USED:

A   Asia
AF   Africa
AN   Antarctica
AU   Australia
C   China
E   Europe
I   India
NA   North America
S   Siberia
SA   South America

FIGURE WM.1 (p. 366, top)
WARMER CLIMATE FEATURES.
Rainfall and permanent atmospheric circulation cells for *warm season* in both hemispheres as a proxy for past warmer global intervals. For illustrative purposes, summer data for both hemispheres are plotted together; difference between humid and arid chosen as 25 centimeters (10 inches) of rainfall. Note pole-to-equator temperature differential is only about 20°C during warm times and subtropical high-pressure cells are correspondingly relatively large. This relationship has a major effect upon prevailing winds and wind-driven ocean currents. (Contrast size of arid regions with Fig. WM.2; also consult Figs. 11.25 and 16.30.)

FIGURE WM.2 (p. 366, bottom)
COLDER CLIMATE FEATURES.
Rainfall and permanent atmospheric circulation cells for *cold seasons* in both hemispheres as a proxy for past colder global intervals. For illustrative purposes, winter data for both hemispheres are plotted together; difference between humid and arid chosen as 25 centimeters (10 inches) of rainfall. Note pole-to-equator temperature differential is 30°C during cold times and subtropical high-pressure cells are correspondingly relatively small. This relationship has a major effect upon prevailing winds and wind-driven ocean currents. (Contrast high-pressure cells and size of arid regions with Fig. WM.1; also compare Fig. 14.18.)

FIGURE WM.3 (p. 367, top)
MID-CENOZOIC (EARLY MIOCENE) PALEOGEOGRAPHY showing that continents had approximately their present sizes and positions and that continental glaciation had begun in polar areas (see Fig. WM.4 for continent identification). Surface marine currents (arrows) were essentially like today's except in the Panama region. Circum-Antarctic current first fully developed due to the final break between South America and Antarctica. The tropical sub-tropical climate belt was still broader than today by about 15° of latitude. The Alpine-Himalayan mountain systems of Eurasia were just being created by continental collisions. Circum-Pacific arcs and mountains also were experiencing new tectonic activity. (Compare Fig. WM.11.)

FIGURE WM.4 (p. 367, bottom)
LATE MESOZOIC (CRETACEOUS) PALEOGEOGRAPHY showing land areas, shallow-marine currents (arrows), and very broad tropical-subtropical climate belt. Tethyan tropi-

cal marine faunal province characterized the entire seaway between northern and southern continents. Distinct faunas characterized northern (Boreal) and southern (Austral) seas. Note evaporite deposition on margins of newly separated Gondwana continents in widening new oceans as well as young mountains on continental margins experiencing active subduction. (Compare Fig. WM.12.)

FIGURE WM.5 (p. 368, top)
JURASSIC PALEOGEOGRAPHY showing land areas, shallow-marine currents (arrows), and very broad tropical–subtropical climate belt. Pangea was just beginning to break up, creating several new seaways. Tropical marine Tethyan province was fully developed with distinct faunal provinces to the north (Boreal) and south (Austral). Note evaporite deposition on margins of several continents, especially where newly separated (as in Gulf of Mexico). Note new mountains along western margins of North and South America.

FIGURE WM.6 (p. 368, bottom)
LATE PALEOZOIC (CARBONIFEROUS) PALEOGEOGRAPHY showing land areas, shallow-marine currents (arrows), and relatively narrow tropical-subtropical zone. Tethys Sea was forming as Gondwanaland collided with the northern continents and as they themselves were colliding with each other (note new Appalachian, Hercynian, and Ural Mountains). Ice caps had formed on parts of Gondwanaland and locally in Siberia. *Panthalassia* is the name sometimes used for the pre-Jurassic ancestral Pacific Ocean. (Compare Fig. 15.21 for Permian time when Pangea was completely assembled.)

FIGURE WM.7 (p. 369, top)
MIDDLE PALEOZOIC (DEVONIAN) PALEOGEOGRAPHY showing land areas, shallow-marine currents (arrows), a very broad tropical marine reef belt, widely scattered evaporite deposition, and distinct northern and southern faunal realms. Note Acadian and other mountains and glaciation in South America.

FIGURE WM.8 (p. 367, bottom)
EARLY PALEOZOIC (CAMBRIAN) PALEOGEOGRAPHY showing small land areas, shallow-marine currents (arrows), and tropical marine (*Archaeocyatha*) belt. With so few large landmasses, atmospheric and marine circulation would be simpler than today; trade-wind–driven currents in the tropics should have been particularly prominent.

WARMER CLIMATE

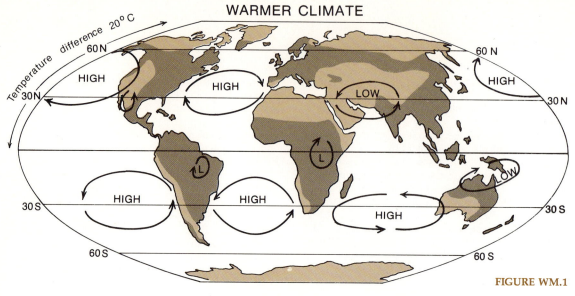

FIGURE WM.1

COLDER CLIMATE

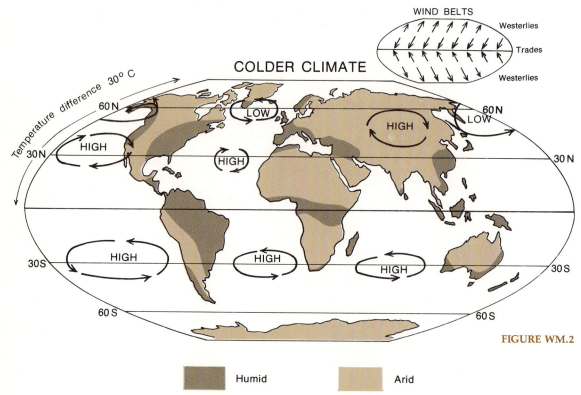

FIGURE WM.2

Humid Arid

MID-CENOZOIC
Early Miocene (20 million years ago)

FIGURE WM.3

LATE MESOZOIC
Cretaceous (80 million years ago)

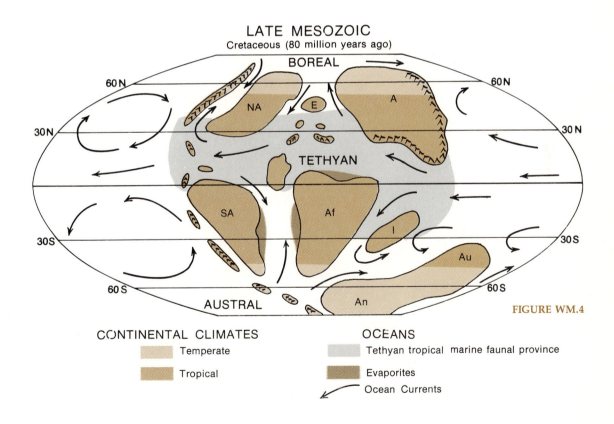

FIGURE WM.4

CONTINENTAL CLIMATES

- Temperate
- Tropical

OCEANS

- Tethyan tropical marine faunal province
- Evaporites
- Ocean Currents

MID-MESOZOIC
Jurassic (160 million years ago)

FIGURE WM.5

LATE PALEOZOIC
Carboniferous (300 million years ago)

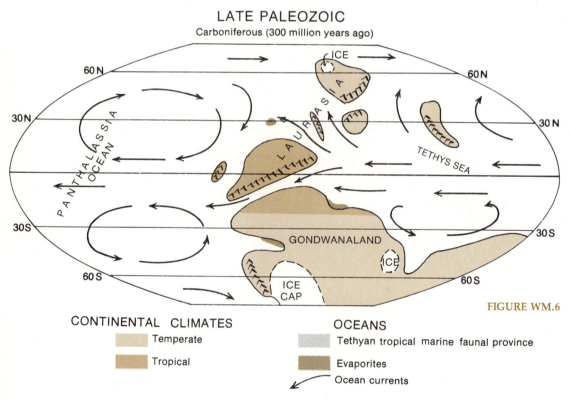

FIGURE WM.6

CONTINENTAL CLIMATES

- Temperate
- Tropical

OCEANS

- Tethyan tropical marine faunal province
- Evaporites
- Ocean currents

MID-PALEOZOIC
Devonian (370 million years ago)

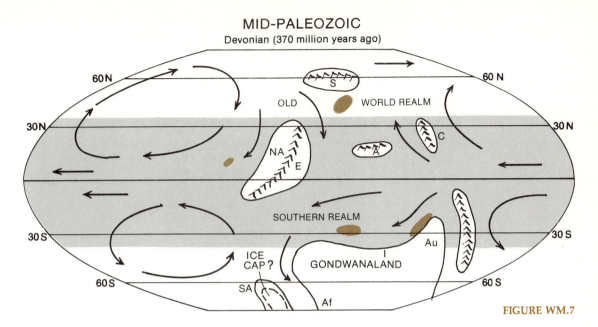

FIGURE WM.7

EARLY PALEOZOIC
Cambrian (500 million years ago)

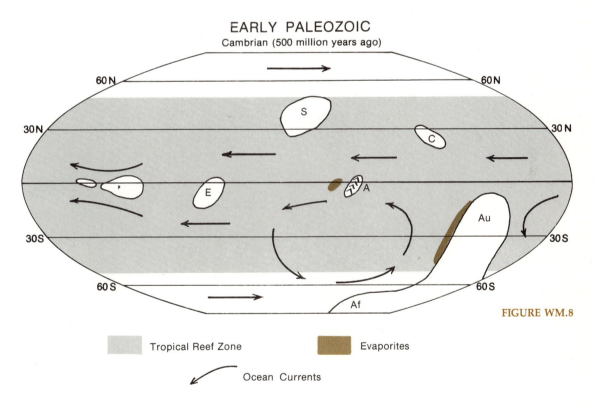

FIGURE WM.8

Tropical Reef Zone

Evaporites

Ocean Currents

FIGURE WM.9
Prepaleozoic And Earliest Paleozoic Isotopic-Data Provinces of basement rocks of continents (middle Paleozoic and younger mobile belts shown blank, emphasizing their general overlapping relationship with older mobile-belt patterns). Note complexity of discordances between provinces. Quality of data varies enormously, being least complete in South America and Asia. (*From many sources, e.g., Cahen and Snelling, 1966, The geochronology of equatorial Africa; Compston and Arriens, 1968, Canadian Journal of Earth Sciences; Gastil, 1960, 21st International Geological Congress; Holmes, 1965; Hseih, 1962; Hurley et al., 1967, Science, v. 157, pp. 54–542; Jenks, 1956, International Geology Review.*)

FIGURE WM.10

World Mobile Belts Of The Past 700 Million Years (Eocambrian and younger). Early Paleozoic belts also appear on Fig. WM.9. Note disjunctive belts (those cut off at continental margins) and their apparent matching counterparts on opposite continents in several cases. Note also modern island arcs and oceanic deep trenches. (*Bartholomew's Nordic Projection; used with permission.*)

MID-CENOZOIC
Early Miocene (20 million years ago)

LATE MESOZOIC
Late Cretaceous (80 million years ago)

Transforms   Subduction zones

Spreading centers

FIGURE WM.11

EARLY MIOCENE GLOBAL TECTONICS showing inferred lithosphere plate configurations with spreading, subduction, and transform margins. Continental outlines are shown for reference. Active collisions along Alpine–Himalayan subduction zone had almost destroyed the Tethys Sea. Note active Circum-Pacific subduction. (Compare Fig. WM.3.)

FIGURE WM.12

LATE CRETACEOUS GLOBAL TECTONICS showing inferred lithosphere plate configurations with spreading, subduction, and transform boundaries; generalized continental outlines shown for reference. Note opening of new Atlantic and Indian oceanic zones and rifting of microcontinents from northern Africa. Position of Southeast Asia not clear at this time. (Compare Fig. WM.4.)

MID-MESOZOIC
Jurassic (160 million years ago)

LATE PALEOZOIC
Carboniferous (300 million years ago)

FIGURE WM.13

JURASSIC GLOBAL TECTONICS showing inferred lithosphere plate configurations with subduction and a few definable spreading and transform margins. Generalized continental outlines shown for reference. Supercontinents Laurasia and Gondwanaland were just starting to break apart, and a radial Jurassic–Triassic basaltic dike swarm (short lines) formed around the North Atlantic basin as it first began to open (possibly due to a new rising mantle plume). Also note beginning of rifting in northern Africa. Subduction characterized entire Pacific margin of Pangea at this time. Southeast Asia and other blocks now on Pacific margin may have been part of Gondwanaland. (Compare Fig. WM.5.)

FIGURE WM.14

LATE PALEOZOIC (CARBONIFEROUS) GLOBAL TECTONICS showing inferred lithosphere plate configurations with subducting margins; spreading zones cannot be designated for pre-Jurassic time. Continental collisions at this time were forming the supercontinent Pangea and producing one of the greatest periods of mountain building (e.g., Appalachian, Ural, Hercynian, Tasman, and Gondwanian orogenies). Indochina and other blocks may have been part of Gondwanaland. The *Panthalassia* ocean floor must have undergone a great deal of spreading and subduction as Pangea was assembled. (Compare Fig. WM.6.)

MID-PALEOZOIC
Devonian (370 million years ago)

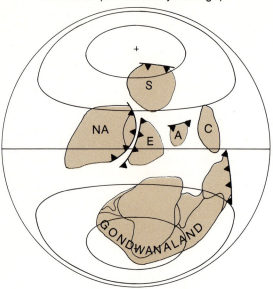

EARLY PALEOZOIC
Cambrian (500 million years ago)

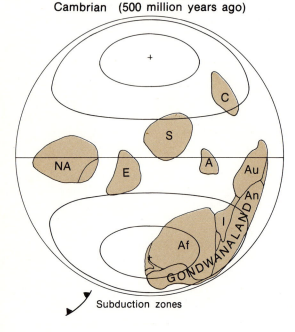

Subduction zones

FIGURE WM.15
MIDDLE PALEOZOIC (DEVONIAN) GLOBAL TECTON-ICS showing inferred lithosphere plate configurations with their subduction margins; spreading zones cannot be designated for pre-Jurassic times. Collision of North America and Europe produced Acadian-Caledonian orogenies; subduction beneath eastern Australia produced orogeny there. (Compare Fig. WM.7.)

FIGURE WM.16
EARLY PALEOZOIC (CAMBRIAN) GLOBAL TECTONICS showing inferred lithosphere plate configurations; spreading zones cannot be designated for pre-Jurassic times. Gondwanaland either had formed or was just forming. Late Cambrian time seems to have been tectonically quiet, thus very little mountainous terrain is shown. (Compare Fig. WM.8.)

References for Map Reconstructions

Gordon, W. A., 1973, Marine life and ocean surface currents in the Cretaceous: Journal of Geology, v. 81, pp. 269–284.

Gordon, W. A., 1974, Physical controls on marine biotic distribution in the Jurassic Period: from Ross, C.A., (ed.), Paleogeographic provinces and provinciality: Society of Economic Paleontologists and Mineralogists Special Publication No. 21.

Hallam, A., (ed.), 1973, Atlas of palaeobiogeography: Oxford, Oxford Press.

McElhinny, M. W., 1973, Palaeomagnetism and plate tectonics: London, Cambridge University Press.

Morel, P., and Irving, E., 1978, Tentative paleocontinental maps for the early Phanerozoic and Proterozoic: Journal of Geology, v. 86, pp. 535–561.

Phillips, J. D., and Forsyth, O., 1972, Plate tectonics, palaeomagnetism and the opening of the Atlantic: Bulletin of Geological Society of America, v. 83, pp. 1579–1600.

Scotese, C. R., et al., 1979, Paleozoic base maps: Journal of Geology, v. 87, pp. 217–277.

Smith, A. G., and Briden, J. C., 1977, Mesozoic and Cenozoic paleocontinental maps: London, Cambridge University Press.

Van der Voo, R., and French, R. B., 1974, Apparent polar wandering for the Atlantic-bordering continents: Late Carboniferous to Eocene: Earth Science Reviews, v. 10, pp. 99–119.

Ziegler, A. M., et al., 1979, Paleozoic paleogeography: Annual Reviews of Earth and Planetary Sciences for 1979, v. 7, pp. 473–502.

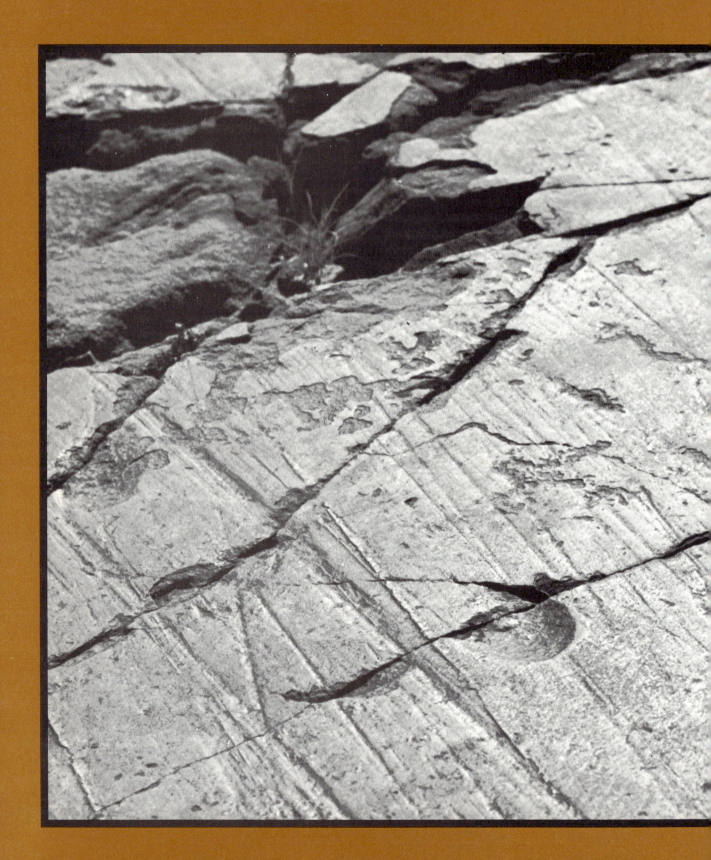

15

PANGEA
Its Makeup and Breakup

And so these men of Indostan
Disputed loud and long,
Each in his own opinion
Exceeding stiff and strong,
Though each was partly in the right
And all were in the wrong!

*The Blind Men and the
Elephant* (Ancient Hindu
fable, John Godfrey Saxe)

FIGURE 15.1
Striated glaciated surface or pavement beneath Carboniferous Dwyka Tillite, Nooitgedacht, near Kimberley, South Africa; surface was cut on Prepaleozoic metamorphosed volcanic rocks. *(Courtesy J. C. Crowell)*

15

In Chap. 6 we presented a brief introduction to the concept of large-scale continental displacements, and we noted some of the evidence from the five continents that today lie south of North America and Eurasia. Now we are ready to examine that evidence more fully and to trace the Paleozoic and Mesozoic history of the southern supercontinent *Gondwanaland* as we have already done for North America and Eurasia, which together comprised *Laurasia* (or Euramerica).

The five southern or Gondwana continents have several peculiarities as compared to the northern or Laurasian ones. Much of the formers' perimeters are represented not by marginal mobile belts active sometime since Cambrian time as in the north, but rather by truncated cratonic edges (see Fig. WM.10). Most of the coastlines look like places where the continents were somehow broken off—pieces of a jigsaw puzzle that once fit together to form a single gigantic craton. The perfect match of Africa and South America was, of course, a major clue that inspired the first continental reconstruction proposed in 1858 by the European, Antonio Snider. For this and other equally compelling reasons, the southern continents have long provided a stronger case for continental displacements than have the northern ones. As with the blind men trying to describe the elephant, the way continents seemed to develop has depended greatly upon where one was standing.

As we shall see, largely nonmarine rocks of late Paleozoic and early Mesozoic ages are strikingly similar on all five southern continents. They also contain nearly identical fossils, which implies a close relation among the five until the late Mesozoic when widespread flooding by the sea occurred and physical histories as well as organisms began to diverge. Important differences between late Paleozoic floras of Laurasia and Gondwanaland (see Fig. 14.38) suggest a barrier of some kind between, but it must have been a climatic rather than a physical one (see Fig. 6.12) because nonmarine fossil reptiles show strong similarities on all continents. Finally, the distribution of late Paleozoic glacial deposits also suggested a polar grouping of southern continents (Fig. 15.1). The evidence points to physical connections among all the present continents in Permo-Triassic time, forming the super-supercontinent *Pangea*. Total land area of the earth was about the same as today—a rare event in earth history with tremendous implications for both climate and the evolution of land life. Paleomagnetic evidence and isotopic dating accumulated since 1955 added great strength to the old continental-drift idea, and the still newer knowledge from the deep seas has clinched major continental displacements through time. Both rock paleomagnetism from continents and sea-floor magnetic anomalies have provided independent time scales for studying those displacements.

Ambiguities and incompleteness of the geologic and paleontologic data for years assured a lively and colorful debate between contending opinions on drift (see Fig. 6.13). Advocates sometimes have been drugged by the most flimsy evidence, such as the seasonal flight path of the Arctic tern as it zigzags from side to side of the Atlantic Ocean in its migration between the northern and southern polar regions. It has been alleged that this weird habit reflects the indecision of a poor confused species whose ancestors began migrating before the continents separated and not knowing what else to do, continued to try frantically to keep both shores in sight over the ages as the continents continually sailed apart. A similar argument has been proposed to explain the seemingly pointless

FIGURE 15.2
Prepaleozoic banded iron deposits and shales (Brockman Iron Formation) in strata about 2.1 billion years old yet still flat-lying. Groups of laminae in this formation can be correlated for tens of kilometers. Colonial Asbestos Mine, Wittenoon Gorge, Hammersley Ranges, Northwest Australia. *(Courtesy D. E. Ayres.)*

migration of large sea turtles across entire ocean basins to breed. Perhaps, after all, the most compelling biologic evidence for drift is the alleged similarity on opposite sides of oceans of modern species of the lowly earthworm, a creature totally incapable of transoceanic travel.

GONDWANA HISTORY

PREPALEOZOIC

As in the north, the patterns of Prepaleozoic rocks in the southern five continents are very complex. There are many criss-crossings of old mobile belts as well as many overprintings of successive metamorphisms. Some of the oldest rocks known (3.5 to 3.0 b.y.) occur in southern Africa, and Russian geologists report an isotopic date approaching 4.0 billion years from the Antarctic craton. Old dates also occur in central and western Africa (2.9 to 2.7 b.y.), western Australia (3.5 to 2.7 b.y.), southwestern India

(2.5 b.y.), and northeastern Brazil (2.1 b.y.)(see Fig. WM.9). Banded iron formations formed widely (Fig. 15.2), the world's largest gold deposits formed in Precambrian conglomerate in South Africa, and important copper deposits accumulated in Eocambrian strata in Africa (Fig. 15.3). At present it is impossible to interpret the relationships among these continents for Prepaleozoic time; apparently they were more or less separate units until the early Paleozoic.

EARLY PALEOZOIC

The first event that can be correlated clearly among the southern continents was the Eocambrian glaciation (see Fig. 11.2). The glacial evidence, however, tells us little about relative positions of those continents, and the paleomagnetic data are not reliable. Regardless of continental positions, between 800 and 700 million years ago, important new tectonic patterns developed on all the southern continents. A series of mobile belts began that were to persist through early Paleozoic time in much of Africa and in southern Australia (Figs. 15.3, 15.4 and WM.10). Presumably these belts were formed by the convergence of lithosphere plates, which culminated in collisions of a

FIGURE 15.3
Tightly folded strata of Eocambrian age in Katanga arc of the Damara mobile belt, Chambishi copper mine of northern Zambia, south-central Africa. (see Fig. WM.10 for location of "Pan-African" Damara belt). *(By R. H. Dott, Jr.)*

mosaic of continental cratons to form the Gondwana supercontinent by no later than Cambrian or Ordovician time (Fig. WM.16). The collisions are inferred from so-called *Pan-African* isotopic date provinces (600 to 400 m.y.) on all southern continents except India (Figs. WM.9 and WM.10). Apparently a sliver of present eastern North America was then a part of northwestern Africa and South America, and apparently that margin of Gondwanaland was experiencing subduction of the old proto-Atlantic plate (see Fig. 12.25). In late Paleozoic time, recall, Gondwanaland collided with North America and Europe, resulting in the final Appalachian and the Hercynian orogenies (Figs. WM.14 and 14.27). Meanwhile, apparently a microcontinent made up of present southern South America, part of West Antarctica, and probably also New Zealand converged upon southern Gondwanaland where mobile belts (Pampean, Cape, and Tasman; see Fig. WM.10) remained active throughout the Paleozoic. The South African drift advocate, Alexander Du Toit, mentioned in

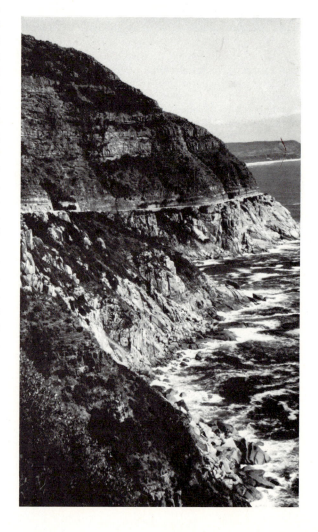

FIGURE 15.4
Basal unconformity at road level between Ordovician–Silurian Table Mountain Sandstone of the Cape mobile-belt and underlying Cambrian Cape Granite (550 m.y.), 5 kilometers south of Cape Town, South Africa. The granite formed during widespread early Paleozoic "Pan-African" orogenesis. Paleozoic Table Mountain Sandstone has thick, cross-stratified quartz sand derived from the southern African craton. *(By R. H. Dott, Jr.)*

Gondwana glacials
known to Du Toit

Inferred extent of
Gondwana Rocks

Assumed ice flow directions

Structural trends

FIGURE 15.5

Du Toit's reconstruction of the *Gondwanaland supercontinent* during late Paleozoic time based upon congruence of shorelines, matching of structural features in ancient basement rocks (lines), and late Paleozoic mobile belts—here combined in what Du Toit called *Samfrau geosyncline* (from the combination of letters from the names of three continents). Apparently a microcontinent compressed the Samfrau belt against main Gondwanaland in Permo-Triassic time. After reconstructing, Du Toit plotted distributions of Gondwana rocks and especially glacial phenomena. Heaviest lines at left represent same boundary on both continents between two major isotopic date provinces. (Compare Figs. 15.9 and WM.9). *(Adapted from Du Toit, 1937.)*

Chap. 6, used these zones of late Paleozoic orogeny in perfecting his now-famous reconstruction of Gondwanaland (Fig. 15.5). His *Samfrau geosyncline* was finally crumpled either by collision of the postulated "Samfrau microcontinent" with Gondwanaland proper or by a special kind of low-angle subduction under this part of Gondwanaland (Fig. 15.6). The resulting Permo-Triassic mountain building was named *Gondwanan orogeny* by Du Toit. Typical clastic wedges accumulated in front of the mountains raised by it (Fig. 15.7B).

Early Paleozoic epeiric seas covered most of the northern Afro-Arabian, northern Indian, western South American, and much of the Australian and Antarctic cratons. Typical early Paleozoic mature quartz sandstones accumulated on cratons as in North America (Fig. 15.4). Devonian seas were the most widespread (see Fig. 13.24), and their faunas in the south were distinct from those of the north (Fig. WM.7). During Late Ordovician time, a large ice sheet covered much of North Africa, which was then located at the south pole according to paleomagnetic data (Fig. 12.25). Silurian and Devonian till-like deposits also occur in central South America (see Figs. 13.24 and WM.7). By Carboniferous time, much of the Gondwana supercontinent had become land, and collisions with the North American–Eurasian supercontinent (Laurasia) produced land connections among all the present continents (Fig. WM.6). In Devonian time, Gondwanaland had shifted northward suddenly about 60 degrees of latitude so that the pole lay in central Africa (Fig. 15.8). Then it shifted about 40 degrees laterally

SAMFRAU MICROCONTINENT

FIGURE 15.6

Two different hypotheses to explain the Gondwana orogeny in the intracontinental Samfrau geosyncline, which has long been a puzzle. *Upper hypothesis* invokes collision by a microcontinent made up of parts of southern South America, Falkland Islands, West Antarctica, and New Zealand (see Fig. WM.14). *Lower hypothesis* calls upon very low-angle subduction to physically couple together subducting and overriding plates for a width of 1,000 kilometers. Presumably coupling would transmit most of the compressional strain to the Cape Mountains. *(Lower adapted from B. E. Lock, Geology, v. 8, pp. 35–39.)*

FIGURE 15.7

Development of the Cape mobile belt. Prior to A, the Cape Granite formed (see Fig. 15.4) and then Table Mountain Sandstone and other middle Paleozoic marine strata accumulated in the mobile belt. *A:* During late Paleozoic time, continental glaciers spread south to the belt where thick till (Dwyka) was deposited. *B:* After glaciation, widespread Mesozoic Gondwana nonmarine strata accumulated in a clastic wedge. *C:* In Triassic and Jurassic time, the belt was upheaved with cratonward thrusting followed by later Mesozoic deposition in normal-fault basins (right) related to initial breakup of this part of Gondwanaland.

in Carboniferous time to bring Australia near the pole by Permian time; in the Triassic Period, the supercontinent moved northward away from the pole.

Because much land area lay at high latitudes in late Paleozoic time, large ice sheets formed on upland areas. Permo-Carboniferous glacial deposits are known on all five of the southern continents. Together with the similarity of present continental margins, the distribution of these deposits provided one of the criteria used for early reconstructions of Gondwanaland (Fig. 6.12). Their presence on either side of the present equator and the absence of similar-aged tillites at high present latitudes in North America and Europe was baffling, to say the least, in terms of present geography. Moreover, in several places the ice seemed to have moved onto

the continents from present deep-ocean basins (e.g., southeastern Australia and eastern South America; Fig. 15.5). The *Dwyka Tillite* of South Africa contains boulders with Lower Cambrian archaeocyathid fossils (see Fig. 9.9) that are unknown in Africa but which could have come from Antarctica (Fig. 15.5). To these criteria, Du Toit added structural features (Fig. 15.5). Then in the 1960s, the new clues for refitting the jigsaw puzzle provided by extensive paleomagnetic studies (Fig. 15.8) and isotopic dating (Fig.

A. PRESENT

B. PALEOZOIC FIT

C. MESOZOIC LATITUDES

—— Africa   —— S. America   — — — Australia

FIGURE 15.8

Paleomagnetic reconstruction of southern continents. *A:* present positions of South America and Africa and the *relative positions* of the south pole as determined paleomagnetically from rocks on each continent. *B:* Note close match of restored zig-zags in respective apparent Paleozoic "polar wandering" paths for Australia and South America with Africa's. It is more convenient to draw *relative* pole shifts, but actually the continents moved. C: Mesozoic latitudes; those indicated for Australia and India do not agree exactly with geologic data. (*After McElhinny, 1967,* UNESCO Symposium on Continental Drift, Montevideo.)

15.9) gave new precision and confidence to the restorations. Deep-sea drilling in the 1970s has added yet more refinement, especially in discovering submerged continental basement that "fills" a former hole in the reconstruction between Africa, Antarctica, and South America. Greatest remaining uncertainty is the proper position of the Pacific side of Antarctica within restored Gondwanaland.

THE GONDWANA ROCKS

Late Paleozoic and early Mesozoic strata of the southern cratons occur in a distinctive sequence that is amazingly similar on all five now widely separated continents. These rocks are collectively known as the *Gondwana rock succession.* (The name Gondwana is derived from an ancient tribe in India.) When Gondwana rocks were first discovered a century ago, their age was unknown. The principal fossils belonged to a then

strange, and as yet undated flora. Because the relative geological time scale was developed in Europe in largely marine strata, the Gondwana succession could be dated only by relating its largely nonmarine fossils to some more familar, well-dated marine index fossils. This first became possible in northern India where lower Gondwana tillites are interstratified with Upper Carboniferous and Lower Permian fossiliferous marine strata (Fig. 15.10). Elsewhere, upper Gondwana strata were found associated with fossiliferous marine Jurassic ones, and so the total sedimentary succession was dated as Carboniferous through Jurassic. More recently, it has been shown that the top of the succession extends into the Cretaceous in some places. Once Gondwana fossils were dated in terms of the marine index fossils of the European standard time scale, they themselves could then be

On the following page:
FIGURE 15.9

Reconstruction of Atlantic border continents using best computer-fit of margins. Some late Prepaleozoic and Paleozoic mobile belts defined by isotopic dating are added to show how well reconstruction could explain many disjunctive geological features. Of great interest is the fact that long-supposed *marginal mobile belts* become *bilateral intercratonic* ones in such a restoration (see Fig. 14.27). Arrows indicate shedding of coarse, red clastic sediments from middle Paleozoic mountains. Brown squares indicate Grenville basement beneath Appalachian belt. (*Reconstructed base map adapted from Bullard, 1965,* Royal Society of London Symposium on Continental Drift)

Ages of Belts

Trans-Atlantic Tertiary
volcanic province

"HERCYNIAN" Late Paleozoic - Triassic

"CALEDONIAN" Silurian - Devonian

"PAN AFRICAN" Eocambrian - Cambrian

"GRENVILLE" Late Prepaleozoic

Ancient Cratons
(1 billion yrs. old)

Dispersal of Devonian clastic deposits
(mostly red)

FIGURE 15.10

Diagrammatic cross section showing how largely non-marine Gondwana rocks were dated in India by relating them to a few interstratified marine tongues of Himalayan (Tethyan) facies containing fossils referable to European standard geologic time scale shown at right (UC—Upper Carboniferous).

used to correlate among different Gondwana localities soon to be explored on all southern continents (Fig. 15.11).

On the five southern continents, the Gondwana succession has prominent glacial tillites in its lower part (Fig. 15.12). These distinctive, unsorted layers overlie scratched surfaces on old rocks abraded by boulders frozen in ice (Fig. 15.1). Many erratic boulders in the tillites also show scratches characteristic of glacial materials (Fig. 15.13). Such deposits in India and South Africa were the first pre-Pleistocene tills to be recognized—and only two decades after proposal of the theory of Pleistocene continental glaciation in Europe in 1840. Although some tillites occur within marine sequences, most of them are interstratified with predominantly nonmarine strata containing important coal seams (Fig. 15.11). Associated with the coals are many fossil plants belonging to an assemblage collectively referred to as the *Glossopteris* flora mentioned in Chap. 14, so named for a seed fern which bore elaborate fruits (Fig. 15.14).

The younger Gondwana rocks include chiefly nonmarine sediments, including conspicuous red beds and wind-blown sands (Fig. 15.15). In some areas, such as eastern Africa, northern India, and South America, they intertongue with marine deposits. Finally, on all continents except Australia, the Gondwana succession is capped by thick basalt flow-rocks and sills (Fig. 15.11). On the southeastern Australian island of

FIGURE 15.11

Distribution of Gondwana rocks showing their remarkable similarity on all five southern continents. *(Bartholomew's Nordic Projection; used by permission.)*

DISTRIBUTION OF GONDWANA ROCKS

FIGURE 15.12
Lower Gondwana (Dwyka) tillite, Port St. Johns, South
Africa. This, the most famous ancient glacial deposit in
the world, formed both on the African craton and in the
north edge of the Cape mobile belt (see Fig. 15.7). *(Courtesy J. C. Crowell.)*

Tasmania, Late Triassic or Jurassic basalt dikes
are present, but no extrusive rocks. In South
Africa and Antarctica, the Gondwana basalts
are of Jurassic age; in South America they are
chiefly Early Cretaceous; while in India 3,000
meters of basalt are Late Cretaceous and early
Cenozoic in age. All these rocks form immense
basalt plateaus deep within cratons where such
rocks are not normally encountered. Like Prepaleozoic and Triassic ones in North America and
Permian ones in Siberia, these basalts are somewhat peculiar, being poor in olivine and sometimes containing minor quartz unlike oceanic

FIGURE 15.13
Gondwana glacial deposits (Talchirs Tillite), India. Striated
cobble of a quartzite whose possible source is only known
today 200 kilometers from the resting place of this erratic.
(Courtesy P. K. Ghosh, Geological Survey of India.)

FIGURE 15.14
Fruits of *Glossopteris* within which seeds matured much as in an apple today. The fruits attached near bases of leaves. Preserved in association with coal seams in South Africa (scale is in centimeters). *(Courtesy E. P. Plumstead; with permission of South African Association for Advancement of Science.)*

basalts. They were fed by countless fissures now filled by dikes.

The famous diamond mines of South Africa have been excavated in a very unusual igneous rock called *kimberlite* (for Kimberley, South Africa), which also occurs in Arkansas, Siberia, Brazil, and Canada. The rock is a magnesium-rich ultramafic species (see p. 94). It is typified by a brecciated chaotic texture, and contains a great variety of inclusions of rocks through which the pipelike masses of kimberlite rose as they blasted their way up through the crust. Especially significant are very dense constituents indicative of high pressure, including the

FIGURE 15.15
Drakensburg Mountains in Royal Natal National Park, South Africa. The escarpment is capped by resistant basalts, which overlie wind-deposited white sandstone cliffs, both of which are Jurassic and comprise the upper Gondwana rock succession here (See Fig. 15.11). *(By R. H. Dott, Jr.)*

FIGURE 15.16

Erect petrified tree stump 32 centimeters in diameter and showing 36 annual growth rings; found in Gondwana strata on Terrace Ridge, the Ohio Range of Horlick Mountains, Antarctica (a few hundred miles from the South Pole). Together with Mesozoic and Cenozoic plant fossils it attests to an intermittently milder past climate than now. (*Photo by W. E. Long; furnished by J. M. Schopf.*)

diamonds, whose densities are 3.1 to 3.5. These dense materials point to derivation at depths below the crust and rapid transport to the surface, which is borne out by textures suggesting explosive emplacement. In Africa, diamonds occur in Cretaceous sandstones, and some pipes penetrated Late Jurassic rocks; therefore the kimberlites are assumed to have been emplaced in Early Cretaceous time, that is, soon after the formation of the nearby Gondwana basalts. Seemingly they provide further evidence of unusual Mesozoic crustal disturbances.

THE *GLOSSOPTERIS* FLORA

How could a uniform land flora be dispersed across wide oceans? Twenty species of leaves found in Antarctica are common also to India now across the equator and far away; yet *Glossopteris* seeds were too large to be windborne. The distribution of the *Glossopteris* flora does not overlap that of the late Paleozoic lycopsid flora of North America, Europe, and North Africa (see Fig. 14.38). Because of its intimate association with glacial deposits, the presence of well-developed seasonal growth rings in petrified tree trunks (Figs. 14.39 and 15.16) and great abundance of leaves at most fossil localities (Fig. 15.17), the *Glossopteris* assemblage is interpreted to have been a temperate-climate deciduous flora. The lycopsid flora, on the other hand, apparently was a tropical-climate one, which is consistent with paleomagnetic data (Fig. WM.6). The

FIGURE 15.17
Glossopteris leaf apparently chewed by insects; leaves as well as fruits must have provided food for animals. (*Courtesy E. P. Plumstead with permission of South African Association for Advancement of Science.*)

Glossopteris flora flourished over wide, swampy low areas on the perimeter of glaciated regions. Interlayering of *Glossopteris*-bearing strata with several tillites indicates considerable climatic fluctuation and implies that the flora was adapted to a wide range of ecologic conditions. The flora persisted into Triassic time long after glaciation had ceased, but by the Jurassic Period, *Glossopteris* had vanished to be replaced by gingkos, cycads, and various new seed ferns (discussed more in Chaps. 16 and 17), which reflects a Mesozoic warming of climate as Gondwanaland moved away from the pole and began to break up (see Fig. WM.5). Besides these new plants, a warming and drying of climate over much of Gondwanaland is suggested also by evaporites, dune sands, red beds, and abundant cold-blooded vertebrates.

ANIMAL FOSSILS

An unexcelled variety of Permo-Triassic amphibian, reptile (including the best-known mammal-like forms; Fig. 15.18), fish, and invertebrate fossils occurs in Gondwana strata. Some fossil leaves even show evidence of having been eaten by insects (Fig. 15.17). A small fresh- or brackish-water reptile called *Mesosaurus* (see Fig. 15.23) is unique to South Africa and Brazil and marks the Carboniferous-Permian boundary. Presumably it could not have crossed the wide Atlantic, and so was taken years ago as additional evidence for existence of the reconstructed Gondwanaland. While many other Gondwana land vertebrates also are common to the southern continents, they have much in common with their northern counterparts. Unlike the *Glossopteris* flora, vertebrates clearly could move easily between the northern and southern supercontinents. For example, African Permo-Triassic vertebrates have about as much

FIGURE 15.18
Cynognathus, a therapsid mammal-like reptile from the Gondwana Sequence (Karoo series), South Africa. (*Courtesy American Museum of Natural History.*)

FIGURE 15.19
Gondwana strata (Beacon Sandstone) with black basaltic sills (Jurassic Ferrar Dolerite) exposed along glaciated walls of Finger Mountain near McMurdo Sound at west edge of the East Antarctic craton. *(Courtesy R. F. Black.)*

in common with Eurasian, North American, and Siberian ones as with those of other Gondwana continents. And Jurassic dinosaurs in eastern Africa were identical with those of Wyoming! Clearly land animals managed to disperse north and south more readily than did plants, which perhaps is not surprising in view of the mobility of animals. Their wide distribution also requires that the herbivorous animals were adaptable to a diet of either lycopsid or glossopterid vegetation.

ANTARCTICA—A TRIUMPH OF PREDICTION

Two-thirds of Antarctica (East) is a craton, whereas a youthful mobile belt marks its Pacific (West) margin; an early Paleozoic *Trans-Antarctic mobile belt* underlies the western edge of the present craton (Figs. 15.5 and WM.10). Since 1909, when Cambrian fossils were discovered at the west edge of the craton, early Paleozoic strata of the Trans-Antarctic belt have been studied intensively. Isotopic dating shows that the belt was deformed near the end of Cambrian time (500 million years ago). Granitic batholiths were emplaced and extensive metamorphism occurred as in Australia, Africa, and South America, making it another *Pan-African* belt.

Early Paleozoic mountains were leveled by erosion, and then a Devonian epeiric sea flooded the region as in the Brazilian and Australian shields (Fig. 13.24). Devonian marine fossils present have close affinities to those of Australia, South America, and the Cape belt of South Africa, all of which comprise the *Southern Faunal Realm* (Fig. WM.7). A very widespread and distinctive sequence of cross-stratified sandstones with important coal seams overlies the Devonian rocks (Figs. 15.11 and 15.19). Important fossil plants were discovered only 100 kilometers from the pole during a 1901–1904 British expedition, and by the ill-fated Scott expedition of 1910–1912. On the return down Beardmore Glacier from the South Pole, Scott's men collected many rock samples, which were found with their frozen bodies at their last camp. Specimens were returned to London, but the plants were misidentified as lycopsids. Years later, reexamination proved that the fifth southern continent also contained the *Glossopteris* flora! Among the specimens were samples also of basalt described as forming thick, black bands in the sandstones along the walls of Beardmore Valley (Fig. 15.19). Yet another characteristic of the Gondwana rock clan was present. Naturally, these early geologic revelations from the frozen continent were greeted with eager anticipation, especially by Southern Hemisphere geologists and by botanists, who were puzzled by presence of coal and fossil trees at such high, hostile latitudes.

Relatively few new discoveries occurred until after World War II. The International Geophysical Year or I.G.Y. (1957–1958), an imaginative venture in worldwide scientific cooperation, gave impetus to a new wave of scientific explo-

FIGURE 15.20
Spectacular fold structures in Paleozoic strata, Sentinel Mountains, West Antarctic mobile belt. This area represents westernmost known occurrence of Gondwana rocks in Antarctica; beneath them is a thick marine sequence much as in the South African Cape mobile belt (see Fig. 15.7). *(Courtesy Campbell Craddock)*

ration on a scale so large that Antarctica appeared in danger of overpopulation. As a result of more extensive explorations, many more Gondwana plant localities from Carboniferous through Triassic ages have been found, and detailed restorations of river flood plains, coal swamps, coastal deposits, and glacial phenomena have been made. Gondwana rocks were also found beyond the craton where they were caught in mountain building (Fig. 15.20), as in the *Cape Belt* at the southern tip of Africa (Fig. 15.7). Mesozoic basaltic rocks also have been found over an immense area, although their main concentration is around Beardmore Glacier. Isotopic dating indicates that these igneous rocks are about 160 million years old (Jurassic), or about the same as those on surrounding continents (Fig. 15.11).

Even by 1930, nearly all elements typical of Gondwana sequences elsewhere were known save evidence of glaciation and nonmarine vertebrate fossils. Similarities seemed strong

enough that the great South African geologist, A. L. Du Toit, surmised that Antarctica was a full-fledged Gondwana continent, and so it must have been glaciated and probably also had been inhabited by land animals. It was 30 years before his prediction was proved true, however. During the I.G.Y., New Zealand geologists found a probable tillite, and in 1960, Ohio State University personnel discovered another example in a complete sequence of Gondwana rock types, including coal and *Glossopteris* remains, only 200 miles from the South Pole. In 1967 an amphibian jaw bone, and in 1969 reptilian bones, both of Triassic age, were found, proving that cold-blooded land animals as well as forest trees inhabited the continent 210 million years ago! At least one of the land-dwelling reptiles was identical with Triassic forms long known on other southern continents. The similarity of East Antarctica to the other four Gondwana continents became firmly established (Fig. 15.11).

PALEOGEOGRAPHY AND PALEOCLIMATE OF PANGEA

Figure 15.21 shows a paleogeographic reconstruction of all of *Pangea* for late Carboniferous-Permian time. Early continental-drift theorizers postulated a single huge polar ice cap, which spread radially outward from the center of Gondwanaland (Figs. 6.12 and 15.21). In Chap. 6 we noted also that the Wegener-Köppen late Paleozoic restoration of the continents placed the north pole in the north Pacific Ocean, which would make it a warmer pole (Fig. 6.12). Probable glacial deposits of late Paleozoic age recently recognized in northeastern Siberia suggest that there was, in fact, some northern glaciation at high latitudes symmetrical with that of Gondwanaland (see Fig. WM.6). This is consistent with the temperate-climate fossil flora of northeastern Asia (see Fig. 14.38). Wegener and Köppen drew the equator through the northern (lycopsid) coal belt, which already was considered tropical on paleobotanical grounds (see p. 331). Based upon evaporite deposits and presumed fossil dune sands, midlatitude deserts like today's were postulated for both hemi-

Sea   Swampy lowlands   Ice   Mountains
(much coal)

FIGURE 15.21
Late Paleozoic reconstruction of Pangea—a compromise of all available data. Ice never covered the entire white area at once. Note marine embayments on Gondwanaland, which must have furnished moisture for glaciers. Probable glacial-marine deposits also exist locally in Siberia near the north paleopole (see Fig. WM.6). (S—south pole; N—north pole.)

spheres (see Fig. 6.12). Everything seemed to fit, and for drifters the most heartwarming recent development has been the almost perfect agreement of paleomagnetic results for ancient latitude with the old 1920s restorations (compare equators in Figs. 6.12 and 15.21).

American geologist A. A. Meyerhoff has pointed out that the total area covered by Gondwana glaciation was three times as large as the North American Pleistocene ice cap. A single landmass so large would have been too dry in its interior for the nourishment of such a huge cap. In Pleistocene time, for example,

Siberia was certainly cold enough, but it was too dry to have had a significant ice cap. In addition, he notes that in southwestern Africa and in Brazil, tillites occupy ancient valleys as much as 1,000 meters deep. This indicates that considerable topographic relief existed and that the glaciers might have formed on high, cool plateaus even if the continents were located at low latitudes just as glaciers are found today on high equatorial mountains in South America, eastern Africa, and New Guinea.

Stratigraphic evidence indicates that glaciation of all of Gondwanaland was not simultaneous. It began first in South America in Silurian or Devonian time and continued there into Carboniferous time. It was confined to the Carboniferous in Antarctica and Africa, but spanned part of Carboniferous and part of Permian time in India and Australia (Fig. 15.11). Paleomagnetic evidence confirms also that Gondwanaland was drifting as a unit across the South Pole (Fig. 15.8); thus the centers of glaciation shifted as this occurred. Therefore, we need not postulate a single ice cap at any one time that was larger than Pleistocene ones (and as shown in Fig. 15.21). The glacier nourishment problem is further alleviated by the recognition in recent years that epeiric sea embayments existed within Gondwanaland (Figs. 15.21 and WM.6). These areas of water would have been important sources for evaporation of moisture to feed the interior ice cap areas.

Northern Australia provides proof that even the northern fringe of late Paleozoic Gondwanaland really was cold. Permian brachiopod shells from a marine layer just above a tillite have been subjected to a geochemical technique for determining the temperature of the sea water in which the shells were formed. The analysis involved precise determination of the ratio of two isotopes of oxygen contained in the calcium carbonate of the shells. That ratio is related to the original water temperature, as explained more fully in the next chapter. The results indicate a Permian sea-water temperature of only 7°C, which supports the Gondwanaland restoration.

Another climatic puzzle raised by Meyerhoff is the presence of Permian evaporites near tillites in Bolivia and Brazil. Although these two types of deposits suggest opposite temperature conditions, there is no reason to suppose that they formed at exactly the same time. Instead they reflect marked changes of climate within Permian time, which is not surprising because glaciation ceased early in South America as that part of Gondwanaland drifted north.

The *Glossopteris*-bearing coal deposits in Gondwanaland, like the northern lycopsid-bearing coals, formed largely in broad, coastal swamps as indicated by their interstratification with fossiliferous marine and brackish-water strata. Because of their higher latitude, the Gondwana swamps were simply more temperate and more variable in climate than the tropical lycopsid ones. Clearly the glaciers flowed down to sea level and out onto continental shelves at many places, such as the northern Australia locality mentioned above (Fig. 15.21). Presence of apparent warm-temperature indicators such as evaporites and cold-blooded reptiles need not be alarming, for today tree ferns grow within a mile of glaciers in New Zealand, and ice extends down to areas with apple orchards in Norway. The point is that if glaciers flow from their cool source areas into warm areas more rapidly than melting occurs, they will persist there in spite of warmth. As noted in Chap. 14, the presence of red beds (Fig. 15.11) probably proves wet and dry seasons more than warm temperature, although other evidence makes warm conditions seem probable, too.

At least six distinct tillite zones are known in Africa, and more than two dozen have been claimed in Australia. Clearly the glaciers advanced and retreated repeatedly, but it is impossible to tell how many times. Apparently the *Glossopteris* flora was adapted to a variable cool-temperate climatic zone surrounding the ice caps. As the glaciers retreated, the flora expanded poleward, and vice versa. Considering the long time during which large glaciers existed on Gondwanaland (and by analogy with the Pleistocene), we conclude that worldwide sea level must have fluctuated up and down many times as the ice shrank and expanded. All coastlines around the world—even in the tropics—would have experienced transgres-

sions and regressions, and it seems more than coincidental that the most conspicuous repetitive or cyclic sedimentation in the entire geologic record occurs in strata of Carboniferous and Early Permian ages in many parts of the world— exactly the span of the main Gondwana glaciation (see Chap. 14). While late Cenozoic glaciation has lasted about 20 or 30 million years so far, late Paleozoic Gondwana glaciation spanned almost 100 million years.

Antarctica shows the most dramatic evidence of climatic changes through time of any place on earth. Today only lichens, algae, and a few puny grasses (and no truly land animals) grow there, but dense forests with sizable trees (Fig. 15.16) have grown there at several times in the past. Moreover, in Permo-Triassic time, vertebrate animals roamed the landscape only 100 miles from the present South Pole. Clearly the continent had a much milder climate during the past than now, even though glaciers as well as forests also existed there during the late Paleozoic. Major temperature fluctuations characterized late Paleozoic time, as indicated by the presence of cold-blooded animal remains and coal seams interstratified with tillites.

The late Paleozoic polar region apparently suffered greater oscillations of climate than did the equatorial land region, just as in Pleistocene time. Whereas in much of Gondwanaland the tillites alternate with coal and reptile-bearing strata, North America and Europe show no evidence of significant climatic fluctuations— only a gradual long-term increase of aridity of the land through Permian and Triassic times. Meanwhile, a tropical sea called *Tethys* (named for one of the mythical Titans, a daughter of Uranus) separated eastern Gondwanaland from eastern Laurasia (Fig. 15.21), and it nurtured a very homogeneous fauna from Permian through Eocene times.

SOME ASPECTS OF PANGEAN BIOGEOGRAPHY

With all of the world's continents sutured together into Pangea, the oceanic and atmospheric circulation would have been far different from what we see today (see Figs. WM.1, WM.2, and WM.6). Solar radiation and surface reflection of the earth, together with those circulation patterns, affected the climate in late Paleozoic time. We have seen glacial evidence that proves that both the southern and northern high latitudes were cold, while between lay the tropical *Tethys sea* with a rich and diverse fauna. The Permian marine faunas reflect the steep pole-to-equator temperature gradient (compare Fig. WM.2) with very cold waters in high latitudes. In the Asiatic part of Laurasia, faunas consist of very few species but individuals of large size. In Gondwanaland, for example Australia, again we find a typical cold-water fauna with few species. These two faunas are separated by the low-latitude *Tethyan fauna* with its hundreds of species, many of which can be traced east and west for thousands of kilometers. This warm-water fauna and its descendants existed through the Mesozoic Era until Eocene tectonic activity began to destroy the continuity of the seaway. Today, only relics of that once glorious ocean remain as the Caspian, Black, and Mediterranean seas, as discussed in Chap. 17.

Valentine and Moores have summarized the factors regarding the relationship between the assembled continents and faunal diversity. When Pangea was assembled, the climate became strongly seasonal with a steep latitudinal temperature gradient. In the oceans, there were few barriers, so that species were widespread, and hence there was moderate species diversi-

ty. The only barrier as such was the colder temperature of the water at higher latitudes, which excluded many species that could tolerate only warm water; the result was lower diversity. But, just as today, some species could tolerate a wide range of temperatures and even of salinity variations, and these were more cosmopolitan (widespread). As Pangea began to break up during the Mesozoic, atmospheric and oceanic circulation were altered, and the climate became more uniform and warmer. Distinctive biotas formed that were separated by barriers. Widening deep-ocean basins between the separating continents were barriers to some marine organisms just as they were to land-dwelling ones. Species diversity increased, particularly in shallow shelf seas, as stable but separated environments became established. Diversity also increased on the now-separated continents. More of the consequences of these major changes will be discussed in Chap. 16.

During Permo-Triassic time, vast stretches of lowlands were present on Pangea with homogeneous floras and faunas traceable over great distances. The diversity of land reptiles was quite low compared to the rich Cenozoic mammals after the break-up of Pangea. The mammal-like reptile *Lystrosaurus* illustrates the wide distribution of the land animals during this time, for it is known from many localities on all five southern continents (Fig. 15.22). Its present disjunct (disconnected) distribution also serves as one of the most convincing pieces of evidence for an Antarctic-Africa-India connection. As

FIGURE 15.22
The mammal-like reptile *Lystrosaurus* from the Triassic of Antarctica. *(Reconstruction by Margaret Colbert in E. H. Colbert, 1965, The Age of Reptiles, Norton.)*

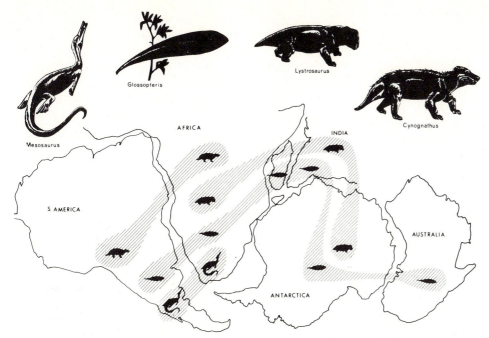

FIGURE 15.23
Reconstruction of Gondwanaland, showing the recorded
distribution of key Permian and Lower Triassic genera,
including *Lystrosaurus*. *(From Colbert, 1973. Copyright© 1973
by Edwin H. Colbert. Reprinted by permission of the
publishers, E. P. Dutton.)*

mentioned earlier, the remains of *Lystrosaurus*
were discovered in 1969 in Antarctica, and it
proved to be identical with species found in
South Africa and India (Fig. 15.23). This reptile
was a land dweller, and could not have swum
across any wide expanse of water. Previously
discovered vertebrates, such as the small aquatic reptile *Mesosaurus,* commonly were discounted as having no relevance to the question of
former land connections (Fig. 15.23).

Disjunct faunas can come about in two ways.
One is that a living fauna became widely
distributed, but owing to some later environmental change along its range, the distribution
was broken into disconnected patches. The
second type of disjunction is caused by continents drifting apart to break up a former continuous range. Either the living organisms ride on
a continent, which becomes a kind of *Noah's
Ark*, or already-fossilized remains ride on a

continent, which is more like a *Viking Funeral
Ship* (see Chap. 16). In the former case, they
may have adapted and evolved as their ark
sailed through different climatic zones, or they
may have become extinct en route if they moved
into severely adverse conditions. Naturally the
fossils floating on Viking Funeral Ship continents would not be affected by such environmental changes. In either case, the same genera
or species will then be found on different
continents and will be considered disjunct as in
the case of *Lystrosaurus* (Fig. 15.23).

Lystrosaurus is not only found in the southern
continents, but also in eastern China. There is
no evidence that China was directly connected
to Gondwanaland, and some critics have seized
upon this as an argument against continental
drift. We assume, however, the *Lystrosaurus* had
a very broad geographic range that later became
disrupted. (The alligators of today show similar
disjunction between China and the Americas.)
Lystrosaurus probably migrated around the west
end of the Tethys Sea via a land bridge between
Gondwanaland and Laurasia during Permo-
Triassic time (Fig. 15.21).

FIGURE 15.24

Map restoring relative past paleomagnetic south pole positions for all continents except Asia. Medium-weight dashed arrows trace pole migrations *relative to continents* for Gondwana and Euramerica in early Paleozoic; S-Cl records their collision in middle and late Paleozoic time. Heavy arrow traces pole migration for all Pangea until breakup in early Mesozoic time (M). Light dashed arrows record separation of the different continents since then. Although we know that it was the continents that moved, this type of map is useful for illustrative purposes. (*After McElhinny, 1973, Palaeomagnetism and plate tectonics, Cambridge; used with permission.*)

THE GREAT PANGEA BREAKUP

After more than 200 million years of peaceful union, the outbursts of middle Mesozoic basalt forewarned of trouble in Gondwanaland. Before the end of the Jurassic Period, serious separation proceedings were underway, and by the end of the Cretaceous, so many links had been broken that an ultimate complete split became inevitable. Strife and unrest even extended to that northern family, where North America and Eurasia also split up before the end of the Mesozoic Era. Antarctica maintained ties with Australia and South America longer, but even these links broke by mid-Cenozoic time. Of all the former continental unions begun in Paleozoic time, only that of Europe and Asia has

survived to the present. Recently Africa (Arabia) and India have joined with Eurasia but the fate of these bonds is as yet uncertain, and meanwhile there is trouble brewing within Afro-Arabia where the Red Sea rift began forming about 30 million years ago. And so it goes, although drawn mysteriously together many different times, continents have had difficulty maintaining stable associations for long periods.

LAURASIA

Figure 15.24 shows the positions of the *South Pole relative to the continents* for the past 500 million years (this is equivalent to holding continents fixed and seeing how the pole appears to have migrated). This map conveniently summarizes paleomagnetic evidence both for the times of supercontinent assembly and for breakup. Beginning of formation of the supercontinent Pangea in mid-Paleozoic time by the collision of Laurasia (or Euramerica) with Gondwanaland is shown by the convergence of their respective *apparent polar wandering paths* (the S-C1 pole position of Fig. 15.24). The coincidence of pole paths forming the heaviest line reflects the existence of Pangea moving as a single unit until the early Mesozoic (M pole position), and its path shows the *relative apparent movement of the pole* across Gondwanaland during late Paleozoic time as discussed above in connection with glaciation. It is the supercontinent that did most of the moving, of course, but this type of portrayal is useful for illustration purposes. Note that the cessation of glaciation in Late Permian time coincided with the relative movement of the south pole to a warmer oceanic position (to M). The light dashed arrows reflect the breakup and divergent relative motions of the separate continents going their own ways (but beginning at different times) in the Mesozoic.

Late Triassic faulting and basaltic igneous activity in Europe, North America, South America, and northwestern Africa forewarned the opening of the central Atlantic basin. From Nova Scotia south to Florida and in northwestern Africa, very thick Upper Triassic and Jurassic nonmarine red beds and interstratified basal-

FIGURE 15.25
Initial breakup of a supercontinent by rifting and sea-floor spreading based upon the North Atlantic and Gulf of Mexico basins. *Late Triassic:* Block faulting, swarms of basaltic dikes, volcanic eruptions (black), and nonmarine sedimentation. *Jurassic:* Initial seafloor spreading and first marine deposition characterized by evaporites. *Cretaceous:* Normal marine deposition including reefs (R); Tethyan fauna now established in Gulf. (Compare Fig. 14.26.)

main embryonic ocean rift zone, evaporites were the first marine deposits (Fig. 15.25). After burial under younger sediments, the lower-density evaporites tended to rise isostatically as intrusions into the overlying strata (such features are discussed more in Chap. 17).

Ocean-floor magnetic anomalies and deep-sea-sediment ages indicate that actual opening of the central Atlantic between North America and Africa began in Early Jurassic time (about 180 m.y. ago; Fig. WM.13). On the other hand, magnetic anomalies indicate that Europe did not separate from Greenland until the beginning of the Cenozoic Era (65 m.y. ago; Fig. WM. 11), which is consistent with the presence of widespread basalts of that age in northern Britain (including Ireland's famous Devil's Causeway) and southern Greenland. Greenland, however, began separating on the west side from the North American mainland in latest Cretaceous time (about 80 m.y. ago; Fig. WM.12). Small Jurassic-Cretaceous igneous bodies, which include some eroded volcanic necks in New Hampshire, near Montreal, Quebec, in Newfoundland, and a series of now-submerged volcanic seamounts extending east from New England on the Atlantic sea floor, apparently formed along fractures activated during the early opening of the ocean basin. The Arctic Ocean doubled in size and began exchanging water with the Atlantic. We see, therefore, that the breakup of Laurasia had a complex history, which was anything but synchronous.

TETHYS SEA

The relationship between Europe and Africa in early Mesozoic time was very different than today's. Part of Iran may have been attached to Arabia, and Turkey and Greece apparently were part of North Africa. Italy, Sardinia, and Corsica were all attached to southern France and eastern Spain (Fig. WM. 13). Northwestern Africa and Spain remained more or less attached until Jurassic time, but the Tethys ocean basin lay between Mesozoic Eurasia and the remainder of Afro-Arabia and India (Fig. WM.13). The present Mediterranean Sea has long been consid-

tic flow rocks and sills occur in downfaulted basins (Fig. 15.25, also 14.23 and 14.24). Both the faulting and the basaltic igneous activity point to a new, extensional behavior after 400 to 500 million years of compressional tectonics. This basaltic eruption and faulting on at least four different continents soon after the Paleozoic era ended is the first hint of the beginning of the breakup of Pangea and the opening of the present North Atlantic basin. Basaltic dikes form a radial pattern around the North Atlantic basin (Fig. WM.13), which must reflect the initial rupture—perhaps caused by a large, hot mantle plume (compare Fig. 7.8). Within the

FIGURE 15.26
The Dolomite Alps of northeastern Italy (near Cortina d'Ampezzo) were carved by Pleistocene glaciers largely from Tethyan Triassic reefs. These rocks apparently were rifted from northern Gondwanaland to collide with southern Europe during the mid-Cenozoic Alpine orogeny. (By R. H. Dott, Jr.)

ered a remnant of the old Tethyan basin, which shrank as India and Arabia collided with Asia. As is shown in Figs. WM.11 and WM.12, however, the Mediterranean was formed instead by spreading behind the Italy-Yugoslavia-Greece-Turkey blocks as they were torn from Africa and driven north to collide with Europe. The floors of the Black and Caspian Seas may be the only remnants of the original Tethys that have not been crushed and shredded by continent-continent collisions. Cyprus apparently is a portion of an old Tethyan ocean ridge squeezed and elevated during Cenozoic motions of Africa toward Europe.

The Tethys seaway, extending from the Caribbean region all the way east to present Indonesia, was originally recognized and named because of a remarkable homogeneity of Mesozoic marine faunas. The rich assemblages include a greater diversity of species of corals, sponges, ammonoids, and other molluscs than to the north or south. Some forms, such as stromatoporoids, the exotic rudistid clams (see p. 438), and certain microfossils, are known only from the Tethyan region. All these fossils occur within a vast expanse of thick carbonate rocks, which were deposited in shallow shelf environments on the margins of the supercontinents facing the Tethys. Reefs are prominent in these rocks (Fig. 15.26). (The prolific oil fields of Arabia produce petroleum from Tethyan limestones of Jurassic age.) The very diverse marine fauna, the presence of widely scattered evaporites, and the types of fossil plants that lived on adjacent lands all point to a warm climate;

clearly the entire Tethyan realm lay in the subtropics and tropics.

The shelves on both sides of the Tethys ocean basin were originally connected (Triassic-Jurassic) at the west end through Spain and northwestern Africa as well as over the restored positions of Yugoslavia, Greece, Italy, and Turkey as shown in Figs. 15.21 and WM.13. By Jurassic time, there was also a shallow marine connection westward to the present region of the northern Gulf of Mexico and Central America (Fig. WM.5). These interconnections of shallow seas allowed the wide migration of the distinctive *Tethyan fauna.* Deep, wide oceans pose almost insurmountable barriers to the dispersal of shelf faunas because most bottom-dwelling invertebrates have a larval stage too short to survive long transoceanic voyages. Therefore, it does not seem mere coincidence that early Mesozoic Tethyan shelf faunas are very homogeneous from southern North America eastward to southern Asia. As Africa and North America separated, however, later marine faunas became more distinctive, apparently due to a widening and deepening Atlantic barrier.

As Laurasia became restive and began to break up, Triassic volcanism not only occurred in eastern North America and northwestern Africa, but also on the northern (European) side of the Tethys. In Early Jurassic time, when the central Atlantic had actually begun opening, the western Tethys also became fragmented. Because Europe remained coupled with northern North America, a major transform-fault system must have developed between Africa and Europe. Apparently it passed through northwestern Africa and southern Spain, perhaps slicing that region into several elongate blocks (Fig. WM.13). A spreading center or centers apparently formed simultaneously in the western Tethys and began to break up the former continental shelf there (Figs. 15.27 and WM.12). First, Yugoslavia, Greece, and Turkey (and perhaps also Iran) were torn from the shelf to begin their eastward journeys as microcontinents. Soon the Italian block was also torn away and sent eastward. These fragments in their turn were transported northeastward to collide

eventually with southern Eurasia (Fig. WM.11). The culmination of these collisions occurred in mid-Cenozoic time with the great Alpine-Himalayan orogenic upheaval resulting from the collisions of Afro-Arabia and India with Eurasia, to be discussed in Chap. 17.

During late Mesozoic time, several new small-ocean basins formed in the western Tethys by the rifting of microcontinental blocks from the main continents. Deep-water sediments were deposited in them—the very same deep-water sediments originally regarded a century ago by European geologists as typical geosynclinal strata (see Chap. 6). These sediments, which include cherts, fine mudstones, and rhythmically stratified *flysch*—dominantly turbidity current deposits (like Fig. 14.22), were deposited upon ophiolite igneous sequences considered by most geologists now as having formed originally at several spreading sites in the new microocean basins within the Mesozoic Tethys. Some of the small-ocean basins were closed through subduction followed by collision between continental fragments as suddenly as they had started opening—like so many ice cubes jostling about in a punch bowl.

The Mesozoic history of the Tethys was so complex as to be perhaps forever indecipherable in a detailed way. To some extent, this complexity must be due to the fact that such a large mass of continental crust was fragmented into so many microcontinents by transform faulting and that local sea-floor spreading then shifted those continental fragments to and fro as overall Atlantic spreading directions and rates changed.

GONDWANALAND

Besides the Mesozoic basalts of eastern North America, northwestern Africa, and parts of interior Gondwanaland, all of which reflect pulling apart of lithosphere plates along countless fissures to allow mantle material to rise, early-warning signals of impending separation also include the formation of narrow marine embayments along the sites of eventual new continental margins. In Middle Jurassic time, transgression began in eastern Africa and Mad-

FIGURE 15.27

Evolution of the north African margin of the Tethys seaway during Jurassic time (bottom to top). Note progressive fragmentation of this edge of Gondwanaland as sea-floor spreading broke up the Tethyan margin. Former shallow-water carbonate rocks were downfaulted, and deep-water Jurassic sediments were deposited in the troughs. *(From Bernoulli and Jenkyns, 1974, in R. H. Dott, Jr., and R. H. Shaver, eds.,* Modern and ancient geosynclinal sedimentation; *used with permission of Society of Economic Paleontologists and Mineralogists.)*

agascar, and by Cretaceous time, transgressions were occurring on all sides of Africa, South America, and at least three sides each of India and Australia. Important evaporite deposits discovered by recent exploration for petroleum indicate that most of these embayments did not at first have free access to the open ocean (Fig. 15.25). But the fact that in all cases normal-marine Late Cretaceous or early Cenozoic strata succeeded the evaporites reflects broadening of the basins as the present continents broke apart and new deep-ocean basins such as the Atlantic were formed between them by sea-floor spread-

ing. Block faulting in Late Triassic time in eastern North America, in the Cretaceous in eastern Brazil and in western Africa (Fig. 15.28), as well as in South Africa (Fig. 15.7C), and in early Cenozoic time in western India contributed to the early development of these basins. Sea-floor magnetic anomalies, age patterns of deep-sea sediments, and paleomagnetic data from the continents together provide the best clues to the initial opening and subsequent spreading histories of the new ocean basins, and thus of continental separation. Such analysis indicates that, as already noted, North America and Gondwanaland began separating to form the central Atlantic basin in Late Triassic time (Fig. 15.25). India began separating from Africa in Jurassic time, while South America split from Africa in the Cretaceous Period (Fig. 15.29). An Australia-Antarctic unit probably began separating from its other Gondwana partners in Jurassic or Cretaceous time, but remained together as a kind of siamese-twin continent until early Cenozoic time when Aus-

FIGURE 15.28
Diagram of west Africa showing typical faulted margin resulting from opening of the present South Atlantic basin beginning in Cretaceous time. Evaporites deposited initially were buried deeply, and then formed intrusive salt ridges and mounds (domes) shown in black. Similar features are being found beneath the margins of most continents facing the Atlantic and Indian Oceans. *(From Beck, 1972, in* Australian Petroleum Exploration Association Journal, *v. 12, pp. 7–28.)*

tralia split for warmer latitudes (Fig. WM.12). At least a slender connection existed until early or middle Cenozoic time between southernmost South America and West Antarctica. At that

FIGURE 15.29
Rio de Janeiro harbor and Sugar Loaf Mountain, Brazil. The picturesque rounded stone monoliths of Rio were eroded from early Paleozoic granites very similar to many along the west coast of Africa; these granites formed during Pan-African orogenesis (e.g. Cape Granite, Fig. 15.4). *(Courtesy Pan American Airways.)*

time, the formerly continuous Andean orogenic belt was dismembered by large-scale faulting to form the Scotia Sea and island arc between the two continents (Fig. WM.11).

SUMMARY

Impressive in the south is evidence of widespread early Paleozoic (Pan-African) mountain building, which is little known in the northern continents. An exception is eastern New England, Nova Scotia, and Newfoundland, which were originally part of Gondwanaland. This event produced much land area in Gondwanaland, followed in middle Paleozoic time by epeiric sea transgressions of Australian, North African, South American, and part of the Antarctic segments of the supercontinent. Present India, two-thirds of Africa, and probably much of East Antarctica remained land. Plateaus or mountains may have characterized large regions, as indicated by buried ancient, deep valleys. Continental glaciation began earliest in South America (middle Paleozoic), and South Africa (early Carboniferous), and continued latest in Australia (Late Permian). Glaciation on such a scale probably provided the mechanism for simultaneous cyclic sedimentation in the northern continents as a result of worldwide fluctuations of sea level as ice expanded and contracted.

The most significant relation among southern continents is the striking similarity of largely nonmarine Paleozoic and Mesozoic rock sequences on all five of the continents now so widely separated by deep seas. Besides glacial deposits, the unique *Glossopteris* flora, land vertebrate fossils (including mammallike reptiles), and basalts are prominent. These characteristics have for two generations provoked questions about the structure of the earth's crust. How could the *Glossopteris* land flora be dispersed from one continent to another across 5,000 kilometers of salt water? And how could cold-blooded animals and luxuriant forests grow within the Antarctic Circle? Even if somehow the climate were warm enough there, could trees adapt their photosynthesis to 6 months of light and 6 months of darkness? And how could continental glaciers develop at low elevation on both sides of the present equator, especially when simultaneously North America and Europe, now at high latitudes, clearly were enjoying warm, humid conditions and lacked glaciers? Also, why do several important structural zones seem to vanish at present coastlines? Wegener and Du Toit provided answers to these questions long ago, and their restorations of Pangea stand today as basically correct in spite of an enormous volume of new, sophisticated data that have appeared since 1955. This is an enormous tribute to the human intellect.

Early Mesozoic faulting and basaltic eruptions in eastern North America and northwestern Africa (and even in Norway and in far-off Siberia) apparently attest to the initial breakup of Pangea, which was soon to lead to the birth of the present Atlantic basin. During Triassic and Jurassic times, the southern continents were practically all land. Immense outpourings of basalt accompanied faulting and the formation of diamond-bearing intrusions; in India and South America, eruptions continued into Cretaceous time. Late Jurassic–Cretaceous transgression affected all continental margins, but to varying degrees, as Pangea was gradually dismembered and new ocean basins formed between them. At lower latitudes, evaporites typically were deposited first before open circulation of sea water became possible. Stratigraphic and paleomagnetic evidence indicates that the North Atlantic, Gulf of Mexico, Mediterranean, and Indian basins began opening first (Jurassic), followed by the South Atlantic, Davis Strait between Greenland and North America (Cretaceous), and then the Atlantic between Greenland and Europe as well as the Pacific between Australia and Antarctica (Eocene). The very last tenuous connection, an isthmus between South America and Antarctica, was severed in Miocene time.

Land faunas were very homogeneous over Pangea's vast area, reflecting easy access to all parts of the supercontinent. After Jurassic time, however, the similarities diminished steadily as the different continents became increasingly isolated. A distinctive, homogeneous, tropical shallow-marine Mesozoic fauna developed in the Tethys seaway between Eurasia and Africa and spread westward to southern North America by Jurassic time. As the ocean basins widened, however, Cretaceous and Cenozoic faunas on opposite sides differed more and more.

Key Concepts

apparent polar wandering path
breakup of Pangea
Cape Belt
disjunct faunas:
   Noah's Ark displacement
   Viking Funeral Ship displacement
Dwyka Tillite
glacier nourishment
Glossopteris
Gondwanaland
Gondwana Rock Succession
Gondwanan orogeny
ice-rafted pebbles
initial evaporite deposition
kimberlite (diamond-bearing)
Laurasia
Lystrosaurus
mammallike reptiles
Mesosaurus
paleopole positions relative to continents

Pan-African orogeny
Pangea
Samfrau geosyncline
Samfrau microcontinent
Southern Faunal Realm
Tethyan fauna
Tethys Sea
Trans-Antarctic mobile belt
worldwide cyclic transgressions and regressions

Readings

Adie, R. J., (ed.), 1972, Symposium on Antarctic geology and solid earth geophysics: Oslo, Universitetforlaget.

Boucot, A. J., Johnson, J. G., and Talent, J. A., 1969, Early Devonian brachiopod zoogeography: Boulder, Geological Society of America Special Paper 119.

Briden, J. C., Smith, A. G., and Sallomy, J. T., 1971, The geomagnetic field in Permo-Triassic time: Geophysical Journal of Royal Astronomical Society, v. 23, pp. 101–117.

Brown, D. A., Campbell, K. S. W., and Crook, K. A. W., 1958, The geological evolution of Australia and New Zealand: London, Pergamon.

Bushnell, V. C., and Craddock, C., 1970, Geologic maps of Antarctica: Antarctic Folio Series, Folio 12, 23 plates; New York, American Geographical Society.

Childs, O. E., and Beebe, B. W., 1963, Backbone of the Americas: Tulsa, American Association of Petroleum Geologists Memoir 2.

Clifford, T. N., 1967, The Damaran episode in the Upper Proterozoic-Lower Paleozoic structural history of southern Africa: New York, Geological Society of America Special Paper 92.

Colbert, E. H., 1973, Wandering lands and animals: New York, E. P. Dutton.

Crowell, J. C., and Frakes, L. A., 1970, Phanerozoic glaciation and the causes of ice ages: American Journal of Science, v. 268, pp. 193–220.

Doumani, G. A., and Long, W. E., 1962, The ancient life of the Antarctic: Scientific American, September, pp. 169–184.

Furon, R., 1963, Geology of Africa: Edinburgh, Oliver & Boyd. (English translation by A. Hallam and L. S. Stevens)

Jenks, W. F., (ed.), 1956, Handbook of South American geology: New York, Geological Society of America Memoir 65.

Kay, M., (ed.), 1969, North Atlantic—geology and continental drift: Tulsa, American Association of Petroleum Geologists Memoir 12.

Kummel, B., 1970, History of the earth (2d ed.): San Francisco, Freeman.

Kurtén, B., 1969, Continental drift and evolution: Scientific American, v. 220, no. 3, pp. 54–64.

McElhinny, M. W., 1973, Palaeomagnetism and plate tectonics: Cambridge, Cambridge.

Meyerhoff, A. A., 1970, Continental drift: implications of paleomagnetic studies, meteorology, physical oceanography, and climatology: Journal of Geology, v. 79, pp. 1–51.

Pearson, B., 1978, Climate and evolution: London, Academic Press.

Plumstead, E. P., 1962, Fossil floras of Antarctica, Geology 2, in British trans-Antarctic expedition 1955–1958: Science Reports.

Seyfert, C. K., and Sirkin, L. A., 1979, Earth history and plate tectonics: New York, Harper & Row.

Smith, A. G., and Hallam, A., 1970, The fit of the southern continents: Nature, v. 225, pp. 139–144.

United Nations, 1971, Tectonics of Africa: Earth Sciences Series, v. 6, Paris, UNESCO.

Valentine, J. W., and Moores, E. M., 1972, Global tectonics and the fossil record: Journal of Geology, v. 80, pp. 167–184.

Wadia, B. N., 1953, Geology of India (3d ed.): New York, St. Martin's.

16

THE MESOZOIC ERA
Age of Reptiles and Continental Breakup

Lives of great men all remind us
We can make our lives sublime,
And, departing, leave behind us
Footprints on the sands of time.

Longfellow,
Psalm of Life, (1838)

FIGURE 16.1
Triassic reptile tracks unearthed in 1966 during excavation for a new state building in Hartford, Connecticut. The discovery was so phenomenal that the site for the building was changed and the excavation reserved as a public park (Dinosaur State Park)—a rare yielding by "progress" *(Courtesy John Howard.)*

16

Marked changes in life occurred after the great extinctions in late Paleozoic time. Between 65 and 230 million years ago, on land, especially, new innovations appeared, including dinosaurs in Late Triassic time, true feathered birds and skinned, flying or gliding reptiles, as well as primitive mammals, in the Jurassic Period. These true mammals were descendants of the earlier mammal-like reptiles. Lycopsid trees declined, while ginkgos, cycads, and conifers flourished. Flowering plants appeared in late Mesozoic time, with ancestors of *Magnolia*, *Eucalyptus*, *Sassafras*, oak, poplar, fig, and willow prominent in Cretaceous forests. Insects underwent a new burst of evolution at about the same time, suggesting a cause-and-effect relationship with plant evolution. Pollen-transporting insects short-circuited the plant reproductive cycle by their selectivity of flowers. This reduced the gene pool size for specific plants, accelerating evolution and adaptive specialization of both plant and insect. In the seas, too, new faces were evident, and many ecologic niches were vacated by extinction of hordes of Paleozoic invertebrates. Only a few corals had survived, but they diversified and expanded greatly. Molluscs underwent great evolutionary diversification as they adapted to a wide spectrum of habitats. Fish continued to flourish, but were challenged by ammonoids and a host of swimming reptiles that readapted to the aquatic habitat.

In 1822 an English physician and amateur paleontologist, Gideon Mantell, called upon a rural patient in Sussex. His wife discovered some large Jurassic bones in the yard while she was waiting. After considerable study, a skeleton was assembled and named *Iguanodon*. The discovery aroused much attention, but in spite of its fame, the zoological affinity of *Iguanodon* was not recognized quickly. None other than Georges Cuvier, leading comparative anato-

mist, finally guessed that it was a member of an extinct group of huge reptiles, later named dinosaurs (terrible lizards). Since then, dinosaurs have stirred the human imagination more than any other geologic phenomenon. Indeed, it is probable that their bones, together with those of giant Pleistocene mammoths, inspired early myths about giants. Today their powers of stimulation still are great, as evidenced by browsing in any well-stocked children's library. Finds over the years, such as fossil dinosaur eggs, polished stones thought to represent gizzard gravel, and, most recently, petrified dinosaur stomach contents and nests of baby dinosaur skeletons, all have served to prevent extinction of interest in ancient monsters.

In 1835, only 13 years after the first discovery of the bones of *Iguanodon*, yet still 7 years before the term "dinosaur" was coined, giant footprints were found on Upper Triassic flagstones being laid on the streets of Greenfield, Massachusetts. The tracks immediately drew attention and triggered a rash of speculation and fantasy. Discovery of bones of the barely extinct giant Moa bird in New Zealand lent credence to the most popular belief that the tracks were made by some overnourished Triassic fowl (as yet, no one knew that birds appeared later). Dramatic support for the existence of such creatures was provided at a scientific meeting by famous English geologist W. E. Buckland, who "exhibited himself as a cock on the edge of a muddy pond, making impressions by lifting one leg after another." So lively was debate over footprints that it provided a great stimulus to American interest in fossils. By the end of the nineteenth century, a dozen localities in the Connecticut Valley had yielded Triassic bones that clearly were reptilian, including some of the earliest dinosaurs. The controversy was settled, with birds the losers, though the oldest true fossil birds (Jurassic) soon were to be discovered in Germany. What was a lively controversy is now so long forgotten that reminiscent vestiges are to be found only in nineteenth-century literature, for example the famous passage quoted above.

Collection and study of fossil vertebrate animals have given rise to some of the best romantic lore in the history of geology. The most colorful period was around the turn of the twentieth century when exploration of the American West was revealing untold wealth of virgin fossil localities, many of which yielded hitherto completely unknown ancient monsters. The men who collected them were at least as colorful as their finds. It was a heroic drama on the American geologic stage, and the stars were cast well for their roles.

Two of the most flamboyant participants were E. D. Cope and O. C. Marsh, who scoured arid western hills for reptile and mammal bones. At first, these two pioneers were harassed only by occasional unfriendly Indians, but later—and far more dangerously—by each other in a jealous gun-toting feud that eclipsed even the earlier Sedgwick-Murchison altercation in Wales. They falsified collecting locations to keep them secret, and telegraphed descriptions of newly discovered skeletons back east in order to beat the other to publication. Cope even bought a scientific journal in order to publish his articles uninhibited, and to have a ready medium for editorial diatribes against Marsh. As a sample of their affection for one another, consider this commentary by Marsh on Cope's extensive studies of an extinct mammal: "This mythical *Eobasileus,* under Professor Cope's domestication has changed its character more rapidly than Darwin himself ever imagined for the most protean of species. Surely such an animal belongs in the Arabian Nights and not in the records of science." And here is Cope's rejoinder: "Marsh's reputation for veracity among his colleagues is very slight, in fact he has none. . . . He has never been known to tell the truth when a falsehood would serve as well. . . . He has no friends save those who do not know him well."

GENERAL SETTING OF MESOZOIC NORTH AMERICA

Although Mesozoic life was greatly changed, sedimentation and tectonism in North America were no different from the preceding Permian Period. Marine deposition was continuous in much of the Cordillera, and red sediments continued to accumulate over the craton. In the Appalachian belt, as we already have seen in Chap. 15, rifting of North America from Gondwanaland began late in the Triassic Period. In the Cordilleran region, disturbances were intensifying after the relatively mild Antler orogeny.

When the second edition of this book appeared in 1976, we felt confident in asserting that a relatively simple and obvious volcanic arc formed in Permian time and persisted until the late Mesozoic from southern Alaska to Central America. This seemed to us the very clearest example of an ancient arc system! But in the brief span of five years, to quote Mark Twain, "researchers cast much darkness on the subject, and if they continue we shall soon know nothing at all about it." Beginning with the Late Jurassic, however, the situation is not so hopeless. It was then that the culminating compressional mountain-building events began; we group them for simplicity under the single term *Cordilleran orogeny.* They continued through Eocene time to be followed by extensional rifting in late Cenozoic time, which is discussed in Chap. 17. The Cordilleran orogeny terminated subsidence and marine deposition in most of the Cordilleran belt. In Cretaceous and Cenozoic time, patterns of subsidence and sedimentation changed; a long, deeply subsided zone in the present Rocky Mountain area developed just east of the former subsiding belt and *within the edge of the craton!* Farther west, complex local basins and mountainous blocks dotted the Cordillera itself; near the Pacific Coast, marine embayments persisted until late Cenozoic time. In the Arctic, marine Cretaceous sedimentation occurred in northern Alaska and northern Canada.

PERMO-TRIASSIC OF THE CORDILLERAN BELT

SUSPECT TERRANES

Records of Permo-Triassic volcanism are, indeed, scattered all around the Pacific perimeter as we asserted in both prior editions of *Evolution of the Earth*. The correlation of these records once seemed perfectly straightforward, but subsequently it was noted that there are important differences of stratigraphy, faunas, and/or volcanic rock types in different zones or terranes that are of essentially the same age and, in some cases, even adjoin each other. With the advent of plate tectonic theory, workers reexamined these regions with new eyes and saw that most had tectonic boundaries, so might be far from their original relative positions. Recently the apt term *suspect terrane* has been given to any such crustal block whose birthplace is suspect; they represent tectonic accretions to the continent. Figure 16.2 shows the major suspect terranes for northwestern North America and the adjacent northeastern Soviet Union. Needless to say, their boundaries and histories are still slightly controversial! Please fight any feeling of panic as you view this figure, for our purpose is merely to suggest the possible complexity of orogenic belts seen more and more as collages of accreted old arcs and microcontinents. (We ourselves throw up our hands with Mark Twain at the confused state of knowledge of regions like this.)

FIGURE 16.2

Suspect terranes of the Pacific margin of North America and adjacent Soviet Union presented only to indicate the complexity of and uncertainty about the Cordilleran belt. Each crustal block—an old arc or microcontinent—has distinct stratigraphy, fauna, and/or volcanic rock type. Apparently all have been displaced from their original locations and jammed together at different times; but they were together as shown by Cretaceous time. Truncation of margin in United States by transcurrent faulting probably carried Stikinia northward to Canada in Triassic(?) time, *(Adapted from J. W. H. Monger, 1975, Geoscience Canada, v. 2, pp. 4–9; R. A. Schweickert, 1976, Nature, v. 260, pp. 586–591; D. L. Jones, et al. 1977, Canadian Journal of Earth Sciences, v. 14, pp. 2565–2577; M. Churkin, Jr., and J. H. Trexler, Jr., 1979, Geology, v. 7, pp. 467–469.)*

Long before plate tectonics, a striking faunal anomaly was recognized in the Cordillera where Permian Tethyan fusulinid types characterize a long zone of limestones associated with volcanic rocks extending from Alaska south to California (Fig. 16.2). Their nearest counterparts today lie in Japan and southeastern Asia. Before plate theory, this anomaly was explained by migration of the fauna from Asia via warm currents (see Fig. WM.6). This is still one reasonable hypothesis as witnessed by penguins living happily today on the equatorial Galápagos Islands, which they obviously reached by following a cold current that extends up the west coast of South America (see Fig. 17.47). But plate tectonics forces us to consider the staggering alternative that a microcontinent containing these fusulinids—either a *Noah's Ark* or a *Viking Funeral Ship* as described in Chap. 15 (p. 395)—may have sailed thousands of kilometers from southeastern Asia or Gondwanaland to collide with North America in Mesozoic time. Indeed, recently a hypothetical *Pacifica continental mass* something like mythical Atlantis has been postulated as a possible spawning ground for many of the suspect terranes of the present

Pacific margin. If Pacifica existed at all, the idea is appealing to us that it was a part of Gondwanaland and also may have been the source of southeastern Asia (see Fig. WM.14). Biologists attempting to explain modern-day floral and faunal disjunct distributions in these terranes also like this concept.

Striking similarities of stratigraphy coupled with paleomagnetic evidence suggest that two of the slivers now in westernmost Canada, *Stikinia*, and an unnamed smaller one in adjacent southeastern Alaska, dominated either by Paleozoic and/or early Mesozoic arc terranes, may have been moved north—presumably by strike-slip faults like the present San Andreas—from northwestern California through about 15° of latitude. Note at the bottom of Fig. 16.2 that the continent was truncated in early Mesozoic time. Slivers apparently were removed, of which Stikinia probably was one. Even more dramatic is paleomagnetic and stratigraphic evidence that a terrane called *Wrangellia* (Fig. 16.2) may have been displaced between 35° and 65° northward from some equatorial region. It has a unique sequence of Middle and Upper Triassic limestones overlying oceanic basalts

FIGURE 16.3
Chitistone Valley, Wrangell Mountains, Alaska, showing a large syncline with oceanic Triassic basalts overlain by Upper Triassic massive limestone and shale. These rocks are part of the suspect terrane *Wrangellia*. Glacier-shrouded peaks in background (right) include Pleistocene volcanoes. (*Courtesy Gene L. LaBerge.*)

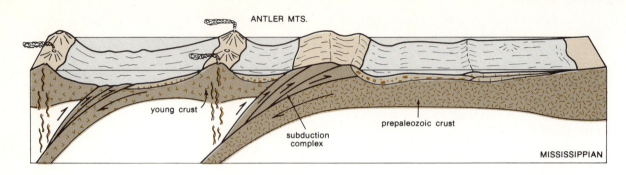

ANTLER MTS.

young crust

subduction
complex

prepaleozoic crust

MISSISSIPPIAN

PERMIAN

SONOMA MTS.

Tidal Flats

EARLY TRIASSIC

WIND

EARLY JURASSIC

CALIFORNIA   NEVADA   UTAH

0   300   600 km

0   300 mi

FIGURE 16.4
Plate tectonic hypothesis for Mississippian to Jurassic development of western margin of North America (see Fig. 16.2 for line of cross sections). Antler and Sonoman orogenies are interpreted as results of successive arc collisions followed by reversal of subduction in Late Triassic time. Note that Sonoma Mountains and arc are part of Sonomia of Fig. 16.2. Eastward subduction produced an Andean-like arc on the edge of the continent, beginning in Late Triassic time. Note Navajo Sandstone deposition at lower right. (*Adapted from M. Churkin, Jr., 1974, in* Society of Economic Paleontologists and Mineralogists Special Publication 19; *R. A. Schweickert, 1976,* Nature, *v. 260, pp. 586–591; F. G. Poole and C. A. Sandburg, 1977, in* Paleozoic Paleogeography of Western U.S., *pp. 67–86; R. C. Speed, 1979, unpublished.*)

(Fig. 16.3), which probably originated on some extinct ocean ridge far to the south. Wrangellia is bounded by other suspect terranes like Stikinia that contain Triassic and upper Paleozoic rocks with different stratigraphy and arc-type volcanics. These latter terranes formed in different tectonic environments than Wrangellia and are also out-of-place. Wrangling over Wrangellia and its neighbors will doubtless go on for some time, and one of the important issues will be their times of collision with each other and of emplacement as part of North America. Most of them seem to have been in their present positions by no later than Cretaceous or early Cenozoic time, but we would like to know the dates more accurately.

SONOMAN OROGENY

In several areas from Nevada to Alaska, there is evidence of major deformation near the Permo-Triassic boundary. Upper Triassic strata were deposited unconformably upon a variety of Middle Permian and older rocks, and they contain pebbles of those as well as granites and ultramafic rocks apparently emplaced during Permo-Triassic time. In Nevada the *Sonomia terrane* was emplaced at this time by overthrust faulting, and this major event has been named the *Sonoman orogeny.* Figure 16.4 illustrates a plausible evolution of the Cordilleran belt across northern California and Nevada from late Paleozoic to Jurassic time. Here we suggest that both the earlier Antler and the Sonoman orogenies

had a similar cause, namely arc collisions in which eastward thrust faulting reflected development of so-called *subduction complexes* in front of the advancing arcs. This origin also was suggested in Chap. 12 as the cause of thrusting during the Taconian Orogeny in the Appalachian belt (see Fig. 12.19). Analogies with the forearc regions of several modern volcanic archipelago-trench systems strongly support this interpretation.

LATE TRIASSIC AND EARLY JURASSIC VOLCANIC ARC

Sonomia and some of the suspect terranes in Canada apparently were emplaced before Late Triassic time when a fairly coherent volcanic arc system formed along the truncated western margin of North America (Fig. 16.4). Figures 16.5 and 16.6 show that, regardless of exact positions of all of the suspect terranes, there certainly was widespread accumulation of volcanic rocks (Fig. 16.7) as well as abundant volcanic-rich conglomerate and graywacke (Fig. 16.8) and local carbonate banks and reefs. Many of the sandstones accumulated in relatively deep water from turbidity currents. The continent margin as a whole was subsiding profoundly, producing deep-water areas at the same time that rapid extrusion of lavas and buildup of carbonate banks and reefs by organic activity produced islands and shoals. Behind the arc to the east lay a small, shallow ocean basin similar to those behind the modern west-

On the following pages:
FIGURE 16.5
Upper Triassic sedimentary facies. Environments and rock types were extremely variable in the western Cordillera, where up to 7,000 meters of Triassic strata are known. As noted in text, some of the Cordilleran areas were not yet in present positions. Note that much limestone formed in the volcanic zone of the Cordilleran belt. (See Fig. 11.7 or endpapers for symbols and sources.)

FIGURE 16.6
Late Triassic paleogeography. In spite of uncertainties about suspect terranes in the Cordillera, approximate character is portrayed. Note Africa just separating from America, but Europe still attached.

some red beds

170

190

Red beds

evaporites

R ?

UPPER TRIASSIC

SEDIMENTARY FACIES

(Karnian)

(about 190 m.y. ago)

0   1000 Miles

0   1000 Kilometers

LATE

TRIASSIC

PALEOGEOGRAPHY

(about 190 m.y. ago)

0   1000 Miles

0   1000 Kilometers

FIGURE 16.7
Ellipsoidal structures in Late Jurassic porphyritic lavas, southwestern Oregon. Submarine volcanic rocks such as these typify Mesozoic sequences along the Pacific Coast. *(By R. H. Dott, Jr.)*

ern Pacific ones (see Fig. WM.10). Some volcanic ash blew southeastward into this sea and onto the low coastal plain farther east (Figs. 16.4, 16.6).

TRIASSIC OF THE CRATON

Red beds continued to form widely over the craton as well as in the Appalachian and East

Greenland mobile belts during the Triassic Period. Continental land area remained very large even though average elevation was reduced. Mountains raised earlier in Colorado and New Mexico were lower and became mostly buried by nonmarine Triassic sediments (Fig. 16.6). Sediments were transported by rivers flowing westward across an immense alluvial plain to the Cordilleran Sea; red beds accumulated on the plain. Forests covered at least the highlands, as evidenced by petrified logs (Fig 16.9). Triassic red beds also have yielded important animal fossils, including remains of amphibians, dinosaurs, and other reptiles. Freshwater molluscs and fish point to some swamp and river environments.

Along the western margin of the craton, Triassic red sediments grade into typical marine gray shales and limestones of the eastern part of the Cordilleran mobile belt. Complex facies variations as well as an abrupt westward thickening of Triassic strata characterize the cratonic margin. It was long assumed that the widespread red beds to the east in the entire Rocky Mountain region were nonmarine deposits. But

FIGURE 16.8
Microscopic photograph of an immature sandstone composed of andesitic rock fragments such as is typical of Mesozoic and Cenozoic strata deposited along the Pacific margin of the Cordilleran belt in proximity to active volcanic arcs. (Coaledo Formation, Coos Bay, Oregon; average grain size is 0.5 millimeters.) *(By R. H. Dott, Jr.)*

FIGURE 16.9
Prostrate late Triassic tree trunks at Petrified Forest National Park, Arizona (Chinle Formation). Large coniferous trees were transported by floods from uplands, then the wood tissue was replaced by silica and, occasionally, by uranium minerals. The same strata also contain cycads and tropical ferns. Volcanic ash is abundant here. *(By R. H. Dott, Jr.)*

in Wyoming and Utah, thin marine limestone and gypsum layers are interstratified in the red beds, forming tongues of western facies penetrating eastern red strata. A considerable amount of the associated red strata represents marine lagoonal and tidal flat deposits. Marine conditions existed, at least intermittently, on the western part of the craton where periodic transgressions extended at least 600 kilometers eastward.

JURASSIC DESERT AND RETURN OF THE EPEIRIC SEA

As the Triassic Period drew to a close, sedimentation changed markedly over western North America. A vast blanketing mass (approximately 40,000 cubic kilometers) of very well-sorted sand was deposited along the entire west edge of the craton in the United States (Fig 16.10). The Late Triassic and Early Jurassic sands consist of 90 percent quartz. They are strikingly similar to the widespread, mature Paleozoic

quartz sandstones of the craton (see Chaps. 11, 12, and 14).

Where did so much sand come from? Relative purity and roundness of the grains point to derivation from older sandy sediments through recycling. With much sandstone present in upper Paleozoic and Triassic strata of the western part of the craton, there is no problem of designating potential source rocks. Paleocurrent data indicate general southerly transport of the sand (Fig. 16.10). This points to derivation from the north, chiefly from the craton in north-central United States and adjacent Canada where a widespread pre-Middle Jurassic unconformity indicates that upper Paleozoic and Triassic strata indeed were being eroded. Derivation from that region would be consistent with the general pattern of tilting of the entire craton accompanied by westward shift of loci of most active sedimentary accumulation and attendant gradual westward stripping of strata that formerly covered most of that region (see Fig 14.4).

Large-scale cross stratification is a prominent trait of the Jurassic Navajo and other sandstones in Utah (Fig. 16.11). This spectacular feature was attributed long ago to the preservation of lee faces of dunes perhaps as much as 200 to 300 meters high; this origin is verified by other, more subtle features as well. The Navajo also provides evidence that the continent had now drifted north into the westerly wind belt (Fig. WM.5). We believe that sand derived from the

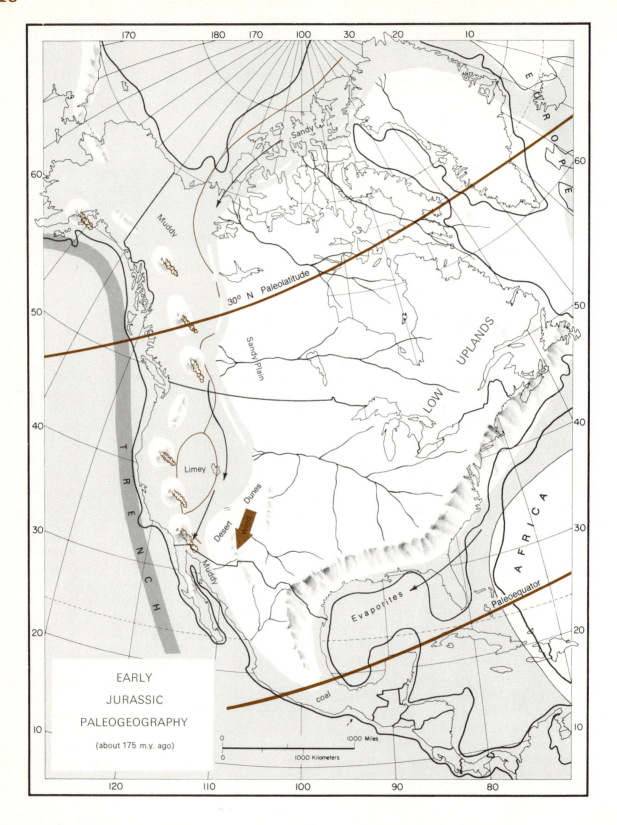

EARLY

JURASSIC

PALEOGEOGRAPHY

(about 175 m.y. ago)

FIGURE 16.10
Late Early Jurassic paleogeography; note beginning of encroachment of the sea along the western edge of the craton with deposition of Navajo dune sands next to the large embayment. Note Africa moving away, but Europe still attached to North America.

craton was transported south parallel to an oscillating eastern shoreline as is the rule today on shallow continental shelves (Fig. 16.4). Some of the sand apparently was deposited on the shelf, but onshore winds also produced coastal dunes, upon which dinosaurs left telltale footprints. Rare land plants and fossil burrows probably formed by beetles also are present.

FIGURE 16.11
Large-scale cross stratification in the Navajo Sandstone near Zion National Park, Utah. Details of the stratification indicate deposition in wind-formed dunes, some of which must have been of the order of 200 meters high. (*By G. Kocurek.*)

Interdune pond deposits and deformed cross strata suggest a rather high water table in the dune field, and, with onshore winds, coastal fog may have characterized the region as it does the apparently similar modern Namibian Desert of southwestern Africa (Fig. 16.12).

Thin carbonate rocks with marine fossils are interstratified locally with the upper sandstones, and they are overlain by widespread limestone, shale, and evaporite deposits. Beginning in Middle Jurassic time, the sea advanced widely over the western part of the craton from the present Arctic and northern Pacific regions. Quartz sand deposition was followed by conditions in the northern Rocky Mountain–Canadian Plains region like Paleozoic epeiric seas. Late Jurassic time marked the last significant carbonate and evaporite deposition anywhere on the craton. Middle and Upper Jurassic

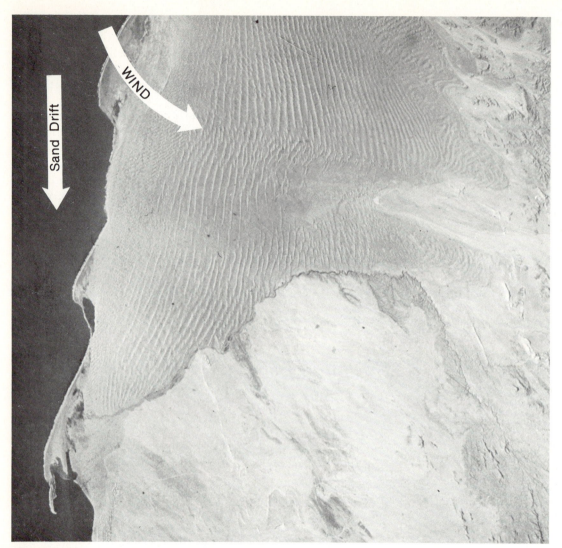

FIGURE 16.12
Coastal sand dunes of the Namibian Desert, Southwest
Africa, produced by cool, dry onshore winds. This region
is a close analogue for the setting of Lower Jurassic Navajo Sandstone of Utah. Note that dunes cease abruptly at a
river because wind is unable to blow sand across the
valley. Curved sand spits along coast are produced by
marine longshore transport of sand from top to bottom.
Fog shrouds the coastal dunes 200 days per year. For
closer comparison with the Navajo in Fig. 16.10, photo is
inverted (that is, North is actually at bottom of photo).
*(Gemini photo 65-45579 taken from about 200 miles above the
earth; Courtesy NASA.)*

strata lap across a widespread major unconformity and are younger toward the center of the
craton. As transgression proceeded, normal
marine conditions with better circulation prevailed, and evaporite deposition practically
ceased in the west, though it continued in the
Gulf of Mexico region (Fig. 16.13). Some workers believe that Jurassic salt was deposited
across the entire present Gulf of Mexico, which
was just beginning to form as a new, small-
ocean basin through the separation of Gondwanaland from North America (see Fig. 15.25).
Nearly every type of epeiric sea deposit dis-

cussed before formed in the Late Jurassic sea. Limestones rich in fossil fragments, oölites, and algal material, fossiliferous shales, and cross-stratified glauconitic sandstones all are prominent, but complexity of the facies variations almost defies analysis.

Most of the Appalachian belt was above sea level and being actively eroded at this time, though areas around the present Gulf of Mexico experienced immersion. Extensive, thick carbonate rocks as well as evaporites formed in Texas, Mexico, and Cuba. The epeiric sea must have joined with waters encroaching northward from the present Gulf of Mexico region (Fig 16.14) to produce a more or less continuous seaway stretching to the present Arctic. But there was some ecologic barrier such as water temperature, for Jurassic faunas of the Pacific Coast and Canadian Rocky Mountain regions have affinities with Asian ones (Boreal), and differ from those of the Gulf region, which were like those of the tropical Tethys seaway located in southern Europe and Africa (see Fig. WM.5).

A DINOSAUR GRAVEYARD

Latest Jurassic deposits of the Rocky Mountain region form a famous nonmarine sequence called the Morrison Formation. It is best known for its treasure of dinosaur skeletons. The strata also contain invertebrate and plant fossils, which, together with the dinosaurs, prove that it was a nonmarine sequence. The Morrison is made up of a varied assemblage of pastel-colored shales, sandstone, and rare conglomerate (Fig 16.15). Volcanic eruptions were still important farther west, and considerable ash blown from that region occurs in the Morrison. Though plant fossils and coal are rare, very large, herbivorous-dinosaur skeletons indicate that vegetation was abundant; oxidation almost completely destroyed any plants missed by the herbivores. This reasoning points to a moderately humid climate. Particularly rich concentrations of fossil dinosaur skeletons in a few localities, such as Dinosaur National Monument, Utah, suggest that unwary beasts gathered and became mired at watering places.

Then floods carried and buried carcasses downriver.

Morrison sediments were deposited as the short-lived Jurassic epeiric sea retreated northward and river and swamp deposits pushed forward to form an immense alluvial plain extending north and east from Arizona to southern Canada. Lands were beginning to be raised in the Cordillera, especially in western Arizona, thus the Morrison constitutes a clastic wedge like those of the Appalachian region. It caused the sea to retreat into present Canada and toward the present Gulf of Mexico, but this regression was short-lived, for the sea returned to the Rocky Mountain region again during Early Cretaceous time.

THE LATE MESOZOIC AND EARLY CENOZOIC CORDILLERAN OROGENY

Over most of the Rocky Mountain and Gulf Coast regions, Cretaceous strata overlie Jurassic ones with no discontinuity, but farther west in the Cordilleran belt, they overlie a variety of deformed rocks, reflecting onset of major mountain building and deep erosion (Fig 16.16). The culminating Cordilleran orogeny spanned Late Jurassic through early Cenozoic time. An entirely different tectonic behavior, distin-

On the following pages:
FIGURE 16.13
Upper Jurassic sedimentary facies. Patterns in the Rocky Mountain region are simplified here due to many rapid shifts of sediment types. Isolated evaporites (E) in north-central United States (Iowa) long were considered Permian, but fossil pollen shows them to be Mesozoic. (See Fig. 11.7 or endpapers for symbols and sources.)

FIGURE 16.14
Late Jurassic paleogeography. Transgression of the western craton is shown as more extensive than thought previously; note also the importance of continuing volcanism in the western Cordilleran belt. Africa has moved far enough away to open the Atlantic basin to normal marine circulation and entry of Tethyan fauna; note possible barrier reef along southeastern margin of continent. (compare Fig. 15.25). Most of the continent now lay in the westerly wind belt (see Fig. 6.18).

UPPER JURASSIC

SEDIMENTARY FACIES

(Oxfordian)

(about 145 m.y. ago)

1000 Miles

1000 Kilometers

MEXICAN GEOSYNCLINE

Evaporites

Evaporites

170 180 170 100 30 20 10

Muddy Bottom

Bottom

Sandy

60

60

EUROPE

Sandy

50

30° N. Paleolatitude

50

Sandy

Sand

Plain

Limey

L

40

Muddy

40

Limey

E

wind

Limey

BARRIER

REEF

E

30

Limey

30

Shoal

AFRICA

Sandy

Evaporites

Limey

20

POSSIBLE

20

Limey Bottom

Limey

Paleoequator

TRENCH

LATE

JURASSIC

PALEOGEOGRAPHY

(about 145 m.y. ago)

1000 Miles

0

0

1000 Kilometers

120 110 100 90 80

FIGURE 16.15
Combined facies and paleogeographic map for the nonmarine, latest Jurassic Morrison and Kootenay Formations. Both coarse facies and paleocurrent data indicate highlands to the west; isotopic dating and unconformities along present Pacific Coast also indicate beginnings of *Cordilleran orogeny*. Westerly winds blew volcanic ash eastward.

FIGURE 16.16

Relations of igneous intrusions and two angular unconformities resulting from Late Jurassic (Nevadan) and mid-Cretaceous phases of *Cordilleran orogeny* respectively in southwestern Oregon. (F = index fossils.)

FIGURE 16.17

Stratigraphic diagram of Cretaceous facies across the present Rocky Mountains of the Utah-Colorado region showing coarse clastic tongues extending eastward into marine shale facies, and indicating cyclic transgressive-regressive events reminiscent of Pennsylvanian times. *(After R. H. Dott, Jr., 1964, Kansas Geological Survey Bulletin 169.)*

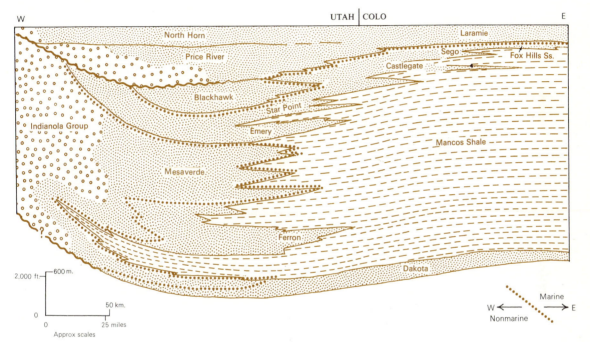

guished by faulting, has characterized western North America since the Oligocene.

The eastward spread of nonmarine, varicolored Morrison sediments reflects the first major influence in the Rocky Mountain region of the Cordilleran orogeny, but Lower Cretaceous strata show much greater effects. All Cretaceous sediments in the eastern Cordillera coarsen westward and thicken in the same direction to form a clastic wedge as much as 10,000 meters thick (Figs. 16.17 and 16.18). This wedge result-

On the following pages:
FIGURE 16.18

Middle Cretaceous sedimentary facies and isotopic dates for granitic batholiths. Note faults in craton (black lines). (See Fig. 11.7 and endpapers for symbols and sources.)

FIGURE 16.19

Middle Cretaceous paleogeography at the time of maximum worldwide transgression. Westerly winds blew volcanic ash widely over the epeiric sea. Seawater temperatures are from oxygen-isotope studies. Note that northward drift of the continent has placed equator and Africa off the map; Europe had not yet separated.

"MIDDLE" CRETACEOUS

SEDIMENTARY FACIES

(Albian-Cenomanian)

(about 100 m.y. ago)

(Maximum Transgression)

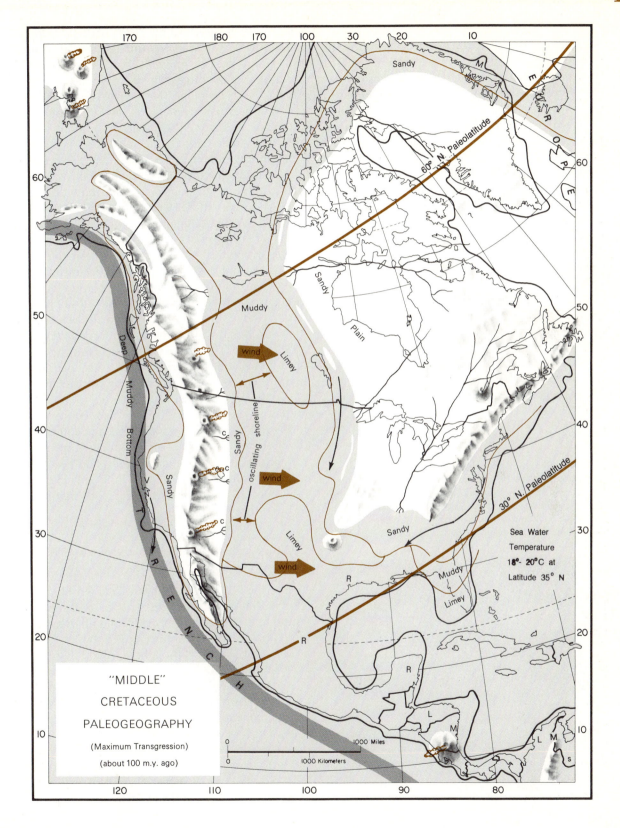

170   180   170   100   30   20   10

Sandy

60° N. Paleolatitude

E U R O P E

60

50

Sandy
Plain

Muddy

Deep
Muddy
Bottom

Limey

wind

oscillating shoreline

Sandy

Sandy

wind

C

C

C

Limey

wind

Sandy

30° N. Paleolatitude

Sea Water
Temperature
18°- 20°C at
Latitude 35° N

R

R

Muddy

Limey

T R E N C H

R

L

M

L   M

S

"MIDDLE"
CRETACEOUS
PALEOGEOGRAPHY

(Maximum Transgression)

(about 100 m.y. ago)

0   1000 Miles

0   1000 Kilometers

120   110   100   90   80

ed from erosion of mountains within the mobile belt (Fig. 16.19). More direct effects of mountain building are evident farther west, where regional metamorphism and igneous activity were widespread. Upheaval began during Jurassic time near the Pacific Coast and then spread eastward. After local Early Jurassic uplifts and volcanism (Fig. 16.10), widespread deformation began in Late Jurassic time (sometimes called the Nevadan Event). Granitic batholiths began to form (Fig. 16.15), but most of the isotopic dates are Cretaceous (75 to 100 m.y.) (Fig. 16.18). Widespread Late Cretaceous granitic pebbles, feldspar-rich sandstones, and unconformities reflect rapid uplift and exposure of many batholiths by erosion.

In the Rocky Mountain region, there is more evidence of Late Cretaceous (Sevier) and early Cenozoic (Laramian) than of earlier deformation. Thick, coarse conglomerates and sandstones attest to vigorous erosion of Cretaceous highlands within the mobile belt to the west as well as to suddenly accelerated subsidence of the cratonic margin, probably due to heating

and possibly westward flow of subcrustal material as the Cordilleran Mountains rose. In central Utah, coarse gravel 3,000 meters thick accumulated on huge alluvial fans close to the mountainous highland (Fig. 16.20). Meanwhile, volcanic activity erupted in western Canada, Montana, and along the Mexican border as well as in the Arctic (Fig. 16.18).

Beginning in latest Cretaceous time, the easternmost edge of the former mobile belt, which had continued to subside profoundly during deposition of thick Cretaceous strata, was crushed eastward against the craton just as strata of other geosynclines had been during earlier orogenies. Cordilleran mountain building had swept eastward like a wave so that in early Cenozoic time, a series of complex, low-angle thrust faults carried immense slabs of rock eastward over one another along a zone extending from Mexico to northwestern Canada (Figs. 16.21 and 16.22). As in older belts, the thrust-fault zone today marks the former cratonic margin (Fig. 16.23). Most of the suspect terranes of Alaska and western Canada (Fig. 16.2) were emplaced by now and were also being deformed. The mobile belt at this stage could be described as having been turned inside out; its former zone of greatest subsidence had become the axis of maximum uplift while subsidence now became concentrated along both margins of the old belt.

CAUSE OF OROGENY ON ANDEAN-TYPE PLATE MARGINS

Unlike most other orogenies discussed before, the Cordilleran did not involve the collision of continents. Instead, deformation and magma generation resulted only from the subduction of a Mesozoic Pacific lithosphere plate. Figures WM.12 and WM.13 show the probable circumstances affecting the western edge of both North and South America. As Pacific lithosphere was thrust beneath the continents, earthquakes must have been generated frequently in the downgoing plate as is the case beneath the modern Andes Mountains of South America. Selective melting of either that plate or lower crustal material (or both) by heat generated by

FIGURE 16.20

Upper Cretaceous conglomerate (Price River Formation), Maple Canyon State Park, central Utah. Coarse fragments of Lower Paleozoic and Prepaleozoic quartzites were deposited on alluvial fans at the foot of the mountains raised in western Utah and Nevada during the Cordilleran orogeny. A sandy coastal plain lay east of the gravels (see Fig. 16.19). (*By R. H. Dott, Jr.*)

FIGURE 16.21

Sentinel Range, northeastern British Columbia, along the Alaskan Highway. The Canadian Rocky Mountains were carved by rivers and glaciers from immense sheets of Paleozoic and Mesozoic strata thrust-faulted eastward against the craton during early Cenozoic time (see Fig. 16.22). In this view, Devonian rocks have been folded and repeated by at least two thrusts; to the right of photo Silurian strata have overridden Triassic. *(Courtesy L. R. Laudon.)*

underthrusting produced andesitic magma, which rose to the surface to form a volcanic arc on the edge of the two continents. A deep trench formed near the present coast of both

continents like the Peru-Chile trench today. Subduction of the Pacific plate presumably began no later than Triassic time when andesitic volcanism began in both the Cordillera and in the Andes. As subduction progressed, deformation of the arc rocks and the edges of the two continents occurred intermittently, but gradually intensified until it culminated in Late Cretaceous time with the uplift of entire mobile belts.

North America has changed dramatically since Mesozoic time, therefore we look to South America as a model for *Andean-type plate margins*. Beneath the Andes Mountains in Peru and Bolivia today, seismic evidence indicates that the subduction zone dips very gently eastward. This special situation is explained by a very fast spreading rate at the nearby East Pacific Rise (15 to 18 centimeters per year). As a result, the

FIGURE 16.22

Generalized structure section through the Canadian Rocky Mountains showing complex thrust faults and the Rocky Mountain Trench, a possible rift or graben feature (P— Paleozoic; M—Mesozoic). Compare Fig. 16.23.

TECTONIC MAP FOLLOWING
CORDILLERAN REVOLUTION
(Middle Cenozoic)

Coastal Plain Provinces

Marginal thrust fault zones

Anticlinal intracratonic
   mountain ranges

Granitic batholiths

(Late Cenozoic faulting omitted)

Colo.
Plateau

CRATON

Late Mesozoic and Early Cenozoic Batholiths

Middle and Late Paleozoic Batholiths

Middle Paleozoic Batholiths

faults

faults

0   1000 Miles

0   1000 Kilometers

downgoing slab is younger, warmer, and thus more buoyant than usual for oceanic lithosphere and is being subducted more rapidly. Consequently, the slab slides in at such a low angle that it interacts directly with the base of the overlying continental lithosphere across a zone 700 to 800 kilometers wide. Partial coupling of the two plates in such a wide zone inevitably produces unusual effects upon the overlying continent. These include great initial subsidence followed by exceptional deformation and magmatism even within the edge of the craton hundreds of kilometers from the plate margin. The same phenomena characterized western United States across a zone more than 1,000 kilometers wide (Fig. 16.23) during Late Cretaceous through Eocene time, which suggests a similar cause (Fig. 16.24).

The Andes, Cordillera, and all other Mesozoic-Cenozoic mountains that rim the Pacific basin contain immense volumes of granitic rocks (Fig. 16.23), and the late Mesozoic marks one of the greatest periods of granite formation in all of earth history. Why should this be so? We believe that it reflects rapid spreading and subduction in the Cretaceous Pacific. Recall from Chap. 8 (p. 164) that current thought attributes to subduction all plutonic granitic rocks as well as extrusive andesite and rhyolite of similar chemical composition. The first plate tectonic ideas had granitic magmas derived from the same source as those volcanic ones; the coarser intrusive rocks simply crystallized from magma that had not made it to the surface. More recently, however, as discussed in Chap. 8, the granitic magmas have been considered to form by melting of deep crustal rocks such as old volcanic, sedimentary, or granitic ones. The necessary heat, however, is still attributed to subduction below, and fast, low-angle subduction must heat and melt vast amounts of crust

FIGURE 16.23
Tectonic map of North America at the end of the Cordilleran orogeny (end of Eocene) and prior to late Cenozoic upheavals. Note widespread batholiths and thrust-fault zones along inner margins of all mobile belts. Colorado Plateau and nearby intracratonic ranges (colored lines) reflect unusual mobilization of the western craton.

(Fig. 16.24). Today we see many small, clearly intrusive bodies formed by magma that rose to relatively shallow levels in the crust. Recall that molten or partially molten magma is less dense than solid crust, therefore it rises—probably as giant tear-shaped masses. Where erosion has cut to greater depths, granitic rocks generally show much more complex relationships with surrounding, profoundly metamorphosed rocks. Especially around the margins of the large granitic batholiths like those of the Sierra Nevada of California and Coast Range of Canada, there are banded rocks with granitic and schistose layers so intimately alternating that the batholith boundaries are difficult to delineate (Fig. 16.25). Clearly the temperature was great enough here to cause thorough recrystallization of all rock types. Such regions must be close to the sources of melting to form the granitic magmas that rose to shallower depths, and they seem to typify the root zones of arcs exposed by complete erosion of a volcanic edifice several kilometers thick. The Mesozoic arc system of western North America suffered this fate. No actual Mesozoic volcanoes are left, but instead we see vast exposures of granitic batholiths and intimately associated metavolcanic and metasedimentary rocks (Fig. 16.23). Therefore, such batholiths provide our most important clue to the existence and position of countless ancient volcanic arcs that have long since been eroded to their roots. They also tend to have important ore deposits associated with them.

CRETACEOUS TRANSGRESSION AND SEDIMENTATION

WORLDWIDE TRANSGRESSION

Worldwide transgression occurred during the Cretaceous Period, for marine rocks of this age unconformably overlap older rocks widely on practically every continent. Maximum dousing, which submerged about one-third of the present land area of the earth, occurred near the middle of the period, roughly 100 million years ago. "High water" occurred at slightly different

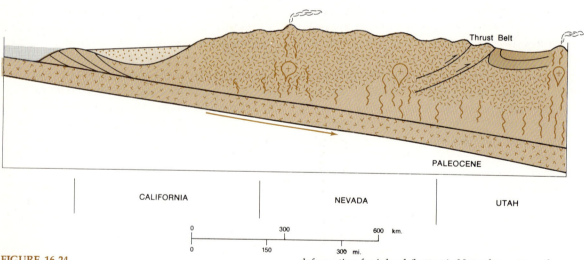

FIGURE 16.24

Development of late Mesozoic–early Cenozoic Cordilleran orogeny due to subduction beneath an Andean-type plate margin (compare Fig. 16.4). Note inferred flattening of dip of subduction zone as rate of sea-floor spreading increased in Late Cretaceous and early Cenozoic time. As in the Andes today, this seems to explain magmatism and severe deformation far inland (bottom). Note also westward expansion of subduction complex and forearc basin (now seen in California west of Sierra Nevada). *(Adapted from W. R. Dickinson and D. R. Seely, 1979, Bulletin of the American Association of Petroleum Geologists, v. 63, pp. 2–31; and other sources.)*

FIGURE 16.25
Banded granitic and metamorphic rock from a late Mesozoic batholithic complex in northern Cascade Range,
Washington. Light-colored sills and pods of quartz diorite
alternate with darker hornblende gneiss. Everything was
intensely contorted by plastic flowage during mountain
building. (*Courtesy D. F. Crowder, U.S. Geological Survey;
from Crowder, 1959,* Bulletin of the Geological Society of
America, *v. 70, plate 7; by permission of the Geological Society
of America.*)

times in different regions, depending upon local
tectonic and sedimentary conditions. In broad
terms, an essentially simultaneous worldwide
relative rise of sea level occurred, followed by
general worldwide regression down to the present time. The Cretaceous flood affected North
America profoundly, producing the last epeiric
sea over the western craton and inaugurating
formation of thick prisms of sediments now
beneath marginal coastal plains and continental
shelves.

Because there is no evidence of Cretaceous
glaciation, a structural cause for this worldwide
transgression must be sought. Acceleration of
sea-floor spreading is known to have occurred
during the Cretaceous. Presumably this caused
the enlargement of ocean ridges, Circum-Pacific
mountain building, and wholesale continental
breakup. All of those changes collectively must
have been sufficient to have displaced enough
sea water to cause the transgression.

EFFECTS IN THE PACIFIC COAST REGION

Along the western margin of the Cordilleran
tectonic land, the Cretaceous shoreline oscillated somewhat, but effects were less pronounced
than on the cratonic side. The western coastline
was steeper, and local structural disturbances
resulted in local transgressive and regressive
complications; while transgression occurred in
one area, regression may have prevailed only
100 kilometers away. A narrow continental shelf
received marine gravel, sand, and mud deposits, while farther offshore in deeper water,
clastic sediments accumulated through the action of submarine sliding and turbidity currents
(Fig 16.26). Much of the region subsided very
rapidly and more than 2 or 3 million cubic
kilometers of sediments eroded from the Cordilleran Mountains accumulated there. Submarine
lavas were erupted intermittently along the
western edge of the continent in what was
probably a deep submarine trench.

The Pacific Coast must have looked much like
the present west coast of South America with
high mountains directly adjacent to a trench,
but details of Cretaceous paleogeography there
are poorly known because of complex Cenozoic
structural disturbances that greatly obscured
the record. Important local unconformities
within the Cretaceous sequence attest to recurring tectonic disturbances; underthrusting of
oceanic crust beneath a narrow continental shelf
apparently occurred continually during Cretaceous time, creating a chaotic mixture of sedimentary, volcanic, and ultramafic rocks (Fig.
16.24). Upper Cretaceous sediments along the
Pacific Coast and interior Alaska contain more
coal and other nonmarine deltaic and shoreline
deposits than do Lower Cretaceous ones, thus
marking a time of considerable filling of former
deep-marine basins and embayments during an
interval of temporary tectonic quiescence immediately after the Cordilleran orogeny had culminated.

EFFECTS ON THE CRATON

The craton suffered flooding both from Arctic
and Gulf regions, with merging of waters over
the present Plains area in mid-Cretaceous time.
Marine strata of this age extend from the Gulf
Coast to the Arctic and from Minnesota to
western Wyoming. On the east side of this

FIGURE 16.26
Very evenly stratified Late Cretaceous or early Cenozoic graded graywacke sandstones and mudstones (*flysch*) considered to have been deposited by turbidity currents in moderately deep water near San Pedro Point, San Francisco peninsula, California (compare Fig. 14.22). (*By R. H. Dott, Jr.*)

seaway, a widespread transgressive, sandy shoreline facies developed. Farther west, the epeiric sea lapped almost at the foot of the Cordilleran Mountains near the middle of the period, but marine strata became less and less widespread through the Late Cretaceous times as regression commenced. By the end of the period, the sea had retreated to the present Plains region; during early Cenozoic time, the waters parted and drained completely from the craton both northward and southward. The great clastic wedge of Cretaceous strata on the western part of the craton (Fig. 16.17) is reminiscent of the Pennsylvanian coal-bearing wedge of the western Appalachian region. Conglomeratic facies grade cratonward into thick, massive, cross-stratified tan sandstones containing coal seams (Fig. 16.27). The total volume of clastic sediments exceeds 3 million cubic kilometers. These in turn pass eastward into widespread black shales with thin limestone layers and

zones of altered volcanic ash blown from the west. Gradually sand was deposited farther and farther eastward during Late Cretaceous time, forming a classic regressive facies pattern (Fig. 16.17). The conglomerates and coal-bearing portions of sandstones are largely nonmarine, whereas the shale and some sandstone are marine. Therefore, we see that the Cretaceous coal-bearing strata have another trait in common with Pennsylvanian coal sequences, namely repetitive or cyclic sedimentation. The intertonguing of marine and nonmarine strata represent wide oscillation of the shoreline due to (1) worldwide fluctuations of sea level, (2) spasmodic uplifts in the Cordillera, or (3) possibly cyclic climatic changes that affected weathering, erosion, and deposition of sand in a rhythmic fashion.

As the shoreline oscillated over the region, large deltas and associated swamps shifted with it as during Pennsylvanian time on the eastern craton. Sand-barrier islands paralleled the shores of marine embayments so that the coastline must have looked much like that of Texas and Louisiana today. Abundant vegetation grew in swamps, and gave rise to widespread coal seams, which constitute North America's second largest coal reserve, which, because of a low sulfur content, may now become our most important one. "Terrible lizards" wallowed in the swamps and roamed the alluvial plain that extended back to the Cordilleran Mountains. Besides coal, Cretaceous strata have long been very important sources of petroleum trapped by folds and faults formed during the last phase of the Cordilleran orogeny.

Black, organic-rich shale was the predominant Cretaceous deposit on the central part of the craton, and Cretaceous black muds also are found widely in deep-sea cores. Such sediment is so widespread and characteristic—a virtual Cretaceous trademark like the red beds of the

Permo-Triassic—that it demands some special explanation. The situation is reminiscent of the similar black shales of Late Devonian and Early Mississippian ages (Chattanooga) discussed in Chap. 13. An attractive recent explanation for the Cretaceous shales calls upon the very warm and uniform global temperature of late Mesozoic time, which was due largely to the location of both poles in open oceans as well as to widespread transgressions that increased total water area (see Figs. WM.4 and WM.5). One effect of this temperature regime would be that sea water could take up less oxygen because gases are less soluble at higher temperatures. Another probable effect would be sluggish deep circulation because of the more uniform temperature of sea water than now. The net result apparently was that much of the ocean was poor in oxygen, thus much organic material could be preserved in muds on foul bottoms that were not susceptible to oxidation. Where buried deeply and heated sufficiently after deposition, the resulting black shales became excellent potential source rocks for petroleum.

FIGURE 16.27

Lower Cretaceous coal-bearing (black) strata at left and Upper Jurassic Morrison Formation at right in huge roadcut on Interstate Highway 70 just west of Denver, Colorado. The cut is across the prominent "Dakota hogback" ridge. Such ridges are formed by Lower Cretaceous Dakota Sandstone around the flanks of most Rocky Mountain ranges because this sandstone is more resistant to erosion than overlying and underlying shales. There is a geologic walkway with signs along each side of this spectacular roadcut. (*By R. H. Dott, Jr.*)

EFFECTS IN GULF AND ATLANTIC COASTAL PLAINS

By Late Jurassic time, the entire Appalachian mobile belt system from Mexico to Newfoundland had been eroded to a low-lying surface, and the present Atlantic and Gulf of Mexico basins had begun to form (see Fig. 15.25). Jurassic transgression brought marine deposition over the southwestern end of the belt and during Cretaceous time, owing to still greater transgression, marine strata unconformably overlapped farther to cover at least half of the old eroded belt (Fig. 16.19). Cretaceous sediments must have extended still farther inland originally than now, but erosion has stripped them from large areas of the Appalachian Mountains and southern plains. A possible giant barrier reef of Late Jurassic–Early Cretaceous age has recently been suggested from drilling on the Atlantic continental shelf. It is of great interest for potential petroleum entrapment. (see Fig. 16.14).

In the southern United States, quartz-bearing sandstones are common in Lower Cretaceous strata, whereas shale and limestone are more important in the Upper Cretaceous. In Mexico, limestone deposition occurred widely throughout most of the period (Fig. 16.18). Beneath the Atlantic coastal plain, Cretaceous strata are of moderate thickness (1,000 to 2,000 meters), but along the Gulf Coast, thicknesses are about 4,000 meters. The Gulf Coast region was subsiding much more rapidly, but sedimentation kept pace, as most known Cretaceous sediments are richly fossiliferous shallow-marine sands, muds, or carbonates. There is evidence of some nonmarine deposition, too. As subsidence occurred beneath the Gulf Coastal region, worldwide transgression also was taking place so that younger marine Cretaceous strata tended to lap farther cratonward until the latest part of the period. Local structural warping and erosional truncations occurred, and there was considerable faulting. Small igneous intrusions (some containing diamonds) were emplaced in widely scattered areas, and a few volcanic vents in Texas, Arkansas, Louisiana, and Mexico erupted considerable ash.

Paleomagnetic evidence suggests that the paleoequator was approaching its present relative position by Late Cretaceous time (Fig. 16.19). This would have placed the Gulf Coast region still at a nearly tropical latitude. It is significant that marine invertebrates were abundant and diverse in the shallow sea there, and that carbonate rocks were of great importance; organic reefs also developed widely.

LATE MESOZOIC PALEOCLIMATOLOGY

PALEONTOLOGIC AND SEDIMENTARY EVIDENCE

Emergence and elevation of most continents in late Paleozoic and early Mesozoic times produced a diversified "continental" climate characterized by considerable aridity. With a much larger proportion of the earth's surface covered by water during late Mesozoic time, we would expect that more solar radiation would be absorbed by the water, and heat would be efficiently distributed poleward by currents, producing an overall warm, mild, "oceanic" climate with ice-free poles. Lack of evidence of late Mesozoic glaciation and widespread distribution of reptiles and mild-climate plants all across the continent strengthen this belief. Fossil plants, which had evolved toward a modern aspect by Late Cretaceous time, indicate mild-temperate to subtropical conditions over most continental areas. For example, conifer and ginkgo forests grew in Franz Josef Land, now at 80°N. Ancestors of breadfruit trees, laurels, *Magnolia*, and *Sequoia* (redwood), which could not stand freezing temperatures, thrived in western Greenland at 70°N. (See Chap. 17 for discussion of the evolution of flowering plants.) Moreover, Cretaceous dinosaur footprints recently have been discovered on Svalbard (Spitsbergen), whose present latitude is 77°N (10 degrees north of the Arctic Circle). This frigid land today enjoys an average annual temperature range from −20 to +5°C! While cold-blooded animals can stand occasional frost and mild freezes, it is inconceivable that they could survive extended periods of subfreezing temperatures unless they hibernated in protected

FIGURE 16.28

Relation of oxygen isotopes to seawater temperature in calcium carbonate shells formed by modern animals; ^{18}O decreases relative to ^{16}O as temperature increases. (*After Epstein et al., 1951,* Bulletin of The Geological Society of America, *v. 62, p. 424; by permission of the Geological Society of America.*)

the sea water in which organisms grew, such that ^{18}O in shells decreases relatively as temperature increases (Fig. 16.28). From experimental studies, chemical relationships controlling this variation are understood sufficiently so that $^{18}O:^{16}O$ ratios in shells can be used for paleotemperature determination. Paleotemperature is estimated through analysis of ancient fossil shells under the assumption that identical chemical relationships controlled oxygen-isotope composition of ancient as well as of modern shells. Determinations in 1951 of $^{18}O:^{16}O$ ratios in Jurassic belemnoids from 57°N latitude (equivalent to Scotland or southern Alaska) points to an average annual seawater temperature of 14 to 20°C (54 to 68°F), or roughly 15°C warmer than is typical at that latitude today. The quantitative oxygen-isotope data, therefore, confirm qualitative fossil and sedimentary evidence pointing to a warm Jurassic climate at midlatitudes. Cretaceous fossils from North America, western Europe, and Russia yield results (Fig 16.29) that confirm mild ocean temperatures on the order of 20 to 25°C in

places. But hibernation would have been next to impossible for giant dinosaurs. On the other hand, it has been suggested recently that dinosaurs could, in fact, regulate their body temperatures to some degree.

Although evaporites were important in Jurassic times, suggesting high evaporative potential at many latitudes, their general absence from Cretaceous sediments suggests relatively humid conditions (or more open-marine circulation) over North America in late Mesozoic time. Cretaceous evaporites did form on the edges of Africa, however (see Fig. WM.4).

OXYGEN-ISOTOPE PALEOTHERMOMETERS

It has been shown by analysis with the mass spectrometer that relative proportions of the oxygen isotopes ^{18}O and ^{16}O, in calcium carbonate marine shells vary according to temperature of

FIGURE 16.29

Latitudinal paleotemperature gradients for shallow-marine water at two separate times late in the Cretaceous Period. For comparison, the modern gradient based upon fossils from North America, Europe, and Russia is shown (compare Fig. WM.1). Note cooling trend from 80 to 70 million years. (*Adapted from Lowenstam and Epstein, 1959,* El Sistema Cretacico (Primer Tomo), *pp. 65–76; 20th International Geological Congress.*)

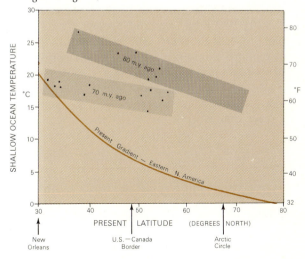

present middle latitudes from 30 to 70°N (present coordinates). Today, comparable surface ocean temperatures are characteristic of such coasts as Florida, Mexico, Central America, and northwest Africa (i.e., roughly between the low latitudes of 5 and 32°N). Even in east-central Greenland, now at a latitude of 72° north, the sea was a balmy 17°C during mid-Cretaceous time or about 15°C warmer than today. By extrapolation from oxygen-isotope studies of Greenlandic and Russian fossils, it is estimated that Cretaceous north polar water was no colder than 10 to 15°C.

If in Late Cretaceous time, North America occupied approximately its present position with respect to pole and equator, and if there was little complication of ocean-current patterns, then it would be possible to reconstruct a rather detailed paleoclimatic map from the above temperatures and what is known of the paleogeography. Geologic and oxygen-isotope evidence indicates that total earth climate was milder and more uniform than today, and the fossil record suggests that North America had largely subtropical conditions. Global conditions can be approximated as in Fig. WM.4, and from this picture we can make a more detailed interpretation of North America shown in Fig. 16.30. Most of the continent would have been within the westerly wind belt, so that moist air would have blown over its western part. The Cordilleran Mountains were well watered by westerly winds, and areas farther east could receive moisture from the epeiric sea east of the mountains and from the Gulf of Mexico. Uniformity of Cretaceous and Eocene plants indicates a lack of sharp climatic zonation of the continent. By analogy with modern continents at similar latitudes, it is assumed that there were distinct seasonal variations of temperature and precipitation, but these cannot be fully assessed.

MESOZOIC ANIMAL LIFE

THE INVERTEBRATES

The great wave of Permian extinctions was not followed immediately by replacement with a new and varied fauna. Early Triassic marine faunas are not well known because epicontinental seas did not spread widely until Late Triassic time. Thus few early shallow-water habitats are known. The earliest fossiliferous Triassic strata contain some ammonites (Fig. 16.31) and a few species of clams. Both the latest Permian and earliest Triassic marine bottom faunas are represented by a few conservative genera, but the species are represented by large numbers of individuals. These highly unusual faunas are similar to those found today in hypersaline lagoons. No single satisfactory cause has been advanced to account for the late Permian extinctions, because both land and marine organisms were affected. Some major groups such as the fishes were not altered at all.

During the Triassic Period, the ammonites diversified and filled a wide range of environments. Some 400 genera evolved. For unknown reasons, almost all these became extinct at the close of the period save a few conservative groups that were evolutionary *cul-de-sacs*. A single family of ammonites remained to give rise to a vast and complex group of 1,200 genera that dominated the Jurassic and Cretaceous Periods. So rapid was the evolution and extinction of subgroups that Mesozoic ammonites serve as the best guide fossils known for worldwide correlation. As a result, the history of life and paleogeography is perhaps better known for the Mesozoic than for any prior segment of time.

Other molluscs show quite a different pattern. Clams and snails gradually took on a modern appearance by addition of new, mostly still-living families. Clams took up residence in the sediments as burrowers, and diversified quickly. Some characteristic Mesozoic families evolved and waned to extinction at the era's end,

FIGURE 16.30

Hypothetical Late Cretaceous paleoclimatic map. Dominance of westerly winds is indicated by volcanic ash fallout; land climate from fossil plants and meteorologic theory; seawater temperatures from oxygen-isotope studies; marine faunal migrations from paleobiogeography; and probable latitudes from paleomagnetic evidence.

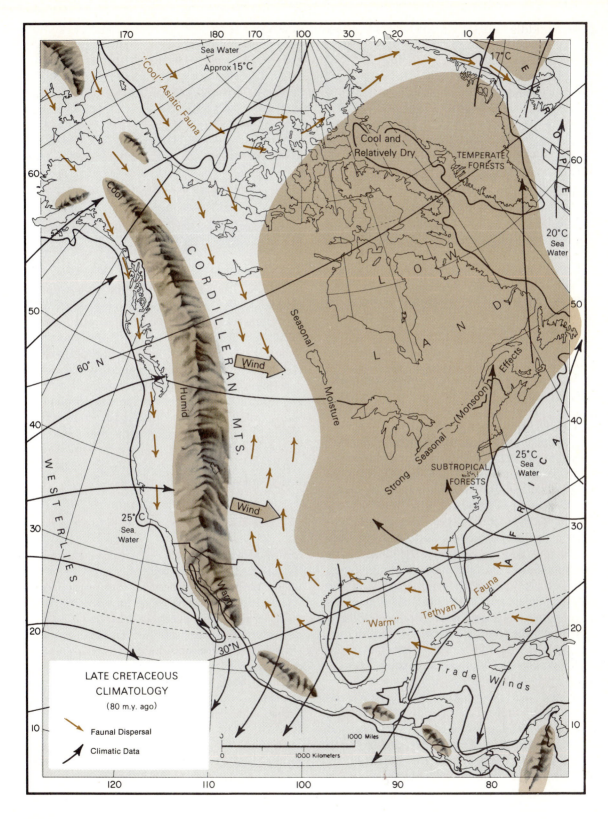

170   180   170   100   30   20   10

Sea Water
Approx 15°C

"Cool" Asiatic Fauna

17°C

Cool and
Relatively Dry

TEMPERATE
FORESTS

60

20°C
Sea Water

Cool

C O R D I L L E R A N   M T S.

Seasonal Moisture

L O W   L A N D

50

Wind

40

Humid

60° N

Strong   Seasonal   (Monsoon)   Effects

SUBTROPICAL
FORESTS

25°C
Sea Water

Wind

W E S T E R L I E S

25°C
Sea Water

30

Warm

Tethyan   Fauna

A F R I C A

30

"Warm"

30°N

20

T r a d e   W i n d s

20

LATE CRETACEOUS
CLIMATOLOGY
(80 m.y. ago)

→ Faunal Dispersal

→ Climatic Data

1000 Miles

0

1000 Kilometers

120   110   100   90   80

FIGURE 16.31
Mesozoic ammonoids. *Left:* Most Triassic ammonoids are characterized by having subdued ornamentation and ceratitic sutures; however, by Late Triassic time ammonitic-type sutures became more common. *(Photograph by G. R. Adlington.) Right:* With the exception of a few Cretaceous pseudoceratitic patterns, all post-Triassic genera have ammonitic sutures. Most evolution and variation in the latter involved development of ribs and nodes. *(Courtesy U.S. National Museum.)* By Cretaceous time, families arose having members that became uncoiled.

such as *Inoceramus* and *rudistid clams*. Another important mollusc found in many Mesozoic faunas is the curious cigar-shaped, internal skeleton belonging to belemnoids, a group related to modern squids.

As noted in Chap. 14, the diverse and common· upper Paleozoic productids and other brachiopods became extinct at the close of the Permian. Only a few Brachiopod groups, such as ribbed, triangular rhynchonellids and the smooth egg-shaped terebratulids, remain as minor elements of the fauna. The lacy Bryozoa came into prominence during the Cretaceous, and their skeletons became contributors of detrital carbonate sediments in lagoonal deposits.

We saw that during the latter part of the Paleozoic Era there was an appreciable decline both of reef and nonreef-forming rugose and tabulate corals. Rugose corals became extinct in Middle Permian time and it was not until the Middle Triassic that a new group of corals (scleractinians) became prominent (Fig. 16.32).

Since Late Triassic time, the latter have been the principal reef formers; however, it was not until the Cretaceous Period that reefs acquired a fully modern aspect. We are unsure if the scleractinians were derived directly from rugose corals or if they were derived from some form that lacked a skeleton (e.g., the modern anemones). Another important reef-former in the Cretaceous was the corallike, massive, rudistid clam which created huge reef tracts along the margins of the equatorial Tethyan belt as far west as the American Gulf Coast. These reefs now serve as very important oil reservoir rocks (see Fig. 16.33).

THE VERTEBRATES

Mammal-like Reptiles

By far the most important preserved record of Early Triassic vertebrates is found in the Gondwana nonmarine red-bed sequence in South Africa (Fig. 15.11). In addition to the interesting depositional history of this important Permo-Triassic sequence and its *Glossopteris* flora, it contains the richest and most varied reptile fauna known for this age. One of the most important reptilian groups is the mammal-like forms, the therapsids, represented by large types thought to be the ancestors of the mammals (Figs. 15.18 and 15.22). The mammal-like reptiles had anatomical characteristics that suggest the possession of body-temperature regulating capabilities.

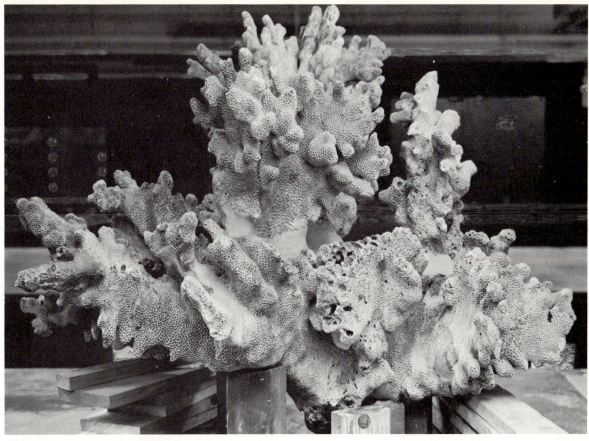

FIGURE 16.32
A colonial scleractinian coral from the Miocene of eastern
Virginia. *(Courtesy U.S. National Museum.)*

Although the mammal-like reptiles persisted
into Jurassic time, we are unsure of the exact
origin of the mammals. They probably arose
from a group called the therapsids. The first

FIGURE 16.33
A reconstructed cross section of a rudistid clam reef com-
plex in the Cretaceous of the Caribbean region. *(From
Kauffman and Sohl, 1974, courtesy Birkhauser Verlag, Basel.)*

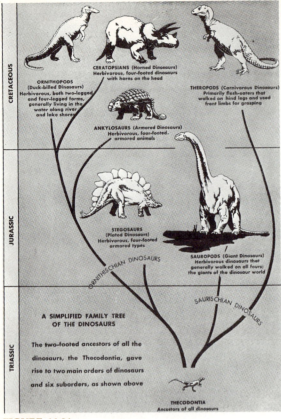

CRETACEOUS

CERATOPSIANS (Horned Dinosaurs)
Herbivorous, four-footed dinosaurs
with horns on the head

ORNITHOPODS
(Duck-billed Dinosaurs)
Herbivorous, both two-legged
and four-legged forms,
generally living in the
water along river
and lake shore.

THEROPODS (Carnivorous Dinosaurs)
Primarily flesh-eaters that
walked on hind legs and used
front limbs for grasping

ANKYLOSAURS (Armored Dinosaurs)
Herbivorous, four-footed
armored animals.

JURASSIC

STEGOSAURS
(Plated Dinosaurs)
Herbivorous, four-footed
armored types

SAUROPODS (Giant Dinosaurs)
Herbivorous dinosaurs that
generally walked on all fours;
the giants of the dinosaur world

ORNITHISCHIAN DINOSAURS

SAURISCHIAN DINOSAURS

A SIMPLIFIED FAMILY TREE
OF THE DINOSAURS

The two-footed ancestors of all the
dinosaurs, the Thecodontia, gave
rise to two main orders of dinosaurs
and six suborders, as shown above

TRIASSIC

THECODONTIA
Ancestors of all dinosaurs

FIGURE 16.34
A simplified family tree of the dinosaurs. *(Courtesy American Museum of Natural History.)*

true mammals appear in Late Triassic strata, and were rather small, rodentlike forms known only from isolated teeth and jaws; not one complete skull is known. There is very little evidence regarding their life-style, but the usual hypothesis calls for them to have been small, nocturnal forest dwellers living in a cool climate and having a varied diet of plant and animal food. They probably were warm-blooded as were their immediate therapsid ancestors. R. T. Bakker has suggested that early mammals were nocturnal just as modern marsupials are. The mammalian eye compared to the reptilian eye, for example, is modified for maximum efficiency at night. The earliest mammals had three-cusped teeth, but soon there appeared a more massive multicusped tooth apparently more

efficient for plant eating. Rarity of Mesozoic mammalian fossils possibly reflects the dominance of reptiles. It was not until the beginning of Cenozoic time that mammals "explosively" radiated to become the most successful land dwellers of all geologic history.

The Dinosaurs

The most important reptiles in the Triassic were the Thecodontia, a varied group with bipedal tendencies. It is from thecodonts that dinasaurs arose sometime during Late Triassic time (Fig. 16.34). Not all thecodonts were bipedal. The phytosaurs, for example, were a long-snouted predaceous form, which occupied a niche to be filled by crocodiles in the Jurassic. Bipedal forms, such as *Ornithosuchus,* had long muscular hind legs and a slender, flexible tail used as a balancing organ; front legs were shortened and it obviously was an excellent runner (Fig. 16.35). The earliest dinosaur order (there is no formal name *dinosaur* in reptilian classification), the Saurischia, is represented in Upper Triassic rocks by forms such as hollow-boned *Coelophysis* (Fig. 16.35), which resembled the thecodonts in being a bipedal runner. Numerous sharp teeth of this form indicate that it was an efficient carnivore. The primitive coelurosaurs gave rise to the great carnivorous saurischians of late Mesozoic time. The name Saurischia means lizard hips and indicates the most important characteristic of the group (Fig. 16.36). In many regions, such as the eastern United States, saurischians were very abundant during the Triassic, judging from extensive tracks they left (Fig. 16.1). These tracks and also bones were found in lake sediments, such as those found in the Triassic Newark Series in New Jersey across the Hudson River from New York City. There are thick deposits with garlike fish and a rich fauna of insects such as crane flies, waterbugs, march flies, cockroaches, and even the garden pest, the lace wing.

The coelurosaurs, such as *Coelophysis,* which include the smallest (dog-sized) dinosaur, continued into the Cretaceous, when they were represented by a small creature, which strongly resembled an ostrich with a long, slender neck and a small head. The teeth were replaced by a

FIGURE 16.35
A Triassic thecodont reptile, *Ornithosuchus* (left). This group is thought to have given rise to the saurischian dinosaurs like *Coelophysis* (right). *(Courtesy American Museum of Natural History.)*

horny, bill-like structure. In fact, many of the small bipedal dinosaurs apparently behaved rather like some of the large flightless birds of today.

Relatively small *saurischian dinosaurs* of the Triassic evolved into dominant and sometimes terrifying monsters in the Jurassic Period. One group retained primitive bipedalism and carnivorous habits. They are represented in the Jurassic by giant *Allosaurus* (Fig. 16.37), which preyed on the gargantuan herbivorous sauropods (Fig. 16.34). By Cretaceous time, *Tyrannosaurus* and *Gorgosaurus* appeared and climaxed the trend toward large size. *Tyrannosaurus* skeletons as long as 15 meters, standing almost 6 meters in a walking position, are known. Front legs were reduced in number of digits and were so tiny that they probably were completely useless. The skull, in contrast, was huge in proportion to body size. It was armed with a vicious set of curved, serrated teeth as much as 15 centimeters in length. Large holes reduced the weight of the huge skull.

When most children hear the word dinosaur, they think of great herbivorous Sauropoda of the Jurassic. These solid-boned quadrupeds retained some ancestral features, such as a small head, long tail, and rather short front legs. By Middle Jurassic time, they reached their zenith

in attaining the largest size and weight of any land animal known. *Apatosaurus* (Fig. 16.37) and *Diplodocus* found in the Morrison Formation in Wyoming and Colorado were massive (up to 36 metric tons). Culmination of this trend in size was one Jurassic genus over 25 meters long. Except in the Southern Hemisphere and India, sauropods are rarely found in Cretaceous strata; and no known later ones achieved the amazing proportions of the Jurassic forms. The *ornithischian dinosaurs* (bird hips; Fig. 16.36) are known from only several genera in the Upper Triassic; they are about 1 meter long and are bipedal with short forelimbs. Before the 1960s, they were known from only a few fragments from South Africa. They are now known from almost every continent, but were not very important or common until Jurassic time.

Ornithischian dinosaurs never attained the giant size of the saurischians, but they became almost unbelievably bizarre (Fig. 16.38). Even the most conservative suborder, the ornithopods, are represented by such forms as the hadrosaurs, a common Upper Cretaceous group, which was bipedal with a duck-billed skull adapted for feeding on water plants (ornithopods of Fig. 16.34). The hadrosaurs were well adapted for swamp dwelling because they possessed webbed appendages; some had nasal tubes that swept back up to the skull roof. Recent discoveries in Montana of "nests" of juveniles and adults of hadrosaurian dinosaurs suggest that they had an extended family life, possibly similar to that of elephants. The had-

FIGURE 16.36

The difference in the pelvic girdles of the two dinosaur orders. Top: *Ornitholestes*—note that in the order Saurischia the pelvic girdle is arranged in a triradiate pattern that provided the giant hind-leg muscle with a firm attachment. Bottom: *Camptosaurus*—note that the pelvic girdle in the order Ornithischia is birdlike in that the two lower pelvic bones extend anterior-posterior forming a tetraradiate pattern with the upper bone. *(Courtesy American Museum of Natural History.)*

FIGURE 16.37
Jurassic saurischian dinosaurs: *Top: Allosaurus,* a carnivorous dinosaur. *Bottom: Apatosaurus,* a herbivorous dinosaur. *(Courtesy American Museum of Natural History.)*

FIGURE 16.38
Ornithischian dinosaurs. *Top: Triceratops* (Cretaceous). *Bottom: Stegosaurus* (Late Jurassic). *(Courtesy American Museum of Natural History.)*

rosaurs were the most diverse and abundant Late Cretaceous terrestrial herbivores, composing 75 percent of the biomass of local faunas. They are found in both hemispheres, and probably arose in Asia.

Though derived from bipedal thecodonts, the majority of ornithischians were quadrupeds primarily with front legs more fully developed than in saurischians (Fig. 16.34). They were exclusively herbivores, and in all genera, teeth were reduced or absent in the front of its mouth which was in the form of a beak.

While bipedal hadrosaurs appeared to be fairly mobile, a prime necessity with carnivores lurking about, the majority of the ornithischian quadrupeds were ponderous. As in most slow-moving organisms, some type of protection was developed and three suborders solved this problem in different ways. The Ceratopsia developed a very large bony head and neck shield formed of the upper facial bones. The head, with its bony shield and turtlelike beak, occupied almost a third of the entire length of the beast (Fig. 16.38). Almost all had a median horn over the nasal area, and the most famous of the group (*Triceratops)* had an additional pair over the eyes. The ankylosaurs, another exclusively Cretaceous group, were entirely covered by heavy dermal plates accompanied by long, bony spines—not very appetizing fare (Fig. 16.34). Another important suborder, the stegosaurs, are primarily found in Jurassic rocks; the most famous of these is *Stegosaurus* (Fig. 16.38) from Morrison strata. Most conspicious features of this group were large spines and triangular plates extending down the back and terminating in a vicious-looking spiked tail. The stegosaurs also are noted for having two "brains"—a very small regulation model in its snakelike head, and another, much larger nerve center at the base of the spine. The latter supposedly controlled the spiked tail. The large plates were undoubtedly for protective purposes, but recent studies have shown that they have a complex structure that probably also served as cooling fins to regulate body temperature (see Fig. 14.42).

The dinosaurs are the most common and therefore the most important land vertebrates of the Mesozoic, and much is known of their evolution, their habitats, and their distribution. Because they are diverse, they occupied varied habitats ranging from that of small birds to very active carnivores. Even the largest of the sauropods were probably excellent runners with a gait similar to the modern elephant. This new concept contrasts with an older one, which argued that, owing to their size, they spent much of their time in lagoons or other shallow water to aid in supporting their weight. The concept of rapid movement involves a further debate regarding the possibility that the dinosaurs were warm-blooded. R. T. Bakker believes that they had to be warm-blooded because the structure of their bones, which is more mammal-like, suggests that they had the ability to regulate their internal temperature. In addition, their predator-prey ratios are high like living large mammal communities rather than reptiles, suggesting a high rate of metabolism possible only with warm-bloodedness. He further observed that, although dinosaurs lacked hair (proven by a specimen from Wyoming with mummified skin), they still could have been warm-blooded because large tropical mammals of today, such as the elephant, rhinos, and hippos, are practically hairless and yet they maintain a constant body temperature.

The dinosaurs underwent gradual adaptive radiation during the Mesozoic Era (over 230 genera are known; Fig. 16.34). At least 70 genera roamed the lands and browsed in near-shore waters during the Late Cretaceous, and then all were gone, utterly and completely, in spite of science fiction novels. Speculations about the cause of dinosaur extinction are legion; we simply do not know the cause.

Finally, the first true birds (*Archaeopteryx*) were found in the Jurassic Solenhofen Limestone of southern Germany in 1861. They represent a classic case of a transitional group in that their skeletons and large teeth strongly resemble small bipedal dinosaurs, but they had many bird features, such as feathers and an expanded brain case. We know that dinosaurs also laid eggs, which had a shell structure almost identical with that of modern birds' eggs (Fig. 16.39). The discovery of *Archaeopteryx* (Fig. 16.40) only three years after *On the Origin of Species* appeared provided the first important example of a *missing evolutionary link* intermediate between two major groups, which helped to support Darwin's theory of natural selection. Specimens fortuitously were preserved in fine-grained limestone in which impressions of feathers were clear; otherwise *Archaeopteryx* might be classified as a dinosaur. In fact, recent studies have shown that the birds are so close to dinosaurs that a new classification has been proposed. The Ornithischia, Saurischia, and Aves (birds)

RESTORATION

FIGURE 16.40
Archaeopteryx, a toothed bird, the oldest known of all birds, from the Jurassic Solenhofen of Bavaria. It comprises an important missing link between reptiles and birds. (*Courtesy American Museum of Natural History.*)

FIGURE 16.41
Swimming reptiles. *Top: Ichthyosaur* from the Lower Jurassic of Germany. *Bottom: Plesiosaur* from the Lower Jurassic of England. *(Courtesy American Museum of Natural History.)*

would be subclasses of a new class Dinosauria proposed by Bakker. This would formalize the concept of warm-bloodedness for all of these groups.

Some Other Reptilian Adaptations

Two other important radiations occurred during the Mesozoic—the invasion of the aquatic and aerial environments by reptiles. Ichthyosaurs (Fig. 16.41) appeared in the Triassic Period, flourished during the Jurassic, but became extinct well before the end of the Cretaceous. Their limbs were reduced to finlike appendages, and the head resembled that of a swordfish, having large eyes and narrow, elongated toothed jaws. Another group, the plesiosaurs (Fig. 16.41), are better known because of their impressive size (15 meters long). Apparently all were active predators.

The earliest-known and most-primitive reptile to become airborne was a gliding form that lacked feathers. It is known from a single Late Triassic specimen found by high-school boys on an afternoon collecting trip to a New Jersey quarry opposite Manhattan. Much more effective animals in the air were pterosaurs, which were able to fly (barely). The most interesting characteristic was the wing structure, which was supported solely by the fourth digit enormously extended and covered by a wing membrane that also attached to the body. Three other digits were short and clawed. Pterosaurs, or "flying reptiles," are found in marine strata of Jurassic and Cretaceous ages, and probably were fish eaters.

There have been several ideas about how primitive birds achieved flight, and, with flight, the occupation of a whole new adaptive zone—the air. The most obvious idea is an extension of a gliding hypothesis, that the birds may have been cliff or tree dwellers—leaping from branch to branch and gliding to the ground. At the same time there was a selective reduction in body weight and an increase in body surface area. Another idea is that the early birds were, like their ancestors the dinosaurs, high-speed bipedal runners that began to elongate their forelimbs for catching prey and through a series of stages increased the forelimb surface to act as insect-catching nets. This would permit sufficient surface for a wing structure to later become functional, according to John Ostrom. Under this hypothesis, the original adaptive value of feathers was for body-heat retention, and only later did their value for flight come into play. They serve both functions in modern flying birds.

THE LATE CRETACEOUS EXTINCTIONS

Within a short period of time at the close of the Cretaceous, a major biologic crisis occurred. The dinosaurs on land, the rudistids and other large-sized clams on the sea floor, the diverse ammonites, the belemnites, coccolithids, other planktonic plants and animals, as well as the marine reptiles, all vanished in less than a million years, possibly only a few thousands of years, similar to the event at the close of the Permian. There are a number of theories to explain this extinction. Most are based on some type of catastrophe, such as an extraterrestrial event like a supernova sending to earth lethal doses of radiation at a time when the magnetic field was weakening during a magnetic reversal. A recently advanced hypothesis of this type would have the Arctic Ocean completely isolated in the Late Cretaceous and gradually becoming brackish to fresh. When rifting began between Greenland and Norway 65 million years ago, the lighter Arctic waters gushed out, eventually covering all the ocean surfaces with cold, nearly fresh water. This caused the extinction of planktonic organisms, which, in turn, triggered a depletion of world oxygen (plant plankton are major contributors to the oxygen supply). The mixing of the cold water would cause a lowering of temperatures of both the land and oceans, which supposedly would cause the extinctions. This type of explanation tends to account for some phenomena, but fails to account for others that also were associated with extinctions. For example, the suggested extinction of dinosaurs by climatic change does not stand up against the evidence that they were warm-blooded. Moreover, oxygen-isotope evidence indicates only a modest cooling at the end of the Cretaceous anyway (Fig. 16.29). Finally, most theories fail to account for the fact that smaller reptiles, such as turtles and crocodiles, as well as the mammals, did not become extinct (see p. 551 for more examples and further discussion).

Early in the 1970s, Valentine and Moores showed how continental drift and continental collisions affected paleogeography and world climate. This has served to bring a number of ideas together in what is called the *regression model* to explain mass extinction, which, for the present, seems to account for many of the puzzling, often contradictory, phenomena associated with the end of the Cretaceous. Briefly, their idea, which was touched upon in Chap. 15, is that when continents come together (as with Pangea in Permo-Triassic time), there would be a period of regression. During regression the climate would become sharply zoned (see Fig. WM.6), and there would be fewer shallow-water ecologic niches available as the

epicontinental seas were drained. There would be a corresponding increase in the number of terrestrial habitats appearing. The breakup of Pangea means that there would be the beginning of a transgression and a gradual increase in floral and faunal diversity on land, as noted in Chap. 15.

There are several explanations for why transgressions and regressions occur. One concept was developed by Hays and Pitman, who noted that there was a rough correspondence between the area covered by epicontinental seas in the Late Cretaceous and Cenozoic times and the volume (size) of the midocean ridges. Hence sea-floor spreading in early Late Cretaceous time caused an increase in midocean ridge volume, which, in turn, raised sea level and displaced water onto continents, resulting in transgression, some effects of which were discussed already. As sea-floor spreading slackened, the ridges became less active, cooled, and sank somewhat by the beginning of the latest Cretaceous. This sinking caused the seas to retreat back into the ocean basins.

What are the effects of regression in general and on the latest Cretaceous in particular? During most of the Mesozoic, the major, diverse tropical marine fauna lived in the equatorial, globe-encircling Tethyan seaway (see Figs. WM.4 and WM.5). As epicontinental seas withdrew toward the end of the Cretaceous, the Tethys became narrower. As more land was exposed, total earth reflectivity increased, causing a cooling tendency, as discussed in Chap. 15. Also warm Tethyan currents were obstructed from moving poleward and seasonality increased (see Fig. WM.3). On land, cool-temperate forests expanded southward. Presumably there was less rainfall and less runoff from the land into the oceans, resulting in a depletion of nutrients, which may have caused a drop in marine plant plankton production. In fact, deep-sea core studies do show a sharp reduction of plant plankton called coccoliths (Fig. 16.42) at the end of the Cretaceous. Since plant plankton is a major source of free oxygen for both the atmosphere and the oceans, any depletion should have a marked effect on the entire ecosystem and could well account for the

FIGURE 16.42
Electron micrograph of plates of a modern coccolith from Atlantic Ocean (scale in micrometers; 1 micrometer = 0.001 mm). Such plates make up most of the famous European Cretaceous chalk deposits. Coccoliths are unicellular plants that float in surface waters. *(Courtesy Andrew McIntyre.)*

near extinction of planktonic Foraminifera and all of the ammonites and belemnites. Of course, with the withdrawal of the seas from the continents, there was a sharp reduction in the shallow-marine ecologic niches, causing the extinction of a number of families of Bryozoa, echinoderms, and bivalves. At the same time, there was an expansion of terrestrial habitats, which quickly began to be filled by early Cenozoic time with mammals.

Cold, heavy, polar water sank and formed deep-water currents, which moved along the bottom of the ocean basin toward the equator. Differential mixing along continental margins caused *upwelling currents*, which very likely killed off the rudistid clam reefs that ringed the tropical margins of the continents.

During the past 10 years, the internationally sponsored Deep-Sea Drilling Program (DSDP) and other major efforts to probe the ocean depths have revolutionized our concepts of earth history. In particular, the study of planktonic organisms such as radiolaria, coccoliths, and globigerinid Foraminifera in deep-sea cores, which are thought to have an uninterrupted record free of stratigraphic unconformities, has

shown patterns of climatic changes. These studies have shown that during the early and middle part of the Late Cretaceous in the deep seas there were profound changes in planktonic diversity with periodic increases and decreases, which appear to be correlated with minor extinctions and radiations of both marine and terrestrial vertebrates as shown by Fischer and Arthur. These integrated studies, particularly of the Cenozoic, are yielding important clues to the effects of the gradual deterioration of the climate culminating in the Pleistocene glacial catastrophe to be discussed in the next two chapters.

SUMMARY

Marine life underwent profound changes during Triassic time after late Paleozoic extinctions, but land life was not so markedly affected. Fish, ammonites, and other molluscs flourished. Land plants flourished and diversified; coniferous trees, together with cycads and gingkos, predominated, but flowering plants appeared near the end of the era. Evolution of the flowering plants is discussed in Chap. 17. Reptiles underwent tremendous diversification with the reign of mammal-like reptiles, dinosaurs, and swimming and flying reptiles. First true birds appeared, having evolved apparently from some dinosaur. Primitive mammals appeared during the middle of the era.

The tropical Tethyan marine fauna spread westward from the Old World as the Gulf of Mexico formed in Jurassic and Cretaceous time. As the Atlantic basin widened, latest Mesozoic and Cenozoic faunas differed more and more on opposite sides of the ocean.

Unlike life, sedimentation in the Mesozoic Era involved fewer innovations. Triassic facies represent a direct holdover of Permian patterns with very widespread red beds on the craton and marine deposition confined to the west. The Permo-Triassic Sonoman orogeny disturbed most of the western Cordilleran belt, and several suspect terranes apparently were accreted to the continent then. Subsequently, the development of an arc with eastward subduction produced a paleogeography resembling the present Japanese region. Red-bed deposition nearly ceased on the craton in Early Jurassic time, when an immense blanket of pure quartz sand, derived from erosion of Triassic and Paleozoic sandstones in the central craton spread over the west edge of the craton. There were both shallow-marine sands like those of early Paleozoic time, and widespread wind dunes on land. In Middle Jurassic time, wider transgression of the western craton commenced. Initially, important evaporites formed in several basins (especially around the Gulf of Mexico), but by Late Jurassic time, normal, shallow-marine deposits prevailed.

Near the end of the Jurassic, the sea retreated both north and south as the nonmarine, dinosaur-rich Morrison Formation was deposited over the Rocky Mountain region. These changes reflected onset of the Cordilleran orogeny farther west. Near the Pacific Coast, severe deformation accompanied by batholith formation began in Late Jurassic time, caused by the eastward underthrusting of oceanic crust beneath the continent. A great wave of deformation spread eastward, culminating in mid-Cretaceous time, and finishing with eastward overthrust faulting at the edge of the craton in early Cenozoic time. Apparently this reflects a flattening of the angle of subduction and an acceleration of Pacific sea-floor spreading. By mid-Cretaceous time there was an immense mountainous land from which at least 5 or 6 million cubic kilometers of material were eroded in Cretaceous time. Simultaneously, the west edge of the craton underwent rapid subsidence accompanied by deposition of a vast Cretaceous clastic wedge with great coal-bearing tongues of sandstone grading eastward into marine shale. These same deposits soon were to be involved in Cenozoic thrusting.

Major changes of ocean-ridge spreading caused sea-level changes. A great worldwide transgression affected North America in Cretaceous time, producing the last inundation of the craton. Reduction of oxygen in the seas resulted in black shale deposition. After maximum flooding near the middle of the period, regression occurred and marine conditions in the cratonic interior terminated in the Paleocene Epoch. Short-period tectonic or climatic cycles produced patterns of repetitive sedimentation in western North America much like those of Pennsylvanian time farther east.

Paleontologic and oxygen-isotope evidence point to a mild, rather uniform Mesozoic climate over North America. Dinosaurs left immortal footprints from far-north Svalbard (Spitsbergen) to Texas. Late Cretaceous ocean surface temperatures were everywhere from 10 to 20°C warmer than now, even in the Arctic! Such mild conditions were to continue for another 25 million years with no sharp change evidenced at the beginning of Cenozoic time. This allowed vegetation similar to that of the Cretaceous to persist through early Cenozoic time, yet terrestrial dinosaurs became extinct and marine ammonoids and swimming reptiles followed suit in the seas. Nearly simultaneous extinction at sea and on land is difficult to explain. The sole cause was not climatic, as has been stated frequently, nor is postulated increased cosmic radiation adequate. Most likely subtle ecologic factors, such as unrecorded changes in the food chain, were responsible.

The Cordilleran region bears striking similarities to eastern Asia as impressive as the eastern North American Paleozoic similarities with Europe. Asiatic Mesozoic similarities include many nearly identical molluscan fossils, dinosaurs, and plants, as well as parallel developments of Permo-Triassic (Sonoman) and late Mesozoic (Cordilleran) orogenies. As in the Cordilleran belt, the greatest granite batholiths of east Asia formed in late Mesozoic time.

Key Concepts

Andean-type plate margin
Archaeopteryx
Cordilleran orogeny
Cretaceous black shale
Cretaceous extinctions
Cretaceous transgression
dinosaurs
    Ornithischia

Readings

Arkell, W. J., 1956, Jurassic geology of the world: New York, Hafner.

Bakker, R. T., 1971, Dinosaur physiology and the origin of mammals: Evolution, v. 25, pp. 636–658.

Bond, G., 1976, Evidence of continental subsidence in North America during the Late Cretaceous global submergence: Geology, v. 4, pp. 557–560.

Casey, R., and Rawson, P. F. (eds.), 1973, The boreal Lower Cretaceous: Liverpool, Geological Journal, Special Issue No. 5.

Colbert, E. H., 1955, Evolution of the vertebrates: New York, Wiley. (Paperback edition, 1961, Science Editions, No. 099-S)

Eicher, D., and McAlester, R., 1980, History of the earth: Englewood Cliffs, Prentice-Hall.

Fischer, A. G., and Arthur, M. A., 1977, Secular variations in the pelagic realm: Soc. Econ. Pal. Min. Special Publication 25, pp. 29–50.

Hallam, A. (ed.), 1973, Atlas of palaeobiogeography: Oxford, Oxford.

Hays, J. D., and Pitman, W. C., 1973, Lithospheric plate motion, sea level changes and climatic and ecologic consequences: Nature, v. 246, pp. 18–22.

Howell, D. G., and McDougall, K. A. (eds.), 1978, Mesozoic Paleogeography of the western United States: SEPM Pacific Coast Paleogeography 2.

Kauffman, E. G., and Sohl, N. F., 1974, Structure and evolution of Antillean Cretaceous rudist framework: Verhandl. Naturfor. Gesell. Basel, Bd. 84, pp. 399–467.

Logan, A., and Hills, L. V. (eds.), 1974, The Permian and Triassic systems and their mutual boundary: Calgary, Alberta Society of Petroleum Geologists, Memoir 2.

McKee E. D., 1954, Stratigraphy and history of the Moenkopi Formation of Triassic age: Geological Society of America Memoir 61.

————, et al., 1956, Paleotectonic maps of the Jurassic system: U. S. Geological Survey Miscellaneous Publications, Map I-175.

————, et al., 1959, Paleotectonic maps of the Triassic system: U. S. Geological Survey Miscellaneous Publications, Map I-300.

Middlemiss, F. A., Rawson, P. F., and Newall, G., 1971, Faunal provinces in space and time: Liverpool, Geological Journal, Special Issue No. 4.

Millison, C., 1964, Paleoclimatology during Mesozoic time in the Rocky Mountain area: Denver, Mountain Geologist, v. 1, pp. 79–88.

Moullade, M., and Nairn, A. E. M., 1978, The Phanerozoic geology of the world—II: the Mesozoic: New York, Elsevier.

Ostrum, J., 1979, Bird flight: how did it begin?: American Scientist, v. 67, pp. 46–56.

Price, R. A., and Doeglas, R. J. W. (eds.), 1972, Variations in tectonic styles in Canada: Geological Association of Canada Special Paper No. 11.

Smith, A. G., 1971, Alpine deformation and the oceanic area of the Tethys, Mediterranean, and Atlantic: Bulletin of the Geological Society of America, v. 82, pp. 2039–2070.

Spencer, A. M. (ed.), 1974, Mesozoic-Cenozoic orogenic belts: London, The Geological Society, Special Publication No. 4.

Stearn, C. W., Carroll, R. L., and Clark, T. H., 1979, The geological evolution of North America (3d ed.): New York, Wiley.

Stokes, W. L., 1973, Essentials of earth history (3d ed.): Englewood Cliffs, Prentice-Hall.

17

CENOZOIC HISTORY
Threshold of the Present

The earth is not finished,
but is now being, and will
forevermore be re-made.

C. R. Van Hise
(1898)

FIGURE 17.1
Rare view of southeast face of Mt. Everest (9,000 meters) showing metamorphic rocks at bottom. *(Courtesy L. N. Ortenburger.)*

17

Figuratively speaking, the story of early geologic exploration of the Cordillera is as colorful as the scenery. For many years, the geologist was not safe there without a gun. Moreover, in the United States, unlike Canada, his activities were not supported consistently for many years. Several state geological surveys were organized in the East as early as the 1830s, but for years all of the West was under the direct care of the federal government. After the Louisiana Purchase in 1803, one expedition after another was sent out to probe the unknown. Explorations accelerated after the Civil War, and several competitors politicked officials to support "their" survey. Finally the United States Geological Survey was founded in 1879—nearly four decades after its Canadian counterpart.

Important geologic discoveries were made in rapid succession. The Grand Canyon had been explored, and several pioneer geologists contributed new concepts of erosion based upon the first systematic observations of arid landscapes. Structures in the arid West were more clearly exposed than in any regions previously studied. Valuable mineral resources were discovered and studied intensively, ultimately leading to development of a revolutionary theory of metallic ore formation. Untold numbers of rich fossil localities also were discovered, many of which yielded hitherto unknown extinct vertebrate animals. As incomparable scenic beauty was discovered, geologists became instrumental in publicizing and working for preservation of such marvels as Yellowstone, the continent's first national park. Notable in this endeavor were F. V. Hayden and two associates, famous landscape artist Thomas Moran and pioneer photographer W. H. Jackson.

The magnificent mountains and canyons explored by early geologists in western Canada and the United States were formed largely in Cenozoic time. In the last chapter, we examined the Cordilleran orogeny, which terminated a long subsidence history, and produced structures typical of ancient mobile belts (see Fig. 16.23). There was one outstanding peculiarity, namely the almost unique reactivation in the western United States of a large part of the craton. Unusually high uplifts of relatively simple anticlinal structures were formed by vertical fault movements of blocks of the Prepaleozoic basement.

Completion of our analysis of Cordilleran history as well as that of other parts of the world, especially the Alpine-Himalayan belt (Fig. 17.1), in this chapter will show that structural disturbances of great magnitude occurred with surprising rapidity in Cenozoic time, thus pointing to an extremely mobile condition. Intense fragmentation of the crust accompanied by unusually widespread volcanism characterized the late Cenozoic. In marked contrast, the other margins of North America were relatively tranquil tectonically as shown by strata of the coastal plains and broad continental shelves. Finally, we shall review the history of Cenozoic life over the past 65 million years with respect to changing geography and climate as all continents emerged from the seas to assume their present configuration.

It is important to realize that the Cenozoic record is more complete than that for earlier times, therefore we are able to read that record more finely. Both the Paleogene (42 m.y.) and Neogene (23 m.y.) Periods were equivalent to a typical pre-Cenozoic Period, but the smaller subdivisions (epochs), which we more commonly use in discussing Cenozoic history, represent much shorter time intervals than have generally been used in previous chapters. The Eocene and Miocene Epochs were each about 20 million years long, the Paleocene and Oligocene about 10 million, and the Pliocene and Pleistocene only 2 or 3 million years each.

FIGURE 17.2
Teapot Rock near Green River, Wyoming, containing Eocene lake deposits (Green River Formation) composed of laminated "oil shales" under a massive, dark deltaic sandstone, which occupies channels cut into the shales. (This photograph was taken by pioneer photographer W. H. Jackson, 1869, who had to coat glass plates with emulsion in a tent before each exposure.) *(Courtesy U. S. Geological Survey.)*

CENOZOIC CORDILLERAN HISTORY

THE ROCKY MOUNTAINS

Early Cenozoic or Paleogene

Between the early Rocky Mountain uplifts elevated during Paleocene and Eocene time (see Chap. 16) lay low areas, which became dumping grounds for sediment eroded from the high, young mountains. These intermontane basins were filled largely with river and lake deposits (Fig. 17.2). At many places the strata contain remarkably well-preserved plant, mammal, fish, reptile, and insect fossils. Volcanic ash is abundant throughout, attesting to great eruptive centers over most of the West (Figs. 17.3 and 17.4).

In the Rocky Mountain and Great Plains regions, Paleocene and Upper Cretaceous strata are conformable, except near ranges. There was no appreciable change in types of sedimentation except that coal, which formed intermittently during Cretaceous regressions, now was deposited even more widely. About the end of Paleocene time, North America's last epeiric sea retreated to the Gulf of Mexico. The last preserved vestige of marine conditions is found in North Dakota, where Paleocene marine strata escaped later erosion (Fig. 17.3).

On the following pages:
FIGURE 17.3
Lower Cenozoic lithofacies of North America. The majority of rocks shown in the Cordilleran region are nonmarine, intermontane sedimentary and volcanic deposits. Note great volcanic outpourings in Greenland and Iceland. (See Fig. 11.7 or endpapers for symbols and sources.)

FIGURE 17.4
Early Cenozoic paleogeography of North America. Note that the configuration of the continent for the first time was approaching that of today, but climate still was milder. Europe had just separated, to produce Greenland–Iceland volcanism. Eocene oil shales formed in the large lakes.

BRITISH ARCTIC VOLCANIC PROVINCE

Edge of Craton

60

65

59

150   48

70   49

Marine Paleocene

Williston Basin

Faulting and Volcanism

Edge of Craton

(Sea retreated south)

Ultramafics

Faults

75

Edge of Craton

LOWER CENOZOIC

SEDIMENTARY FACIES

(Paleocene-Eocene)

(about 50-60 m.y. ago)

1000 Miles

1000 Kilometers

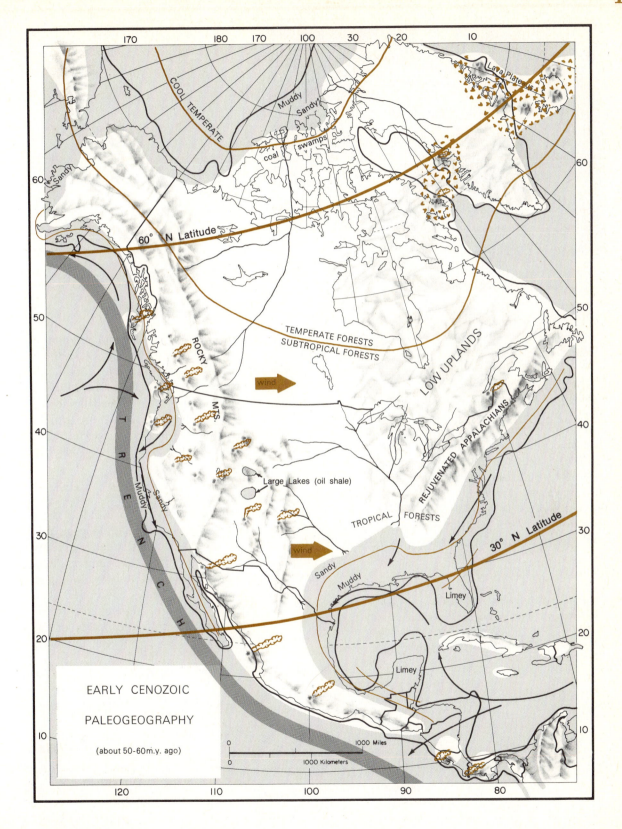

COOL TEMPERATE

Muddy

Sandy

coal

swamps

Sandy

60° N Latitude

60

50

TEMPERATE FORESTS

SUBTROPICAL FORESTS

ROCKY

MTS.

Lava Plateaus

LOW UPLANDS

REJUVENATED APPALACHIANS

wind

40

Large Lakes (oil shale)

TROPICAL   FORESTS

30

30° N Latitude

wind

T
R
E
N
C
H

Muddy

Sandy

Sandy

Muddy

Limey

20

Limey

EARLY   CENOZOIC

PALEOGEOGRAPHY

(about 50-60 m.y. ago)

0   1000 Miles

1000 Kilometers

FIGURE 17.5
Badlands topography eroded in varicolored, soft Eocene river and lake deposits of the Wasatch Formation, Bryce Canyon National Park, Utah. (*Courtesy R. B. Doremus.*)

During the Eocene Epoch, basin-filling continued, but with gradual diminution of coal. Pink, yellow, and red silts and shales are especially characteristic of the Eocene, and typically are eroded into picturesque badlands topography (Fig. 17.5). The most unusual and economically important deposit is the famous *oil shale* in Wyoming, Colorado, and Utah (Fig. 17.2). It originated in two huge lakes (each larger than Salt Lake; Fig. 17.4) in which organisms of many kinds flourished. Microscopic plankton were so abundant periodically that their remains enriched the accumulating fine muds with an organic compound called kerogen, which can be distilled to yield petroleum-like compounds. The lake deposits also contain freshwater fish skeletons and other fossils. On uplands surrounding the Eocene lakes, a lush forest of redwood and other trees thrived. By ecologic analogy with modern counterparts, the fossil flora points to uniformly mild, humid temperate conditions in marked contrast with the present climate (Fig. 17.4).

The last phase of the Cordilleran orogeny occurred during the Eocene Epoch. By this time, the mountains stood a few thousand feet high. At several scattered localities, Paleocene and Eocene strata have been tilted steeply and are overlain unconformably by late Eocene or Oligocene gravels (Fig. 17.6), and at a few localities, Paleocene and Eocene gravels have been overthrust by Paleozoic and Prepaleozoic rocks. Most of the structural movement east of the thrust zone (Fig. 16.23) was vertical, involving

FIGURE 17.6
Restored cross section showing structural and stratigraphic relationships that allow close dating of tectonic events associated with formation of the Rocky Mountains in western Wyoming. (*Adapted from Dorr, 1958, Bulletin of the Geological Society of America; and Eardley, 1962, Structural geology of North America.*)

FIGURE 17.7

Monoclinal structure at Flaming Gorge on north flank of the Uinta Mountains, northeastern Utah. Prepaleozoic basement rocks were raised and thrust outward, carrying Paleozoic and Mesozoic strata upward to form the nearly right-angle fold. The Green River cut a deep canyon directly across hard rocks of the range during late Cenozoic rejuvenation of the Rocky Mountains. Major John Wesley Powell navigated this canyon en route to explore the Colorado River in 1869, one year before the photo was taken by pioneer photographer W. H. Jackson. *(Courtesy U. S. Geological Survey.)*

large basement blocks and producing chiefly simple flexures in overlying strata (Fig. 17.7), but overthrust faults along some ranges. Small granitic plutons and volcanism (Fig. 17.8) also attest to structural unrest. Widespread volcanic centers far east within the craton have long been puzzling. They are now attributed to subduction along a very low-dipping zone so that arc volcanism extended across an unusually wide area (see Fig. 16.24). Heat so produced may have made the crust above unusually mobile.

Late Cenozoic or Neogene

Basin filling was largely completed during late Eocene and early Oligocene time. Miocene strata overlap beyond older ones to rest unconformably upon eroded pre-Cenozoic rocks (Fig. 17.9). The net result of erosion and filling was to smooth the topography, and, as uplift and erosion slackened for perhaps 5 or 10 million years, the landscape became relatively stable. Besides filling intermontane basins, Oligocene and Miocene sediments also were spread in a great apron eastward onto the plains to form a clastic wedge sequence (Fig. 17.10). The Great Plains deposits contain some of the world's finest mammalian fossils. Associated plants show that grasslands were replacing forests at lower elevations, reflecting a drying of climate, which resulted partly from the rise of mountains that blocked the flow of moist air from the Pacific, but mainly from a worldwide climatic change.

Beginning near the end of Miocene time and continuing to the present, almost the entire Rocky Mountain and adjacent regions were warped up as a large unit, producing a profound effect upon erosion. In the central Rockies, considerable faulting and volcanism occurred again. Uplift rejuvenated all streams and rivers by steepening their gradients, causing a long period of downcutting (Fig. 17.8). Early Cenozoic basin-fill sediments were eroded and

FIGURE 17.8
Yellowstone Falls and Canyon, Yellowstone National Park, Wyoming. Rock in the Canyon is altered early Cenozoic volcanic material, which gave the region its name. Volcanic activity occurred throughout Cenozoic time here, which is part of a vast Absaroka volcanic plateau, and is represented today by thermal springs and geysers. Seismic as well as thermal evidence suggests that Yellowstone overlies a hot plume of magma.

ranges were partially exhumed (Figs. 17.9 and 17.11). Most dramatic has been the cutting of deep gorges superimposed directly across huge structures in hard, pre-Cenozoic rocks (Fig. 17.7). The most famous example among dozens, the Grand Canyon, was formed during late Cenozoic time as the Colorado Plateau was raised vertically. But why this vertical up-warping on a scale covering several states? One idea is that the postulated low-angle subduction discussed above may have heated the lithosphere beneath the Colorado Plateau and surrounding ranges sufficiently to cause a simple upward expansion. There also must be some relationship to whatever processes led to wide-

spread block faulting farther west in late Cenozoic time, which is discussed later.

THE CENTRAL CORDILLERA

Early Cenozoic (Paleogene)

As the Rocky Mountains and Colorado Plateau were evolving, the axis of the mobile belt farther west was undergoing a somewhat different history. Mountains formed in late Mesozoic time still dominated the region during the early Cenozoic. They sloped westward to the Pacific Ocean, where thick marine Cenozoic strata accumulated (Fig. 17.4). Present Sierra Nevada, Cascade, and Coast Range Mountains did not yet exist, and so rivers flowed directly west to the ocean.

Early Cenozoic sediments are poorly represented in the central Cordillera because much of the region was high and undergoing active erosion. Volcanic activity is known, however (Fig. 17.3). Famous early Cenozoic fossil floras with palms and associated mammal fossils in

RECONSTRUCTED STAGES IN THE DEVELOPMENT OF
THE MEDICINE BOW MOUNTAINS, WYOMING
By S. H. Knight
1953

FIGURE 17.9
Structural and erosional development of the Rocky Mountain ranges of the Wyoming–Colorado–New Mexico region, showing effects of Cordilleran orogeny (1–4), basin filling (5), and late Cenozoic upwarping and rejuvenated erosion (6). *(After S. H. Knight, 1953,* Wyoming Geological Association 8th Annual Guidebook; *by permission of S. H. Knight and Wyoming Geological Association.)*

Oregon and Washington indicate a warm, humid subtropical climate, as do mid-Cenozoic *bauxite* deposits there (bauxite) is an aluminum-rich product of extreme weathering). Meanwhile, warm-temperate hardwood and redwood forests grew in Alaska.

Late Cenozoic Plateau Basalts
In the Oregon-Washington-Idaho region, an unusual accumulation of basalt began during Oligocene and continued locally into Pleistocene times (Fig. 17.10). Swarms of dikes represent fissures related to late Cenozoic faulting. Each lava flow spread rapidly out over very

large areas (Fig. 17.12). The total region affected exceeds 300,000 square kilometers in the Columbia and Snake River regions. Flow after flow spilled out from fissures, each successive one spreading over its predecessor. Former valleys were filled with lavas totaling as much as 4,000

On the following pages:
FIGURE 17.10
Upper Cenozoic lithofacies, reflecting great increase of volcanism and widespread deposition in intermontane basins. Note major faults in continent and beneath Pacific Ocean. (See Fig. 11.7 or endpapers for symbols and sources.)

FIGURE 17.11
Late Cenozoic paleogeography showing latitudes essentially same as today's. Regression had progressed almost to the present position of the shoreline, and climate was approaching that of today. Basalt plateaus and other volcanic materials became widespread in the west, as did block faulting, which formed most modern ranges and valleys.

UPPER CENOZOIC

SEDIMENTARY FACIES

(Miocene)

(about 20 m.y. ago)

LATE
CENOZOIC
PALEOGEOGRAPHY
(about 20 m.y. ago)

Hot, dry-climate flora

FIGURE 17.12
Columbia River basalt near Mitchell, north-central Oregon, showing typical flow-on-flow character of the plateau lavas erupted from fissures to cover thousands of square miles. A fault cut the flows and tilted the block at right.

FIGURE 17.13
Air view looking west at eastern escarpment of the southern Sierra Nevada, California; Mt. Whitney (named for California's first State Geologist) left center skyline; Alabama Hills, town of Lone Pine, and Owens River in foreground. More than 3,000 meters of topographic relief are shown. The Sierra and Alabama Hills escarpments, westernmost of the Basin and Range faults, have been active from late Miocene to present. *(Courtesy P. C. Bateman and the U. S. Geological Survey.)*

meters in thickness until vast basalt plateaus without volcanic peaks were built up (Fig. 17.11).

Late Cenozoic Block Faulting

Renewed structural disturbances began in middle Cenozoic time in the central Cordilleran region, as in the Rocky Mountains, and have continued to the present. But here they were much more severe, involving chiefly block faulting (Figs. 17.13 and 17.14). In Nevada, southeastern California, western Utah, southern Arizona, adjacent Mexico, and even along the Rio Grande valley in New Mexico, parallel northerly trending faults produced alternating narrow ranges and valleys, giving rise to the name *Basin and Range Province*. The general structural configuration is that of a series of parallel horsts and grabens (Fig. 17.15).

Fragmentation of the crust of western North America was the hallmark of late Cenozoic time—beginning in the late Oligocene. Faulting provided paths of escape of magma from the mantle. Flows spread over downfaulted valleys and lapped against ranges. Erosion of ranges produced sediments that also were dumped

FIGURE 17.14

Small fault scarp (light line on slope) produced by the 1915 Dixie Valley earthquake in northwestern Nevada. Such recent movements prove the fault origin of basins and ranges in the central Cordillera of the United States.

into the valleys. Renewed faulting offset the lavas and sediments, and new volcanic outpourings then buried older, faulted rocks. Thus were late Cenozoic block faulting and accompanying volcanism superimposed upon all of the older complex structures in the mobile belt (Fig. 17.15). Faulting also disrupted earlier drainage.

FIGURE 17.15

Diagrammatic sections showing history of the Basin and Range region; Cordilleran thrust faulting and folding (*A*: Late Mesozoic to early Cenozoic) with block faulting superimposed (*B*: Late Cenozoic or today).

For example, rivers that previously flowed west to the sea across the site of the Sierra Nevada were beheaded when the present range was faulted up in latest Cenozoic time. Nearby Death Valley came into being at the same time by unusual depression of a fault block.

PACIFIC COAST REGION

Early Cenozoic (Paleogene)

The present coast ranges were sites of marine deposition during early Cenozoic time. Paleocene and Eocene strata appear to represent a continuation of conditions prevailing in Late Cretaceous time. Thick marine graywacke sandstone and shale sequences occur, representing turbidity current deposition beyond a narrow continental shelf. Along the Eocene shoreline, deltaic coal-bearing strata also accumulated widely (Fig. 17.16). Apparently marine Cenozoic strata along much of the Pacific Coast, like Cretaceous ones before, encroached westward to produce a sort of "continental accretion" by thick sedimentation on oceanic crust followed by uplift. The underlying crust, however, has not acquired continental characteristics—at least not yet.

San Andreas Fault System

Today most people have heard of the famous San Andreas fault system of western California (Figs. 17.17 and 17.18). Many have some notion of its relation to recent earthquakes and of the threats of bisection of a winery near Hollister and of the Berkeley campus football stadium; nearby Stanford University is not safe, either, because the main San Andreas zone passes within a mile of the campus (see Fig. 1.6). But the full geologic significance of the San Andreas system of faults has been appreciated only in the past decade. The San Andreas is but one in a family of northwest-trending faults slicing diagonally across the California and southern Oregon coast ranges, and the Baja California Peninsula (Fig. 17.19).

Many years ago it was argued from offsets of fences and roads during historic quakes that the San Andreas had suffered largely *lateral or*

FIGURE 17.16
Contorted strata beneath a deltaic channel in late Eocene deposits, Coos Bay, Oregon. Sudden cutting of channel caused cohesive muds to collapse and fold. Coal-bearing strata were deposited then along a retreating shoreline characterized by widespread swamps as early Cenozoic sediments filled embayments along the Pacific Coast. *(By R. H. Dott, Jr.)*

FIGURE 17.17
Radar image (not a photo) of the San Francisco Peninsula, showing the San Andreas and related faults delineated by straight valleys. Pacific Ocean to lower left; San Francisco Bay at top (note International Airport runways bordering Bay). Expansion of greater San Francisco and the construction of huge housing tracts directly athwart the fault zone foretell future earthquake damage far in excess of that of 1906. *(NASA photo No. 67-H-1362; by permission of NASA).*

transcurrent rather than vertical movement. Measurable historic movements averaged about 1 centimeter per year. Recently it has been suggested that right lateral movement[1] of the San Andreas system may have been on the order of 500 kilometers and may have spanned more than 100 million years. The evidence for such staggering conclusions comes from offset of peculiar rock types. The most dramatic reve-

[1]So named because as one looks across the fault from either side, the opposite block has moved to the right relative to the block on which the observer is standing (it is the same no matter on which side one stands). Such faults also are termed right-handed or dextral, and strike-slip or transcurrent faults.

FIGURE 17.18
San Andreas fault zone of western California. Intricate folds in Pliocene strata within the fault zone near Palm- dale, California. In areas where rocks in the fault zone were more brittle, intense crushing is characteristic. (*Courtesy K. O. Stanley.*)

A. EOCENE

B. LATE MIOCENE

C. LATE PLIOCENE

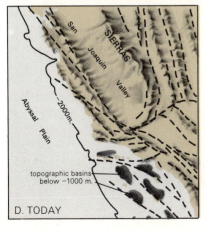

D. TODAY

FIGURE 17.19
Paleogeographic and tectonic maps showing Cenozoic evolution of California Coast Ranges. Intense faulting (dashed black lines) and rapidly changing uplift and subsidence, and marine to nonmarine conditions, have characterized this region up to the present time, (*Adapted from Hoots et al., 1954, in* Guide to the geology of Southern California; *Reed, 1933,* Geology of California.)

lation of all is the suggestion that movements along the system have torn Baja California away from the Mexican mainland, creating a scar or chasm in the crust. Oceanic material now floors the Gulf of California, while continental crust occurs on either side (Fig. 17.10).

What of the antiquity of the fault system? If, indeed, there have been hundreds of kilometers of displacement over a long period of time, then on our maps we should restore the western side of the fault southward to its original position before drawing its paleogeography. But how far back in time would we have to go to find the inception of faulting? Fences, roads, and wineries show only recent movements. Pliocene and Pleistocene strata on either side of the fault suggest a modest movement on the order of a few kilometers. The clearest suggestion of great offset is found in Miocene rocks of the Coast Ranges. In the southern San Joaquin Valley (Fig. 17.19), the thickest Miocene strata are next to the San Andreas fault. This fact, together with facies data, suggests that they were chopped off. Across the fault, one has to seek the best apparent match of Miocene facies 200 kilometers farther north!

Other evidences of important disturbances include Miocene and Pliocene volcanic rocks in the California Coast Ranges, an angular unconformity within the Miocene sequence, and initial elevation of the Sierras. In Oregon and Nevada there is also evidence that major faulting was most intense in Miocene and later time. But, although Miocene and younger conglomerates show almost continuous influence of local, uplifted fault blocks, early Cenozoic strata of western California appear to be much more homogeneous and to reflect little obvious influence of faulting upon sedimentation. Therefore, it is concluded that the most intensive activity of the San Andreas system began in the Miocene Epoch.

Late Cenozoic (Neogene) Basins

Late Cenozoic sedimentary basins of southern California are shown in Fig. 17.19. Structure of the San Joaquin basin is relatively simple except along its westernmost, faulted margin. Coast Range basins, on the other hand, suffered remarkable "see-saw" tectonics—that is, areas depressed in one epoch to receive thick sediments were upheaved to form ranges in another epoch, thus being converted to sources of new sediments deposited in still other, depressed areas. Local conglomerates and angular unconformities abound.

In the California Coast Ranges, one is most impressed by the extreme youth of the structures, which is reflected clearly by present topography (Figs. 17.17 and 17.19). Most of the anticlines of southern California are so youthful that they still form topographic elevations. Early oil seekers in southern California soon recognized this, and immediately began drilling hilly areas within the sedimentary basins with remarkable success. In recent years, oil interest has shifted offshore, for it is known that similar geology extends seaward (Figs. 17.20). In fact, islands and submerged basins off much of the Pacific Coast represent modern counterparts of the character of the present onshore region a few million years ago (Fig. 17.19C).

FIGURE 17.20

Stationary offshore petroleum drilling platform in Cook Inlet, southwest of Anchorage, Alaska. This is an unusual "monopod" platform designed for stability in very strong tidal currents. Other stationary platforms have three or four legs; floating (ship) platforms also are used. Petroleum occurs here in lower Cenozoic strata. *(Courtesy Marathon Oil Co. and L. C. Pray.)*

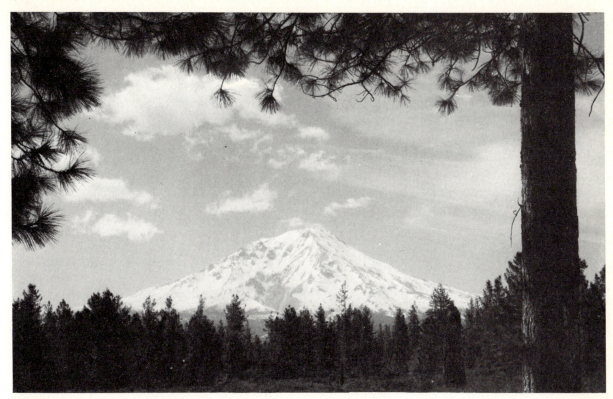

FIGURE 17.21
Mount Shasta, northern California, a Cascade volcanic peak. The Cascade volcanoes were active chiefly during Pliocene and Pleistocene times; however, recent earthquakes and volcanic pyrotechnics attest to continued Cascade arc activity. *(Courtesy California Division of Highways.)*

The Cascade Arc

The last late Cenozoic feature to note is the Cascade Range extending from southern British Columbia southward across western Washington and Oregon into northern California. The *Cascade volcanoes* (Fig. 17.21) (which include Mount St. Helens) represent a new volcanic arc superimposed upon older structures. They were more or less contemporaneous with the young Aleutian and southern Mexico–Central America arcs (Fig. 17.11). Mild folding, local metamorphism, and emplacement of a few small granitic batholiths also occurred along the Cascade trend. Mount St. Helens' awakening in 1980 proves to us that the Cascade arc is not extinct.

A trench-arc system much like the Andean system of present South America (see Figs. 7.15, WM.10) clearly existed along the Pacific margin of North America from Mesozoic through early Cenozoic time, but after the Oligocene no continuous arc-trench can be identified. The Cascade volcanoes represent the only apparent vestige of arc activity between the Aleutians in the far north and the volcanoes of Central America on the south. Some workers have questioned that the Cascade Range is a true magmatic arc, because no deep trench is associated with it today and no deep-focus earthquakes have been recorded. Certainly the volcanoes and their eruptions are of arc type, and the lack of deep earthquakes is consistent with the fact that the volcanoes until recently have been dormant. The lack of a deep trench seems due to the great abundance of sediments being supplied by the Columbia and other rivers draining a vast, humid, and recently glaciated land; sedimentation simply kept the trench

A. Beginning Cenozoic   B. End Miocene   C. Today

FIGURE 17.22

Restoration maps showing evolution of Cenozoic Cordilleran structures (such maps involve restoration, or "undoing" of structures to restore geography). A: The Cordilleran orogeny produced folding, thrust faulting, metamorphism, and batholiths ("ductile" deformation). Rotation in late Cenozoic time is indicated by relative orientations of Eocene (E) and Miocene (M) paleomagnetic vector arrows (compare A and C. B: Shearing of the Pacific basin northward against the continent produced intense faulting ("brittle" behavior) and staggering volumes of volcanic outpourings. C: Note progressive shift along San Andreas fault and formation of Cascade Range. (*Adapted chiefly from Carey, 1959,* Continental Drift; *Wise, 1963,* Bulletin Geological Society of America; *W. Hamilton and Myers, 1966,* Reviews of Geophysics.)

filled. By contrast, the deep Peru-Chile trench is still empty because adjacent South America is so arid that little sediment is carried to the coast.

LATE CENOZOIC TECTONIC REVOLUTION IN THE PACIFIC

We now have seen much evidence of a widespread and profound late Cenozoic or Neogene structural change in the Cordilleran region of North America. This change in structural habit

began in middle Cenozoic time, that is, in the Oligocene and Miocene Epochs; it continues today. Late Cenozoic time was dominated in the central and western Cordillera from Central America to Alaska by intense faulting and volcanism, whose patterns were discordant with older structures. Farther east, epeirogenic uplift and resultant canyon cutting occurred.

The brittle behavior of western North America in late Cenozoic time (as contrasted with its response during the earlier Cordilleran orogeny) produced unusual heterogeneity. Nevertheless, the seemingly chaotic fragmentation can be shown to be genetically related to disturbances all around the Pacific basin perimeter. The roughly simultaneous onset of such profound changes can hardly be coincidental. As shown in Fig. 17.22, these phenomena suggest change from the compression of the continental margin (as by continuing subduction; Fig. 17.22A) to tearing and bending of the margin as if the Pacific edge were sliding northward against the interior of the continent (17.22B). Apparently the fragmentation extended through western Canada to Alaska (Fig. 17.23). We must look to the northern Pacific sea floor for the explanation of this late Cenozoic tectonic revolution.

FIGURE 17.23
Oblique air photo of the Rocky Mountain trench in eastern British Columbia (looking north). Straightness of trench has invited speculation that it is either a major strike-slip fault zone or a down-dropped feature (see Fig. 17.22B). *(Courtesy British Columbia Government photo No. BC 762:110.)*

FIGURE 17.24
Evolution of the northern Pacific Ocean basin showing progressive collision of North America with ocean ridges (compare Figs. WM.11 and WM.12). Bend in Emperor Hawaiian seamount chain may reflect change of motion of Pacific plate about 40 million years ago at the same time as initial collision of North America with the East Pacific ridge and inception of San Andreas fault. (SA—San Andreas fault; GC—future site of Gulf of California; SF—San Francisco; HI—present position of Hawaiian Islands; numbers—dates m.y. ago). *(Adapted from Atwater, 1970, Bulletin of the Geological Society of America; Grow and Atwater, 1970, ibid.; Dalrymple, Silver, and Jackson, 1973, American Scientist; Larson and Chase, 1972, Bulletin of the Geological Society of America; Winterer, 1973, Bulletin American Association of Petroleum Geologists; Pitman, Larson, and Herron, 1974, The age of the ocean basins, special map by Geological Society of America.)*

The present discontinuous nature of both the East Pacific spreading ridge and its associated magnetic anomalies in the northeastern Pacific (as well as the disrupted trench system) indicates fragmentation of the sea floor as well as of the adjacent continental margin (Fig. 7.15). The cause of all this disruption of formerly continuous tectonic features is now explained as the result of a collision between westward-moving North America and a part of the Pacific ridge system accompanied by the formation of immense transform faults, including the San Andreas, that have since offset large segments of the ridge, the sea floor, and even the edge of the continent itself. Volcanic-arc activity continued only where a spreading ridge segment still lay west of the continent as in Oregon and Washington (Figs. 7.15 and 17.24). From a detailed analysis of sea-floor magnetic anomalies, geophysicist Tanya Atwater concluded in 1970 that what is now northern California first collided with the ridge about 30 million years ago (late Oligocene). As North America continued to move, more and more of the continental margin encountered the ridge, and the *San Andreas transform fault* grew in length (Fig. 17.24). Not later than 5 million years ago (early Pliocene), Baja California was torn away from the mainland and the Gulf of California began to form (Fig. 17.22). In effect, the sliver of the continent now lying west of the San Andreas transform was detached from the American lithosphere plate and has since moved as a part of the Pacific plate. Eventually western California seems destined either to become completely detached as a long island or to find itself joining western Canada.

But how could the *continent-ridge collision* have also caused the varied late Cenozoic disturbances within the continent farther east, especially the block faulting of the Basin and Range region? The most popular view is that a California segment of the ridge was destroyed by the collision, and that the new oblique shearing of the edge of the continent, manifested most clearly along the San Andreas fault system (Fig. 17.22), was resolved mechanically into simple tension within the continent farther east; this resulted in vertical block faulting and basaltic eruptions there. An alternate hypothesis is that the California ridge segment was not destroyed, but was subducted and continues to spread (Fig. 17.25). Yet another idea, called *back-arc spreading*, postulates complex convective eddies above a subducting plate to cause tension in the crust behind an arc (Fig. 17.25). Such tension may even separate an arc from the continent to produce a new, marginal ocean basin like the Japan Sea and possibly the southern Bering Sea behind the Aleutian arc (Fig. 17.25). The arcs may later be crushed back against continents. Such phenomena are believed to have been common throughout earth history.

Meanwhile, large volumes of old suboceanic plates were consumed beneath the Aleutian arc in the northern Pacific and beneath the arcs of the western Pacific (as well as beneath the Americas) (Fig. 17.24). Besides the magnetic anomaly patterns, another clue to Pacific plate motions is provided by deep-sea sediments. Because warm equatorial waters support the greatest populations of planktonic organisms, several times more shells of microscopic creatures such as Foraminifera are deposited on the equatorial deep-sea floor than at higher latitudes. Deep-sea drilling has shown that a zone of thicker, shell-rich sediments can be recognized back to late Eocene time. The thicker zone, which is presumed to mark a past equator position, is successively farther north in older sediments and suggests a northward movement of the Pacific plate beneath the equator of nearly 10 degrees of latitude in the past 40 million years. Tentative paleomagnetic data from Cretaceous volcanic seamounts near Japan and Cretaceous sediments in the center of the Pacific suggest that the Pacific plate has moved northward about 30 degrees in the past 100 million years.

Lines or chains of volcanic islands and submerged volcanic seamounts, such as the Hawaiian-Emperor chain, give still another clue for Pacific motions. The volcanic centers along this chain become progressively younger toward the southeast (Fig. 17.24). It has been proposed that plumes of hot material rise from the base of the mantle and produce volcanic eruptions through the Pacific lithosphere plate above (see Fig. 7.8). It is argued that the plume remains fixed in position for long times, but because the plate was moving continually over the plume, the lines of extinct volcanic centers trace the history of plate motion. Fossils dredged from the northernmost of the Emperor Seamounts are Late Cretaceous. Isotopic dates for volcanic rocks farther southeast indicate ages from 46 million years just north of the prominent bend in the chain to 18 million years at Midway Island, and from 6 million to 0 years for the Hawaiian Islands themselves (again with younger dates southeastward). If the Emperor-Hawaiian chain indeed was formed by passage of the Pacific plate over a hot mantle plume, then the sharp bend in the chain must record a major change of direction of plate motion around 35 to 40 million years ago (Oligocene). Each volcanic center was built from the abyssal sea floor 6,000 meters below sea level in only 3 to 5 million years. The older, northern islands later subsided beneath sea level to form a submarine ridge.

A mid-Cenozoic date for a major change of Pacific plate motion from south-to-north to

FIGURE 17.25

Two alternate hypotheses for the late Cenozoic fragmentation and upwarping of the western United States. *Top*: Low-angle subduction caused widespread andesitic and granitic magmatism accompanied by ore formation (compare Fig. 16.24). *Middle*: Fragmentation and basaltic volcanism resulted from subduction of the East Pacific ridge (I). *or* by back-arc spreading (II), *or* by some other cause (see text). Broad upwarping indicated by dashed lines. *Bottom*: Continued extension in Basin and Range region may produce a new, spreading ocean basin in the future. (SAF—San Andreas fault system).

southeast-to-northwest is very close to the time of initial collision of the American plate with the East Pacific ridge and with a rejuvenation of activity in the Aleutian and all the western Pacific island arcs (Figs. 17.24 and 17.26). It seems to be one more evidence of major mid-Cenozoic tectonic changes that were of global importance. There is considerable evidence, for example, that Japan and several of the other western Pacific island arcs formed from slivers

FIGURE 17.26
Mount Fuji, Japan, one of many late Cenozoic active volcanoes of the western Pacific volcanic belt. Fuji lies along the Fossa Magna rift that cuts central Honshu. *(Courtesy Japan Air Lines.)*

of continental crust torn from mainland Asia and that New Zealand was torn from Australia, resulting in the formation of new small marginal ocean basins behind them. While the separation of New Zealand is thought to have occurred in late Mesozoic time, the separation of Japan and formation of the Marianas and other arcs apparently began in mid-Cenozoic time when the other major tectonic changes occurred in the Pacific (see Fig. 7.15). Local sea-floor spreading behind certain arcs seems to produce the marginal basins, but the ultimate cause of that spreading is not yet proven (Fig. 17.25). In the days when continents were seen as fixed and as always growing larger by accretion, the western Pacific arcs were imagined to be embryonic mobile belts that would eventually become additions to Asia or Australia. It is an indication of how profoundly our views of global evolution have changed that now many of those same arcs are seen as slivers *removed* from continents rather than as new additions.

Other parts of the Pacific perimeter show similar revolutionary tectonic changes. The development of the Caribbean region is still very obscure, but it appears that a major spreading ridge—an extension of the original North Atlan-

tic ridge (Fig. WM.13)—must have separated North and South America from Jurassic to Late Cretaceous time, when a collision occurred (Fig. WM.12) like that in California during the Miocene. Not until Pliocene time, however, did volcanism build a continuous Isthmus of Panama land connection between the two continents. Major transcurrent faults fragmented the Caribbean region during the late Cenozoic, moving Cuba, Hispaniola, and Jamaica eastward. Simultaneously the eastern Caribbean or Lesser Antilles arc formed (Fig. WM.11). Way down south, the last Gondwana separation also was completed in mid-Cenozoic time. Until the Miocene, South America and Antarctica were still structurally connected via the Antarctic Peninsula (Fig. WM.12). Sea-floor magnetic anomaly patterns, deep-sea drilling, and comparisons of land geology show that the separation occurred in early Miocene time (23 m.y. ago; Fig. 17.27). Not only did this complete the physical separation of the Gondwana continents (Fig. 17.28), but it also allowed the full development of a vigorous *Circum-Antarctic ocean current* (see Fig. 17.47). This cold, encircling flow blocks any close approach of warmer waters to the polar continent, which, together with the very strong Circum-Antarctic Roaring Forties westerly winds, isolates it meterologically as well. Without the occasional intrusion of warming air masses, the climate of Antarctica turned glacial, as confirmed by identification of local Miocene tillites in Antarctica and ice-rafted

Seafloor Age: ▨ Cretaceous ▨ Paleogene ☐ Neogene

FIGURE 17.27
Cenozoic history of the southeastern Pacific region showing opening of the Scotia Sea between South America and Antarctica, the last stage in the breakup of Gondwanaland. Apparently Gondwanaland collided with a spreading-ridge *triple junction* at the beginning of Miocene time (23 m.y. ago), somewhat like the continent-ridge collision of California. (All numbers indicate m.y. ago.) *(After R. H. Dott, Jr., 1976, in* American Geophysical Union Monograph 19, *pp. 299-310; W. C. Pitman III, et al., 1974,* Geological Society of America Sea Floor Age Chart.)

pebbles in Miocene deep-sea sediments near the continent.

THE LATE CENOZOIC REVOLUTION OUTSIDE THE PACIFIC REGION

EARLY CENOZOIC EVENTS

Major plate reorganizations also occurred in Cenozoic time outside the Pacific region. Greenland began separating from Europe in the Paleocene or Eocene (60 to 65 m.y. ago) as the northern mid-Atlantic ridge formed. Widely scattered Eocene basalt flows and swarms of dikes mark this event in Ireland, Scotland, Iceland, and Greenland (see Fig. 17.3). On the other side of the world, Antarctica and Australia finally separated and went their opposite ways, and India completed its separation from Africa with outpourings of huge volumes of basalts in northwestern India. The southern continents had been very stable as evidenced by huge erosion surfaces with old lateritic soils. In mid-Cenozoic time, as we have seen, profound

FIGURE 17.28
Seaman's view of Cape Horn (actually an island) on an unusually calm day. The Cape is underlain by the Andean batholith, which is similar in composition and age to the Sierra batholith. This southernmost end of South America represents a scar where the former connection with Antarctica was severed about 23 million years ago. *(By R. H. Dott, Jr.)*

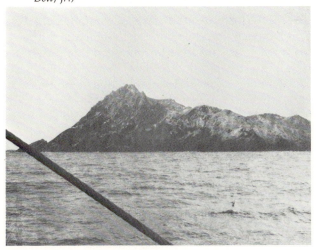

tectonic changes occurred on a worldwide basis. New activity in the Circum-Pacific arcs accompanying changes in the Pacific plate was described above. The great Alpine-Himalayan mountain system was formed, and important rifting with associated volcanism occurred on at least four continents—Africa, Europe, Asia, and Antarctica (Fig. WM.11).

THE ALPINE-HIMALAYAN BELT OF EURASIA

While the entire Pacific margin was experiencing intense deformation and volcanism associated with late Mesozoic and Cenozoic mountain building, the Tethyan region between Gondwanaland and Eurasia continued to experience deformation, which culminated in the upheaval of the Alps, Himalayas, and many other, less-famous ranges of Asia Minor in late Cenozoic time (Fig. 17.1). Old continental crust on the southern margin of Eurasia, which was disturbed previously by the late Paleozoic Hercynian orogeny (see Fig. 14.27), was involved in this Cenozoic upheaval. In addition, however, Mesozoic oceanic (Tethyan) rocks were thrust northward onto the old continental margin and crushed together with the Hercynian basement. As tectonic activity intensified in Cretaceous time, islands were pushed up near the north side of the shrinking Tethys seaway. Rapid erosion of them produced clastic debris that periodically was swept by turbidity currents into the deep-water basins between the islands much as sediments are now being deposited in basins among the many islands of the Aegean Sea between Greece and Turkey. The result was rhythmic alternations of Cretaceous to Eocene sandstone and shale originally named *flysch* in Switzerland (see Fig. 16.26). Less frequently, huge, chaotic submarine slides occurred. As compression increased, the area and height of uplifted terrain increased rapidly. The culmination of compressive deformation occurred in Oligocene and Miocene time, when sedimentation changed to very thick nonmarine gravel and sands as well as some deltaic, coal swamp, and shallow-marine deposits, all collectively named *molasse* long ago in Switzerland. Some of

the world's most complex thrust faulting accompanied by a plastic folding of strata into immense recumbent folds (called *nappes* in Switzerland) characterize the Alps. Similar structures are inferred for older mountain belts as well, but erosion has cut more deeply into those and removed much of the upper parts of the great folds (Fig. 17.29). Alpine granites formed around 30 million years ago, but metamorphism continued locally until 11 million years ago. The great vertical uplift of perhaps 7 to 8 kilometers in the earth's youngest mountains continues today.

In the last chapter we noted that fragmentation of the southern margin of the Tethys seaway accompanied the formation of a series of spreading sites. One result was the formation of a number of microcontinents separated by new, small-ocean basins. It is believed that, as the units moved northward toward Eurasia, there was subduction of considerable oceanic material; andesitic volcanism occurred along the continental margin. Finally, just as F. B. Taylor suggested long ago (see Fig. 6.10), India collided with Asia to produce the Himalayas. Arabia and Iran also collided with Asia to produce the Taurus, Zagros, and other mountains of Asia Minor (see Fig. 17.30), and parts of north Africa collided with Europe to give birth to the Alpine mountain systems (Fig. 17.31). Corsica and Sardinia were torn from Spain and France and rotated counterclockwise to their present positions near Italy. Italy itself had previously been torn from northern Africa. Meanwhile, at least part of Indochina may have separated from Gondwanaland and collided with China in late Mesozoic time. It was in turn nudged by north-traveling Australia in the late Cenozoic (Fig. 17.32). Where two continents collided along the old Tethyan zone, the resulting mobile belt is more or less structurally symmetrical. Old oceanic ophiolite rocks commonly mark a suture zone between the two formerly separate continents, and outward thrust faulting generally occurred toward each craton. While most of the former Tethyan oceanic material was consumed by subduction, some was thrust up onto the sedimentary rocks. Gravity measurements in the Himalayan region suggest that the thickest

LEVELS OF EROSION IN MOBILE BELTS

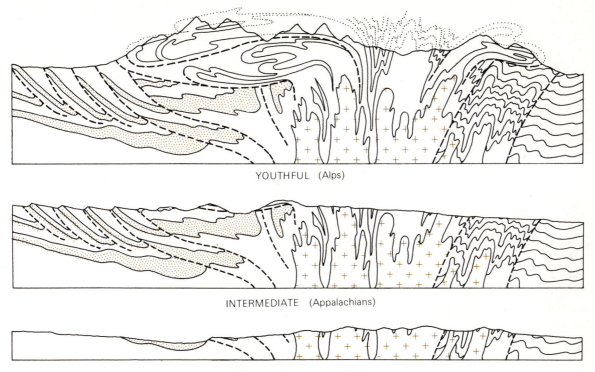

YOUTHFUL (Alps)

INTERMEDIATE (Appalachians)

MATURE (Prepaleozoic Shield)

FIGURE 17.29

Diagrammatic comparison of effects of different levels of erosion on exposed characteristics of mobile belts. *Nappe* folds shown at top may seem unique to the Alpine belt only because it is so youthful and shallowly eroded.

known continental crust in the world probably underlies these highest of all mountains, that is, about 70 kilometers, or twice the normal thickness. It has been suggested that collision of the continents here drove Indian crust beneath Asiatic.

Active subduction is continuing in the northern Mediterranean region and in Indonesia, where oceanic trenches, deep-focus earthquakes, and andesitic volcanism are still active. In both cases, oceanic material is still being underthrust northward (Fig. WM.11). It seems a safe bet that the Mediteranean will eventually be closed by the ultimate complete collision of northern Africa with Europe. In other words, the Mediterranean–Black Sea region seems to

represent the final stage in the destruction of the Tethyan realm, which process has already been completed farther east—from Turkey to Burma.

WHEN THE TETHYS DRIED UP

By early Miocene time (about 20 m.y. ago), northward drift of Africa toward Europe was restricting the western Tethys Sea (Fig. WM.3). Mountain building had closed the Tethys gap of Fig. 17.31 and split the western Tethys into two parts, the *Mediterranean Tethys* and a *Paratethys* ("Subordinate to or beside Tethys") inland sea across the present region of the Black, Caspian, and Aral seas (Fig. 17.33A). The early Miocene fauna was similar in both. Then about 6 to 6.5 million years ago, collision temporarily closed the main connection(s) between the Mediterranean and the Atlantic Ocean. Faunas decreased

FIGURE 17.30
Eroded folds of the Zagros Mountains on the northeastern shore of the Persian Gulf in southern Iran. These mountains are part of the central Alpine-Himalayan mobile belt. Thick Mesozoic and Paleogene strata deposited upon the northern side of Gondwanaland were crumpled into huge folds and thrust faults by collision between the Arabian craton (south) and Asian craton (north). (see Fig. 17.31) *(Taken from Gemini manned space capsule; NASA photo 66-63483.)*

in diversity and became very endemic. Fossil plant pollen indicates that the climate of the region was becoming cooler and drier, which fits with late Miocene evaporites long-known around the Mediterranean basin. Recent deep drilling has shown that these evaporites underlie virtually the entire basin, which seems to require that the entire Mediterranean Sea dried up! But a single event would produce only some 100 meters of salt and gypsum (see Table 13.1), whereas the maximum thickness drilled is a staggering 1,000 to 2,000 meters. Apparently the sea had an intermittent connection to the Atlantic that resupplied water repeatedly; possibly as many as 40 fillings and dryings occurred within only 1 or 2 million years. Meanwhile, the Nile River cut a very deep gorge that later filled up with alluvium as the sea returned in Pliocene time (Fig. 17.33C).

About 5 to 5.5 million years ago, fresh- to brackish-water faunas appeared in several parts of the Mediterranean basin, indicating that fresh water from somewhere caused lakes to form. Investigations northeast of the Mediterranean recently have provided evidence that the then-landlocked Parathys also partially dried up, apparently due to the cutting of channels in the Yugoslavia-Greece region (Fig. 17.33B) that drained some of its water into the Mediterranean. Then about 5 million years ago at the beginning of Pliocene time, the Strait of Gib-

FIGURE 17.31
Possible relation of continental displacements (brown arrows) to sea-floor spreading from the Indian Ocean ridges and bending of mobile belts. According to S. W. Carey, the Tethyan and other mobile belts originally were straight; bends developed due to rotations during continental displacements. Divergences of paleomagnetic vectors (lower map) in Europe (Permian) and in eastern Africa and Arabia (early Cenozoic) seem to confirm rotation. (*Adapted from Carey, 1959; van Hilten, 1964, Tectonophysics; McElhinny, 1967.*)

ralter opened, and normal Atlantic water gushed in to fill the Mediterranean (Fig. 11.33C) and to overflow through the Bosporus narrows into the Black Sea. The Caspian and Aral seas gradually became isolated as salty lakes, however. During Pleistocene glacial episodes, the Mediterranean level dropped below that of the Bosporus, which is less than 90 meters deep.

This converted the Black Sea to a lake again, and the Bosporus to a river. With each interglacial rising of sea level, the Black Sea then became a brackish sea as it is today.

Dramatic as the late Miocene drying up is to the history of southern Europe, it also is claimed to have affected the composition of the entire ocean and, in turn, the climate of the globe. The

Tertiary

Mesozoic

Pre-Tertiary of Indonesia

Paleozoic

Basement

Oceanic Basalts

Volcanics

FIGURE 17.32
Block diagram showing topography and structure of a
portion of the northwestern Australian craton and the
Indonesian arc. The faulted margin of Australia and Jurassic oceanic crust in front of it plunge toward the Indonesian trench. An arc-continent collision is impending. *(After
R. H. Beck, 1974, Petroleum Exploration Association Journal, v. 12, pp. 121; by permission.)*

immense volume of evaporites deposited there
is estimated to have amounted to 5 or 6 percent
of total oceanic dissolved salts. It is even
claimed that the seas were sufficiently less salty
that the polar seas could have frozen, which
would further chill the climate and possibly
contribute to the onset of glaciation in Miocene
times. These claims are extremely speculative,
however.

CRATONIC RIFTING AND MANTLE PLUMES

While Pacific arcs were being rejuvenated and
the *Alpine-Himalayan orogeny* was in progress,
extensive rift faulting accompanied by outpourings of immense volumes of lava affected large

FIGURE 17.33
Paleogeography of southern Europe during Neogene time,
showing destruction of the western Tethys Sea at the peak
of the Alpine orogeny due to collision of Africa with
Europe. *(Adapted from R. Brinkmann, 1960, Geologic evolution of Europe, Verlag-Hafner; L. J. Wills, 1951, A palaeogeographical atlas, Blackie; K. J. Hsü, 1978, Scientific
American, May, pp. 52–63.)*

481

A. 20 MILLION YEARS AGO
(EARLY MIOCENE)

BALTIC SEA

ATLANTIC OCEAN

EUROPE

YOUNG ALPINE RANGES

P A R A T E T H Y S

GIBRALTAR

M E D I - T E R R A N E A N

ASIA

AFRICA

T E T H Y S

0   200   400   600 km.
0   200   400 mi.

NILE RIVER

B. 5-6 MILLION YEARS AGO
(LATE MIOCENE "MESSINIAN CRISIS")

BALTIC SEA

ATLANTIC OCEAN

EUROPE

Lakes 5 m.y. ago

P A R A T E T H Y S

ARAL SEA

?

SPILLOVER CHANNELS

BLACK SEA

CASPIAN SEA

?

GIBRALTAR

M E D I T E R R A N E A N   E V A P O R I T E   B A S I N

(500 - 2000 M DEEP)

ASIA

AFRICA

0   200   400   600 km.
0   200   400 mi.

NILE RIVER

PERSIAN GULF

C. 4 MILLION YEARS AGO
(EARLY PLIOCENE)

BALTIC SEA

ATLANTIC OCEAN

EUROPE

B R A C K I S H   P A R A T E T H Y S

ARAL SEA

LAKES

BLACK   SEA

CASPIAN SEA

GIBRALTAR

BOSPORUS

M E D I T E R R A N E A N   S E A

ASIA

AFRICA

0   200   400   600 km.
0   200   400 mi.

NILE RIVER

PERSIAN GULF

FIGURE 17.34
Mount Erebus (3,794 meters), an "active" volcano on the shore of the Ross Sea, McMurdo Sound, Antarctica; view looking north from Observation Hill with monument to ill-fated Robert F. Scott (British) South Pole Expedition of 1910–1912 in foreground. Erebus formed along a late Cenozoic rift zone. (*Courtesy R. F. Black.*)

parts of the cratons of Africa, southern Siberia (Baikal rifts), central Europe (Rhine graben; see Fig. WM.11), and Antarctica (Fig. 17.34). This volcanism was unusual in that it occurred within, rather than at the edges of, lithosphere plates. Large-scale upwarping (by mantle plumes?) generally preceded rifting. In Africa some plateau areas were raised by 2,000 meters, and then their crests collapsed along normal faults to form huge grabens (see Fig. 8.23). The resulting rift valleys in East Africa have become famous for their remains of early humans.

Most of the African rifting began in Oligocene or Miocene time and continues today. Evaporite deposits were the first prominent sediments to form (as in Fig. 15.25). Associated volcanoes, including Africa's tallest, Kilimanjaro (5,995 meters), produced olivine basalts like those typical of oceanic regions together with peculiar alkaline (calcium- and sodium-rich) lavas (Fig. 17.35). The rifts average 50 kilometers in width, but the Red Sea and Gulf of Aden rifts are twice as wide, and they extend to oceanic depths (2,000 meters below sea level). The latter have oceanic crust beneath their axes and are zones of unusually high rates of heat flow through the crust. Hot solutions beneath the central Red Sea are now creating ore deposits there. Twenty years ago the suggestions by Australian geologist S. W. Carey that these two huge rifts resulted from Africa being literally pulled 200 kilometers away from Arabia, tearing open a long scar in the continental crust was regarded

as outrageous. Now it is universally accepted, and paleomagnetic evidence has confirmed that Africa was rotated 11 degrees counterclockwise away from Arabia (Fig. 17.31). The Jordan or Dead Sea rift discussed in Chap. 1 is narrower than most, and it represents a dominantly strike-slip offshoot from the Red Sea rift. It is most like the much larger Gulf of California. Offset streams and lavas indicate a total of 100 kilometers of movement since Miocene time. Although displacements have been geologically rapid in this region, one might wish that they could be greatly accelerated in order to solve local political problems by creating natural oceanic moats in drowned rifts between contending nations.

COASTAL PLAINS AND CONTINENTAL SHELVES

SUCCESSORS AND PREDECESSORS TO MOBILE BELTS

The history of the earth has now been traced from Prepaleozoic through Cenozoic time. This study has shown that mobile belts form by subduction alone or by collisions. They typically evolved through long histories of mountain building generally accompanied by formation of granitic batholiths to become stabilized finally as parts of cratons. Analysis of continental evolution would be incomplete, however, without a look at the young coastal plain strata and

FIGURE 17.35
Kilimanjaro volcano (5,995 meters), highest point in Africa, looking southeast from Ambosceli Game Reserve, Kenya. Kilimanjaro is one of many volcanoes that formed in East Africa during late Cenozoic rift faulting. (*Courtesy R. F. Black.*)

their extensions beneath the continental shelves, which formed partly over the sites of old mobile belts at present continental margins (see Fig. 16.23). They are especially prominent on passive margins as in eastern and southern United States and in the Arctic. But, although passive today, such margins are always vulnerable to active compressive tectonics should lithosphere plate patterns change, as we know they have many times in the past. Recall that

the margins of North America and those of the Tethys Sea began as passive ones. Thus coastal plain regions are bridges through time of tranquility between episodes of great tectonic activity such as we have just discussed. Remember what Heraclitus said long ago, "There is nothing permanent except change."

The northern margin of the Gulf of Mexico contains beneath it by far the thickest Cenozoic sequence of any of the coastal plains. The strata are almost 12,000 meters thick near the Mississippi Delta, while in Mexico and along the Atlantic Coast, they are only one-fifth as thick. Under the Arctic coastal plain, Cenozoic sediments are still thinner. In all cases the strata dip gently seaward and are thickest near present coastlines (Fig. 17.36). Knowledge of sub-

FIGURE 17.36
Cross section of Gulf of Mexico showing general configuration of Cenozoic continental-shelf and coastal-plain deposits (north), Sigsbee abyssal plain (center), and Yucatan carbonate reef platform (right) (*Adapted from G. I. Atwater, 1959, 5th World Petroleum Congress Transactions; and Ewing et al., 1955, Geophysics.*)

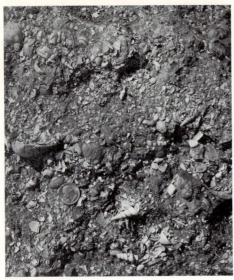

FIGURE 17.37
Miocene, richly fossiliferous, shallow-marine strata in Chesapeake Bay region; these are typical of the Atlantic coastal plain. These strata are being decimated by thoughtless collectors, so they should be protected. (Note coin for scale). *(Courtesy Thomas G. Gibson.)*

merged extensions of the deposits beneath the shelves comes from offshore deep drilling for petroleum and from geophysical surveys. The staggering thickness of sediments beneath the northern Gulf of Mexico shelf was first revealed by gravity measurements, which give an estimate of the depth to the dense basement. Seismology provides insight into the structure of the young sediments themselves and can clarify the nature of the basement.

THE ATLANTIC COASTAL PROVINCE AND APPALACHIAN REJUVENATION

The Atlantic coastal plain is the simplest of our three examples. Cenozoic strata are only about 1,000 meters thick, and clastic sediments predominate; carbonate facies are confined to southernmost Georgia and Florida (Figs. 17.3 and 17.10). No Cretaceous or Cenozoic volcanic rocks are known. During early Cenozoic time, a tropical marine fauna thrived, implying a clear, warm sea. Texture and composition of sands suggest that the adjacent Appalachian region

was low (Fig. 17.4) and undergoing mature weathering. Fossil floras of the southeastern states and the development of bauxite in Arkansas also attest to very warm, humid conditions.

After an Oligocene regression, the sea returned, but brought a rich, cooler-water Miocene fauna (Fig. 17.37). A mid-Cenozoic stratigraphic unconformity also is recognized in the Gulf Coast, Europe, and many other regions as well. Perhaps it reflects indirectly the change of spreading direction in the Pacific, the collision of North America with the East Pacific ridge, and the collision of India with Asia.

In the present Appalachian Mountains, there is conspicuous topographic evidence that, after Mesozoic beveling, regional epeirogenic upwarping much as in the Rocky Mountains caused rivers to reexpose the buried old landscape (Fig. 17.38). As the area rose, gradients of major rivers were rejuvenated, and soft, early Cenozoic and Cretaceous strata were stripped off, exposing the harder, deformed rocks of the old mobile belt. Rivers managed to maintain their flow directions originally established on younger deposits, and their drainage patterns became superimposed upon old rocks as the region slowly rose. Entrenched meanders winding through anomalously narrow valleys and

FIGURE 17.38
Diagrammatic evolution of modern Appalachian Mountains by epeirogenic rejuvenation and entrenchment of rivers. *(After Douglas Johnson, 1931, Stream sculpture on the Atlantic slope; by permission of Columbia University Press.)*

A. End of Triassic

B. Cretaceous

C. Late Cenozoic

D. Today

water gaps in resistant ridges are the results. Tributaries took advantage of relatively softer strata to carve out long, narrow valleys between parallel mountain ridges to produce ridge and valley topography (Fig. 17.38*D*).

GULF OF MEXICO COASTAL PROVINCE

General Patterns

After the Appalachian-Ouachita mountain building ceased in Late Triassic time, sedimentation commenced again in the Gulf Coast region during the Jurassic Period (see Fig. 16.13). As noted in the last chapter, it is now thought that the Gulf of Mexico was formed in Jurassic time as Gondwanaland split from North America. Local structural warping and erosional truncations occurred, and there was considerable faulting. Small igneous intrusions (some containing diamonds) were emplaced in widely scattered areas, and a few volcanic vents in Texas, Arkansas, Louisiana, and Mexico erupted considerable ash. The present lower Mississippi River Valley appears to be underlain by an *aulacogen* (p. 232), that is, a rift that developed perpendicular to the larger rift that was to become the Gulf of Mexico. Occasional earthquakes reflect continuing adjustments along its marginal faults, including the most severe event in United States history, at New Madrid, Missouri, in 1811–1812. As we found in discussing the breakup of Gondwanaland, the first deposits formed were not normal marine ones. Rather they were evaporites and red beds deposited in shallow, restricted lagoons. By the end of Jurassic time, normal marine deposition ensued and continued into the Cretaceous Period. Cretaceous deposits include chiefly limestones and shales with some important reef developments. Cenozoic deposits are composed largely of terrigenous sands and shales with only minor carbonates. Enlargement of the continent, coupled with a gradual change of climate that accelerated erosion, caused this change. The Mississippi and other modern river systems had formed at least by Miocene time. Having large drainage basins, these streams brought immense volumes of sediment to the northern Gulf, and began building great deltas as today.

Following the great Cretaceous transgression, a Cenozoic regression commenced, and the axis of maximum sedimentation continually shifted southward (Fig. 17.36). The continental shelf prograded seaward, for it is but a sedimentary embankment produced by spreading of material in shallow, agitated water. At the same time, subsidence of the crust has occurred by broad, regional, isostatic downbending accompanied by some faulting under the load of such a large volume of material (see Fig. 12.18).

Evaporite Intrusions

The most interesting features of the Gulf Coast province are hundreds of intrusions of evaporite material through Cretaceous and Cenozoic strata (Fig. 17.39). Such intrusions occur over an immense area, even well beyond the continental shelf margin. Pollen contained in the salt in Louisiana indicates a Jurassic and Triassic(?) age for the parent evaporites (see Fig. 16.13).

Evaporite intrusions, better known as salt domes or diapirs, take several forms. The simplest are roughly cylindrical, extending downward to the deeply buried parent evaporite zones. Their tops are domal in shape, and they arch overlying sediments as they rise. Complex faulting also is produced over them; graben blocks are common as are radial faults like fractures produced around an impact hole in glass. Internal structure of the domes is fantastically complex due to continuous, slow plastic flow of the evaporite. Sediments are pierced and steeply tilted along flanks of the intrusions. In many cases, the tops are mushroom-shaped, and it is thought that some of the intrusions are teardrop-shaped, being pinched off at depth. Some domes join in elongate evaporite ridges. The rise of the domes has been intermittent as evidenced by the unconformities and complex facies patterns around them.

The origin of evaporite intrusions is explained by simple physical considerations. The density of a buried evaporite stratum (2.2) is less than that of overlying compacted sediments (2.5 to 2.7 at depths exceeding 3,000 meters). The small density contrast of 0.3 to 0.5 produces an isostatic disequilibrium, and so deeply buried,

GENERALIZED CROSS SECTION
THROUGH
EASTERN LOUISIANA

FIGURE 17.39
Diagrammatic cross section of thick Gulf Coast stratigraphic sequence punctured by salt intrusions (diapirs) or
salt domes due to density disequilibrium of deeply buried,
less-dense Jurassic salt beneath more-dense, younger strata. Salt rises in ridges, cylinders, and "teardrops" until
isostatic equilibrium is achieved. Black crescents indicate
oil accumulations. See also Fig. 15.27. *(After J. B. Carsey,*
1950, Bulletin American Association of Petroleum Geolo-
gists, v. 34, pp. 361–385; used by permission.)

low-density evaporites will rise slowly in
moundlike knobs, first arching, then piercing
overlying material. As upward flow continues,
the intrusions become more elongated and may
pinch off at their bottoms. Upward intrusion
ceases when the evaporite reaches a position of
density equilibrium. A number of domes that
reached the surface were, of course, the first
such structures to be discovered. Subsequently,
because oil was found to be trapped against and
over salt domes, there was a great impetus to
discover buried ones. A breakthrough was
provided by the torsion balance in the 1920s and

gravity meter in the 1930s. The density contrast
between evaporites and sediments was easily
detectable with these instruments so that a rash
of major petroleum discoveries followed their
invention. Seismic methods later were used in
locating deep domes and associated fault structures (Fig. 17.40).

The Gulf Coast region has the largest North
American petroleum reserves, but it also is
important for salt and sulfur. Salt domes have a
curious cap rock composed of carbonate, evaporite, and pure sulfur. Pressure from the upward rise of the intrusions plays a role in the
formation of cap rock through a series of chemical reactions. Bacterial attack of petroleum also
may be important. Calcium sulfate ($CaSO_4$)
from the evaporites probably reacts with carbon
dioxide (CO_2) to form carbonate rock ($CaCO_3$),
releasing the sulfate ion, which in turn could
react with water trapped in the sediments to
produce pure sulfur.

ARCTIC COASTAL PLAIN

Concepts of mobile belts and coastal plain regions already were well advanced when modern geologic studies began in the remote Arctic. So it was with satisfaction that geologists greeted confirmation that the structural symmetry of North America was complete on its north margin. Full documentation of the existence of the Franklin belt proved that almost the entire perimeter of the craton was rimmed by mobile belts of Paleozoic or younger age, and all had similar rocks and structures.

The Arctic coastal plain provides yet another important parallel with the better-known margins of the continent. The Arctic example bears a striking resemblance to the Gulf Coast in that it, too, contains evaporite intrusions. Eroded gypsum domes are very evident over the barren northern islands; source of the evaporite is in upper Paleozoic strata. Naturally the prolific petroleum production associated with Gulf Coast evaporite structures has attracted much interest in the petroleum potential of the Arctic.

LIFE OF THE CENOZOIC

INVERTEBRATES

By the early Eocene, the planktonic plants and animals had once again diversified until near the end of the Oligocene when, yet again, there was a sharp drop in plant plankton productivity, which led to a wave of extinction, particularly of the all-important globigerinid Foraminifera (Fig. 17.41). Both the benthonic and planktonic Foraminifera are utilized by micropaleontologists for stratigraphic correlation and to recon-

FIGURE 17.40
High-resolution seismic cross section of Freeport salt dome off the Texas coast. Vertical scale is in seconds required for reflection of acoustic energy from within the sediments back to the water surface. Modern geophysical and geological methods have led to an unprecedented 70 percent petroleum discovery record off Louisiana and Texas. *(Courtesy Teledyne Exploration, Marine Sciences Division.)*

FIGURE 17.41
The globigerine Foraminifera *Globigerinoides sacculifer* (Brady) from dredging in 2,188 meters of water in the northwestern Indian Ocean. (Enlarged 100 times). *(Courtesy A. H. Bé.)*

struct past environments. The globigerinid Foraminifera are extensively used in Cenozoic and Pleistocene deep-sea stratigraphy, not only for correlation, but as temperature and climatic indicators. Their shells can be used for oxygen-isotope analysis in the same way that mollusc shells were used to work out paleotemperatures for the Cretaceous. The benthonic Foraminifera came to be used for Cretaceous and Cenozoic stratigraphic correlation as the result, in part, of the development of the rotary drilling rig for use in recovering petroleum from Gulf Coast sediments. All other, larger, fossils were ground up during drilling and only the microscopic Foraminifera (and other microfossils such as the ostracods) came up in the drilling mud. These microfossils are useful because they are very abundant in the Cenozoic marine sediments, which are frequently petroleum reservoirs. The benthonic Foraminifera rapidly diversified so that by early Eocene time, some groups had evolved species as large as a 50-cent piece, which formed Eocene limestones used to construct the pyramids of Gizeh.

Except for extinction of the ammonoids and extinctions within a few other classes, the marine fauna appears to have been less affected by events of the end of the Mesozoic. Disappearance of conspicuous organisms such as reef-forming clams (rudistids) and some typical Mesozoic snails and sponges made the Cenozoic marine fauna appear different. There was a rapid expansion of the Foraminifera, corals, advanced snails, clams, echinoids, and crustacea. As the climate deteriorated toward the end of the Cenozoic, the broad faunal provinces of today were formed.

FLOWERING PLANTS

The way the living world appears to us today is largely the result of a breakup of Pangea and the consequent evolution and distribution of the angiospermous (flowering) plants, insects, and mammals. The character of the land surfaces changed radically after the Jurassic. Beginning in Early Cretaceous time, the angiosperms began a period of explosive evolution. Most of today's common plants can be recognized in early Cenozoic floras. This expansion, as mentioned in Chap. 16, was accompanied by a parallel evolution of insects. The final important expansion occurred in mid-Cenozoic time, when prairie grasses appeared, opening many new habitats for grazing mammals. As with the fauna, the flora rapidly expanded and developed provinciality with increasing climatic differentiation. Colder regions were dominated by angiosperms that dropped leaves or otherwise formed hibernation features, while in the tropics most plants were evergreens. The leaves of the earliest-appearing angiosperms had smooth margins and disorganized venation, but during the first—Early Cretaceous—radiation venation quickly became organized, and toothed and lobed margins developed. These primitive angiosperms lived along the margins of streams and estuaries and they quickly evolved into aquatic habitats (lily ponds, etc.) and into forests. The modern grasses first appeared in mid-Cenozoic time.

VERTEBRATES

We now must devote our full attention to the adaptive radiation of the mammals. By latest Cretaceous time, mammals, as seen in deposits of western North America, gained in sheer numbers what they lacked in body size. In addition to the primitive trituberculates, multi-tuberculates (whose names refer to number of tooth cusps), and other groups that became extinct at the close of the Mesozoic Era or in early Cenozoic time, two important groups appeared—the marsupials and the insectivores (see p. A5). It is the shrewlike insectivores that probably served as the stock from which all other mammals arose.

On the Advantage of Being a Mammal

Except for the curious egg-laying, duck-billed platypus of Australia, and the *marsupials* (including our opossum and most native Australian mammals), mammals reproduce by *placental* development so that young are born alive and fully formed, in contrast to most other vertebrates, which lay eggs. In addition, the young are suckled by the mother. This characteristic has great importance, when combined with the much more highly developed mammalian brain and a very long period of prematurity, for it allows the young time to be trained and provides the groundwork for a structured social order. Social order is observed only rarely in lower vertebrates (except possibly some dinosaurs) which must rely on instinctive behavior. In fact, intelligence, or the use of an ever-evolving brain capacity, to a large extent accounts for the success and evolution of mammals during the Cenozoic. The brain and other features of the skull reflect this evolution.

The mammals were and are far more successful vertebrates in terms of environment because they are warm-blooded. Maintenance of a constant body temperature (by a metabolism utilizing food converted to energy and stored energy, and by hair) allows the animal to maintain sustained activity, unlike the reptiles, fish, and amphibia, whose activity is dependent on air or water temperature. Mammalian teeth are differ-

FIGURE 17.42
An opossum, a primitive mammal (marsupial) from Virginia. Note the differentiated teeth (i.e., front vs. rear). *(Courtesy American Museum of Natural History.)*

entiated for biting, chewing, and grinding. This allows them to utilize different types of food and hence to move into a large number of environments.

Early Cenozoic Mammalian Faunas

The earliest Cenozoic mammals were fairly small herbivores not too unlike their Mesozoic ancestors (Fig. 17.42). By middle Paleocene time, however, the picture abruptly changed. The ungulates or hoofed mammals became abundant and are represented by five-toed condylarths.

By Eocene time (Fig. 17.43), the condylarths had achieved sizes of our modern large mammals (see p. A5). Just as in all balanced faunas where herbivores exist, the carnivores are sure to be handy to help stabilize populations. The carnivores are first represented by forms which possessed small hoofs rather than the more typical carnivore claw. Their teeth, good carnivore types, were fairly well-developed bladelike forms that were used to shear flesh like a pair of scissors. The hoofed carnivores became extinct at the end of the Eocene. Other carnivore groups in turn gave rise to cats and dogs by Oligocene time and finally to the hyenas, bears, seals, raccoons, civets, etc., by Miocene time.

In addition to a large number of Paleocene orders that were still extant in the Eocene, many new orders came into existence. Two of these

FIGURE 17.43

North American Eocene mammals. Note the early horses in right center of the painting by J. H. Matternes. (*Courtesy U.S. National Museum.*)

important new orders are the artiodactyls, represented today by pigs, giraffes, deer, and camels; and the perissodactyls, having as modern descendants the rhinos, horses, and the rarely encountered tapir. Two interesting early perissodactyl groups include a curious type that had large claws instead of hoofs and was fairly large sized, and another type which was rhino-like (*Uintatherium*) and well endowed with horns (Fig. 17.43). Both of these groups had molars with fused cusps with the dentine forming a double V pattern well adapted for grinding up plant food.

Horses

The most important of the odd-toed ungulates (perissodactyls) of the Cenozoic from our point of view is the horse. There are two reasons for

our interest: one is that they have a continuous fossil record and are fairly numerous as vertebrate fossils go, and the other reason is that we can interpret the evolution over the course of the era better than in most other groups. Our friendly Victorian "bishop-eater," T. H. Huxley, first observed an evolutionary trend in the horses, but it was an American, H. F. Osborn, who first documented the history in a thorough way (Fig. 17.44). Intensive collecting over the past 30 years has shown that the picture drawn by Osborn is a gross oversimplification. Since our treatment must be brief, we will follow his simple picture, recognizing that the trends are not always consistent within groups.

In general, there is a progressive evolutionary pattern involving a whole series of related character complexes in the main phylogenetic line. As in most observed evolution, there is an increase in general body size from the earliest dog-sized horse of the early Cenozoic to the horse-sized horse (*Equus*) of today. It is impor-

FIGURE 17.44

H. F. Osborn's concept of evolution of the horse, considered to be a simplification today. (*Courtesy American Museum of Natural History.*)

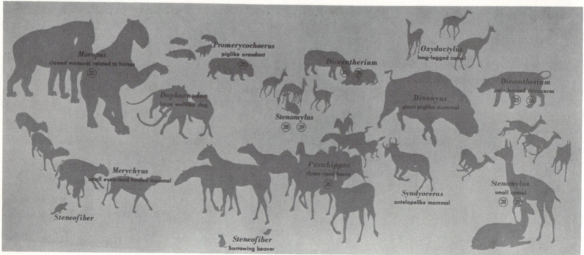

FIGURE 17.45
North American Miocene mammals. Note the large,
clawed perissodactyl (*Moropus*) on left; early horses in the
lower center are three-toed. (*Painting by J. H. Matternes,
courtesy U.S. National Museum.*)

tant to know that this was not a slow persistent
increase because the rates of change differed in
the various evolutionary lines of descent.

Another relatively persistent evolutionary
trend involved the foot and leg complex. The
basic mammalian foot consists of five digits. The
early Cenozoic horse already had displayed the
trend for a reduction of the number of digits
because it possessed four toes on the front legs
and three on the rear legs. By late Eocene and
Oligocene time there were three toes on all legs;
the middle toe was elongated with the nail
developed into a hoof, and the side toes were
reduced. By Miocene time, the middle toe
became even more elongated, taking over the
principal walking and running function (Figs.
17.44 and 17.45). This trend was accompanied

by a change of the molar teeth from a low-crowned cuspate tooth to a high-crowned, almost flat-topped tooth, which became increasingly elongate and with a more complex enamel pattern through time. With this change there was a deepening of the jaws to accommodate the longer, constantly growing teeth and a lengthening of the face. Most of these changes are thought to be in response to a change of diet, and thus a change in habitat, from a leaf-eating and probably forest-dwelling (browsing) animal to an open prairie (grazing) animal. At this time (the early Miocene), siliceous prairie grasses had evolved, thus providing a new food source. The deep roots of such grasses fit them unusually well for semiarid climates. Since the horse was becoming a highly efficient runner, it was possible for it to move out into the open country and to escape predators by running rather than hiding. This hypothesis is a simplification, but floral and faunal associations tend to suggest that it has a ring of plausibility.

The enormous number of horse fossils found in the western intermontane basins indicate that most of the evolution of the horse took place in North America. The first major invasion of the Eurasian land mass through Siberia probably occurred in Miocene time when a more primitive leaf-eating horse stock appeared there. It was not until the Pliocene that true grass-eaters became established in the "Old World." By late Pliocene or early Pleistocene time, the modern horse, *Equus*, had evolved, and surviving Pliocene genera became extinct. In Eurasia, *Equus* became highly successful and spread rapidly. However, in the Western Hemisphere the horse became extinct, for no known reason, and was brought back to North America in the 1500s by the Spanish.

Other Mammals
By Miocene time, the diverse odd-toed (perissodactyl) ungulates (Fig. 17.45) began to decline, mostly by the extinction of primitive groups. The more successful even-toed ungulates (artiodactyls) began to expand at their expense. Most of the larger mammals of the mountains, plains, and forests of today, such as the llamas, camels,

deer, and the domesticated sheep, goats, and cattle, attest to the success of their multi-stomached ruminant digestive system (the pigs and their relatives lack this unique digestive process).

Together with the even-toed ungulates, rodents and, to a lesser extent, elephant groups became diverse from middle Cenozoic on, although they declined sharply in the Pliocene-Pleistocene. The elephants have a fairly well-documented phylogeny, and showed some interesting adaptations, particularly during Oligocene to early Pliocene times (Fig. 17.46). The main line, which led to the elephants, mastodonts, and mammoths, rapidly became large-sized and developed tusks of remarkably different shapes and sizes originating from either or both jaws, indicating markedly different habitats.

Cenozoic mammalian faunas of the world are sufficiently well documented to enable us to establish migration routes and patterns for reconstructing their zoogeography. The background and understanding of these patterns are highly important in our analysis of the coming of humans.

SOME ASPECTS OF CENOZOIC BIOGEOGRAPHY

PROBLEMS OF DISPERSAL

The biogeography of the recent flora and fauna of the world is the result of a complex history resulting from isolation, evolution, contraction, dispersal, the unique events of Pleistocene glaciation, and, above all, the appearance of the most destructive of all known species through the history of life—man. Contraction of the geographic range of a widespread group is the result of intervening extinctions due to changes of environmental factors, such as the rapid climatic alterations in the Pliocene and Pleistocene. Such contractions resulted in many scattered distributions of living forms.

More than 20 years ago, American biologist and vertebrate paleontologist G. G. Simpson recognized that much of the distribution of

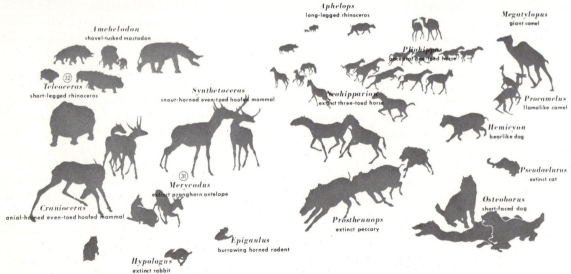

FIGURE 17.46
North American Pliocene mammals. Note the shovel-tusked elephant in the upper left of the painting by J. H. Matternes. (*Courtesy U. S. National Museum.*)

Cenozoic land faunas and floras could be explained as the result of migrations from one land mass to another via *corridors* such as *land bridges* (like the Isthmus of Panama) or *island chains* (such as Indonesia), and combinations of subsequent *isolation* (as on islands) and *contraction of geographic ranges* due to adverse climatic changes or the appearance of competitors. However, most of the biogeographic distributions must be reviewed with continental drift firmly in mind. Latitudinal changes, which affect light and temperature, are most noticeable in the Cretaceous and early Cenozoic in terms of distributional patterns. The most primitive biotas tend to be those that have been most isolated. His explanation still seems to account for the curious *disjunctive* (isolated) distribution, for example, of the camels in north Africa and southern Asia far from their cameloid relatives (llama, etc.) in South America. The Cenozoic fossil record shows that camels formerly were widespread over Asia and North America in between, and so there must have been a continuous geographic range in Pliocene time. Since then, however, the range contracted due to extinctions in North America and Asia.

Presence of modern oceanic island biotas far from continents, as on the Galápagos Islands (see Chap. 5), proves long-distance dispersal. Peculiarities of these biotic communities, however, indicate that the distribution is selective and somewhat accidental. New Zealand, for example, has an "unbalanced" fauna replete with birds, but totally lacking in indigenous mammals, turtles, snakes, and freshwater fishes. There is a very high percentage of endemic plants and animals resulting from lengthy isolation of ancestral groups derived from Australia-Antarctica in the Cretaceous and from subsequent periodic additions such as birds, butterflies, bees, and some plants during the Cenozoic. Similarly, Madagascar has a few amphibians and reptiles, one shrew, one primate, one carnivore, and some rodents. Evidence indicates that Madagascar separated from Africa in early Eocene time. Though the fauna is clearly of African origin, no larger mammals or reptiles are present.

Dispersal across water of many birds, some reptiles, and insects is easily envisioned, as is that of light-weight plant seeds adapted for long-range wind transport; all these are common on islands. But only rare, very improbable introductions of other plant and animal forms have occurred since Jurassic time, probably by bird transport of seeds, or even less-probable rafting of seeds and small land animals on drifting logs, icebergs, or (recently) ships. When G. G. Simpson pointed out the importance of *chance or "sweepstakes" dispersals,* he noted that if the chances of an event happening were say one in a million in any 1 year's time, it would be raised to a 63 percent chance in 1 million years, and to 99 percent in 10 million years (equivalent to post-mid-Miocene time). Thus a seemingly impossible event (by human standards) may become possible, while the improbable becomes inevitable given enough time! While this is dazzling mathematics, it overlooks the added improbabilities that enough individuals will be dispersed *and* find the proper ecologic niche in which to survive *and* reproduce to establish viable populations. Probability of success in a new location is considerably lower than that of freak arrivals, and probability of successful dispersal of whole communities of many co-adapted plants and animals is still smaller.

Not all distribution problems can be resolved by invoking drift or migration. In the Paleocene and Eocene in the suspect terranes of Alaska (such as Wrangellia; Fig. 16.2), at latitudes of about 60°N there are broad-leafed evergreen floras clearly of a tropical rain forest. Note that these terranes had docked in their present positions by Late Cretaceous or early Paleocene time (p. 410) and probably have not had much latitudinal change since then. The main problem in explaining these tropical floras in high latitudes is the lack of adequate sun to drive the ecosystem's photosynthesis to produce dense forests. Paleobotanist Wolfe has suggested that, to bring the sun into high latitudes, the earth's axis of rotation was less inclined (perhaps 15°). We do not particularly like the idea of shifting axes of rotation to explain floral distribution, but then the story of the suspect terranes is just beginning to whet our appetites.

SOUTHERN TEMPERATE FOREST

Out of some 20 groups of trees and shrubs making up the southern temperate forests, two trees, the southern beech (*Nothofagus*) and a southern conifer (*Araucaria*), together with associated insect communities, are especially interesting (Figs. 17.47 and 17.48). Today they live only south of the equator, and are closely adapted to cool, humid climatic conditions. Yet

Nothofagus /////   Araucaria ▨

Cold currents →   Penguins 〰

FIGURE 17.47

Distribution of certain distinctive Southern Hemisphere organisms that suggest Antarctica as an important former land bridge. Distribution of penguins is explained easily by patterns of cold currents; early Cenozoic fossil penguins are known in New Zealand, Patagonia, and Antarctica, suggesting origin in the south. Distribution of *Araucaria* and *Nothofagus* trees seems to require land-bridge connections among southern continents until Late Cretaceous time (both trees are now extinct in Antarctica; *Araucaria* also in New Zealand and South Africa). Dashed line is Australia-Asiatic biogeographic boundary called Wallace's Line. *(Adapted from Murphy, 1928,* Problems of polar research, American Geographical Society Special Publication No. 7; *Gressitt et al., 1963; Darlington, 1965,* Biogeography of the southern end of the world.*)*

FIGURE 17.48

Straits of Magellan from Bahia Fortesque, southern Chile (Chilean naval vessel in middle ground). This region is characterized by the southern beech (*Nothofagus*) forest community, nearly identical today in Chile, New Zealand, Australia, and New Guinea (see Fig. 17.47). The beech forests were publicized by Charles Darwin after *H.M.S. Beagle* passed through the Magellan Straits (1832–1833). Note effects of very strong winds upon tree shapes. *(By R. H. Dott, Jr.)*

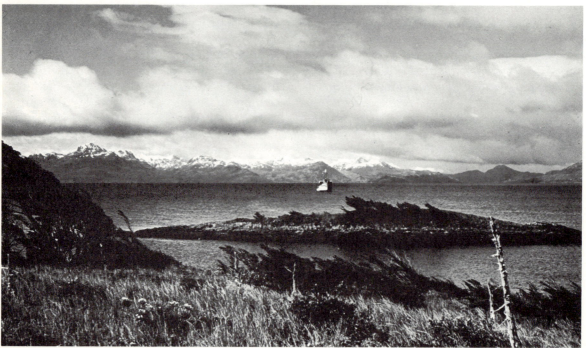

they are widely separated (i.e., disjunctive) across seemingly insurmountable water barriers. *Nothofagus* seeds typically are windborne only a few hundred meters, even during gales; the tree has failed to populate islands only a few tens of kilometers offshore from New Zealand. On the other hand, *Nothofagus* forests lie in the strongest wind zone on earth. Experiments show that in sea water, which is detrimental, the seeds soon become water-logged and sink. Yet *Nothofagus* logs are known to drift from Chile to Tasmania. Could seeds have been transported on drifting logs (note currents in Fig. 17.47)? It is judged unlikely that birds dispersed these particular trees because oceanic species rarely frequent the *Nothofagus* forests.

The Cenozoic record in the northern Antarctic Peninsula contains marine and terrestrial fossils, including giant Miocene penguins. Important plant fossils include both *Araucaria* and *Nothofagus*. These and related plants have lived since Cretaceous time in widely scattered Southern Hemisphere lands (Fig. 17.47). Their presence in Antarctica leads to two important conclusions. First, some means of biologic dispersal was possible among Australia, New Guinea, New Zealand, Antarctica, and South America in Cretaceous or early Cenozoic time; in all probability, Antarctica served as a land bridge. Since then, however, these lands have become more ecologically isolated. Secondly, the climate, at least in the relatively low-latitude Antarctic Peninsula, still was temperate enough until mid-Cenozoic time (Miocene) for forests to thrive. Continental glaciation then brought to extinction all indigenous land life in Antarctica.

The most primitive species of *Nothofagus* are found in Chile and these are closely allied to the Northern Hemisphere beech *(Fagus)*. This observation, along with the *Nothofagus* distribution noted above, has been used to develop the idea of migration "tracks." The *Nothofagus* track begins in North America and swings through South America and other southern continents. The dinosaurs and later the marsupial mammals also followed this track. Botanist Melville and others would reverse this track, at least insofar as the *Nothofagus* distribution is concerned.

They believe that the angiosperms originated in southwest Gondwanaland and that *Nothofagus* traveled Noah's Ark fashion on fragments of the lost continent of Pacifica as they drifted northward (see p. 409).

To carry the idea of a southern Gondwana origin further, the floras found in northern India clearly were related to southern continent ones until Eocene time when they became mixed with types derived from Laurasia. This coincides with the docking of India onto the shores of Asia; again the angiosperms rode a Noah's Ark. The Australasian plate also served as a Noah's Ark, for we find a mixture of northern and southern floras in southeast Asia where there are a large number of very primitive angiosperm plant groups.

SOME CENOZOIC ANIMAL DISTRIBUTIONS

When Simpson first proposed his explanation of Cenozoic biogeography, continental drift was not yet respectable. He attempted to explain all distributions of Cenozoic mammals using the Bering and Panamanian land bridges as well as one Cretaceous or early Cenozoic "sweepstakes" migration of marsupials to Australia from Asia. Now that drift is better accepted and the timing of separations of different land masses is well established, we can improve upon his story considerably. For example, there is no known fossil record of marsupials in southeastern Asia, and we now know that Australia was far from Asia until late Cenozoic time (Fig. 17.49). Therefore, it seems less plausible to derive Australia's fauna from Asia as Simpson suggested. The distribution of post-Paleocene mammals is best explained by Simpson's methods rather than by a continental drift model because the northern hemisphere continents were essentially in their present relationship and there were relatively small latitudinal shifts thereafter. The pre-Paleocene breakup of a continuous ecologic zone due to continental drift created a number of interesting convergent evolutionary patterns. For example, four unrelated groups of mammals in the southern conti-

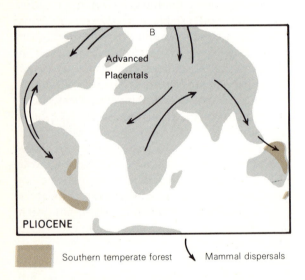

Southern temperate forest Mammal dispersals

FIGURE 17.49
Suggested major dispersal patterns of mammals from Cretaceous to present, showing probable role of continental displacements both in isolating (for example, marsupials in the south) and in allowing mixing at different times via land bridges and island stepping stones. Also shown is disjunctive southern temperate (*Nothofagus*) forest, which illustrates contraction of its range. (B—Bering bridge, which lies just above map.) (*Adapted from Raven and Axelrod, 1972,* Science, *v. 176, pp. 1379–1386; Fooden, 1972,* Science, *v. 175, pp. 894-898; Tedford, 1974, in* Soc. Econ. Paleo. and Mineral Spec. Publ. No. 21, *pp. 109–126; and unpublished sources.*)

nents became specialized for ant and termite eating (Fig. 17.50). Plate motions have rearranged past dispersal routes and Simpson could not have hypothesized *Noah's Arks* and *Viking Funeral Ships* (see p. 395) because his world was one of stable continents.

As noted above, the ancestral southern temperate forest community apparently dispersed over South America, Antarctica, Australia, New Zealand, New Guinea, and New Caledonia in Cretaceous time when these land masses were all still grouped in a supercontinent (Fig. 17.49). New Zealand separated from Australia about 80 million years ago (Late Cretaceous), Australia and Antarctica remained together until 45 to 50 million years ago (late Eocene), and the Antarctic Peninsula was still connected with South America until mid-Cenozoic time (Fig. 17.27). Apparently early land mammals and birds also dispersed together with the southern forest. Large flightless birds similar to the ostrich reached New Zealand by mid-Cretaceous time. Their relatives also lived in South America, South Africa, and Australia. Of greatest interest, however, are the marsupial mammals. who first appear in mid-Cretaceous strata of North America and in the Upper Cretaceous in South America. Marsupials as well as the placental mammal groups called edentates (such as the armadillos, sloths, and anteaters of today) and ungulates (hoofed forms) presumably reached South America from the north. By Eocene time, the marsupials had begun to diversify, some becoming carnivores, others ratlike. The ungulates and edentates (Fig. 17.50) also underwent considerable evolution, which resulted in a rich and unique fauna, including forms that resembled elephants, camels, horses, rabbits, and pigs. None of these groups is known outside South America, indicating isolation there since early Cenozoic time. Monkeys and rodents were added to this fauna from the north by middle Cenozoic time, but the New World monkeys are unrelated to those in Africa. It is thought that these forms might have "island hopped" from Central America. Both primates and rodents were already well established in North America and Europe.

It is in Australia that the marsupials have had their glory, for there they have developed the most fascinating and strangest examples of faunal isolation. Only two native groups of mammals are known, the monotremes (duck-billed platypus; see p. A4) and the marsupials. Unfortunately the fossil record of mammals in Australia is very poor. There is a gap from Triassic to Oligocene, which makes it difficult to establish the origin of the fauna. The monotremes, though very primitive forms, are not known as fossils prior to the Pleistocene, and marsupial fossils first appear in Oligocene strata. The record does show, however, that the marsupials underwent a great adaptive radiation to fill many unoccupied ecological niches. There are even more different marsupials here than in South America, including forms that resemble placental dogs, cats, wolves, rabbits, squirrels, pigs, rodents, bears, and horses of other continents. Others, such as the kangaroos, have no counterparts among either placental mammals or marsupials on other continents, but they do fill a similar ecologic niche as the placental ungulates elsewhere. New Zealand, on the other hand, which separated from Australia near the end of the Cretaceous Period, had no native mammals until humans arrived.

Although we cannot rule out an origin for the marsupials in Australia or Antarctica, the fossil record suggests that they probably originated instead either in North America and migrated into South America in Cretaceous time, or vice versa. From South America, apparently they migrated via Antarctica to Australia (Fig. 17.49).

Exactly when they arrived "down under" is not clear, as noted above, but presumably it was before late Eocene time when these two continents separated from one another. The absence of any marsupials from New Zealand may indicate that that island had already separated from Australia before their arrival. Although there are other factors that also could explain such an absence, the lack of snakes in New Zealand may also indicate isolation of the island before snakes evolved in the latest Cretaceous. New Zealand, together with New Caledonia and Norfolk Islands (between New Zealand and Australia), illustrates a commonly observed situation whereby the most (or longest) isolated biotas tend to be the most primitive and least diverse.

Meanwhile the placental mammals had also appeared and were flourishing in the north by the beginning of the Cenozoic Era. Whether they originated in Eurasia or North America is not known, but early placental ungulates and edentates noted above entered South America at the beginning of the Cenozoic, where they became isolated with the marsupial immigrants by separation of the Americas. In Pliocene time, the building of the Panamanian isthmus by volcanism allowed a new influx of more advanced placentals from the north, including camels, horses, deer, and some carnivores (Fig. 17.49). Meanwhile, ground sloths, armadillos, opossums, and porcupines moved north over the new land bridge (Fig. 17.49). One of the most puzzling questions concerning Cenozoic mammalian distributions is why the early placentals did not reach Australia. They were abundant in North America and Eurasia throughout the era, and they entered South America together with the marsupials as just noted. Perhaps climate or some other environmental factor permitted the early, probably less-specialized marsupials to filter across the southern access route, but prohibited more specialized placentals from being fellow travelers.

In Africa, unlike South America and Australia, only placental mammals are known. They show no relation to groups found in South America, which probably reflects the relatively

EQUATOR

ANT BEAR
(EDENTATA)

PANGOLIN
(PHOLIDOTA)

AARDVARK
(TUBULIDENTATA)

FIGURE 17.50

Convergent anteaters. Four unrelated ant-eating mammals developed elongated jaws without teeth to accommodate long tongues specialized to pick up ants. They also developed front feet that are clawed or otherwise modified for breaking into ant colonies. The Australian spiny anteater is the only other relative of the primitive duck-billed platypus, while the others belong to separate mammalian orders that are very specialized. The edentates include sloths and armadillos; the aardvark and pangolin and their fossil ancestors are the sole members of separate orders. *(From B. Kurtén, 1969,* Continental drift and evolution; Scientific American, *v. 220, no. 3, pp. 54–64, Copyright© 1969 by Scientific American, Inc.; used by permission).*

early separation (Cretaceous) of those two continents. The earliest known African mammalian fauna is of late Eocene age. It contains ancestral elephants and other forms that still dominate the African fauna today. Madagascar received an influx of primitive placentals from Africa, but after the Eocene they seem to have become isolated from the mainland fauna. After permanent northern contact was established by the collision of Africa and Eurasia in Miocene time, there were some important interchanges of mammals (Fig. 17.49). Camels, elephants, and lions are well-known examples. The Australia-New Guinea land mass first encountered southeastern Asia (Indonesia) in Miocene time. Extensive uplift of the edge of the southern plate created a host of new ecologic niches, and there was a prompt influx of tropical Malaysian plants and animals by island hopping and rafting. This influx of tropical flora and fauna made possible for the first time some mixing of primitive

SPINY ANTEATER
(MONOTREMATA)

tween the Asian plate and Australian plate, which had only recently collided. Besides a deepwater moat at Wallace's Line, there has been little time for the intermixing of land animals through filters or other migratory means.

India presents its own peculiar puzzles. The mammal record there is from Eocene to Recent. But the Eocene mammals are most akin to those of Eurasia and North America rather than, say, Africa! Yet, geologic and oceanographic evidence suggests that the continent was far to the south of Asia and had already separated from Gondwanaland. In short, it would seem to have been isolated by the Eocene. So we have an unexplained problem—how could northern mammals have gotten to India so early? In Miocene time, when Afro-Arabia collided with Iran and Turkey, African forms migrated across Asia Minor into India and tropical southeastern Asia (Fig. 17.49). From this Miocene fauna developed the Indian elephant and lion. Still later, other immigrants, including monkeys, also came in.

The puzzles presented by mammalian fossil distributions remind us of the following favorite quotation, which serves as a reminder that we do not have all the answers yet:

> If observations agree with our theory, that is nice, but if they do not, that is interesting. (An Anonymous Wise Man)

Australian and more advanced Asiatic animals and plants.

Such a mixture is not the case for the distribution of mammals and to a lesser extent birds and reptiles, which are sharply separated by a boundary which runs through the islands of southeastern Asia called the *Wallace Line* (Fig. 17.47) because it was first recognized by A. R. Wallace, the coauthor with Darwin of the first theory of natural selection. To the north and west of the line there is a very diverse tropical fauna, while to the south there are only a few species of monkeys, a deer, a porcupine and, of course, the great Australian marsupial fauna (only one marsupial is known from north of the Wallace Line). Undoubtedly this great and sharp separation represents the boundary be-

SUMMARY

In this chapter we have examined the development of western North America since the culminating Cordilleran orogeny. The central Cordillera underwent vigorous erosion during early Cenozoic or Paleogene time, while thick marine sedimentation continued along the Pacific margin. In present northwestern United States, immense outpourings of basalt began in mid-Cenozoic time. Simultaneously a profound change in structural habit affected all of western North America. It was characterized by brittle fragmentation of the crust and extreme volcanic activity. Block faulting created the Basin and Range Province from Oregon into Mexico. The

Sierra Nevada was raised at the west edge of that region, and major transcurrent faulting also commenced (or at least intensified) in the present California Coast Ranges. Faulting in western Oregon and Washington led to late Cenozoic or Neogene eruptions that built the majestic Cascade volcanoes, which represent a new volcanic arc. Large-scale fragmentation of western North America is the result of movement of the eastern Pacific basin against the west edge of the continent after collision of the continent with a portion of the East Pacific spreading ridge. Since then, the San Andreas system has acted as a transform fault. Baja California and the California Coast Ranges have been moved north along the San Andreas system up to 200 kilometers in the past 25 million years.

While the Cordilleran region has suffered severe and rapid tectonic changes during late Cenozoic (Neogene) time, the eastern Cordillera, and much of the remainder of the continent, have been gently upwarped. Rejuvenated rivers coupled with Pleistocene glaciation produced the present Rocky Mountain landscape. The modern Appalachian Mountains were topographically (but not orogenically) rejuvenated as well, causing entrenchment of large, meandering rivers whose courses were established in Cretaceous or early Cenozoic time.

We have seen that, after a last major Cretaceous transgression, regression characterized the Cenozoic Era. But not only has the continent become larger in area, its average elevation also has increased due to mountain building and widespread epeirogenic upwarping. Increase of area and elevation—i.e., greater continentality—naturally has resulted in more erosion and therefore accelerated clastic sedimentation both on and adjacent to the continent. Pleistocene glaciation, a subject of the next chapter, accelerated such sedimentation even more.

The rest of the world also experienced a tectonic revolution. Regression of the sea was worldwide as all continents were elevated and the world sea-floor-spreading regime underwent important changes beginning in mid-Cenozoic time. The change of direction of motion of the Pacific plate, formation of several western Pacific arcs by the tearing away of microcontinent slivers from Asia, the upheaval of the great Alpine-Himalayan mountain systems by continental collisions, and important rifting of at least four continents are the major results. The rifting and spreading changes seem destined to bring about a new phase of continental drift. Indeed, this already is in progress between northwestern Africa and Arabia (the Red Sea–Aden rifts).

Coastal plains and continental shelves represent tranquil accumulations of sediments on relatively passive crust at the edges of continents. Most of the structural deformation in them is due to intrusions of low-density evaporite masses. Relative thickness of strata within the plains and shelves is simply a function of the magnitude of sediment delivered by particular rivers and available for lateral spreading by marine processes. What subsidence of the crust has occurred on the shelves seems due only to isostatic loading and compaction. Rather than being sites of inevitable future accretions of continental crust, the shelves are simply marine accumulations on passive coasts (or trailing edges) following separations of the continents in Mesozoic time. They are, however, always susceptible to later deformation if such a margin becomes an active compressive one as has happened repeatedly.

In the previous chapter, we demonstrated that warm temperature characterized most of the Northern Hemisphere about 90 million years ago (Late Cretaceous). Evidence from early Cenozoic plant fossils shows a persistence of mild conditions, perhaps even with a slight warming. Paleobotanist Ralph Chaney showed in 1940 that humid, subtropical, and mild-temperate climate plants covered large areas of North America. Latest Cretaceous or early Cenozoic fossil redwoods (*Sequoia*) occur at present latitudes of 70°N in Alaska, Greenland, and Siberia, and 78°N in Canada and Svalbard (Spitsbergen); palm trees grew as far north as 50°N latitude. Moreover, climatic zonation suggested by plant distributions was closely parallel with the present equator, conforming with paleomagnetic evidence that the continent had essentially its present relation to the pole and equator. Floras very similar to Asiatic ones spread widely over the continent. Seas were slightly larger, the continent was somewhat lower, and climatic zones were wider than today. By late Cenozoic time, the climatic zones, as suggested both by land and marine fossils, were narrower and were shifting southward. Cooler temperatures are indicated for the first time. A trend toward greater aridity had developed, especially in the southwestern United States and adjacent Mexico as indicated by shifts of humid- and arid-climate plant communities. A contributing factor was rapid elevation of the Cascade Range and Sierra Nevada near the Pacific Coast. Early Cenozoic floras were so uniform as to suggest no sharp east-west climatic differentiation, but by late Cenozoic (Neogene) time, westerly winds were disrupted, creating aridity to the east (rain shadow). A global deterioration of climate, however, was more important. In the west, grasslands enlarged in area and forests shrank to moist river beds and high, cool uplands. A number of plants known today only in Asia (e.g. *Ginkgo*) vanished from western North America at about this time, and the ranges of others contracted greatly. For example, as recently as Pleistocene time, redwood and Douglas fir trees grew on islands off southern California 200 kilometers farther south than they now grow. A parallel change from browsing to grazing habits occurred among many mammals as an adaptation to changing environment as the grasslands spread. Some important mid-Cenozoic extinctions of mammals may have resulted from failures to adapt, but on the whole, mammals proved well suited for late Cenozoic changes of topography, climate, and vegetation.

Evolution of modern land plants and animals was greatly influenced by the breakup of Pangea and the separation of its component continents. Some groups, such as the marsupials, underwent unique evolution after being isolated early in South America and Australia. The placentals, however, had free run of Eurasia, Africa, and North America, but for the most part reached South America and Australia very late. Climatic and other environmental changes in late Cenozoic time have reduced former much broader geographic ranges of many organisms, leaving widely separated disjunct outposts (e.g., camels of Africa-Asia and cameloids of South America) due to extinctions over intervening territory. The final character of the modern flora and fauna resulted from events of the Pleistocene ice age, during which evolved the most adaptable and destructive species of all time—man.

Key Concepts

Alpine-Himalayan orogeny
angiosperms
aulacogen
back-arc spreading
Basin and Range Province
bauxite
block faulting
Cascade volcanic arc
Circum-Antarctic ocean current
continent-ridge collision
contraction of geographic ranges
disjunctive
evaporite intrusions
globigerinid Foraminifera
land bridges or corridors
late Miocene drying of Mediterranean
lateral or transcurrent fault
marsupial versus placental mammals
Neogene fragmentation
Nothofagus (southern beech) forest
oil shale
Paratethys Sea
plateau basalts
regional upwarping and canyon cutting
San Andreas transform fault
sweepstakes dispersals
Wallace's Line

Readings

Armentrout, J., et al. (eds.), 1979, Cenozoic paleogeography of the western United States: SEPM Pacific Section Symposium Volume 3.

Atwater, T., 1970, Implications of plate tectonics for the Cenozoic tectonic evolution of western North America: Bulletin of the Geological Society of America, v. 81, pp. 3513–3536.

Beck, C. B., 1976, Origin and early evolution of the angiosperms: New York, Columbia.

Brinkmann, R., 1976, Geology of Turkey: New York, Elsevier.

Burke, K., and Whiteman, A. J., 1973, Uplift, rifting, and the break-up of Africa, in Continental Drift, Sea Floor Spreading and Plate Tectonics: New York, Academic.

Colbert, E. H., 1955, Evolution of the vertebrates: New York, Wiley. (Paperback edition, 1961, Science Editions, No. 099-S)

Coleman, P. J. (ed.), 1973, The Western Pacific island arcs, marginal seas, geochemistry: Report of the 12th Pacific Science Congress, Canberra, 1971: New York, Crane, Russack and Co. and Univ. of Western Australia Press.

Crowell, J. C., 1962, Displacement along the San Andreas Fault, California: Geological Society of America Regional Reviews, Special Paper 71.

Curtis, B. F. (ed.), 1975, Cenozoic history of the Southern Rocky Mountains: Boulder, Geological Society of America, Memoir 144.

Fooden, J., 1972, Breakup of Pangaea and isolation of relict mammals in Australia, South America and Madagascar: Science, v. 175, pp. 894–898.

Gansser, A., 1964, Geology of the Himalaya: London, Interscience.

Girdler, R. W. (ed.), 1972, East African rifts: Amsterdam, Elsevier.

Gobbett, D. J., and Hutchison, C. S., 1973, Geology of the Malay Peninsula: New York, Wiley-Interscience.

Hamilton, W., 1978, Tectonics of Indonesia: United States Geological Survey Professional Paper 1078.

Himalayan and alpine orogeny, 1964, Part XI of Report of 22nd International Geological Congress: New Delhi, India.

Hsü, K. J., 1972, When the Mediterranean dried up: New York, Scientific American, December; and 1978, When the Black Sea dried up: Scientific American, May.

Karig, D. E., 1971, Origin and development of marginal basins in the western Pacific: Journal of Geophysical Research, v. 76, pp. 2542–2561.

Kingma, J. T., 1974, The geological structure of New Zealand: New York, Wiley.

Kurtén, B., 1972, The age of mammals: New York, Columbia.

Larson, R. L., and Chase, C. G., 1972, Late Mesozoic evolution of the western Pacific Ocean: Bulletin of the Geological Society of America, v. 83, pp. 3627–3644.

Lemoine, M., 1978, Geological atlas of Alpine Europe: New York: Elsevier.

McKee, Bates, 1972, Cascadia—the geological evolution of the Pacific Northwest: New York, McGraw-Hill.

McKenna, M. C., 1973, Sweepstakes, filters, corridors, Noah's Arks, and beached Viking Funeral Ships in paleogeography: in Implications of continental drift to the earth sciences, v. 1, pp. 295–308: New York, Academic.

Melville, R., 1973, Continental drift and plant distribution, in Implications of Continental Drift to the Earth Sciences, v. 1: New York, Academic.

Minato, M., Gorai, M., Hunahasi, M., 1965, The geologic development of the Japanese Islands: Tokyo, Tsukiji Shokan.

Murray, G. E., 1959, Geology of Atlantic and Gulf Coast province of North America: New York, Harper & Row.

Raven, P. H., and Axelrod, D. I., 1972, Plate tectonics and Australasian paleobiogeography: Science, v. 176, pp. 1379–1386.

Ross, C. A. (ed.), 1974, Paleogeographic provinces and provinciality: Society of Economic Paleontologists and Mineralogists Special Publication No. 21.

Rutten, M. G., 1969, The geology of western Europe: Amsterdam, Elsevier.

Raup, D. M., and Stanley, S. M., 1978, Principles of Paleontology (2d ed.): San Francisco, Freeman.

Sestini, G., 1970, Development of the northern Apennines geosyncline: Sedimentary Geology, v. 4, no. 3–4.

Sharp, R. P., 1972, Geology field guide to southern California: Dubuque, W. C. Brown.

Smith, R. B., and Eaton, G. P., 1978, Cenozoic tectonics and regional geophysics of the western Cordillera: Boulder, Geological Society of America, Memoir 152.

Spencer, A. M. (ed.), 1974, Mesozoic-Cenozoic orogenic belts: London, The Geological Society, Special Publication No. 4.

Sugimura, A., and Uyeda, S., 1973, Island arcs: Japan and its environs: Amsterdam, Elsevier.

Van Andel, T. J., et al., 1975, Cenozoic history and paleoceanography of the central equatorial Pacific: Boulder, Geological Society of America, Memoir 143.

Winterer, E. L., 1973, Sedimentary facies and plate tectonics of equatorial Pacific: Bulletin of the American Association of Petroleum Geologists, v. 57, pp. 265–282.

Wolfe, J. A., 1981, Vicariance biogeography of the angiosperms in relation to paleobotanical data: in Vicariance Biogeography: A Critique: New York, Columbia.

18

PLEISTOCENE GLACIATION AND THE RISE OF MAN

The more it
SNOWS-tiddely-pom,
The more it
GOES-tiddely-pom
The more it
GOES-tiddely-pom
On
Snowing.

A. A. Milne,
The House at Pooh Corner, 1928
(By permission of E. P. Dutton and Co.
Inc., and Methuen and Co., Ltd., 1928)

FIGURE 18.1
Adélie penguins near
McMurdo Sound,
Antarctica. Penguins
eat small marine ani-
mals abounding in the
cold waters; the birds
are delicately adapted
to their peculiar niche,
but easily could be-
come extinct if land
predators swarmed
over their rookeries.
*(Courtesy National Sci-
ence Foundation; photo
by Cartsinger.)*

18

James Hutton was among the first to publish (1795) a suggestion that Alpine glaciers in the past had been much more extensive than today. His reasoning was inspired by reading of famous occurrences of erratic boulders on hilltops in the Swiss Plain near Geneva, and on the slopes of the Jura Mountains to the north, more than 50 kilometers from the nearest modern glacier. This was a lucky bit of Huttonian intuition, but it is a curious irony that he himself never recognized the abundant evidence of glaciation in his Scottish homeland.

The glacial interpretation of erratic boulders was overlooked in the tempest over granite and basalt and because of the curious state of thought about features now routinely attributed to continental glaciation. As was pointed out in Chap. 3, the Noachian Flood dogma hung as a pall over early nineteenth-century thought. The Diluvialists continued to regard erratic boulders scattered over northern Europe as having been drifted in by the Flood (see Fig. 3.3). In northern Scotland, a famous set of conspicuous parallel benches along a valley, the Parallel Roads of Glen Roy, also were interpreted as having been formed by the Deluge pouring down a narrow defile; in fact, they are glacial lake beaches.

First full recognition of former glaciation on a continental scale by huge, vanished ice sheets was in Norway in 1824 by J. Esmark, though the great German writer Goethe apparently anticipated Esmark somewhat earlier. Esmark's suggestion, too, was largely overlooked, and it was a Swiss engineer (Ignace Venetz-Sitten) who, beginning in 1821 and joined about 1830 by a friend (Jean de Charpentier), confirmed earlier speculation that Alpine glaciers once extended out of the mountains onto the Swiss Plain, leaving *moraines* and *erratic boulders* far north of present glaciers. A prominent Swiss biologist, Louis Agassiz, felt compelled to see what manner of heresy these men were promulgating. To his own surprise, he became convinced in 1836

that they were correct. By 1840 Agassiz himself became the leading champion of continental glaciation by gathering evidence of a widespread cover of ice over nearly all of northern Europe, including Britain. Presence of peculiar rock types among the erratic boulders of northern Europe proved that most of the ice had not come from the Alps, but had flowed south from Scandinavia, the only place where such rocks were known to occur. In 1846 Agassiz joined the faculty of Harvard University, and was elated to find evidence of similar continental glaciation in New England.

By midcentury, other geologists were becoming interested in continental glaciation. Patterns of ice flow were being recognized:

> The rock scorings are the trails left by the invader. Their character should reveal the nature of the icy visitant as tracks reveal the track maker. (T. C. Chamberlin, 1886)

And in the midwestern United States, as well as in Scotland, several levels of glacial till superimposed upon one another and separated by obvious buried soils containing plant and animal remains were differentiated. There had been not one but multiple advances and retreats of the ice. It had already been suggested in 1842 that a major, worldwide lowering of sea level must have accompanied the past expansion of glaciers; with recognition of multiple ice advances about 1875, it became apparent that sea level must have fluctuated several times. And in 1865 the isostatic depression of the crust by ice was inferred in Scotland. Finally, only 20 years after Agassiz began his glacial researches, evidence of a much more ancient (late Paleozoic) glacial episode was discovered in India. (See Fig. 15.13.)

The *Pleistocene Epoch* represents a truly unique interval in geologic history. Not only did it encompass the Great Ice Age, but also during its brief 2- to 3-million-year span, man evolved both biologically and culturally to take his preeminent position on the ladder of life. Because of its recency, the record of the Pleistocene can be read in fascinating detail. The presence of man makes Pleistocene study the most interdisciplinary facet of geology. Anthropologists, climatologists, botanists, zoologists, and even biblical scholars are very much concerned with latest Cenozoic history.

In most minds, the Pleistocene is synonymous with glaciation, and we have stressed that glaciations have been truly rare events in history. We also have taken pains to show that the typical climate of the past was warmer and more uniform than that of the last 15 to 20 million years. Therefore, neither climatically nor topographically is the present a perfect key to the past! Definition of the limits of the Pleistocene Epoch is by no means simple. It is obvious that, solely on the basis of glaciation, polar and alpine parts of the earth would be classed as still in the Pleistocene today (Fig. 18.2), whereas

midlatitudes emerged from the ice age between 10,000 and 11,000 years ago. Similarly, ice caps must have begun forming earlier in polar areas than elsewhere, and so neither beginning nor end of glaciation provides wholly satisfactory defining time limits. Lyell suggested a biologic definition of the Pliocene-Pleistocene boundary in Italy. Pleistocene marine strata were defined as containing only marine species that still are living, while Pliocene ones contained a few extinct forms. Problems arise, however, in trying to extrapolate this datum from shallow-marine deposits, where it was defined, to deep-marine and to nonmarine successions. Isotopic dating places this boundary between 2 and 3 million years ago, which is more than twice as old as the long-popular, but poorly based, figure of 1 million years. The *Pleistocene-Holocene* boundary is best defined either as the end of the last rapid rise of sea level between 6,000 and 8,000 years ago or as the time of deglaciation of the midlatitudes about 10,000 years ago.

DIVERSE EFFECTS OF GLACIATION

GENERAL

With nearly one-third of the present land area of the globe covered by roughly 43 million cubic

FIGURE 18.2
Glacier-enshrouded Antarctic Peninsula near Chilean scientific base Gabriel Gonzalez Videla. This region, still very much under Pleistocene climatic conditions, lies within a Mesozoic and early Cenozoic mobile belt closely related to the Andes Mountains of Chile. Glaciation began in interior Antarctica in middle Cenozoic time, but temperate forests existed near this site until the Miocene. (*Courtesy M. Halpern.*)

MAXIMUM GLACIATION AND LOWERED SEA LEVEL

major
ice caps

seas

warm
ocean currents

flow
of moist air

FIGURE 18.3
Maximum extent of Pleistocene glaciers and corresponding minimum level of seas together with warm ocean currents that were major sources of moisture for nourishing ice caps (prevailing moist air currents in broad arrows). Note that very cold Siberia was little glaciated because it is too dry. (*Adapted from Antevs, 1929, Bulletin of the Geological Society of America; Wills, 1951, A palaeogeographical atlas, Blackie & Son Ltd.; Goode's world atlas, 1964; Bartholomew's Atlantis Equal Area Projection map used with permission.*)

kilometers of ice during maximum Pleistocene glaciation (Figs. 18.3 and 18.4), it is not difficult to imagine that profound effects resulted. Practically no corner of the globe—not even the deep seas—escaped at least some influence of Pleistocene climate (Fig. 18.5); paleobotanical studies show that tropical land areas were least affected. Direct influences of glacial ice, such as its erosive power, deposition of erratic boulders and moraines, and the like, are well known, but certain other phenomena deserve special mention.

ISOSTASY

Depression of the crust by the load of ice caps up to 3,000 meters thick, and the crustal rise or *rebound* (Fig. 18.6) upon their melting provide convincing proof of the general validity of the principle of isostasy and of the ability of mantle material to flow plastically as the crust is warped epeirogenically (see Chap. 6). Actual evidence of postglacial rebound consists primarily of raised ocean and lakeshore features such as beaches. Around the Great Lakes and the northern Baltic Sea, for example, such raised features are not level, but are tilted southward. This reflects thicker ice at the northern ends of these areas, thus greater crustal depression followed by greater rebound. Rebound continues today, indicating that, by ordinary human standards, it is a slow process spanning 10,000 years or more. If Greenland and Antarctica were to be deglaciated, they too would rise. Figure 1.8 shows differential rise of the Baltic region.

LAKES

Glaciation had a marked effect upon drainage patterns and produced many temporary lakes. Preglacial drainage valleys were gouged by the ice to leave large depressions subsequently filled with water; the Finger Lakes of New York are examples. The Great Lakes basins are of similar origin, and had a complex history during advances and retreats of the ice. Huge postglacial Lake Agassiz—four times as large as Lake Superior—formed temporarily in the southern Canadian Plains when the ice began to

retreat (Fig. 18.4). At various times it drained into the Mississippi River (Fig. 18.6), later into Lake Superior, and finally into Hudson Bay. Similarly, the Great Lakes spilled over and drained in various directions as the ice retreated (Fig. 18.6). As lower outlets emerged from beneath the ice, former drainage routes were abandoned.

Large lakes formed from three to five times in the northern Rocky Mountains. Immense terraces of coarse gravel and sand over 30 meters high line the upper Columbia River valley in Canada, attesting to staggering volumes of meltwater. In western Montana, a large basin filled several times to form Lake Missoula, which was as much as 300 meters deep (Fig. 18.7). About 18,000 years ago its ice-and-moraine dam in northern Idaho suddenly broke. A wall of water rushed across eastern Washington with incredible velocity. This gigantic flood scoured channels and deposited immense gravel ripples 10 meters high over a large part of the Columbia Plateau, now called channeled scablands because of the peculiar topography left by the flood. Lake Bonneville in Utah and a smaller lake on the Snake River overspilled their lava dams almost simultaneously about 30,000 years ago to create a giant flood down the Snake River (Fig. 18.7). In all of these floods, flow velocities probably reached 80 kilometers per hour (50 miles per hour); huge boulders were tossed about as if they were tiny sand grains. Such floods lasted only days or a few weeks at most. Wherever great volumes of water were released, canyons were deepened rapidly, and some were abandoned after the ice disappeared (Fig. 18.8).

During much of Pleistocene time, accumulation of water in low depressions also formed large lakes far from continental glaciers. Most notable were those of the Great Basin region of western United States (Fig. 18.7). Today, under a warmer and more arid climate, only small, saline bodies, such as Great Salt Lake, Pyramid Lake, and playa or alkali flats remain as remnants of once much larger and far more numerous lakes. These were formed in closed basins between the youthful block-faulted ranges of the Utah-Nevada-California region (Fig. 18.9).

GREENLAND CENTER

BERING LAND BRIDGE

BAFFIN CENTER

CORDILLERAN ICE SHEET

KEEWATIN CENTER

LABRADOR CENTER

SUBPOLAR TEMPERATE

Lake Agassiz

CASCADES

SIERRAS

TEMPERATE SUBTROPICAL

PANAMANIAN LAND BRIDGE

PLEISTOCENE

Maximum Glacial Advance

and

Minimum Sea Level

Direction of ice movement

Hot, dry-climate flora

0   1000 Miles

0   1000 Kilometers

FIGURE 18.4
Pleistocene paleogeography of North America showing maximum ice advance and retreat of the sea, immediate postglacial Lake Agassiz, Cascade and other volcanoes, and late Pleistocene desert floras in the southwest. Joining of Cordilleran ice caps with main Canadian-Shield-cap in western Canadian plains blocked migrations of organisms to and from Asia during glacial advances. Note Bering and Panamanian land bridges. (Contrast climatic patterns with Fig. 17.11.)

The largest were glacial Lake Bonneville in Utah and Lake Lahontan in western Nevada, each of which covered up to 50,000 square kilometers and was as much as 300 meters deep. Lake levels fluctuated enormously several times as climate oscillated (Fig. 18.10). Such playas are important sources of rare elements like boron and lithium.

WIND EFFECTS

With much meltwater pouring from wasting glaciers near the end of glacial episodes, tremendous quantities of sediment were dumped in front of receding ice fronts. Rivers became so sediment-choked that they formed mosaics of interlaced or braided channels. Wind then worked over the sediment bars along such channels, winnowing out fine sand and silt. Huge dune fields were formed, as in the Sand Hills of Nebraska, and large areas of fine dust

(loess) accumulated near major river valleys (Fig. 18.6). As the ice front retreated farther, vegetation became reestablished, and wind effects gradually decreased.

WORLDWIDE CHANGES OF SEA LEVEL

Worldwide fluctuation of sea level in response to glacial oscillations was one of the most profound phenomena of the Pleistocene. *Submerged beach ridges* far out on continental shelves indicate a maximum reduction of sea level on the order of 100 to 140 meters. Elephant teeth 25,000 years old have been dredged by fishermen from more than 40 sites as far as 130 kilometers off the Atlantic coast and in water as much as 120 meters deep. Higher-than-present sea levels are recorded by *dead coral reefs* on several oceanic islands. Melting of all existing glaciers, especially those of Antarctica with 90 percent of present ice, would raise sea level about 65 meters above present level; this would flood nearly 15 percent of existing land area!

Major drops of sea level occurred at least four times, with the greatest about 40,000 years ago. Short-term oscillations were superimposed upon more subtle, long-term worldwide Cenozoic regression of the sea. It is interesting to note that the last rise (Fig. 18.11) operated to produce universal transgression in opposition to the regressive tendency caused by isostatic rebound along many high-latitude coasts. Hud-

FIGURE 18.5
Iceberg-rafted cobbles dropped onto the sea floor 4,000 meters below sea level between Antarctica and South America; such cobbles have been carried as far as 3,000 kilometers from Antarctica. Similar ancient rafted pebbles also are known (e.g., Fig. 8.25). (*Official NSF photo, USNS Eltanin, Cruise 10; courtesy Smithsonian Oceanographic Sorting Center.*)

Late Pleistocene in North America

Wind Deposits (mostly loess)

Glacial Lakes

Spillways (active at different times)

—150M— Postglacial isostatic rebound (meters)

FIGURE 18.6

Some late Pleistocene phenomena associated with retreat of the Wisconsinan ice sheet (approximately 10,000 to 15,000 years ago). Ice-front lakes overflowed via temporary spillways, but these were not all active simultaneously. Forest zones at the time of glacial maximum are shown at the bottom; these shifted northward as ice retreated. Main areas of wind-blown silt (called loess) surround major glacial meltwater drainage systems. Compare Fig. 18.4. *(After Hough, 1958, Geology of the Great Lakes; King, in Wright and Frey, eds., 1965, The Quaternary of the United States; Whitehead, 1973, Quaternary Research, v. 3.)*

son Bay, most of which lies less than 600 meters below sea level today, was drowned by rapid rise, but it appears destined to become largely land as crustal rebound continues (Fig. 18.6). Here is another example of the interaction of local versus universal effects operating on different time scales.

Sea-level changes had widespread effects extending from well inland, on the one hand, to the deep seas, on the other. Reduction of sea level lowers effective base level of major rivers. This causes downcutting, which produces *river terraces* that correlate with former *marine beaches* and deltas (Fig. 18.12). In coastal regions, vast expanses of former shallow seas were laid bare, in several cases forming important *land bridges* for migration of organisms between adjacent continents or islands. Finally, lowering of sea

FIGURE 18.7

Pleistocene lakes and giant floods in western United States. Lake Missoula was dammed by ice, lakes Bonneville and American Falls by lava flows, and the two westernmost ones by debris temporarily choking tributary valleys. All others resulted from drainage into closed valleys in the Basin and Range region. Each of those rose during colder episodes and fell during warmer ones, due mainly to changing evaporation rates. *(After Feth, 1961, U. S. Geological Survey Professional Paper 424-B; Weiss and Newman, 1973, The channeled scablands of eastern Washington; U. S. Geological Survey pamphlet.)*

so that rivers swollen many times during early melting of the ice could carry vast quantities of sand to the very edge of the shelf. Ordinarily, most such sediment would be trapped on the inner shelves, but with little or no submerged shelf, much sand could escape to the deep seas. It is thought by most authorities that gushing, muddy meltwater torrents stimulated the cutting of *submarine canyons* that nick most continental shelves, though the deeper ends of the canyons extend far below even the lowest glacial sea level. Below that level they must have been cut by submarine currents and mud flows. Once cut, the canyons provided funnels for continuing export of sediment from the shelves onto abyssal plains by turbidity currents.

FIGURE 18.8

Ouimet Canyon near Nipigon, Ontario, on the north shore of Lake Superior. Nearly 200 meters deep, this gorge was completely abandoned after the deglaciation of southern Canada. Apparently this was one temporary outlet of glacial Lake Agassiz (see Fig. 18.4). (Person in upper left corner provides an idea of scale.)

level made possible a much greater-than-normal delivery of relatively coarse clastic material to the deep seas. The total rate of sedimentation throughout large portions of the ocean basins must have accelerated greatly. The greatest drops of sea level carried the shoreline out to continental shelf margins (Figs. 18.3 and 18.4)

FIGURE 18.9
Great Salt Lake, Utah, with Stansbury Island in the foreground and Wasatch Mountains behind. The lake is a briny relict of much larger Pleistocene Lake Bonneville. Notch on side of mountain in foreground marks a high shoreline of Lake Bonneville (arrows). Evaporation due to warmer postglacial climate has produced many smaller saline lakes and playas (alkali flats) in the Basin and Range region (see Fig. 18.7). Most modern evaporites are forming in such nonmarine settings (see Fig. 13.23). Playas may contain economically important alkali deposits. (*Courtesy L. J. Maher.*)

FIGURE 18.10
At least nine former beach levels of glacial Lake Lahontan visible on perimeter of Pyramid Lake, northwestern Nevada. Lake Lahontan, like other Pleistocene lakes in the west, rose at least four times and then shrank in response to Pleistocene climatic oscillations (see Fig. 18.7).

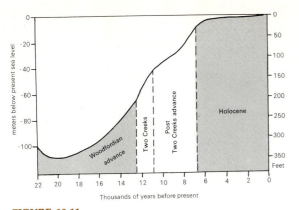

FIGURE 18.11
Curve representing the last rise of sea level to its present position in late Pleistocene time. Note comparison of episodes with Table 18.1.

tion, than for earlier episodes. Full understanding of Pleistocene history requires correlations of deposits, first among glaciated areas and then with unglaciated ones. The same principles and problems encountered in correlation and interpretation of more ancient strata apply.

Perfection of *carbon 14 isotopic dating* methods about 1950 provided a breakthrough for late Pleistocene chronology. Deposits from widely separated localities containing wood, charcoal, bone, or calcareous shells could be dated and correlation was thus enhanced. Even carbon-bearing deep-sea sediments could be correlated with the glacial standard time scale established on the continents. Now it became possible to compare in detail the effects of glacial conditions on the continents with conditions in the oceans. At the same time, ^{14}C dating became very

PLEISTOCENE CHRONOLOGY

As mapping of glacial features progressed in North America and Europe, the different glacial and interglacial deposits were differentiated and named. In central North America, four major glacial advances were recognized long ago, and in Europe, from three to five advances, depending upon location. Apparently each major glacial episode was in turn made up of lesser oscillations of the ice. Details are, of course, much clearer for the last, or Wisconsinan glacia-

FIGURE 18.12
Effects of sea-level changes on river valleys and coast lines. *Left:* Initial drop of sea level caused cutting of main valley and formation of now-submerged shore features; a rise, then a second fall, and a second rise also are recorded. The final rise produced a branching delta, present beaches, and valley alluvium. *Right:* Recent drop in sea level caused downcutting into all previous valley alluvium, seaward migration of crescentic delta-bar, and left several high terrace levels reflecting earlier events (terraces numbered in order from oldest to youngest). (*Adapted from* Scientific American, The Bering Strait land bridge, *W. G. Haag, January, 1962; Copyright© 1962 by Scientific American, Inc.; used by permission.*)

FIGURE 18.13
Idealized representation of typical sequences from four different types of regions showing different records of glacial and interglacial episodes in each as well as means of correlating among them. (A—glacial advance, R—glacial retreat; C—cold; W—warm; numbers represent ¹⁴C dates.)

important to archaeologists studying the last 65,000 years of human history. Attempts continually are under way to extend isotopic dating back farther in time from 65,000 to 100,000 years, at which age the K–Ar method becomes applicable. Protactinium 231–thorium 230 dating is being used for deep-sea clays back to 300,000 years ago. *Fission-track dating* is also used (see Chap. 4), as is dating by the *degree of alteration* (hydration) of the volcanic glass obsidian, which is not an isotopic method. Obsidian alteration dating is new, but offers promise for glass fragments in volcanic ash and in certain human artifacts. Dating by comparing *magnetic polarity reversals* in Pleistocene deposits is also used. But even with numerical dating

becoming increasingly available and more reliable, it still is necessary to correlate Pleistocene strata by other, more conventional means as well. Figure 18.13 shows diagrammatically some ways that correlations are accomplished among different types of Pleistocene sequences. In deep-marine sediments, temperature-induced *reversal of coiling direction* of foraminiferal shells (Fig. 18.14), together with *oxygen-isotope* analyses (described in Chap. 16), provide bases for constructing temperature fluctuation curves that can be used for correlation together with ¹⁴C dating of the same shells (Fig. 18.13). Changes in polarity of the earth's magnetic field leave an imprint on magnetically susceptible minerals in sediments. Using the standard time scale of magnetic polarity reversal events, these can now be correlated from place to place (see Fig. 7.13).

Several persistent and obvious Pleistocene *volcanic ash* layers occur over western North

FIGURE 18.14
Contrast of coiling directions of microscopic shells of planktonic Foraminifera *(Globorotalia truncatulinoides)* from deep-sea Pleistocene sediments of the North Atlantic; right-hand coiling (left) occurs in waters warmer than 8 to 10°C, whereas left-hand coiling (right) characterizes cooler temperatures. *(After D.B. Ericson et al., 1954,* Deep sea research; *by permission of Pergamon Press.)*

America, and they provide invaluable datums for correlation. Some of the best known of these extend through the Great Plains from El Paso, Texas, to North Dakota. Most of the widespread ash layers were derived either from Yellowstone or from the Cascade volcanoes. Careful geological detective work has shown that the ashes are mineralogically distinctive enough so that their sources can be pinpointed accurately (Fig. 18.15). The final eruption of Mt. Mazama in Oregon, which created Crater Lake (Fig. 18.16), was witnessed by Indians, for its ash buried some sandals that have been dated isotopically as 6,700 years old. Lavas interstratified with glacial tills may provide indirect means for dating glaciations. For example, K–Ar dates on lavas in the Sierra Nevada date an interstratified till as 3 million years old, probably the oldest known Cenozoic glacial deposit in middle latitudes.

Table 18.1 shows a comparison of the classic stratigraphic classifications of Pleistocene records for central North America and Europe as evolved through the application of procedures illustrated above. The chronologies, of course, apply strictly only to the glaciated areas where they were first established. Not many years ago, the Pleistocene Epoch was assumed to be encompassed at most within a bare 1 million years. But deep-sea cores and K–Ar dating now suggest that it was at least twice that long, that is, 2.0 to 2.5 million years. This revelation has required some drastic revision of concepts of subdivision and correlation of the older divisions.

As ^{14}C dating increased in the sixties, it became apparent that the record for the last, or *Wisconsinan*, glacial episode was much more complete than for earlier ones. Perhaps this should not have been surprising, for repeated erosion both by ice and glacial rivers together with oxidation would inevitably degrade the older deposits. But the degree of disparity of the two records was unexpected. In the seventies, doubts increased about the earlier episodes (Illinoian, Kansan, Nebraskan). Did the deposits for these represent a complete record of glacial-interglacial events as long assumed? Were those events of equal duration? In short, what are the numerical ages of those events? Widespread volcanic ash called Pearlette in the Plains region was found to include four or five different ash layers rather than a single one as long believed (Fig. 18.15). This threw many correlations into doubt, and recent dating using fission-tracks, obsidian alteration, and magnetic polarity reversals suggests that most of the earlier half of the Pleistocene Epoch may have no record whatsoever in the glaciated region of North America (Table 18.1). Therefore, the classic glacial sequence now must be questioned, for neither the numbers nor the exact numerical dates of pre-Wisconsinan events are clear.

The deep-marine sequences should provide a more complete record of alternations between colder and warmer intervals, and from that record it appears that there were many more than the classic four Pleistocene midlatitude

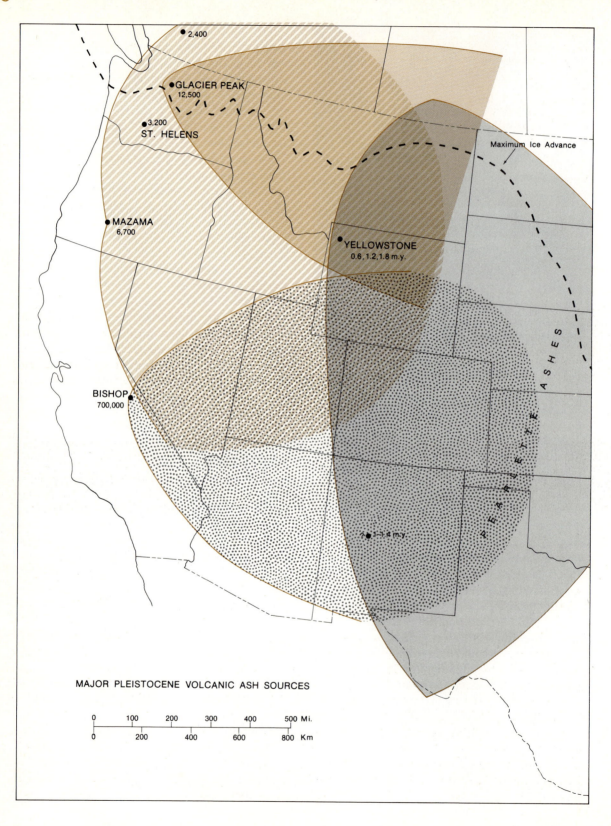

• 2,400

●GLACIER PEAK
12,500

● 3,200
ST. HELENS

●MAZAMA
6,700

●YELLOWSTONE
0.6, 1.2, 1.8 m.y.

Maximum Ice Advance

BISHOP●
700,000

P E A R L E T T E   A S H E S

●1.4 m.y.

MAJOR PLEISTOCENE VOLCANIC ASH SOURCES

0	100	200	300	400	500 Mi.
0	200	400	600	800 Km	

FIGURE 18.15

Major sources and fallout patterns of Pleistocene volcanic ash in western North America. Ash layers provide marker datums for correlation of Pleistocene deposits between widely scattered areas, especially from unglaciated into glaciated country. Glass and mineral fragments in the ash are numerically datable. (*Adapted from Wilcox, in Wright and Frey, eds., 1965*, The Quaternary of the United States.)

glacial episodes (Fig. 18.17). For polar regions we expect more, and the record verifies glaciation back to Miocene time. But what about lower latitudes? What is now needed are reliable correlations between well-documented deep-sea sequences and the classic glacial sequences on land (compare Table 18.1 and Fig. 18.17). While

FIGURE 18.16

North across Crater Lake, Oregon, showing north-south alignment of Cascade volcanoes in the distance. Crater Lake formed when former Mt. Mazama suffered a cataclysmic eruption. Immense volumes of ash were blown hundreds of miles eastward by prevailing westerly winds (see Fig. 18.15). The top of the mountain then collapsed, forming a caldera, which filled with water. Ash from Mt. Mazama and other Cascade volcanoes provides important marker layers for correlation of late Pleistocene deposits over many western states.

progress has been made, it will be awhile before the correlations can be fully accepted, for the marine record has its problems too. It tends to be compressed because of the relatively slow rate of sedimentation in the deep seas, which makes fine discrimination difficult. Also, the oxygen-isotope data are not straightforward. For example, the $^{18}O/^{16}O$ ratio is sensitive to ocean salinity (saltiness) as well as to temperature. Therefore, some of the variation observed reflects overall changes of oxygen-isotope composition of sea water due to episodic influxes of fresh glacial meltwater and not just temperature. With care, it is possible to adjust for this effect. Another problem is the selective dissolving of shells deposited on the deep-sea floor where calcium carbonate is unstable. The shells most affected are those of animals that lived at shallower—and warmer—depths, therefore loss by solution results in an error that makes the apparent temperature appear too cold. Finally, even if we can accept each marine oxygen-isotope cold interval as valid, we cannot be

TABLE 18.1
Classic Pleistocene stratigraphic classification standards for glaciated North America and Europe (numbers indicate years before present where known). (After Black, and others, 1973; Boellstorff, 1978)

North America (Upper Mississippi Valley)				Europe (Alps and Southern Baltic Regions)
HOLOCENE EPOCH		(0–7,000 or 10,000 years ago)		HOLOCENE EPOCH
Wisconsinan glacial	Greatlakean advance (8,000–11,000) *Twocreekan retreat* (11,800–12,500)		Würm glacial	Younger Dryas advance (10,000–11,000) *Allerod retreat* (11,000–12,000)
	Woodfordian advance (12,500–22,000) *Farmdalian retreat* (22,000–28,000)			Older Dryas advance (13,000–20,000) *Riss-Würm retreat* (20,000–30,000+)
	Altonian advance (28,000–70,000+)			Riss advance (30,000–60,000+)
Sangamonian interglacial				*Mindel-Riss interglacial*
Illinoian glacial				Mindel glacial
Yarmouthian interglacial				*Gunz-Mindel interglacial*
Kansan glacial				Gunz glacial
Aftonian interglacial (0.6–0.7 m.y.)				*Donau-Gunz interglacial*
Nebraskan glacial *volcanic ash* (2.2 m.y.)				Donau glacial
Pre-Nebraskan glacial				
PLIOCENE EPOCH		(2.0–2.5 m.y.)		PLIOCENE EPOCH

(Note: PLEISTOCENE EPOCH labels the vertical span from below HOLOCENE EPOCH through above PLIOCENE EPOCH on the left margin.)

certain which ones actually resulted in significant glaciation (Fig. 18.17).

WHAT HAS CAUSED GLACIATIONS?

CLIMATOLOGIC BACKGROUND

By 1850 or so, the principal phenomena associated with Pleistocene glaciation had been recognized. Naturally, speculation about the cause of glaciation began, but even a century later, there is still much doubt. Before discussing several alternate working hypotheses, it is important to state some important climatological background.

First, *continental glaciations have been rare events in earth history*. Besides the well-known late Cenozoic ice ages (0 to 20 m.y. ago), there was widespread glaciation in the southern continents in late Paleozoic time (230 to 350 m.y. ago), in South America in Siluro-Devonian time (around 410 m.y. ago), and in North Africa during the Ordovician (about 450 m.y. ago), as well as the very widespread one in Eocambrian time (circa 700 m.y. ago). Prepaleozoic episodes are not so well documented (see Chap. 8). The time period between the recognized great glacial events is irregular, varying from 50 to 1,000 million years, and so it is difficult to claim that glaciation is a regularly cyclic phenomenon of the earth. Because of its rarity, glaciation must have required a very special combination of conditions.

A second important observation is that all

COLDER ←——→ WARMER

MAGNETIC REVERSAL TIME SCALE

Possible Glaciations

FIGURE 18.17

Late Pleistocene standard marine paleotemperature curve (left) based upon oxygen-isotope analyses of calcium carbonate in microfossil shells from deep-sea cores of three oceans. Magnetic polarity measurements on the same cores (right) and limited isotopic dating of cores provide a time scale. The only two well-dated divisions of the classic North American midlatitude continental glacial sequence, Wisconsinan and Aftonian, are shown. The remainder of that sequence (e.g., Illinoian, Kansan, Nebraskan of Table 18.1) cannot yet be correlated with the marine curve, and there is no assurance that during each cold interval shown here there was glaciation at midlatitudes. Note that for the last 800,000 years, cold intervals had a periodicity of about 100,000 years; from then back to about 1.4 million years, the period was about 40,000 years. (J—Jaramillo brief normal polarity event.) *(Adapted from Emiliani and Shackleton, 1974, Science, v. 183, pp. 511–514, and Shackleton and Opdyke, 1976, Geological Society of America Memoir 145, pp. 449–464.)*

available evidence, incomplete as it is, indicates that *the "normal" climate during the past 1 billion years was milder than that of the last 40 million years.* Present lower-latitude regions offer closest modern analogies with typical climate of the past, but even they were not unaffected by late Cenozoic cooling.

The third point, really a corollary to the second, is that *past climatic conditions typically were not only milder but also more uniform over the earth.* Although the overall mean temperature of the earth cannot have changed greatly, temperature and moisture gradients were less abrupt, and climatic zones tended to be broader. Nonetheless, because of the geometry of the earth-sun system, polar areas must *always have been relatively cooler* than lower latitudes, although it does not follow that there always was ice at the poles. Polar areas would be coldest when continents lay at or near them because land has a lesser heat capacity than does water (see Fig. 11.26). Conversely, poles would be relatively milder whenever open, well-mixed seas lay there. This important paleoclimatic principle was appreciated and discussed at length by Lyell over 100 years ago.

The fourth point to note is that *magnitude of overall average yearly temperature differential between a normal and a glacial climate is only on the order of* 10°C. That is to say, as noted in Chap. 5, the earth enjoys a very critical temperature position in the solar system and has a sensitive thermal budget. It is estimated that a drop on the order of only 4 to 5°C from the present mean annual temperature could cause renewal of continental glaciation (the present must be considered as interglacial rather than truly normal, however). This is confirmed by $^{18}O/^{16}O$ ratios and coiling directions of Foraminifera from deep-sea cores (Fig. 18.17), which indicate that the average sea-surface temperature 18,000 years ago—near the peak of the last glacial advance—was only 2.3°C less than now; maximum differentials were 6 to 10°C less. Such modest differences suggest that the earth's heat budget is very sensitive, which in turn might lead one to wonder why glaciations have not been more common. The answer must be that a rather stable thermal equilibrium has character-

FIGURE 18.18

Global temperature-decline curves for latest Cretaceous and Cenozoic times based upon fossil plants from North America and Europe (upper curve) and oxygen-isotope studies of shells from deep marine sediments (lower curve). *(Adapted from Dorf, in Nairn, eds., 1964,* Problems in paleoclimatology; *Savin, et al., 1974,* Geological Society of America Bulletin, *v. 86, pp. 1499–1510.)*

ized most of geologic time, and, because the total heat budget has varied only within a narrow range, it follows that glaciation requires not wholesale change in kind but only in degree!

LATE CENOZOIC CLIMATIC DETERIORATION

Some evidence for the decline of overall temperature was presented in the last chapter. Long-known land-fossil evidence of cooling is supplemented with morphologic and oxygen-isotope paleotemperature data from calcareous shells in marine sediments. Figure 18.18 shows the general trend of land and sea temperatures for the past 70 million years. The striking shifts of plant communities in arid southwestern United States and Mexico introduced in the last chapter continued into Pleistocene time (compare Figs. 17.4, 17.11, and 18.4). Whatever the cause, it is clear that a steady temperature decline occurred beginning 35 to 40 million years ago, until a critical threshold level was reached when ice caps began to form and grow. Glaciation began at both poles in mid-Cenozoic time. Oligocene and Miocene fossil plants suggest cooling temperatures, and local Miocene tills are known near both poles; some ice-rafted pebbles appear in Miocene deep-sea sediments near Antarctica. D. L. Clark of Wisconsin has shown that the

Arctic Ocean probably has been ice-covered for approximately the past 15 million years. It is inescapable that, by Pliocene time, polar and alpine ice caps were sizable, as evidenced by the appearance of ice-rafted pebbles widely in Pliocene deep-sea sediments surrounding Antarctica (Fig. 18.5) and by the 3-million-year-old till in the Sierra Nevada.

The *critical threshold* necessary for glaciation was reached when annual snow precipitation slightly exceeded summer wastage by melting and evaporation for a number of years. This amounts to saying that if the *permanent snow line* (above which not all snow melts in the summer) descends sufficiently in elevation or latitude, ice can begin to form. The difference between present levels of descent of glaciers relative to that of past glacial episodes shows effects of fluctuation of overall temperature on snow-line changes (Figs. 18.19 and 18.20). Apparently the snow line was lowered an average of from 600 to 1,000 meters during glacial episodes. The snow-line level is critical in the original area of ice-cap formation, but once a glacier becomes large, it may flow to lower elevations or latitudes well below the snow line. If rate of flow is sufficient, the ice front will advance in spite of relatively warm temperatures. This accounts for the seeming anomaly of finding mild-climate fossils very near obvious glacial moraines. A notable example exists on the equator in East Africa, where glaciers from the 5,113-meter-high Ruwenzori Mountains left terminal moraines on the edge of eastern Congo jungles.

FIGURE 18.19

Elevation versus latitude of terminal moraines during maximum glaciation and at present; dark gray band shows chief latitudinal span of major Pleistocene continental ice sheets.

FIGURE 18.20
Grand Teton Range and Snake River viewed across Jackson Hole valley. The Tetons were carved by glaciation from Prepaleozoic metamorphic rocks that had been raised along a late Cenozoic normal fault. Larger glaciers flowed out onto the edge of the valley; only one puny relict glacier remains since the warming of climate ensued several thousand years ago. Prominent river terraces along the Snake River reflect the Pleistocene climatic oscillations.

Fossil crocodiles and giant land turtles occur in Pliocene and Pleistocene sediments of the Plains and southern United States. These were cold-blooded creatures requiring above-freezing temperatures practically all the time. Their distribution suggests relatively warm, moist conditions as far north as South Dakota in early Pliocene time. In late Pliocene, they still ranged as far north as Kansas; the Plains region was more humid than now and had scattered forests. Even during the first two glacial episodes, turtles and crocodiles still lived in the southwestern Plains region, and during interglacial times they again ranged north to Kansas and Nebraska. It is apparent that a really harsh climate did not grip most of the United States until the last (Wisconsinan) glacial advance. That episode actually involved three well-dated fluctuations of the ice (Table 18.1). A number of significant fluctuations within postglacial times also are recorded by fossil and archaeological evidence. The last ice was retreating from midlatitudes by 10,000 years ago, and the warmest postglacial climate occurred between 5,000 and 7,000 years ago. Soon thereafter early human civilization approached its zenith in Asia Minor. After some fluctuations, cooling recurred from about A.D. 1500 to 1880 and from then to 1950 there was a period of amelioration characterized by warmer and drier conditions. Glaciers have retreated since 1890, but cooling again is indicated by recent advances in many mountainous areas. Many fluctuations of climate in historic times have occurred, but they have widely differing time periods. At least the shorter ones seem to be controlled by subtle meteorological changes in atmospheric circulation patterns. We know that, historically, alpine glaciers have been extremely sensitive to such shorter-term changes; therefore, large continental ice sheets must have responded similarly (but on a much larger scale) to longer-period changes.

TYPES OF GLACIAL HYPOTHESES

Two principal phenomena associated with Pleistocene glaciation require explanation. First is the general deterioration of climate down to a threshold for ice-cap accumulation just discussed, and second is the oscillatory or periodic nature of glacial expansion and retreat. Note that major colder and warmer episodes appear to have been rhythmic with a period of roughly 100,000 years (Fig. 18.17). It is natural to assume that both aspects of Pleistocene glaciation were produced by one mechanism, but the great rarity of continental glaciation suggests an *unusual chance combination of several factors*.

Many working hypotheses and speculations have been offered to explain glaciation, but they can be conveniently grouped into a relatively few types as follows:

1 Changes in solar radiation
2 Astronomical or orbital effects involving changes of earth-sun geometry
3 Terrestrial changes affecting net heat budget:
    a By changes of atmospheric transparency (e.g., dust, clouds, etc.)
    b By changes of relative land, sea, or snow areas and/or elevations
4 Variations of heat exchange from equatorial to polar regions due to paleogeographic changes

Hypotheses of the first type, while plausible, are difficult to test because no adequate absolute measure of solar radiation has been made. Moreover, such measurements would be needed for a very long time span. Apparent fluctuations of ^{14}C generation suggest solar fluctuations, but appeal to sunspot cycles seems inval-

id because no corresponding patterns have been verified by analysis of tree rings or annual layers (called varves) in glacial lake sediments. We do not deny the possibility of solar variations as the major cause of glaciation and other climatic changes, but it is fruitful to see if other factors can be discovered that are more readily testable by geologic means.

ASTRONOMICAL OR ORBITAL HYPOTHESES

The second category of hypotheses has received a great deal of attention for more than a century. In the 1860s, a humble but clever Scot named James Croll, who was then working as a janitor in a Glasgow university, proposed that orbital variations discussed below caused the alternations of glacial and interglacial climate. His ingenious hypothesis was largely ignored, however, until in 1920 a Yugoslavian meteorologist, M. Milankovitch, made further calculations of the net solar radiation variations due to three known cyclic parameters of the earth's orbit. These parameters are variations of *eccentricity* of the orbit, *tilt* of the axis, and a *wobble* of the axis due to changing gravitational interactions of the moon and sun on the earth's equatorial bulge (also known as precession of equinoxes or longitude of the perihelion). The periods of these perturbations in the earth's movements are shown in Table 18.2

Because the periods of these separate phenomena are different, there can be only a few times when their effects reinforce one another to

TABLE 18.2
Milankovitch Orbital Factors

Parameter	Variation	Approximate periods
Eccentricity of the orbit (ellipticity)	0.017–0.053	93,000 years
Inclination of the axis (obliquity)	$21\frac{1}{2}$–$24\frac{1}{2}°$	41,000 years
Precession of equinoxes (wobble)	0–360°	21,000 years

cause unusually great or small net solar radiation. For example, if the earth simultaneously had its smallest angular tilt, maximum orbital eccentricity, and also was at its greatest annual distance from the sun during the Northern Hemisphere summer, then a small but significant lowering of net radiation in the north would occur. Some meteorologists believe that the tilt of the axis is of greatest importance because of its effect upon the angle of incidence of solar radiation, and therefore upon net radiation, especially at high latitudes (see Fig. 11.25). Wisconsin climatologist R. A. Bryson believes that the north-south temperature gradient, determined by contrasts in net radiation between poles and equator, is the most significant single climatic parameter. It affects not only net radiation, but also large-scale atmospheric circulation patterns.

In 1922, famous German climatologist W. Köppen revived Croll's hypothesis by suggesting that Milankovitch's radiation curve for the past 600,000 years might match the glacial advances and retreats. Ever since, there has been a great deal of interest in this possible relationship, and a number of hypotheses have invoked it to help explain glaciation. Subsequently the Milankovitch curve has been refined. In the 1960s, Cesare Emiliani pioneered oxygen-isotope paleotemperature determinations for deep-sea sediments (Fig. 18.17). He and others compared the resulting marine cold-warm alternations with the Milankovitch curve and noted a rather close correspondence between their periods of variation. During the 1970s, there was a massive cooperative research program called CLIMAP, which involved dozens of scientists in the investigation of Pleistocene climate using oceanographic data primarily. This program was so successful that it will now be extended back through the entire Cenozoic record. One result of CLIMAP was a sophisticated mathematical analysis of the frequencies of several different parameters observed in deep-sea cores. In 1976 it was shown conclusively that the principal period of variation for the past 500,000 years has had a frequency of about 100,000 years (see Fig. 18.17). This dominant variation seems to be due

primarily to the variations of *orbital eccentricity*, but it is doubtless also modulated or influenced in a complex way by other variations. Less-pronounced periodic features of the marine sequences have periods of 23,000 and 42,000 years each, which are very close to those of the other two Milankovitch factors (Table 18.2). Thus the reality of the Milankovitch effect as the control of Pleistocene glacial-interglacial alternations is strongly supported. Presumably the effect has operated over all of geologic time, yet glaciation has not occurred every 100,000 years in earth history. The total effect is rather small, causing a change of annual temperature of only 10°C or less, so could leave a mark in the historical record *only when some other factors also operated* to reduce overall average temperature. In other words, only when the earth was made susceptible to glaciation by other factors, could the Milankovitch effect operate to cause expansion and contraction of glaciers. Next we shall seek those other factors.

HYPOTHESES OF CHANGING HEAT BUDGET

Albedo

Any change that shifts the balance between incoming and outgoing radiation—the heat budget—would affect overall temperature of the earth. If such changes were of sufficient magnitude, they might contribute to forming (or terminating) a glacial climate. The changes envisioned include variations of *transparency of the atmosphere* either to incoming or outgoing radiation and variations of *reflectivity of the earth's surface*. Viewed from space, the earth presents a variable face, with dark oceanic and forested areas, and light clouds, snow, and deserts. Photographs of the earth taken from space (Figs. 1.4 and 7.7) prove that much solar radiation is scattered, while a great deal is reflected (more properly, reradiated) back to space from the light-colored surfaces. The magnitude of radiation reflected from different surfaces expressed as a percentage of the incoming radiation is termed *albedo*. Albedo can be measured with great accuracy using a radiometer;

TABLE 18.3
Relative earth albedo
  Percentage of total incoming solar radiation reflected

Snow or ice	45–85%
Clouds	40–80%
Grasslands	15–35%
Bare ground; light-colored rocks	15–20%
Forests; dark-colored rocks	5–10%
Water	2– 8%

such measurements show the importance of relative albedo (see Table 18.3). Overall albedo for the present earth is about 30 percent. It is estimated that a 1 percent change of albedo would produce approximately a 1 degree change of average temperature. Put another way, only a 2.5 percent increase of total average annual cloud cover could reduce summer temperatures to their probable level during a typical glacial episode.

Transparency of the Atmosphere

Relative area of clouds obviously affects atmospheric transparency as well as earth albedo, hence overall temperature. Dust and other particles suspended in the atmosphere have similar effects. Climatologists tell us that a mere 7 percent increase of overall atmospheric turbidity would lower the average annual temperature on the order of 1°C. Between 1950 and 1965, average temperature dropped about 0.3°C, and rapidly increasing atmospheric pollution probably has contributed to this decline. Long ago it was suggested that unusual volcanic activity might have brought on glaciation by filling the atmosphere with volcanic dust. It was supposed that the Cenozoic was a time of unusual volcanism, but there is no clear correlation of more ancient volcanism with glaciations. Nonetheless, secondary effects of eruptions do occur. Three months after Krakatoa exploded in 1883 (see Chap. 1), dust from it reached Europe. Instruments in France recorded a sudden drop in solar radiation and a level 10 percent below normal was measured for 3 years. Explosion of

Mt. Katmai, Alaska, in 1912, was followed by a 20 percent radiation reduction in faraway North Africa. Dust particles play still another important climatic role. They act as potential nuclei for ice crystal formation high in the atmosphere, and so may cause cloud formation, which would further increase earth albedo.

More subtle factors that influence transparency are related to gaseous composition of the atmosphere. As was pointed out in Chap. 5, carbon dioxide, water vapor, and ozone all act as filters. Ozone filters out deadly incoming ultraviolet radiation, but molecules of all three gases inhibit escape of infrared and longer-wavelength radiation to space, producing the well-known *greenhouse effect*.[1] If all long-wave radiation from the earth's surface were to escape to space, the earth would be cold and uninhabitable.

Suggestion that changing carbon dioxide content of the atmosphere could be a major factor in climatic change dates from 1861, when it was proposed by British physicist John Tyndall. The carbon dioxide budget is complex. Additions come through igneous and hot spring activity, animal respiration, decay, and combustion. Carbon dioxide is consumed by plant photosyn-

---

[1]Solar radiation is divisible approximately as follows:

	Wavelength (micrometers[*])	Proportions	
X-rays and gamma rays	0.0005–0.01	"Short"	10%
Ultraviolet	0.2–0.4		
Visible light	0.4–0.7	"Medium"	40%
Infrared	0.7–3.0	"Long"	50%
Heat rays	3.0–3,000		

[*]1 micrometer = 0.001 millimeter.

Nearly half of incoming solar radiation is medium-wavelength visible light to which the atmosphere is transparent (see Fig. 5.11). But the earth's surface radiates infrared and long-wave energy toward the sky, to which the atmosphere is not transparent. Water vapor and carbon dioxide molecules are of such sizes as to absorb much of this wavelength spectrum, which causes them to reradiate heat that warms the atmosphere. The same principle causes solar heating of greenhouses and closed automobiles. Glass transmits short-wave solar radiation, but surfaces inside reradiate longer-wave energy to which glass is opaque; thus air inside the greenhouse or auto is heated.

thesis and deposition of carbonate rocks. Unfortunately we cannot estimate accurately changes of past CO_2 content of either atmosphere or oceans, nor is there any firm quantitative basis for estimating the magnitude of drop in carbon dioxide content necessary to trigger glaciation. Moreover, the entire concept of an atmospheric greenhouse effect is controversial, for the rate of ocean-atmosphere-photosynthesis equalization is uncertain. Until these questions are resolved, the carbon dioxide hypothesis cannot be considered satisfactory for explaining glaciation. Much less do we know whether increasing smoke and dust is likely to bring on glaciation or whether increasing CO_2 is more likely to cause melting of existing glaciers (Fig. 18.21).

The Emiliani-Geiss Hypothesis

One interesting glacial hypothesis invokes

changes of total earth albedo to explain the long-term temperature change that triggered glaciation, but couples with that the Milankovitch effect to explain smaller magnitude cyclic temperature fluctuations indicated by the glacial and interglacial episodes. As continents became larger and higher during the Cenozoic Era, authors Emiliani and Geiss argue, total albedo would have increased (Table 18.3). This could produce general temperature decline until the threshold for glaciation was reached about 15 to 20 million years ago in polar areas. Once a critical heat-budget level was reached, the Milankovitch effect then could become significant in varying average temperature slightly, but critically, for expansion and contraction of ice caps over millennia. As was shown earlier, the Milankovitch effect has roughly the same time period as do at least the last five or six cold-warm episodes (Fig. 18.17). Recently it has been shown that ice caps have an inherent tendency for instability, which has also affected their regimes. As caps expand, the wastage or ablation area increases as the ice flows out-

FIGURE 18.21

Which way will future climate go? Will increasing atmospheric dust cause readvance of glaciers or will increasing carbon dioxide cause melting of existing ice and a substantial rise of sea level? . . . Either way we lose the Big Apple. (By R. H. Dott, Jr.)

ward and downward into warmer areas. Meanwhile, the accumulation area has remained unchanged, however. A small ice cap may grow slowly to a critical size and then expand suddenly, but it may then shrink suddenly if it becomes unstable. The most critical factor is the relative elevation of the upper ablation limit versus total size of the ice cap. The paleotemperature curves suggest that glacial episodes did end rapidly.

HYPOTHESES OF PALEOGEOGRAPHIC CAUSES

It was suggested long ago that the great amount of mountain building and enlargement of continents during late Mesozoic and Cenozoic time may have altered atmospheric circulation so as to bring on glaciation. The average elevation of continents is now 800 meters above sea level, whereas at the end of the Mesozoic Era, average elevation probably was closer to 200 meters. American Pleistocene geologist R. F. Flint emphasized that an essentially worldwide uplift of continents in Miocene and Pliocene time may have contributed to the onset of glaciation. Interestingly this large-scale uplift also was accompanied by retreat of seas, therefore an increase of worldwide albedo. Result: a double cooling effect. Mountains and plateaus deflect moist air upward to colder levels and induce considerable atmospheric turbulence. The net result is to cause cloud formation and local precipitation. As already recorded, mountain building dramatically affected local climate in western North America, but, because most large continental ice sheets accumulated on lowland areas far from high mountains, this hypothesis seems of secondary importance. A number of scientists have proposed that geographic factors causing variations of equator-to-pole heat exchange controlled glacial and nonglacial climates. In all probability, such changes also were only of secondary importance. There are several variations of these hypotheses, only one of which will be mentioned.

Geophysicists M. Ewing and W. L. Donn adopted in the 1950s a combination of geographic factors. To explain general Cenozoic decline of temperature, they pointed to the paleomagnetic evidence of continental drift. While the North Pole apparently lay in or near the North Pacific until Mesozoic time, warm, mild climate persisted. But, presumably, as the pole became *thermally isolated* in the nearly enclosed Arctic Ocean, temperatures declined. Meanwhile, Antarctica had migrated over the South Pole, an even more thermally isolated position.

Continental glaciation is thought by Ewing and Donn to have begun first in Antarctica during Pliocene or earlier time and to have chilled the entire globe. Continental glaciation then began on the northern continents through evaporation of moisture from a presumably ice-free Arctic Ocean. Finally, the Arctic Ocean froze over. Ice caps ceased to advance for lack of nourishment, and melting along their southern margins soon caused retreat. As ice melted, total climate would warm, more ice would melt, and so on. But if the Arctic Ocean were reopened by climatic warming and rise of sea level, then it again might provide moisture for new northern ice caps. Presumably glacial and interglacial fluctuations should continue indefinitely until some major change either of pole positions or of polar ocean water exchange occurs.

From archaeological records, we know that between A.D. 800 and 1200, whales could move east along the Arctic coast from Alaska to northern Greenland, where they were hunted by Eskimos. From historic records, we also know that at this same time Scandinavia enjoyed very favorable climate, and the Vikings settled in Iceland, Greenland, and Newfoundland; they met (and were raided by) Eskimos in Greenland. Viking sailors reported nearly ice-free conditions as far north as Svalbard (Spitsbergen). While this suggests a semiopen Arctic coastal condition, it hardly proves a completely ice-free ocean.

Many meteorologists doubt that significant evaporation of moisture could occur even *if* the Arctic Ocean were ice free, and studies of oxygen isotopes and Foraminifera in Arctic Ocean sediments, as well as land plant fossils, all suggest a continuous major ice cover throughout the Pleistocene. Moreover, centers of the ice caps lay well south of the Arctic Ocean (Figs. 18.3 and 18.4), and practically all moisture today comes to these areas from elsewhere—the Gulf of Mexico for North America and the North Atlantic Gulf Stream for Europe.

PLEISTOCENE CLIMATIC EFFECTS UPON LIFE

RANGES OF ORGANISMS

Continental glaciation in the Northern Hemisphere is reflected in Pleistocene marine faunas as well as among land fossils. For example, species living today along the northern California coast ranged only as far north as San Diego during glacial episodes.

Worldwide changes in sea level during Pleistocene time had a great effect on coral reef growth. For example, in the West Indies, fossil coral reefs occur from 1.5 to 5 meters above present low-tide level. These reefs have been dated as interglacial, when water may have stood slightly higher than at present (alternatively, the islands may have risen). Most nearby living reefs are postglacial, being no older than 4,000 to 5,000 years, and form veneers growing on fossil reefs. Furthermore, the biota is unique to the West Indies. During earlier Cenozoic time, the reefs were closely related to Indo-Pacific and Mediterranean (Tethyan) faunas. Alternate cooling, plus many other factors, have caused the recent reef fauna to become more restricted geographically and faunally. Other reefs of the world also show the ravages of sea-level fluctuations, but not as clearly or to the same extent as in the West Indies.

Even in polar regions and in deep-water deposits, glacial and interglacial episodes are seen reflected in the microfossil floras and faunas. Evidence from deep-sea cores shows that new species and extinctions of old ones seem to correlate with alternate periods of normal and reverse polarity of the magnetic field (discussed further in Chap. 19); it is not clear why this should be so.

On land there were even more dramatic effects upon the distribution of plants and animals. For example, large logs, cones, and tree leaves dated as 30,000 years old have been dredged off the west coast of Mexico. Since no such trees grow along this very dry coast today, it is inferred that the climate has become much drier there in the past 30,000 years. In other, widely scattered areas, peculiar relict outposts of cooler-climate floras and faunas are found today, as in many scattered ranges of the southern Rocky Mountains. Cold-adapted species of trees live at high elevations surrounded by vast lower areas with forms characteristic of warmer and drier climate. Many such examples reflect past widespread cool-temperate forests that have shrunk in the past few thousand years leaving these relics, but some of these "islands" may have resulted from wind or bird transport of seeds to isolated mountain tops.

Faunal and floral changes due to glaciation were most extreme near the southern limit of the ice. Here, alternately glacial and nonglacial conditions prevailed. The Great Lakes region provides a convenient example. The last major ice advance (Woodfordian substage) culminated there about 16,000 to 18,000 years ago. Ice retreated from the southern Great Lakes by 12,000 years ago, and cold-climate spruce forests, much like those of central Canada today, reoccupied the region. A famous buried forest at Two Creeks, Wisconsin (about 11,850 years old) gave the name to this warmer interval (Table 18.1). Then the ice readvanced south a short distance to cover part of the northern Great Lakes region from about 10,000 to 11,000 years ago. The Two Creeks forest was overridden, trees were toppled and then were buried beneath till (Fig. 18.22). Spruce forests persisted, however, farther south. About 10,000 years ago the glacial climate regime ended and ice began its last retreat. The ice front paused briefly in the western Hudson Bay region about 8,000 years ago and finally completed its retreat from that region by 6,000 years ago (Fig. 18.23). As the ice left the Great Lakes region, boreal spruce forests migrated northward. By about 8,000 years ago, spruce forests were replaced by pine in the Great Lakes region. From 8,000 to about 1,500 years ago, a relatively dry period set

FIGURE 18.22
Excavation of Two Creeks Forest on Lake Michigan shore, northeastern Wisconsin. Dozens of spruce logs were buried in lake sediments, which in turn were covered by late Wisconsinan glacial till. Some stumps are *in situ*; prostrate logs were driftwood. Carbon 14 dating indicates that the logs are about 11,850 years old. *(Courtesy R. F. Black.)*

in, which is reflected in a change from pine to widespread oak-hemlock forests.

The postglacial climate culminated about 5,000 years ago, but in the past 1,500 years there have been several small fluctuations of climate that are reflected in soils, vegetation, and human and animal activities. Today the north-

ern hardwood forest covers much of the Great Lakes region with boreal spruce forests just touching its northern limits; prairie-oak forest characterizes the southwestern corner of the region. Boundaries between these entities have fluctuated, as is indicated by plant types found in vertical sequences of late Pleistocene and Holocene sediments in lakes and bogs. Wood, leaves, fruits, pollen, animal remains, and soil profiles all are employed in their study (Fig. 18.24). Magnitude of changes suggested by such data is indicated in Fig. 18.25 for Europe where a wealth of evidence is available. As noted earlier, by far the harshest climatic stress

FIGURE 18.23
Contours of carbon 14 dates in thousands of years for late Wisconsin deglaciation of eastern North America. Approximate relative timing and patterns of shrinkage of the last great ice sheet are shown by successively younger dates. Cross-hatching denotes present ice caps. *(Courtesy Bryson and Wendland, 1967,* University of Wisconsin Department of Meterology Technical Report No. 35.*)*

FIGURE 18.24
Microscopic photos of pollen grains from *Ephedra* or "Mormon Tea," a plant found in semiarid regions. Climate must be interpreted carefully from pollen, because some grains such as these are known to be blown hundreds of miles. These grains are 0.05 to 0.07 millimeter long. *(Courtesy L. J. Maher.)*

FIGURE 18.25

(Above and opposite page) Effects of climatic changes on plant communities of Europe during the last maximum ice advance about 25,000 years ago (*A*, above) compared with native vegetation of modern Europe (*B*, opposite page). (*A adapted from Brinkmann, 1960, Geological evolution of Europe, Ferdinand Enke; Wells, 1951, A palaeogeographical atlas, Blackie & Son Ltd.;* B *after* Goode's atlas, *Rand McNally Corp.*).

occurred during the last (Wisconsinan) glacial episode. Cold-blooded reptiles retreated farther south during (and reinvaded less far north after) each advance. Only the Wisconsinan episode was able to relegate them completely to the Gulf Coast region where they now reside.

LAND BRIDGES

Formation of land bridges by glacial lowering of sea level was of paramount importance to the present biogeography of the earth. The most important examples include the Bering bridge between Siberia and Alaska, the North Sea bridge to Britain, the Sunda bridge between western Indonesia and Asia, and the New Guinea–Australia bridge (Fig. 17.49). The Isthmus of Panama, second only to the Bering bridge in biological importance, was formed by structural rather than sea-level changes.

The Bering bridge deserves special mention. Fossil plant and invertebrate evidence indicates that Alaska and Siberia were connected during most of Cenozoic time. Comparisons of terrestrial biotas of all of North America and Asia indicate that two-way traffic occurred through most of the Cenozoic; west-to-east flow was twice as heavy as the opposite. In late Miocene time, the bridge was flooded, terminating connection for several million years. But during glacial advances, Alaska again was biologically more like Siberia because of its isolation from the remainder of North America by a wide ice barrier in Canada (Fig. 18.4). During interglacial high sea levels, however, it showed more affinities with interior North America when an ice-free corridor opened in the western Canadian Plains. The Bering bridge was breached 13,000 years ago and remains so today. The 90 kilometers of water that separate Siberia from Alaska hardly constitute a barrier, but the harsh climate now effectively blocks temperate-dwelling organisms from migrating.

THE END OF AN ERA

As with the close of the other eras, there was (and is) a period of extinction in late Cenozoic time. This is also a period of regression, but in contrast to the end of the Paleozoic and Mesozoic, all continents (except for Europe and Asia) are now separated. One might expect that the extreme climatic variations brought about by glaciation would cause widespread extinction. The picture, however, is not that clear. It was thought that many large mammals became extinct because of the cold; yet as we have seen, the climate almost to the ice edge in some areas was relatively mild, especially before the very last glacial episode. An alternate theory has gained some support. It suggests that humans might have been partly responsible for the extinction of the larger mammals. Over 200 genera of mammals became extinct between Late Pliocene and Recent. In some instances

FIGURE 18.26
Examples of Pleistocene mammals from La Brea tar pits, Los Angeles, California. The pits were sites of both water and tar springs. Animals came to water, occasionally became mired, and sank in quicksand. The tar preserved their skeletons, but churning of the springs disarticulated many skeletons. (*Photo courtesy American Museum of Natural History.*)

these extinctions do not coincide with periods of glaciation. In North America, many extinctions occurred well after the final glacial stage, when fossil pollen indicates that vegetation and climates were similar to prior interglacial times, but after the human hunter arrived.

Earliest Pleistocene mammalian extinctions in Africa also seem to be related to the arrival or development of Stone Age hunters. However, in reviewing actual fossil sites, the record is not clear. In Eurasia, artifacts are found along with large mammal remains; yet only four genera of mammals became extinct during man's existence there. In view of the small population of early humans, it would appear unlikely that they alone could have erased so many mammals during their existence, the later near-extinction of the buffalo notwithstanding.

Mammalian extinction began in the Pliocene with loss in North America of some artiodactyl families and the rhinos. It was among the larger types that we find the most striking extinction patterns. While some artiodactyls, such as the buffalo, survived, we lost giant beavers, the mammoth and mastodon, saber-toothed tigers, camels, horses, and many others. A magnificent cross section of a Pleistocene terrestrial fauna was recovered from the classic Rancho La Brea tar pits of Los Angeles, which are preserved in a park (Fig. 18.26).

The interglacial and glacial stages caused

rapid migrations of organisms, a process which to this day is still occurring. The range of the armadillo has been steadily expanding northward, and only recently it has crossed the Mississippi River. Similarly the opossum and raccoon have migrated north into the Great Lakes and eastern seaboard in recent decades.

THE EVOLUTION OF PRIMATES AND MAN

As we shall see in the final chapter, the most destructive and influential mammal of the Holocene fauna is man. The subject of human origins has been of intense interest ever since the first fragmentary fossil skull was unearthed in the Neander Tal (Valley) near Düsseldorf in 1856. The story is fraught with high emotions and irrational responses, and whole libraries have been written about it; much can be classified as rank speculation. To this day, the pathway leading from a primitive ancestor to modern man and his relatives remains obscure. There is good reason for this mystery, for the fossil record of the human family—the Hominidae—is pathetically small and fragmentary; most of it consists of broken jaws, broken skulls, and teeth. No classification or family tree, no matter how sophisticated, has remained undebated for long. Teams of paleoanthropologists and friends have scoured the earth intensely in the past 10 years, and the best fossil recovered to date is Lucy, an immature female skeleton 40 percent complete, found at Hadar in the Afar Triangle of Ethiopia in 1972. Even with this remarkably exquisite piece of evidence, there is a dispute about whether or not she represents a new species (*Australopithecus afarensis*) or an immature early member of a well-known species (*Australopithecus africanus*). This is a typical problem in classification when new specimens cannot be interpreted accurately by comparing them with other related specimens, which are nearly devoid, in themselves, of characteristics useful to resolve such problems.

To approach the problem of the descent of man, we need a set of characteristics or a definition of humanness. In this discussion we use the term *man* not in a chauvinistic way, but as the common name of the genus *Homo*. Two features which are accepted by most workers are brain size and bipedalism, which are interrelated. Most morphological changes in the hominids can be traced to these features and also to the basic primate character. The human brain is huge in proportion to body size, for example it is almost twice the size of a gorilla's. An elephant weighing several tons might have a brain weighing 10 pounds, while a 150-pound person might have a brain half that of the elephant. Brain size has been linked to bipedalism on the basis that as the hands were freed from locomotion, tool-making occurred, and this, along with hunting, required reasoning and complex neurological hookups giving a survival advantage to those with the largest brains. Bonding of a society followed cooperation in hunting, and from that point languages developed, or so the scenario goes.

The exact origin of the primates is unknown, but they probably were derived from the insectivores, which early shrewlike forms closely resemble. The earliest forms appeared to be arboreal (tree-climbing) lemurs and tarsiers represented in the Paleocene of Europe (Fig. 18.27). By Eocene time, these groups had become widespread and are found as disjunct records in Asia, Africa, and North America. The primates possess many primitive mammalian characteristics; for example, they retained the primitive five-digit feet, which are well suited for arboreal life. One important early-appearing characteristic was the development of stereoscopic vision, which is an advantage for judging distances and is important for forms living above the ground.

Fossil primates are a very rare part of any Cenozoic fauna, mainly because they probably inhabited forested uplands most subject to erosion. After the prosimian primates (tree shrews, lemurs, and tarsiers) experienced a rapid adaptive radiation in Paleocene and Eocene times, they became somewhat restricted and fairly conservative throughout the later Cenozoic. It is in Africa, where Eocene prosimians became well established, that major events of anthropoid evolution evidently occurred.

The earliest true apes have been found in the

FIGURE 18.27

Notharctus, an Eocene prosimian primate (lemur). *Courtesy American Museum of Natural History.)*

Oligocene Fayum beds of Egypt, some 100 kilometers (60 miles) southwest of Cairo, in an area of desert land today on the edge of the fertile and historically important Nile Valley. During the Oligocene, this region was the site of a rich, tropical forest with broad rivers and vast swamps inhabited by crocodiles, primitive elephants, rhinos, and ever-present carnivores.

The apes of the Oligocene and Miocene were arboreal fruit eaters. By the end of the Miocene, two lineages of apes spent at least part of their lives in open country. A group of great apes (dryopithecines) appeared during the Miocene in widely scattered areas; they are known mostly from fragments of the skull. These apes had rows of evenly spaced teeth and other features suggesting that they were related to the hominids (the family which includes man); they probably were early forms that might have given rise to later great apes such as the gorilla.

The most important late Miocene genus is

Ramapithecus, which is the earliest genus included in the hominids. The incisors and other front teeth were small compared to the great apes; the cheek teeth had flat, broad chewing surfaces, and are hominidlike. *Ramapithecus* fossils are found in forested areas with open spaces near rivers and lakes as well as in open grasslands. Probably these animals ate seeds and roots. It is also possible that they ate meat from animals killed by other animals, but they were not hunters. The move from tropical forests to open, temperate grasslands probably occurred 14 to 20 million years ago when India docked against Asia and triggered the formation of the Himalaya, which in turn changed the climate, creating temperate forests and grasslands. *Ramapithecus* was first found in India and is also known from East Africa. The early hominids are distinguished from the apes by head (cranial) features such as dentition, shape of skull, and size of brain, and by the postcranial skeleton (which is rarely preserved).

The similarity between *Ramapithecus* and *Australopithecus* (Fig. 18.28), which appeared in very late Pliocene or early Pleistocene, is close, suggesting to some students that they were on a direct line to man. It is more likely that *Ramapithecus* or some related form may have served as the ancestral stock for both *Australopithecus* and *Homo* (see Fig. 18.30). But the problem is that there is a gap in the record, for *Ramapithecus* apparently disappeared well before the first *Australopithecus* appeared about 4 million years ago.

THE EARLY HOMINID RECORD

The earliest record of the Hominidae is at two sites in eastern Africa. Laetoli is in northern Tanzania some 50 kilometers south of the famous Olduvai Gorge and Hadar is in the Afar Triangle of Ethiopia. Laetoli is the older site with ash beds dated at 3.6 to 3.8 million years, and has yielded only a few fragments of jaw and some teeth from about 25 specimens indicating a true hominid, probably of *Australopithecus,* but Mary Leakey, who discovered the fragments, believes they might be of the genus *Homo.* Even so, not much information can be extracted from the specimens thus far known. The most fascinating find is two sets of human footprints, along with the tracks of a great number of mammals still seen today on the plains of Tanzania. The prints show that the hominids were bipedal with a shuffling gait and a short stride.

At Hadar, which has been dated at 2.6 to 3.3 million years slightly younger than Laetoli, there has been found a very rich treasure trove of hominid specimens, fragments of at least 40 individuals, including Lucy mentioned above. Nearly every type of bone in the body has been found. The nature of the hand and foot bones together with the pelvic bones and knees clearly show these forms to have been bipedal as well. But they were not as erect as the more advanced *Homo erectus* (Fig. 18.29). They had large canine and incisor teeth and a large overhung jaw. A crest is present on the top of the skull similar to that of the apes but the opening for the spine is much more centrally located than in the apes. The brain size is unknown at present, but it apparently was similar to that of later species of *Australopithecus* (500 to 600 cm^3). Lucy was about 20 years old and stood at just under 4 feet. Sexual differences are pronounced in these, as well as all species of hominids.

These early forms have primitive features like the two *Australopithecus* lineages that are found in the next-younger group of fossil sites dated at 2.0 to 2.5 million years at or near the Pliocene-Pleistocene boundary. At a number of sites in south Africa and East Africa, two groups of hominids have been recognized. The first of these was described in 1925 by Dart from an immature skull as *Australopithecus africanus*. This group is called the gracile lineage; gracile refers to a relatively dainty jaw with rather small cheek teeth, no skull crest, a centrally located spinal opening indicating that the spinal column was in a vertical position and hence the gracile individuals were better walkers. The brain was about 600 cm^3 in size. This gracile group has a curious relationship to the contemporaneous robust lineage named *Australopithecus robustus* (or *A. boisei*). *Australopithecus robustus* has a massive jaw with very large molars and premolars indicating a heavier diet. The skull

does have a crest and the brain capacity of 550 cm^3 is roughly the same as the graciles. The skeleton is much larger and the individuals are believed to have weighed in at 100 to 150 pounds. Definite sharp-edged tools have been found with these species by Leakey at Olduvai. Some have suggested that the gracile species simply represents the females, and the robust species the males of the same species. But, insofar as we are aware, they are not found together at a single site—a rather desirable occurrence if that hypothesis should be correct. Leaving out the Laetoli specimens, which Mary Leakey believes to be *Homo*, *Australopithecus* dates from 2.5+ million years to about 1 million years. During that time there was virtually no increase in brain size.

THE LATER RECORD OF THE HOMINIDAE

Among the fossils found by Louis Leakey at Olduvai and dated at 2 million years, is specimen number OH 7 with a dramatically larger brain of 750 cm^3. The skull has no crest and the jaws and cheek teeth are moderately robust. Leakey named it *Homo habilis* (the "handy man" because tools were found with it). Up to that time (1960s) it was believed that *Australopithecus* was the direct ancestor of *Homo*. The problem is that both are found together at Olduvai so then who is the ancestor and who is the descendant (see Fig. 18.30)?

Another *H. habilis* skull was found by R. E. Leakey (the son of Mary and Louis) in the Koobi Fora Formation at Lake Turkhana in East Africa. The rocks were dated just over 2 million years. This skull is even more manlike with a brain of 800 cm^3, more than one-half that of *Homo sapiens*. But in 1975 Leakey found another skull with a brain of 900 cm^3. This casts some doubts on the concept of *H. habilis* and many workers are considering *habilis* as an early-appearing *Homo erectus*.

In Java and China during the 1920s and 1930s, a number of specimens of a tall, large-brained (over 800 cm^3) and small-faced man were found in mid-Pleistocene deposits dated at about 500,000 years. These fossils were given a variety of names but are now placed in the species *Homo erectus* (Fig. 18.29). Legs and associated bones

FIGURE 18.28

Australopithecus, an advanced ape from South Africa thought to be an immediate precursor of the genus *Homo.* *(Photograph of head model courtesy American Museum of Natural History.)*

show that *H. erectus* was, in fact, fully erect. Additional fossils found in Africa are somewhat different, but in any event it is likely that *H. erectus* evolved into modern man, *Homo sapiens.* During the 300,000 years of *H. erectus's* existence, the brain size increased to over 1,000 cm^3, and this was accompanied by an evolution of complex behavior and the development of social patterns. Evidence for these conclusions is based on presence of tools, use of fire for cooking and warmth, construction of stone and wood shelters, and arrangement of these shelters into communities.

The first members of *H. sapiens* were found in Europe in late Pleistocene sites dated at about 300,000 years. They were called Neanderthal men, and their brain size averaged around 1,200 cm^3 (modern man may have brains of about 1,400 cm^3). Thus in less than 1 million years the brain size doubled in *Homo* compared to essentially no brain size increase in *Australopithecus*

during the entire span of about 2 million years. Surely a large brain with all the attendant features is indeed the most important difference between man and all other species of animals. Neanderthal man had heavy eyebrows, as did all earlier hominids, a pronounced chin, and was rather stocky. And despite characterizations as a brutish caveman, Neanderthal man in reality had a well-developed society, as evidenced by complex burial sites and a tool culture. Modern man (Cro-Magnon) first appeared over 50,000 years ago, and like Neanderthal man was a hunter and gatherer. Unlike other species, though, modern man had such features as a high flattened brow, a higher shorter skull, and a more rounded face. Both groups coexisted for perhaps 10,000 years, but modern man, with more highly developed technical skills, arts, and language, dominated and rapidly spread throughout the world.

Because of enormous capacities, man has been able to move into more diverse environments than any other organism by the simple device of constructing an environment if adverse conditions exist. As a result of this geographic ubiqui-

FIGURE 18.29

Home erectus (Pithecanthropus), an immediate ancestor of *Homo sapiens*. *(Photograph of head model courtesy American Museum of Natural History.)*

different skin types evolved, but the strong suggestion is that it was in response to ultraviolet light intensity at different latitudes. Pigmented skin is an effective ultraviolet light filter,

ty, distinct races of man have appeared. Many racial characteristics clearly were adaptive for particular environments. As a single example, we cite skin pigmentation. No one is sure why

FIGURE 18.30

A suggested relationship of Hominid species to each other. This is a family tree that some, but not all workers, would agree upon, and is but one of several possible schemes.

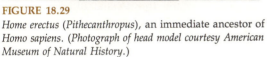

which not only protects delicate tissues but prevents overproduction of vitamin D. It is postulated that, as man moved into higher latitudes where ultraviolet radiation is less intense, there was insufficient light to promote the synthesis of vitamin D that occurs in the skin. Selective pressures operated toward light skin, hence the evolution of white skin.

Most anthropologists believe that one ancestral stock spread from Africa or Asia Minor to other, more or less isolated, regions and then became racially diversified in response to local environmental selective pressures. Alternatively, *Homo erectus* may have given rise independently to *H. sapiens* in several separate areas. In any event, by no later than 10,000 years ago, all major races of modern man had appeared and had occupied their primary distribution areas (i.e., their distributions as of the beginning of historic times).

MAN IN NORTH AMERICA

Large mammals such as the wooly mammoth and reindeer apparently crossed over to America about 100,000 years ago. Mammal-hunting man then gained access to the New World by way of the Bering bridge, and dispersed far southward via the Isthmus of Panama. Human occupations as old as 40,000 years have been claimed, and a skull found at Del Mar, California, recently was dated at 48,000 years. This means that man migrated from southern Asia to California in a mere 50,000 years, for the first *Homo sapiens* known in the Old World is only 100,000 years old. By no later than 10,000 years ago, humans had reached southernmost South America. People had to invent tools, weapons, and clothing to adjust to climate changes as they proceeded north and east across the Bering bridge. Open-plains paleohunters using bifacial, fluted stone weapons became specialized in central and southeastern Europe sometime before 65,000 years ago. The fluted projectile-point culture then spread north and east across Asia, and the people gradually adapted to a cold climate as they dispersed. The culture probably was car-

ried across the Bering bridge between 30,000 and 50,000 years ago. Artifacts show that it spread south into the western United States and to the southern tip of South America (Fig. 18.31). The immigrants were the ancestors of the Plains Paleo-Indians, who developed in isolation during the Wisconsinan glacial advance. This early American culture was distinguished by the use of fluted Clovis and Folsom projectile points, known in the United States at least as early as 12,500 years ago and in South America at least 10,000 years ago.

Additional influxes of humans southward into the New World apparently were impossible until the Wisconsinan ice retreated enough to open once again the corridor from Alaska through western Canada (Fig. 18.31). This apparently occurred between 10,000 and 12,000 years ago, when a new immigration brought the ancestors of the Eskimo and Aleut peoples. These Mongoloid groups remained in the far north, chiefly along the coasts, but they influenced the older, southern Indian culture. For example, fluted hunting points were replaced by unfluted ones in North America about 9,000 years ago. This new stone industry had developed in Europe by 27,000 years ago, but the more primitive American Paleo-Indian tradition persisted here much later due to its isolation from Asia until 12,000 years ago. Prehistoric Indians lived near the southern ice margin and witnessed the last retreat; climatic changes greatly influenced their habits.

The Eskimos, whose ancestors were early Japanese-Korean peoples, developed the greatest adaptation ever to the harsh Arctic climate. Once in Alaska, rather than moving south as earlier peoples had done, they dispersed eastward several times across Arctic Canada, reaching as far as eastern Greenland. In northern isolation, they evolved both to biologic and cultural uniqueness. Meanwhile, other groups evolved distinctively in at least semi-isolation in other parts of the New World. Human migrations and biological as well as cultural adaptations in the Americas, under the influence of the transitory Bering land bridge, fluctuating Canadian ice barrier, and diverse ecologic niches,

Ice   Bifacial, fluted projectile points industries   Unfluted projectile points industries

FIGURE 18.31

Probable history of entry of some early human cultures into the New World, showing importance of glacier histories in western Canada. A primitive hunting culture (represented by fluted Folsom and Clovis projectile points) penetrated into the United States at least 25,000 years ago. New southward immigration was prevented by readvance of glaciers until about 10,000 years ago when the corridor reopened, and a more advanced, unfluted projectile-point culture entered from Asia, soon influencing formerly isolated peoples. (*Adapted from Müller-Beck,* Science, *v. 152, 27 May 1966, pp. 1191-1210; copyright 1966 by American Association for the Advancement of Science.*)

provide outstanding illustrations of the basic principles of organic evolution, migration, and adaptation that we have applied before to lower life forms.

SUMMARY

Pleistocene glaciation represents one of the most dramatic episodes in all of geologic history. Effects of this profound climatic event affected practically the entire earth from the deepest sea floor to the highest mountain peak and from the poles to the equator, and in a great variety of ways. The enormity and rapidity of changes are breathtaking. Practically all kinds of life were affected by it in one way or another. Some even became extinct during the refrigeration, while others, such as the penguin, thrived. Animals with body-temperature-regulating capabilities

should have fared better than cold-blooded creatures, but many mammals also suffered extinction—possibly due to man.

In spite of intensive study for more than a century, the causes of glaciation still are not fully known. Rarity of glaciations suggests that a combination of peculiar conditions must have been required. The appeal of the Emiliani-Geiss hypothesis is its reliance upon earthly mechanisms and its close correspondence to the total record of glaciations. The three best-documented glacial intervals of history—Eocambrian, late Paleozoic, and Pleistocene—all correspond to times when continents were demonstrably much larger than normal. It seems inescapable that the known and measurable effects of total earth albedo and relative shifts of the poles must have been of great importance to ancient climates, and probably were instrumental in lowering temperature to the glacial threshold. The Milankovitch effect seems firmly established as the major control of glacial and interglacial oscillations. Additional factors, such as mountain building and the closing of the Isthmus of Panama to modify interchanges of Arctic and equatorial ocean waters, probably have contributed to glacier changes.

Predicting future climatic trends is almost as difficult as predicting the stock market, which famous banker J. P. Morgan forecast "would in all probability fluctuate." But all indications are that the earth is now in an interglacial episode that will inevitably be followed by further glaciation. Moreover, there seems every reason to believe that glacial and interglacial climates will fluctuate indefinitely until there is either a major transgression or a shift of the continents to bring the poles into open oceanic locations to tip the overall global heat budget into a warmer mode.

Besides dramatic north-south shifts of ranges of both land and marine organisms as glaciers waxed and waned, reefs were affected by rise and fall of sea level. Similarly, wooly mammoths and other land animals roamed freely over what are now submerged continental shelves during lowered sea level. Marine microorganisms were affected by marked evolutionary changes and temperature-related shell modifications. Most dramatic of all, however, was the evolution of hominids, leading to modern man about 50,000 years ago. Through migrations, isolation, and natural selection, adaptive improvements occurred, and *Homo sapiens* resulted. Evolutionary principles have applied to man as well as to lower organisms, at least until the past few centuries, when man began to manipulate the environment profoundly. Some believe that man's further evolution will be chiefly cultural and psychological, but it is difficult, based upon geologic history, to believe that all biologic evolution of man has ceased. It is claimed that the capability to control genes is almost at hand. Thus, although man may soon become the first organism to manipulate its own evolution, so too may man become the first organism to so seriously degrade its environment as to invite its own extinction. In the next chapter we explore these possibilities more fully.

Three monkeys sat in a coconut tree
Discussing things as they're said to be.
Said one to others, "Now listen, you two,
There's a certain rumor that can't be true,
That man descended from our noble race.
That very idea is a disgrace.
Yes, man descended, the ornery cuss—
But, brother, he didn't descend from us."

Anonymous

Key Concepts

climatic oscillation frequency (100,000 years)
critical glacial threshold
dating and correlation:
    Foraminifera coiling directions
    isotopic dating (e.g., ^{14}C, fission-track, etc.)
    magnetic polarity reversals
    obsidian dehydration
    oxygen-isotope ratios (^{18}O/^{16}O)
    volcanic ash
glaciation hypotheses:
    1. solar changes
    2. orbital changes
    3. heat-budget changes
    4. paleogeographic changes
greenhouse effect
Holocene Epoch (6,000 to 10,000 years)
Hominidae:
    Australopithecus
    Homo habilis
    Homo erectus
    Homo sapiens
isostatic rebound
land bridges (Bering and Panamanian)
lakes and spillways
Milankovitch effect (due to eccentricity, tilt, and wobble)
permanent snow line
Pleistocene Epoch (2.5 to 2.8 m.y.)
raised (or drowned) beaches and reefs
reflectivity or earth albedo
river terraces
submarine canyons
thermally isolated poles
transparency of atmosphere
Wisconsinan glaciation

Readings

Bishop, W. W., and Miller, J. (eds.), 1972, Calibration of hominoid evolution: Edinburgh, Scottish Universities Press.

Black, R. F., Goldthwait, R. P., and Willman, H. B. (eds.), 1973, The Wisconsin Stage: Geological Society of America Memoir 136.

Campbell, B. G., 1967, Human evolution: an introduction to man's adaptations: London, Heinemann.

Charlesworth, J. K., 1957, The Quaternary Era (2 vols.): London, Edward Arnold.

Clark, D. L., 1974, Late Mesozoic and Early Cenozoic

sediment cores from the Arctic Ocean: Geology, v. 2, pp. 41–44.

Cline, R. M., and Hays, J. D., 1976, Investigation of Late Quaternary paleooceanography and paleo-climatology (CLIMAP): Geological Society of America Memoir 145.

Dort, W., Jr., and Jones, J. R., Jr., 1970, Pleistocene and Recent environments of the Central Great Plains: Univ. Kansas Special Publication No. 3, Lawrence, Univ. Kansas Press.

Emiliani, C., and Geiss, J., 1957, On glaciations and their causes: Geologische Rundschau, v. 46, pp. 576–601.

Flint, R. F., 1971, Glacial and Pleistocene geology: New York, Wiley.

Harland, W. B., 1971, The Phanerozoic time scale—a supplement: London, The Geological Society Special Publication No. 5.

Hays, J. D., Imbrie, J., and Shackleton, N. J., 1976, Variations in the earth's orbit: pacemaker of the ice ages: Science, v. 194, pp. 1121–1132.

Imbrie, J., and Imbrie, K. P., 1979, Ice ages: solving the mystery: Short Hills, Enslow.

Johanson, D. C., and White, T. D., 1979, A systematic assessment of early African hominids: Science, v. 303, pp. 321–330.

John, B., 1979, The winters of the world: New York, Halsted.

Lamberg-Karlkovsky, C. C., 1972, Old world archaeology: foundations of civilization: Readings from Scientific American: San Francisco, Freeman.

Leakey, R. E., 1976, Hominids in Africa: American Scientist, v. 64, pp. 174–178.

Loomis, W. G., 1967, Skin pigment regulation of vitamin-D biosynthesis in man: Science, v. 157, pp. 501–506.

MacNeish, R. (ed.), 1973, Early man in America: Readings from Scientific American: San Francisco, Freeman.

Matsch, C. L., 1976, North America and the great ice age: New York, McGraw-Hill (Paperback)

Müller-Beck, H., 1966, Paleohunters in America: origins and diffusion: Science, v. 152, pp. 1191–1215.

Pearson, R., 1964, Animals and plants of the Cenozoic era: London, Butterworths.

Pilbeam, D., 1972, The ascent of man: Macmillan series of physical anthropology: New York, Macmillan.

Rankama, K. (ed.), 1965, The Quaternary, v. 1; 1967, v. 2: New York, Wiley.

Ruddiman, W. F., 1971, Pleistocene sedimentation in the equatorial Atlantic: Bulletin of the Geological Society of America, v. 82, pp. 238–302.

Savin, S. M., Douglas, R. G., and Stehli, F. G., 1975, Tertiary marine paleotemperatures: Geological Society of America Bulletin, v. 86, pp. 1499–1510.

Tattersall, Ian, and Eldredge, N., 1977, Fact, theory, and fantasy in human paleontology: American Scientist, v. 65, no. 2, pp. 204–211.

Turekian, K. K. (ed.), 1971, The late Cenozoic glacial ages: New Haven, Yale.

United States Geological Survey, 1973, The channeled scablands of eastern Washington—The story of the Spokane Flood: INF Pamphlet 72–2: Washington, Government Printing Office.

Washburn, S. L., and Delhinow, P. (eds.), 1972, Perspectives on human evolution—2: New York, Holt.

Wright, A. E., and Moseley, F., 1975, Ice ages: ancient and modern: Liverpool, Geological Journal, Special Issue No. 6.

Wright, H. E., and Frey, D. G. (eds.), 1965, The Quaternary of the United States, Review volume for VII Congress of International Association for Quaternary Research: Princeton, Princeton.

19

THE BEST OF ALL POSSIBLE WORLDS?

Life can only be understood backward,
but must be lived forward.

Sören Kierkegaard

FIGURE 19.1
"When you have seen one Redwood tree, you have seen them all," said the then-governor of California during a 1960s controversy over establishment of Redwoods National Park (Mill Creek Road, Jedediah Smith Redwoods, California). *(Courtesy Clyde Thomas and Save-The-Redwoods-League.)*

19

At the end of our long journey through the maze of geologic time, it is appropriate to ask, What are the greatest implications of earth history? We believe that there is a message of lasting value for all, and in this closing chapter, we enumerate what seem to us the few most important and transcendent implications.

From a scientific standpoint, we have attempted to plot a course through the historical maze based upon an objective relating of evidence and the formulation and testing of working hypotheses in a search for general explanations of the development of the physical earth and its life. But, though we have succeeded in formulating a few comprehensive explanations, it should be obvious that the scientific process is never completed, especially in a young science like geology. Hypotheses are transitory; we should nurture them only until we can construct still better ones to explain phenomena. We are not in the least surprised that some of the speculations presented in our first two editions are now outdated. By the time you read this page, perhaps some of the ideas incorporated in this edition will already be out of date.

All progress is initiated by challenging current conceptions. (Bernard Shaw)

It is common to all intellectual pursuits that each question answered tends to raise new questions. This situation is so universally true that we may question if *anything*, strictly speaking, is fact, for so-called facts have an annoying tendency to be reinterpreted without notice. Skeptical minds question the most basic premise of all, namely that there is some inherent order in nature to be discovered, described, and explained. Voltaire once suggested that ''If life has no meaning, man will invent one,'' and existentialist writer Albert Camus, in his essay *The Myth of Sisyphus*, wrestled with the dilemma presented by the seeming chaos of constant change and lack of fixed absolutes. Is there a hopeless futility in the ceaseless quest for understanding, or does the search itself give meaning to human activity? Robert Louis Stevenson already had anticipated Camus with the affirmation that ''To travel hopefully is a better thing than to arrive.''[1]

THREE IMPORTANT MAXIMS

Faced with the reality of uncertainty, it becomes important to ask if there are not at least a few relatively durable geologic tenets. We believe that the study of earth history carries three maxims of far greater significance and timelessness for humanity as a whole than any alleged facts of history or the hypotheses we have developed to explain them. *First*, a study of geologic history necessarily develops *wholly new concepts of time and rates of change*. It reveals man's position on the scale of life history. Instead of having appeared together with practically all other life soon after the origin of the earth, as was assumed until the eighteenth century, man has resided on earth for but a brief 2 to 2.5 million years—only 0.04 percent of geologic time (modern man—an even briefer 50,000 years)! *Second*, the history of earth and life shows forcefully that, indeed, *''there is nothing permanent except change.''* More importantly, however, is that the change is of a special kind. It is not the perfectly cyclic or ''steady-state'' change envisioned a century ago, but rather an *irreversible cumulative evolutionary change*. ''Nature creates ever new forms; what

[1]Camus took Homer's ancient myth as a point of departure for analyzing the seeming absurdity of life's toil symbolized by Sisyphus's futile labors. Sisyphus was banished to Pluto's underworld, where he was made repeatedly to roll a stone to the top of a mountain from which it rolled back down every time. Camus, like Stevenson, concluded that the relentless struggle for the summit gives life meaning enough.

exists has never existed before; what has existed returns not again," wrote Goethe as early as 1781. The *third* major implication follows from a study of evolution through geologic time, which is what our book is all about. This most important maxim is the great *ecologic interaction between both the living and nonliving realms of the earth throughout history*. Because an evolutionary sequence represents a series of unique events, history cannot, after all, repeat itself exactly. While organic evolution has been possible because of random changes called mutations, the *results of evolution were not random*. As we have shown, the particular way that life developed on earth is the result of natural selection operating on mutants through a 3.5- or 4-billion-year series of changes. Involved were changes both in the living and nonliving realms in a particular order and at particular rates. Had some of these occurred in a different fashion, the course of evolution would have been different; yet no one can say exactly how it would have differed.

One of the chief goals of science is *prediction of consequences*, and if, as we contend, a study of earth history truly has relevance to the lives of humans, then we feel obliged to make some extrapolations from past history and to suggest some implications for the future of humans and their environment. We shall proceed by reviewing briefly the evolutionary highlights in earth history as they affected life, and conclude by applying our third maxim to the contemporary human world.

THE EVOLUTION OF THE EARTH

THE IDEA OF EVOLUTION

The concept of evolutionary change undoubtedly is one of the greatest of human ideas, and, as we have tried to show repeatedly, it is as important in the nonliving as in the biologic realm. Evolution, we feel, provides a powerful unifying basis for the study of earth history. Indeed, it should be clear that nature cannot be understood fully without special attention to evolution, which is most completely revealed in the historical sciences. There was little interest in history, however, until the eighteenth century when Pompeii, Mayan ruins, and several other ancient cities were discovered and excavated, and fossils were finally understood. And it was also in the 1700s that people really began to think of natural change in an evolutionary fashion (see Chap. 10). Previously, practically everything was assumed to be fixed in form and in relation to everything else. All life was assumed to have been created nearly simultaneously soon after 4000 B.C.; thus it seemed necessary for Noah to have taken pairs of every life form into the Ark. No extinctions or new creations of species had yet been demonstrated, though Robert Hooke and G. L. L. de Buffon suspected them.

The seventeenth-century cosmogonist Leibniz first preached a principle of grand continuity in nature, which led to a notion called the Great Chain of Being from atoms to God. Allegedly, nature abhors discontinuities; rather, everything is arranged in gradational sequences. Development of calculus by Leibniz and Newton doubtless lent impetus to embryonic evolutionary thought by focusing on the integration of small changes or increments both in space and time. In the nineteenth century, Lord Kelvin, using new principles of thermodynamics, anticipated the modern conception of the physical earth as a dynamic, irreversibly changing sphere. Then in 1896, 37 years after Darwin's momentous *Origin of Species* appeared, discovery of radioactivity provided an important example of *inorganic evolution* through transformation of one element to another, as described in Chap. 4. It also provided an important clue to an overall chemical evolution of the earth in which radioactive decay has played a major role, as was stressed in Chap. 5.

CHEMICAL EVOLUTION OF THE EARTH

We believe that terrestrial evolution began with the aggregation and densification of the proto-earth. Internal heating, especially through radioactivity, facilitated the density differentiation of core and mantle between 4.5 and 4.7 billion years ago, and later the crust. The primitive atmosphere and sea water began accumulating and underwent changes in composition as a result of igneous activity, weathering, and photochemical reactions.

As one result of changes of matter on the earth's surface, life developed—apparently as an inevitable consequence of the evolution of complex carbon-bearing molecules (see Chap. 5). Three or four billion years ago, indeed, this *was the best of all possible worlds for life* in our solar system, for it contained abundant carbon, hydrogen, nitrogen, oxygen, and phosphorus; had a critical surface temperature range such that much water was present in the liquid state; and possessed a strong magnetic field to shield the surface from most cosmic radiation. It did not yet have an ozone shield from ultraviolet radiation, however. It has been said by a famous Russian biochemist, A. I. Oparin, that matter always is changing from one form of motion to another, each more complex and harmonious than before. Life, to him, appears as a very complex form of matter in motion that arose at a particular stage in the general evolution of matter on earth. As we have seen, the origin of life required very special conditions, especially the moderately warm temperatures, high-energy radiation, and lack of abundant free oxygen in the environment.

Subsequent development of life occurred in an irreversibly evolving physical environment. Once photosynthesis became possible, free oxygen, and eventually ozone, began to accumulate slowly in the atmosphere. With the addition of free oxygen, life itself brought about a change in the physical realm that irreversibly precluded its being created anew. Organisms are also important in the cycling of both carbon and nitrogen among the atmosphere, water, and soil. Atmospheric argon evolved continually through decay of radioactive potassium 40. It is the only atmospheric gas wholly independent of life processes. As noted in Chaps. 5 and 8, an oxygen-rich atmosphere is indicated no later than 1.5 billion years ago; by at least 0.7 billion years, marine animals also had appeared, indicating that oxygen respiration had become possible. But neither plants nor animals had invaded the land as yet, though large land areas had existed long, especially at the end of Prepaleozoic time. Only after considerable free atmospheric oxygen accumulated could the important ultraviolet-filtering ozone layer develop fully at a level well above the land surface. Such development may have taken until middle Paleozoic time when land organisms first appeared (see Chaps. 13 and 14). Significantly, mountain building at that same time had formed a variety of potential ecologic niches, and these were quickly filled by a sudden invasion of the land both by plants and animals. Here we see a clear example of physical changes influencing the path of organic evolution, but, conversely, once organisms were established on land, they, in turn, markedly influenced the land by greatly modifying processes of weathering, erosion, and sedimentation.

Increasing mountain building on all continents in late Paleozoic time created more new habitats for expanding land life. As land emerged from the sea in Carboniferous time, great coastal coal swamps formed on a colossal scale. Relative land area and elevation are major factors in the earth's heat budget, as are atmospheric and oceanic circulation. As the continents achieved their largest total area and greatest elevation since Prepaleozoic time, climate changed in response. Permo-Triassic redbed deposits became nearly universal on northern lands, which were then located in low latitudes, while glaciation set in on southern high-latitude ones.

Paleozoic and Mesozoic land animal life was surprisingly homogeneous on most continents, indicating connections among most land areas. Plants were somewhat less uniform, apparently due to climatic zonation. By Cretaceous time, all land organisms began to be more differentiated on present continental units. As the Mesozoic separation of continents occurred, it would have affected distribution and evolution of life, as seems indicated by the fossil record. Presence of several tenuous land bridges has maintained some interchange of land organisms among the different continents up to the present, but Cenozoic land life was far more differentiated than earlier life. In the south, isolation seems to have been more complete and of longer duration than in the north so that more disjunctive distributions of organisms are found there (as discussed in Chaps. 17 and 18).

The late Cenozoic tectonic episode has been more extreme and rapid than geologists appreciated until recently. Since Miocene time, truly profound events have occurred. These included: inception (or rejuvenation?) of all island-arc systems as we know them today, the Alpine-Himalayan mountain building, changes in the ocean-ridge-spreading regime leading to changes of plate motions, and the beginning of large-scale rifting on five continents. As all this occurred, continents have enlarged and risen.

During Cenozoic time, mammals and plants became diversified, and some major extinctions occurred, doubtless influenced by physical changes. Finally, the atmospheric heat budget tipped in favor of Pleistocene glaciation. It was during this geologically recent climatic and structural turmoil that *Homo* evolved in Africa or Asia Minor and within the brief span of 2 million years dispersed throughout all land areas. In the past two centuries, man has modified the earth's surface, atmosphere, and water to such a phenomenal degree as to surely represent the most rapid and potentially catastrophic organic event in all geologic history. Anthropologist Loren Eiseley mused that, while evolution of the human body from that of an aggressive ape is essentially completed, man seems to possess an unfinished mind that "could lead humanity down the road to oblivion."

MASS EXTINCTIONS

Mass extinctions of major groups have been used by paleontologists to mark the close of time divisions (Fig. 19.2). The causes of these extinctions are uncertain. No one hypothesis has stood the test of close scrutiny; therefore, probably we should seek more than one cause. And extinctions were not always instantaneous. The great Permo-Triassic extinction of marine animals began perhaps 10 million years or more before the close of the Permian Period. That extinction was so devastating that it took 15 to 20 million years to evolve a normal, balanced

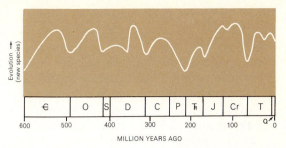

FIGURE 19.2

Graph of evolution and extinction of major groups of animals through time showing unusual crises when mass extinctions occurred, especially at the end of the Paleozoic and Mesozoic Eras and in Pleistocene time. (*Adapted from N. D. Newell,* Crises in the history of life; *Copyright© 1963 by Scientific American, Inc. All rights reserved.*)

fauna in the Late Triassic (see Chaps. 14 and 16). The Permian extinctions involved more than 50 percent of the major groups of marine organisms, and, of course, it is tempting to suggest that emergence of land caused the eradication of many shallow-water environments. But why did some nonmarine amphibia and reptiles also become extinct? The Late Cretaceous mass extinction was not as severe, affecting only about 25 percent of the animal families. However, this extinction again affected both land and sea organisms, but on a highly selective basis (see Chap. 17).

CAUSES OF EXTINCTIONS

Extinctions apparently resulted from failure of species to adapt at a rate equal to or greater than, the rate of environmental change. Extinction hypotheses involve countless ideas ranging from abrupt Cuvieran catastrophism to a kind of evolutionary old age or *overspecialization*.

Probably there are more theories on the extinction of the dinosaurs than on any other group. Suggestions run the gamut from excess atmospheric oxygen to the supposed chill of impending glaciation and even slipped vertebral discs in the oversized monsters. To a seemingly endless list, we offered in our second edition a tongue-in-cheek *asthmatic hypothesis*, which has subsequently come to seem plausible to us. One spring a few years ago when our eyes began to itch and our noses to tingle, we recalled that flowering plants first appeared in late Mesozoic time. Could the appearance of many pollens to which dinosaurs had not previously been exposed produce violent allergic reactions that culminated in wholesale death by hayfever? The new mammals would, of course, also be subjected to the selective stress of pollen, but, seemingly, mutations were producing mammalian strains resistant to the pollens more rapidly than reptilian ones. We might say that mammals were flower children, while blossoms were sprung on the dinosaurs too late in their racial life so that they, poor things, sneezed to extinction.

Some dinosaur groups began dropping out well before the close of the Cretaceous, but up to the end they were still relatively abundant. Both large and very small dinosaurs were present until the end so that ideas that the dinosaurs got too big to survive in the total ecosystem are not satisfactory. Some students have suggested that a pathogenic fungus or some disease wiped them out, but if we take diseases of today as a model, they usually affect only a single species and there is no known species that has been eradicated completely, even by a pandemic. Another theory has considered that mammals ate all of the dinosaur eggs, but again it is difficult to believe this would have been 100 percent effective. Still another concept involved deadly radiation from a supernova or from cosmic rays in lethal quantities during periods of zero magnetism when reversals of the earth's field were occurring. If this were true, why did the radiation selectively eliminate some groups and not others? Surely the mammals, other reptiles, and plants would receive the same dosages; yet they were not noticeably affected. Even more puzzling is the simultaneous extinction of marine forms such as the ammonoids and swimming reptiles, for a moderate depth of water is an effective shield from radiation. Moreover, recent calculations indicate that complete removal of the magnetic field would increase cosmic radiation at the earth's surface only about 15 percent, which is deemed by many authorities inadequate to cause mass extinctions or accelerated mutations. We are not impressed by the idea that an alleged asteroid impact with earth led to the extinction either. As we stated in Chap. 16, none of these catastrophic, single-cause explanations appeal to us, but see Alvarez, et al., 1980.

Mountain building and its relation to climatic change has been cited traditionally as a cause for extinction, which is based on the fact that many unconformities coincide with faunal breaks. Yet some of the most profound faunal discontinuities occurred during times of relative crustal and apparent climatic stability. The Cretaceous extinction occurred during a time of mild climate; significant cooling did not begin until considerably later (see Fig. 18.18). Climate may have played little role in the Pleistocene extinctions as well, for most occurred during interglacial episodes, but it may be that an extinction effect lagged after some climatic cause. A popular view in recent years blames early human hunters for the demise of large Pleistocene grazing mammals like the mastodon in America (see Chap. 18). But why did such beasts survive longer in the Old World if man was the cause, for people had been hunting these much longer than in the New World?

The most likely cause of extinctions is the most difficult to substantiate. The finest balance in nature involves the *ecosystem*; that is, the interaction of organisms in the biological and physical structure of the community. The success of any community is largely dependent upon a few key species of plants and/or animals in the food chain. If any of these species are removed, the effect will be felt immediately through the whole food pyramid of the community—a sort of domino effect. And obviously those organisms that are high on the food chain are most vulnerable. It is possible that a key species, which is completely unknown as a fossil, could have become extinct, causing an entire community to collapse. Some paleontologists even wonder if we are seeking causes where none existed. Extinctions must have been occurring throughout the history of life (extinction simply means that rate of elimination of species exceeds production of new ones). By chance alone, an excess of elimination of species may have affected several major groups more or less simultaneously (but infrequently) without any need for a dramatic explanation—a rare but otherwise not mysterious event.

MAN'S PLACE IN THE ECOSYSTEM

EFFECTS ON OTHER LIFE

Some astronomers estimate that there are about 10^{20} planets in the universe capable of supporting intelligent life. But such life has one chance per planet, for the resources of each body are finite and evolution is irreversible. Man, though not nearly so strong as many other animals, is the cleverest and most adaptable organism that has yet evolved on earth. Through the development of tools and the utilization of stored energy (i.e., fuels), man has more than compensated for any lack of brawn. Tools, however, have also given humans an enormous capability for manipulating and contaminating the environment, leading to man's belief that a complete control of the environment will soon be possible. Indeed, Western man long ago assumed the self-appointed role of predestined master over all of nature. Thus the natural environment came to be thought of as a *commodity* possessed by the human race.

Man has brought about the extinction of hundreds of species of other organisms, for an average of about one species per year over the past century, and now threatens an additional 500 species. By the end of this century, humans may have caused extinctions of tens of thousands of species. Fur coats, for example, take such a merciless toll of the wild cat family (e.g., six leopard skins per coat), that several of the cats are all but extinct because of human vanity. The alligator is suffering a similar fate. Besides overtly killing and inadvertently poisoning organisms, humans also threaten many forms by destroying their habitats—often quite unwittingly. In some cases man has introduced new predators, and in other cases competitors for food. It is estimated that, in the past, the average span for isolated island bird species was about 200,000 years. Since aboriginal man populated islands, the bird species have averaged only 30,000 years, and since Europeans arrived, the span has dropped to only 12,000 years.

Many of the results of man's tampering with natural communities have been quite unexpected. Some species doubtless become extinct without ever being discovered—or even being named! The effects of upsetting ecologic balance are difficult to predict with present limited knowledge. For example, oak forests have increased in area since Anglo-Saxon settlement of the northern Mississippi Valley region, just the reverse of the general pattern of conquest of the frontiers. In this case, cessation of the Indians' habit of regularly burning prairie grass has allowed oak saplings to thrive and mature as never before. Here the settlers had to cut down forests *after* settlement, rather than before, as a consequence of their own activities.

Many changes induced by man have been beneficial to certain organisms. The skunk, opossum, coyote, raccoon, deer, rabbit, cow, robin, dog, and sparrow, to name but a few, have thrived in company with man. And no doubt urban slums and dump grounds have been great for rats. Roadsides also have produced a whole new ecologic niche successfully exploited by rabbits, foxes, pheasants, skunks, ground squirrels, toads, and a host of other birds, mammals, insects, and, of course, so-called weeds. After many years of fire control in western American forests, it is becoming apparent that former periodic burning greatly enhanced the propagation of certain desirable forest trees that do not reproduce well in shady, thickly forested areas. Examples include the *Ponderosa* or western yellow pine, one of our major lumber trees. This serves as an example of unforeseen results of well-meaning actions. While no one is advocating wanton burning, foresters are now letting some natural fires burn themselves out with no interference, and they are doing controlled burning. Some deliberate manipulations of animal communities were motivated only by sentimentality, such as the introduction of the English sparrow into North America and the rabbit into Australia. Serious devastation of rangelands resulted from the latter within but two decades. All this stresses how little we really can predict of the long-term consequences of ecologic manipulation, and also suggests some parallels with the results of natural competition among species in past ecosystems.

Most of the changes cited so far seem relatively innocuous, but many others have been more detrimental and have adversely affected man. Current degradation of the Great Lakes provides an outstanding example. Those lakes are said to hold nearly one-third of the world's freshwater supply, but most are hardly fresh any more; Lake Erie must be described as a gigantic cesspool. Insecticides and herbicides, among other things, find their way into the lakes and then into the food chain—including that of humans (DDT has even been found in Antarctic penguins).

The entry of the parasitic sea lamprey into the upper Great Lakes produced catastrophic effects upon fish populations. Early in this century, the Welland Canal provided access for the lampreys from Lake Ontario around the Niagara Falls barrier to the upper Great Lakes. By 1921 they were found in Lake Erie, by 1937 in Lakes Michigan and Huron, and in 1946 in Superior. They multiplied rapidly in the three upper lakes, and soon decimated the fish populations upon which they are parasitic. Of most direct concern to man were the quick inroads in lake trout fisheries. In Huron and Michigan, large trout production dropped to nothing; Superior's annual production shrank from 4.5 to 0.3 million pounds in only 15 years. Since 1953, a joint Canadian–United States control program has been in effect, which has successfully reduced the lamprey population in Lake Superior so that trout are beginning to increase.

It can be argued that man-caused extinctions of other life forms, as well as acceleration of modifications of the natural environment, simply are contemporary examples of selection in action—"survival of the fittest." According to such a view, history is repeating itself as the human species evolves and expands its domain at the expense of others, producing a colossal *ecologic replacement* analogous to many examples illustrated for past geologic periods. But, although people live more and more in artificially controlled subenvironments of steel and con-

crete, man is still the product of billions of years of evolution in the natural environment. Moreover, *Homo sapiens* always will be dependent upon the mineral and biologic resources of the earth as well as upon its water and atmosphere. Just as plants help to shape their neighbors, man is influenced by other animals, plants, and microbes as surely as individual people are affected by their cultures and intellectual atmospheres. People are part of a fantastically complex ecologic community or ecosystem—unable to extract more from the earth than it can produce, and ultimately subject to selection processes acting upon genetic makeup much as are other organisms. It behooves humans to attempt to maintain an ecological balance that will ensure the well-being of future generations of people as well as of the rest of the system upon which we are dependent. Man, rather than regarding nature as a commodity to be exploited, must think of nature as a complex community of which the human species is both a member and the custodian. It is well, therefore, to remind ourselves in the next section of some of the consequences to *us* of membership in the community of nature.

MAN'S IMPENDING ECOLOGIC CRISIS

FOULING OF THE ENVIRONMENT

Since the first Earth or E-Day in 1970, pollution of the environment has received much publicity, although the problem has been with people a very long time. Ancient Rome was perpetually encased in a smelly dust cloud thanks to heavy horse-chariot traffic. And it is estimated that New York City had 150,000 horses at the turn of the century, and each produced about 9 kilograms (20 pounds) of manure per day. Apparently in those good old days, horses far outdid the dogs of today in littering urban streets.[2] Broadly defined, pollution includes everything

---

[2]Still, pets contribute significantly. The 500,000 dogs in New York City deposit on the streets about 70,000 kilograms (150,000 pounds) of feces and 320,000 liters (90,000 gallons) of urine *each day*.

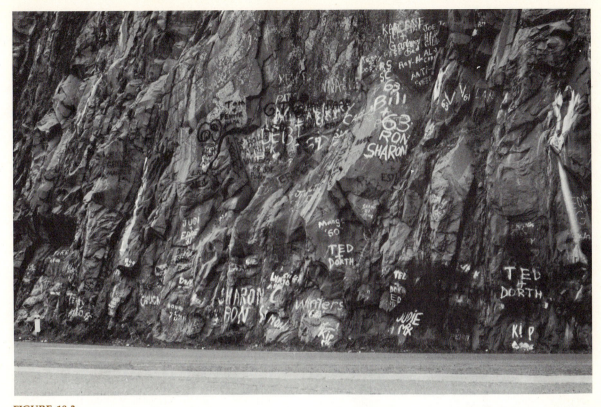

FIGURE 19.3
Roadside art I. A road cut in Cretaceous sandstones on California Highway 128, 35 miles west of Sacramento. Ironically, very old autographs on rocks (as along the Oregon Trail) are prized and protected, but surely wanton autographing by millions of people is intolerable.

from dumping of raw sewage, DDT, and fertilizers into streams to the "uglification" of the landscape with billboards, junked cars, and other solid wastes (Figs. 19.3 and 19.4) and electric utility poles and wires (which, incidentally, might be missed by birds). Even adverse noise from industry, aircraft, vehicles, and television constitutes environmental pollution. The sonic boom generated by fast jet aircraft, for example, is not only an annoyance in urban regions, but has triggered massive rock slides in western national parks.

Unfortunately, indifference toward degradation of the environment still is the rule. Seemingly only a city garbage-collectors' strike can arouse very many people. For example, as reported by *The Times* of London a few years ago, the city fathers of an English coastal resort voted to continue pumping raw sewage into their bay for "that is the cheapest, and the proper method for the resort. We have been doing this for ages past and it can safely be done for ages in the future." But, then, New York City does it, too, to the tune of 750 million liters (200 million gallons) per day into the Hudson River.

In recent years, chemical insecticides, fertilizers, and detergents have caused serious damage to many aquatic ecosystems. Lakes rapidly are becoming fertile with nutrients from sewage plant effluent and dissolved agricultural fertilizers carried by runoff. What once were clear fishing and swimming havens are being converted rapidly to smelly slime ponds as algae thrive. Nitrates from fertilizers ultimately can have an ironic deleterious effect upon humans because, even from drinking water, some nitrate inevitably is taken into the body. Intestinal

FIGURE 19.4
Roadside art II. A "sculpture" made from litter collected by Boy Scouts along four blocks of a typical street in Madison, Wisconsin.

pesticides, for clearly they are of great benefit to man, but consequences of their application are being studied intensively, as well as alternative control measures. For example, there are clever biologic ways to cope with insect pests. Release of sterile males into populations impedes propagation, and introduction of natural predators also can control them. In Australia, after a century of rangeland devastation by European rabbits, introduction of a virus quickly reduced the rabbit population by about 90 percent. Methods such as these are preferable for several reasons. Not only do they pose no threat of poisoning, they also do a more efficient job of extermination *over the long term*. Insects have such short life cycles and propagate so rapidly that chemical insecticides produce unnaturally rapid selection in their populations. Only the fittest survive, and so evolution of hardier strains actually is accelerated by the insecticides, as one can understand from our consideration of mutations and selection. Soon the chemicals become less effective even in dosages that begin to be toxic to man. Unfortunately, chemicals may not be very selective, either: when roadsides are sprayed, wild flowers as well as ragweed are killed—and a brown fire hazard is created as a by-product.

Atmospheric pollution is well known, especially as a result of concern over radioactive fallout and killing smogs (Fig. 19.5). It is a classic irony that the technology developed by the Western civilization spawned in Greece now threatens the famous Acropolis and other priceless ruins in Athens. All of the ancient buildings and statues there were made of marble, which is devastated by acid fog and rain resulting from smog. Effects of acid rain long familiar in industrialized urban areas are now showing up in wilderness regions, too, where they threaten stream and lake ecology. Still there is much ignorance about long-term climatic and biologic effects of the pollutants that man has pumped into the air. Some possible effects have been mentioned before (see Chaps. 5 and 18). Until recent years, the atmosphere, like the oceans, seemed an infinite reservoir into which wastes could be dumped forever without harm. In fact,

bacteria convert it to nitrite, which hinders hemoglobin's ability to transport oxygen in the blood stream, and can cause suffocation of infants.

No objective conservationist arbitrarily advocates complete outlawing of all fertilizers and

FIGURE 19.5
Effects of atmospheric pollution in Los Angeles, California. *Upper*: City Hall on a clear day in 1956. *Lower*: same view on a smoggy day in the same year. A temperature inversion—warm air above a cool layer near the ground—prohibits diffusion of contaminated air into the higher atmosphere. The inversion, seen here at an altitude of about 300 feet, is present approximately 320 days per year. Recent control measures have cut Los Angeles smog back to approximately the 1954 level. (*Courtesy Los Angeles County Air Pollution Control District.*)

both of these reservoirs are closed systems, and so can take on and disperse only finite amounts of pollutants before life becomes adversely affected.

The atmosphere has acquired untold quantities of pollutants from human activities—both agricultural and industrial—much of it as smoke. In addition, combustion has released carbon dioxide and toxic gases in ever-increasing quantities since the industrial revolu-

tion. The most serious culprit is exhaust from internal-combustion engines. Tetraethyl lead released from burning of gasoline affects our nervous systems and can be fatal. In recent decades, the average person's lead content has risen a hundredfold, bringing it near present estimates of the human tolerance level. Lead also is accumulating in the shallow seas, which are our major fishing grounds; apparently it diffuses into the deeper ocean much more slowly than it is being added. Carbon monoxide is a more familiar product of exhaust pipes, and nitrous oxide gives rise through photochemical reactions to the chemicals that cause the ugly color of smog (Fig. 19.5). Both of these are also prominent in tobacco smoke, along with the carcinogenic components. Could man be destined to extinction by a respiratory cause similar to that postulated by us for the dinosaurs?

Besides known toxic effects of gaseous wastes in the atmosphere, gross climatic changes might be triggered by atmospheric pollution. Local meteorological effects are well known around large, industrialized cities, where rainfall is increased by smoke and dust particles; cities receive about 10 percent more rain and fog than do their surroundings. Yet in cities, most rain water runs off because more than half of the ground surface is covered with pavement and buildings, and so infiltration to the ground water table is correspondingly reduced. Also as a result of smoke, dust, and fog, the intensity of visible sunlight in cities is only about 60 percent and ultraviolet radiation about 10 percent as great as in open areas.

Cities alone are not guilty of atmospheric pollution. Rural areas also contribute, especially where low-technology agriculture is prevalent. Widely practiced burning of brush and overgrazing contribute so much smoke and dust that the ground commonly is invisible from a plane in some parts of the world. Worldwide increase of turbidity of the lower atmosphere results in diminished incoming solar radiation, thus a climatic cooling tendency. A 10 percent increase in turbidity causes an estimated reduction of average world temperature of 1°C. From Chap. 18 we recall that reduction of annual temperature of a mere 5 to 8°C over many years would be enough to cause the readvance of ice caps. Possible increase of atmospheric carbon dioxide through combustion also is of concern because of its entrapment of infrared energy radiated from the earth's surface after heating by the sun. Curiously, the latter "greenhouse effect" would tend to warm the climate in opposition to the dust effect. Therefore, it is important to learn the magnitude of both and the rate at which each is operating.

Besides the poisonous and climatic effects of atmospheric pollution, there is growing concern that exhaust from supersonic jet aircraft (the SST) and fluorocarbons from the convenient but unnecessary little aerosol spray can will disrupt the atmospheric ozone layer. This would subject all life forms to dangerous levels of ultraviolet radiation and bring an abrupt end to sun bathing, as well as to life itself. Are we really in such a hurry that we must be able to cross the continent or the Atlantic in a mere 2 or 3 hours less than is now possible? And is it so exhausting to rub on rather than spray your deodorant?

Changes in atmospheric (or oceanic) composition are sluggish things that have a certain momentum; thus they may be very difficult to change quickly even if methods of cleaning were developed tomorrow. Damage produced over more than a century probably cannot be corrected in less time.

Pollution, as serious as it is, is really a symptom—rather than a cause—of our environmental crisis. It reflects the deeper problems of rapid *growth of population* and still-more-rapid *increase in resource consumption*. Only by balancing these factors in the total-demand equation with the earth's finite resources can we avoid calamity.

LIMITS OF THE EARTH'S RESOURCES

Renewable Resources

It is not known just how many people the earth can support. Pessimistic estimates suggest something on the order of 10 billion, which would be achieved in about the year 2015 at the present rate of population growth (Table 19.1). The most optimistic estimates place the figure nearer 100 billion, which easily could be achieved before the end of the twenty-first century. At the present population growth rate, the Malthusian saturation point discussed in Chap. 10 would be reached a mere century from now, but hopefully that rate will be reduced overall as it has been recently in the United States, Japan and several other countries.

Some of the earth's resources, such as lumber and food, are renewable up to a point. Seemingly, any plant-based resource can be regrown,

TABLE 19.1
The growth of human population*
  Note especially the rapidly increasing rate of population growth

Year	World population	Doubling time	Year	U.S.A. population
A.D. 1	¼ billion	——		
1650	½ billion	1650 years	1790	4 million
1850	1 billion	200 years		
			1915	100 million
1930	2 billion	80 years		
1960	3 billion	44 years		
1975	4 billion	40 years	1967	200 million
2000	6–8 billion	< 35 years	2000	300 million
2025	50 billion (?)	(?)		

*After S. Cain, 1967: *Geoscience News,* v. 1, and other sources.

and even water can be cleaned up as it has been in Lake Erie and elsewhere if we are willing to pay the costs. In reality, however, even so-called renewable resources are limited by the earth's ultimate productive capacity, which is the limit of how much and how quickly energy can be supplied to life in usable forms. The total earth energy reservoirs are limited like a bank savings account; how long the savings will last depends upon how rapidly we make withdrawals. But it is even more complicated, because the difficulty of extracting the resources increases continually. In the face of rapidly growing human population pressure, an *International Biological Program* is under way to try to evaluate the total productivity of the earth and the degree of man's dependence both upon native and agricultural vegetation.

The human food chain or pyramid is fantastically complex. Primary producers of food for the chain are, of course, plants, the only organisms capable of manufacturing food. But in the total energy budget of ecosystems, some of the energy stored by plants must be released for recycling in the system. Perpetual recycling affects every single creature. Your body contains carbon that perhaps was used for awhile by a dinosaur, later loaned to some mammal, and possibly borrowed again by some cockroach or dandelion before reincarnation in you. Recycling is accomplished largely by decomposition

to maintain an energy balance. Plants would deplete all carbon dioxide rather quickly from the atmosphere if fire and animal respiration did not recycle it. Certain bacteria could deplete the atmospheric nitrogen in about a million years, but other organisms release it back to the atmosphere as ammonia. Much of the decomposition is accomplished by lowly "pests," which, like it or not, are essential to the whole system. Primary consumers of plant production are herbivorous animals and decomposing microbes. Secondary consumers, in turn, eat them or their products, such as milk and honey, and so on. Organisms are notoriously inefficient, however, and the last link in the chain, such as man, may consume the equivalent of hundreds of millions of pounds of primary plant food annually (see Fig. 9.23).

There is no immediate prospect that man can short-circuit photosynthesis artificially. But if the human population grows unchecked, man may be forced to short-circuit the food chain by eliminating all intermediate animal and plant consumers that are not absolutely essential in the human food chain so as to gain exclusive use of earth resources. Shade trees, garden flowers, and pets would not be permissible. Moreover, diet eventually would be entirely vegetarian, for even now, the total existing human population could not be supported on a largely meat diet. But even an intensive, rigidly controlled agricultural system still would be limited in total productivity by soil fertility, which is only extended by a limited amount and for a limited time through chemical treatment. And we face a serious fertilizer shortage because much of it is manufactured from petroleum. Well over half of the arable land of the earth is now in use, and most of what remains is marginal. The cost of producing food on marginal land is enormous, and production tends to decrease yearly as such land is used—even with fertilizers. This is especially true of soils in the humid tropics where a large portion of the human population resides. In 1950, each additional ton of fertilizer produced 10 tons more of grain, but by 1960 only 8 tons, and in 1980, only about 6 tons. Already conflicts over the use of land for housing and industry versus agriculture are upon us; they surely will intensify in the future. Perhaps we could ease both dilemmas by depending upon algae grown in glass water tanks on the roofs of huge apartment houses or upon bacteria, fungi, and yeast grown under conditions requiring little land and little energy.

Nonrenewable Resources
Ultimate limits of nonrenewable mineral resources, such as ores of the metals, building materials, and the natural hydrocarbon fuels (with all of which geologists have had intimate concern) should be even more obvious than the limits of renewable resources, especially since the 1973 fuel crisis. Perhaps the greatest service anyone interested in geology can perform for society in the next quarter century is to participate in the search for and development of new fuel and other mineral resources. Obviously there will be plenty of work to do.

For many nonrenewable raw materials, supply exceeds current demand by so wide a margin that there may seem to be no immediate problem. But all such resources are finite, and most people have no inkling of the magnitude of our conspicuous consumption. If we had to depend only upon presently known mineral reserves, our industrial society might collapse in a mere two or three decades. On the other hand, estimates of this sort are elusive, because definitions of reserves are based so much upon

economics; ore that is too low grade to be considered today may become workable ore in the future as demand forces economic redefinitions. Lower-grade raw materials, however, require more energy for processing. New York sewer sludge might become an "ore deposit" one day, for it contains a greater percentage of silver, chromium, copper, tin, lead, and zinc than do ordinary sedimentary rocks. But are not new mineral deposits still being generated in the earth? Yes, of course, but at rates far slower than demand. Here we are once again confronted with the human versus the geologic time scale.

When we compare present demands upon nonrenewable resources with population increase, then the picture becomes more sobering. On the average, every member of the human race is increasing usage of energy by 3 percent per year. From 1935 to 1967, world petroleum production grew from 6.5 to 30 million barrels per day; probably within the next decade, requirements will be nearly 60 million barrels per day (one barrel is 42 gallons). Put another way, a 1-billion-year supply for a constant consumption rate will last only half as long at the presently increasing rate. It is estimated that, since 1955, North Americans have burned more mineral fuels than did the *entire world in all of previous history*. Today, North America alone requires five or six times as much fuel on a per-capita basis as the rest of the world (Fig. 19.6). Consumption of metals in the United States is increasing at about double the rate of population growth (Fig. 19.7). Industrially developed societies with roughly 20 percent of the world's population consume at least half of all the earth's annual total mineral raw material production, and a great deal of this is imported from underdeveloped countries. North America alone, with only about 7 percent of the world's population, consumes nearly 30 percent of the mineral raw materials produced by the entire world! From such figures, it is clear that each person in a developed nation puts a much greater stress on world resources than does a counterpart in an underdeveloped country.

The United States has a material wealth one-third greater than the next-wealthiest nations (Fig. 19.6). Will such disparities be tolerat-

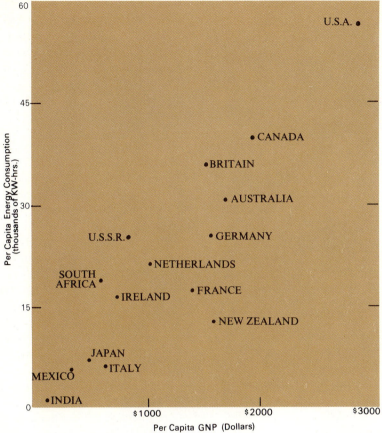

FIGURE 19.6
Relation between per-capita wealth expressed as gross national product and per-capita energy consumption expressed in thousands of kilowatt-hours. Obviously climate also plays some role in these figures; for example, India's energy consumption is low partly because it has a warm climate. (*Adapted from date in Cook, 1971,* Scientific American *and* The Plain Truth, *July–August, 1973.*)

FIGURE 19.7
Graph comparing world population increase, a typical example of the increasing demand for metals (Copper), and increasing demand for petroleum. The latter two rates of increase are even more rapid than that for population. Such exponential increases of rate are the most alarming facets of man's ecologic dilemma. (*Adapted from C. F. Park, Jr., 1968,* Affluence in jeopardy, *and other sources.*)

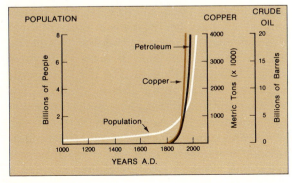

ed? Can we sustain an ever-increasing gross national product, or will we be forced to adjust to a lower material standard of living? In the 1970s, the latter began to happen. If underdeveloped countries should bring their populations under control, their demands for metals and fuels will increase exponentially as they, too, aspire to higher standards of living. With such further increasing rates of demand, surely we would be much closer to ultimate resource limits, especially of certain rare metals, than we now think.

Salvage and reuse of discarded materials will become increasingly necessary, which is as it should be, for we are frightfully wasteful (Fig. 19.4). The average American, who is the world's worst litterbug, disposes of 6 to 8 pounds (3 to 4 kilograms) of solid wastes per day, or nearly 1 ton (0.9 metric ton) per year (and we already

FIGURE 19.8

Bingham open-pit copper mine in the Oquirrh Mountains near Salt Lake, Utah. This is one of the largest open-pit mines in the world and a major producer of copper from low-grade ore disseminated in a late Mesozoic granitic pluton. Low-grade ore of this sort already is the major world source of copper. Still lower-grade ores will be mined in the future for the many metals essential to complex, industrialized societies. *(Courtesy Bureau of Mines, U. S. Department of the Interior.)*

have begun littering the moon, too). As is illustrated by the composition of New York sewer sludge noted above, a fundamental redefinition of "waste" as "resources out of place" is being called for by many authorities because we can no longer afford to waste anything. Development of substitutes already has alleviated some resource pressures and no doubt will become increasingly necessary. Substitutes may have to supersede hydrocarbon fuels before ultimate supplies are threatened for at least two

reasons: first, burning of fossil fuels may become intolerable because of the atmospheric pollution it produces, and second, hydrocarbons may become too precious for lubricants and as raw materials for the chemical industry (e.g., for plastics and fertilizers). As use of fossil fuels declines, nuclear, solar, geothermal, and perhaps even wind and tidal energy will become more important. Of most importance, ultimately, may be the earth's great internal thermal energy. Besides building mountains and displacing continents, as noted in previous chapters, it might be tapped by deep heat wells to drive man's society.

Other serious by-products of the extraction of nonrenewable resources are atmospheric, water, and sound pollution, as well as modification of the landscape through extensive excavations (Fig. 19.8). These ills can be alleviated, however, if society is willing to pay for corrective measures through higher costs; the producer cannot be expected to carry the entire burden of reclamation. The magnitude of excavations and water required to mine and process coal and oil shale in the amounts needed to supplant declining petroleum reserves staggers the imagination, however, and makes one wonder if other energy sources might be preferable.

Nonfuel Minerals

Impending shortages of nonfuel mineral resources have been less publicized than have energy shortages, but they are no less real. Figure 19.9 shows just how dependent you as a citizen of an industrial society are upon such prosaic materials as stone and gravel, salt and clay, to say nothing of the more obvious metals.

Figures 19.7 and 19.10 show how dramatically our appetite for such resources has increased. Most significant is that the increase of demand now exceeds the rate of population growth! Yet another facet of the mineral resource problem is revealed in Fig. 19.10, which shows that, over recent decades, consumption has exceeded domestic production of 18 major minerals in the United States. In short, industrial nations, of which the United States is but one example, are importing more and more of their mineral needs (Table 19.2). Use of substitute materials and recycling of scrap can help to alleviate this undesirable situation, and recycling efforts have been increasingly significant. For examples, roughly half of the recent input of iron, lead, and antimony, and about 25 percent of copper, zinc, and tin used in manufacturing recently in

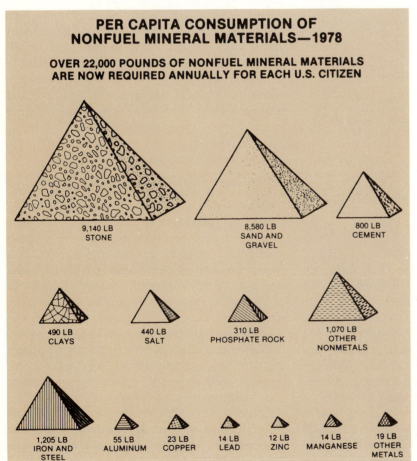

PER CAPITA CONSUMPTION OF NONFUEL MINERAL MATERIALS—1978

OVER 22,000 POUNDS OF NONFUEL MINERAL MATERIALS ARE NOW REQUIRED ANNUALLY FOR EACH U.S. CITIZEN

9,140 LB STONE

8,580 LB SAND AND GRAVEL

800 LB CEMENT

490 LB CLAYS

440 LB SALT

310 LB PHOSPHATE ROCK

1,070 LB OTHER NONMETALS

1,205 LB IRON AND STEEL

55 LB ALUMINUM

23 LB COPPER

14 LB LEAD

12 LB ZINC

14 LB MANGANESE

19 LB OTHER METALS

FIGURE 19.9
Every man, woman, and child has a stake in mineral resources. Over 22,000 pounds of these materials are required annually for each citizen. The total of such minerals used in the United States in 1978 was about 2 billion tons. (*From U. S. Bureau of Mines, 1979;* Status of the Mineral Industries.)

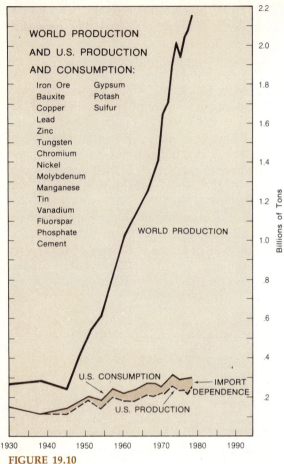

WORLD PRODUCTION
AND U.S. PRODUCTION
AND CONSUMPTION:

Iron Ore   Gypsum
Bauxite   Potash
Copper   Sulfur
Lead
Zinc
Tungsten
Chromium
Nickel
Molybdenum
Manganese
Tin
Vanadium
Fluorspar
Phosphate
Cement

WORLD PRODUCTION

Billions of Tons

U.S. CONSUMPTION

IMPORT DEPENDENCE

U.S. PRODUCTION

1930 1940 1950 1960 1970 1980 1990

FIGURE 19.10

World production compared with United States production and consumption of 18 important minerals. Note the widening gap between domestic production and consumption, making the United States increasingly dependent upon foreign sources. *(Courtesy E. N. Cameron.)*

the United States has been from scrap. Recycling has the added dividend, of course, of alleviating the environmental problems of waste disposal. Many minerals, however, are dissipated during use or are converted to products such as concrete that cannot be recycled.

The importing of some minerals is inescapable because we simply have little or no domestic supply. For example, Table 19.2 reveals why there is so much interest in mining the sea floor for manganese nodules. But in many cases, importation occurs only because the domestic mineral industry does not compete favorably with foreign suppliers. Indeed, some of the domestic companies have themselves gone abroad to obtain ores more cheaply. Most deplorable, however, is the importing of fabricated products, such as construction steel, which we are perfectly capable of manufacturing at home. When we turn from manufacturing things from mineral raw materials to importing finished goods, in effect we are exporting part of our industrial economy and employment opportunities for our own people. There are many reasons why the situation has arisen, but not the least are rising domestic labor costs and declines in grade of ore. Obviously these factors are complex and are not intrinsically bad, but it is important for an enlightened citizenry to recognize the facts. First, industrial societies are utterly dependent upon mineral resources, and, secondly, it behooves them, therefore, to have healthy mineral industries. Only one of several reasons for doing so is that the reliability of some foreign sources is doubtful (consider cobalt, chromium, and tin; Table 19.2). Moreover, it is irresponsible to export all of the socially unpopular aspects of mineral production to other, especially Third World, countries. We should accept our dependence upon such resources and take full responsibility, which means paying the entire social cost, for developing the resources and then rehabilitating the environment as necessary.

ENERGY—THE ULTIMATE LIMIT

Energy is required for all human activities, including utilization of the mineral resources just discussed and the growing, harvesting, processing, and transportation of the food from which our bodies derive energy. Indeed, today there is 10 times as much energy input to the total food system as is gained by man from his food! More alarming, however, is the fact that our increasingly technological food system now requires four times as much energy for approximately the same level of per-capita caloric food energy value as in 1940. Increasing mechanization apparently is a dubious blessing. The U.S. Department of Agriculture has shown that the most energy-efficient agriculture is the small, single-family farm. Because every aspect of life is ultimately dependent upon energy, much

TABLE 19.2
Net import reliance for the United States for some important minerals (percent of consumption for 1978).*

Minerals**	Percent imported	Major foreign sources
Columbium	100	Brazil, Thailand, Canada
Manganese	98	Gabon, Brazil, South Africa
Cobalt	97	Zaire, Zambia, Finland
Aluminum	93	Jamaica, Australia, Surinam
Chromium	92	South Africa, U.S.S.R., Rhodesia, Turkey
Platinum	91	South Africa, U.S.S.R.
Tin	81	Malaysia, Bolivia, Thailand, Indonesia
Nickel	77	Canada, New Caledonia, Dominican Republic
Zinc	62	Canada, Mexico, Australia
Gold	54	Canada, South Africa, U.S.S.R.
Tungsten	50	Canada, Bolivia, Peru, Thailand
Silver	44	Canada, Mexico, Peru
Iron ore	29	Canada, Venezuela, Brazil, Liberia
Copper	19	Canada, Chile, Peru, Zambia
Iron and steel products	13	Japan, Europe, Canada

*Data from *Status of the Mineral Industries,* 1979, United States Bureau of Mines.
**Includes refined metals as well as raw ore in some cases.

attention has been focused upon it in recent years. Every activity must be examined in terms of its energy budget. For example, wise as it may seem to recycle materials, it can be more costly in total energy expended for recycling in some cases than for processing the raw material in the first place. The glass bottle is a case in point. There is no shortage of the raw material silica, and when all hidden energy costs (e.g., transportation) are added to the energy re-quired for reprocessing glass, the result is discouraging. It simply does not make sense to recycle materials that are not rare when to do so costs more in energy than to start from scratch. But reuse of existing bottles saves both energy and money.

Figure 19.11 shows the situation for the fossil fuel resources of the earth. From it we see that about 80 percent of the total fossil fuel reserves that took 500 million years to accumulate will

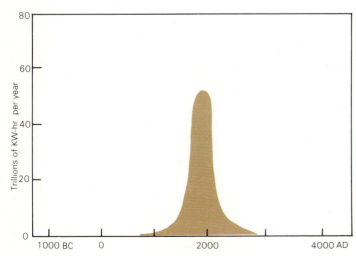

FIGURE 19.11
Exploitation of fossil fuel energy (in kilowatt-hours) from 1000 B.C. to A.D. 5000. According to this estimate, almost the entire earth's reserves will have been consumed in a mere 15 or 16 centuries. *(From Hubbert, 1962, National Academy of Sciences—National Research Council Publ. 1000-D.)*

FIGURE 19.12
Planned deep-water oil-drilling platform about as tall as the tallest skyscraper, the Sears Tower (440 meters). It will be supported with guy wires for use in outer shelf areas of the Gulf of Mexico in water depths of the order of 400 meters. Cost of such a structure will be more than $500 million. *(Courtesy Exxon Corporation, by permission.)*

Clearly, alternative energy sources must be developed. Nuclear energy does not now seem as promising as in the flush of post-World War II atomic optimism. Not only were the Atomic Energy Commission and the power industry naïve in overestimating available nuclear ore reserves, but waste-disposal and safety problems seem to have been underestimated. Perfection of the breeder reactor would help greatly. Even if solar, geothermal, and other new energy sources ultimately can provide our major needs, we shall still be dependent upon fossil fuels for several more decades. And because the demand for these fuels is increasing at such a phenomenal rate, we face serious economic and environmental problems in addition to the fuel pinch itself. Consider these figures. World petroleum consumption is doubling twice as fast as is population (Fig. 19.7). In recent years, the number of automobiles per capita has increased 25 percent in the United States, 50 percent in Britain, and 90 percent in Japan. Yet, the private automobile, which consumes half the total energy expended for transportation, is notoriously inefficient. A municipal bus is about five times more efficient for carrying people than the car, while urban railways are still more efficient. Similarly, the train is 5 times more efficient than a jumbo jet and 10 times more so than an SST. Ironically, the extinct electric street car, which was more efficient and less polluting than the bus, was killed deliberately by industry when the diesel bus came into production after World War II. Equally ironic has been the subsequent construction of costly freeways alongside existing railroad tracks that are falling into disrepair and being abandoned. The freeways have been self-fulfilling prophecies. As more were built, positive feedback produced more cars and suburban sprawl, thus requiring still more freeways. If fuel costs make commuting by auto prohibitive, then we shall rue the day that we let our railroads deteriorate.

Canada and Mexico are well fixed with domestic petroleum reserves adequate for their demands for the foreseeable future, but the energy-hungry United States during the 1980s will have to import approximately half of its needed petroleum. The famous 15-billion-barrel

have been used within a mere 3 to 4 centuries of human history. Considered on the geologic time scale, this represents a catastrophic event. Finding new reserves will help buy time, but it is getting harder and more expensive to do so (Fig. 19.12). Best-informed projections of total petroleum reserves indicate that the peak of possible crude oil production was reached in the United States when the first edition of our book was published in 1971. The peak for natural gas production in the United States was reached about the time our second edition appeared 5 years later. For the world as a whole, these same peaks will probably be reached about the end of the century. The peak for world coal production probably will occur around A.D. 2100 or 2200.

(630-billion-gallon) Prudhoe Bay oil field in northern Alaska, largest in North America, by itself could supply the present total United States demand for but 3 years and the total present world demand for only 6 months. A few more large oil fields like this one doubtless are yet to be discovered, especially beneath continental shelves like the North Sea and around North America (Fig. 19.12), but with any finite, nonrenewable resource, the race between discovery and consumption cannot be won in the long run. Also, we have to determine if it would cost more energy to extract these remote reserves than we would gain from them. In 1973 the Arab embargo on petroleum exports gave us both a warning and an opportunity, but due to political wrangling we muffed our chance to begin getting our energy house in order. Meanwhile, time runs on, and the remainder of the decade of the 80s is critical in developing alternate ways of satisfying our energy needs. The remaining fossil fuel reserves, especially coal, can buy time, but new sources are not developed overnight.

Conservation of energy has not received the impetus that it deserves. Industry, always sensitive to a cost pinch, has done very well by reducing its energy-consumption growth rate to 0.4 percent while still increasing output by about 3 percent since 1973. Recycling of scrap, for example, not only stretches raw ore supplies, but it saves enormous amounts of energy. Savings through manufacturing products from recycled metal rather than raw ore is about 50 percent for steel, 90 percent for copper, and 95 percent for aluminum. Conservation also involves more efficient energy use. Different sources of energy should be tailored to their best uses. For example, electricity should not be used directly for heating, but rather for lighting and running machinery to do other tasks. Authorities such as Amory Lovins advocate both conservation and emphasis upon so-called *soft technology*. The argument is that *hard technology* is less energy efficient and requires huge capital resources at the expense of labor resources. Thus it aggravates not only the energy problem, but also the social problem of unemployment, which is a growing concern. Lovins argues that electricity can be produced more cheaply by smaller plants; they require less investment of capital, can be built faster, which reduces costs due to interest and inflation, and shorter distribution distances reduce costs and energy loss. Also the hot water exhausted from them can be more readily used for local heating than can that from huge, remote plants. Lovins, Steinhart, and many others also advocate more emphasis upon solar, wind, and other alternate energy sources, but view with skepticism the gigantic synthetic fuels proposals. Oil-shale-processing technology seems to be developing favorably. Better insulation of buildings also is urged. Lovins talks of houses so well insulated that body heat, electric lights, and the sun could satisfy all heat requirements. "If one feels chilly in the evening, just have your pet chase a ball every so often to give off a little more body heat."

ALTERNATE FUTURES

What sort of life will our children and grandchildren lead? Will it be a steak-and-convenience-blessed paradise, or mere existence at a bland algae-and-tenement subsistence level? Is concern for lower life forms and for untrammeled open spaces free of billboards and beer bottles mere sentimentality irrelevant to preeminent man of the twenty-first century? We have shown that the only real certainty in nature is change. Among laymen there are two prevalent and opposing views of social and technological change. One, the blindly optimistic view, assumes uncritically that *all* change is good and represents "progress." This has been aptly dubbed the *Gospel of Growth*, and part of its creed reads: "That which can be built must be." On the other hand, the pessimistic or "good-old-days" view holds that most change is bad because it undermines some glorified *status quo*.

All of this reflects in a curious way our culture's definition of "progress," which dates back a couple of centuries to about when medieval times were dubbed the "Dark Ages." Today we see that those ages were not so dark, after all, in either the technological nor the cultural sense. In a similar way, we should be thinking of the future in other than just the two extreme, simplistic options paraphrased above. There should be a quite acceptable middle ground of limited growth. It is as absurd to assume that all technological and environmental changes that man produces, even if masquerading as "advances," are beneficial in the long run as it is to weep, ostrichlike, for a sentimentalized past. Therefore, if change *is* so inevitable as we claim, then surely it behooves mankind to study, plan, and adapt or try to control biological and other environmental changes for the best long-run effects on the total ecosystem. This is very much a personal as well as a social problem. John W. Gardner has discussed in an eloquent manner the importance on the social level of critically analyzing and planning for change. His inspiring treatment should be required reading for any thinking person concerned about the future of man in the biologic as well as the socioeconomic realm. Surely a society clever enough to travel to the moon has not exhausted all potential for intelligent management of its own destiny. Our society has traditionally prized innovation more than it seems to today. Perhaps we need to reinstate that "good-old-days" innovation attribute.

Certainly humans are the most adaptable creatures ever, and perhaps someday we can control the environment absolutely and manipulate it purely for self-convenience, but we doubt it. One inescapable truth, however, is that the amount and location of mineral resources and of agriculturally favorable climatic and soil conditions have already been fixed by geologic history. Faced with the complex ecological interdependence both among life forms and with their physical environment throughout geologic history, together with our limited understanding of the intricacies of ecosystems, it is unwise to plot a course for the future on blind faith. History teaches that evolution is irreversible, and once any creature becomes extinct, it can never return no matter how important it may turn out to have been to peoples' own well-being. This applies even to the lowliest microbes, some of the most benefi-

cial of which probably have not yet even been discovered by man. Examples cited of our unwitting and commonly unwanted degradation of the environment provide ample evidence of the folly of any policy of blind faith that "everything will work out for the best because science will find a way." Famous pioneer conservationist Aldo Leopold once said that the height of ignorance is the person who asks what good is an animal or plant. "If the land mechanism as a whole is good, then every part is good, whether we understand it or not. If the biota, in the course of aeons, has built something we like but do not understand, then who but a fool would discard seemingly useless parts?" Here we have a hard-headed rather than a sentimental reason for the preservation of land and organisms. For example, many nondescript plants and bacteria manufacture substances of great value, such as penicillin. Nearly half of our drugs were first discovered in nature, yet only a small fraction of plants and microbes have been assayed for useful chemicals.

It is a basic principle of biology that *diversity* is the key both to racial survival and to ecologic stability. The more genetic diversity a species has, the less susceptible it is to complete extinction in the face of environmental change. The highly touted green revolution created very productive hybrid crop plants through careful breeding. But those same plants require more care and are far less resistant to disease, pests, and climatic adversity than are the natural strains. Because of their narrow range of genetic diversity (a *monoculture*), the hybrids could be completely decimated overnight. Diversity is also a boon to ecologic communities, for it assures a greater degree of stability for the entire community, which suggests why we should take Aldo Leopold's advice very seriously. Increasing pressure for food to feed the world's burgeoning population will drive us to seek new plant food sources with desirable genetic traits. Wild, native plants are the reservoir from which new genetic strains of crop foods can be developed. So there are sober reasons for us to beware of disrupting natural communities and triggering further extinctions. We need those wild weeds!

FIGURE 19.13
The pressure of a growing population on space. (*Reprinted from* Burdened acres—the people problem, *by R. Wendolin with drawing by Monroe Bush, in* The living wilderness, *spring–summer 1967, edited by M. Nadel. By permission of The Wilderness Society, Washington, D.C.*)

Let us suppose that complete ecologic understanding could be achieved and with it the ability to manipulate safely everything in the environment for the benefit of man. Would the quality of life be acceptable in what would likely be some kind of Brave New World or Orwellian 1984? Can we presume to judge what is best for future generations?

Space is rarely discussed as a resource. In North America today, natural water and landscapes are being lost to development at the rate of one acre per minute! Because of land filling, the area of San Francisco Bay is only two-thirds of what it was 200 years ago. In 1970, there was an average of about 60 persons per square mile of land on earth, though distribution was very irregular (70 percent of Americans live on 10 percent of the available land). At the present rate of population growth, by about the year 2025 there would be 10,000 persons per square mile. Many, if not most, people require contact with the natural environment for their emotional well-being, and so space may turn out to be one of the most precious commodities as population increases. For example, it took more than 40 years for national parks in the United States to tally their first billion visitors, but only 12 years for the second billion (see Fig. 19.13). Already reservations are required long in advance for camping in several of the parks.

Laboratory experiments show that rats become psychopathic when too closely crowded for a long time. Unplanned sociological experiments in urban ghettos suggest that crowding under subsistence-level conditions has adverse psychological effects upon *Homo sapiens* as well, even blurring the distinction from brother rat. If population continues to expand unchecked, the entire human species will be subjected to unknown selective pressures imposed by overurbanization.

SUMMARY—WHAT QUALITY OF LIFE?

Any way that we examine the future of man, even for the geologically minuscule span of the next century, the quality of life will be dependent first and foremost upon food and energy resources. But the scenario really depends upon population growth and the resource demand resulting from it. Problems of food, raw materials, pollution, and space all are secondary to population. Figures 19.7 and 19.13 and Table 19.1 illustrate the magnitude of this problem. It is the phenomenal *acceleration of the growth rate* that is so alarming together with the resulting *acceleration of demands placed upon the environment.* World population will double in a mere 30 to 40 years, achieving 7 or 8 billions by the year 2000. India, with 600 million now, will add at least 200 million more people before 1990—equivalent to the entire 1970 population of the United States! Unless nutritional standards were dropped, even North America would have difficulty feeding 200 million more mouths if added in such a short period. At the time of publication of our second edition of *Evolution of the Earth* in 1976, this fact was just being realized, although a 1974 World Food Conference in Rome seemed unwilling to acknowledge it. In many parts of the world, nearly half of the population is undernourished, and famine seems unavoidable for some countries.

Studies of exponential rates of change in past geologic times emphasize the dramatic impact of man upon this earth in the past two centuries. Consider the phenomenal acceleration of technological changes alone. For at least 2 million years, man was a hunter and gatherer; the agricultural revolution occurred no more than 10,000 years ago; the industrial revolution is but 200 years old; and only within the past century have the electric light, automobile, airplane, television, nuclear reactor, antibiotics, electric toothbrushes, electric matches, computers and pocket calculators appeared. But some people doubt that man is prepared either emotionally or socially to cope with change at still-faster rates. Should we not, however, take a more positive view toward change and try to plan to influence change for the better? We might as well, for change will occur, like it or not.

The population problem is not just the result of increase in absolute numbers of births, but also improvement of birth-survival expectancy and longevity. A little-appreciated consequence of disease control is the perpetuation in the breeding portion of the population, or *gene pool*, of all types, including the mentally and physically deficient, which by natural selection would have been reduced. Together with chemicals and radiation, such perpetuation presents a genetic (thus evolutionary) problem for the race, and at the same time complex sociological

problems. It cannot be ignored, for man like so many of his predecessors on earth, could become overspecialized in this complex, artificial civilization, and perhaps become genetically weakened. On the optimistic side, however, there are predictions of genetic engineering just around the corner that may correct defects and even provide prenatal controls.

As a product of a 3.5-billion-year history of genetic mutations and natural selection by countless changes of environment, and as a member of a highly complex ecosystem, apparently man must reform if a tolerable ecologic equilibrium is to be achieved. And there will be but one chance. For at least the foreseeable future, man will not be omnipotent, but rather will remain very much dependent upon, and a part of, the total environment. Above all, we need to achieve a balance of human population in order to conserve that environment and the present way of life, for unlimited growth disrupts stability of orderly ecosystems. Remember Darwin's early observation that most natural populations tend to be stable in size. As that modern philosopher Pogo put it, "we have met the enemy, and he is us."

Conservation, simply defined, is applied ecology, and it does not mean only to preserve or lock up. About 40 years ago Wisconsin geologist C. K. Leith aptly described conservation as the "balancing of natural resources against human resources and the rights of the present generation against the rights of future ones." Conservation should ensure maximum present and future benefit from resources. Both public and private machinery are required to make it succeed, but inevitably serious conflicts arise between public and private interests. The conflicts involve long-cherished property rights and freedom to exploit private property as one sees fit. Yet, who is to say what interest is the more important? Should the public interest take precedence over individual self-interest and private initiative? It seems self-evident that as societies increase in size and complexity, some individual freedoms must be restricted, and it is presently popular to cry for governmental control of every socioeconomic problem that arises. But, as Orwell warned, one of the greatest threats to the quality of individual lives is the gradual, insidious restriction of personal freedoms by large organizations of all kinds—governmental, military, business, and even educational. Bigness itself may, after all, be the greatest threat to individualism; therefore the development of responsible environmental and of equitable socioeconomic policies without making a brainwashed slave out of the individual is the greatest social challenge facing all nations. Diversity of people is as important socially as it is genetically. We feel that there is a disturbing trend of abdication of responsibility for one's actions no matter what the circumstances. Witness increasingly outrageous lawsuits against governments, for example, by someone who drove into a bridge abutment or stepped over a cliff in a park while drunk. At the corporate level, we hear brave talk about free enterprise one day and pleas for subsidies or other forms of bail-out another day when the economic going gets a little rough. We cannot have it all ways.

It has been observed that ours is the first generation that could foresee a major cultural change within its span. We believe that the decade of the 1970s will prove to have been a major watershed in the history of industrial societies when such a change began. "Post-industrial society" and "stagflation" were added to our vocabulary, and the United States standard of living dropped for the first time since the Great Depression of the 1930s in the face of 100 percent inflation over the past decade. Classical economic theories did not seem to be working very well, and some dubious experiments were tried, such as attempting to balance the books by selling sophisticated armaments abroad. Ironically, seemingly good policies designed to conserve energy and to hold fuel prices down meant less tax revenue to run the government. Our economic ailments are very complex, but much of the problem can be traced to the imbalance of trade, which is aggravated chiefly by the importation of so much petroleum. During the 1970s, the United States increased petroleum imports from 33 to nearly 50 percent. The magnitude of costs for obtaining energy are now so large that they are as mind boggling as geologic time. It is estimat-

ed that by 1985 the United States will be paying over $10 billion annually for its oil imports. A single new deep-water offshore drilling platform to tap new petroleum reserves costs between $120 and $500 million (Fig. 19.12), and one supertanker to carry the oil costs over $70 million. In the face of such staggering expenses, capital must now be regarded as a critical resource, and *capital formation* could become the most severe limit on the acquisition of needed natural resources. It may also force hard choices among competing large projects that can be supported by society, all of which should provide added incentive to diminish our external dependence for resources.

Certainly respect for the natural environment is required on several grounds. Apart from the esthetic and sentimental arguments for conservation, the steepness of the population growth curve is somber warning that already it is later than we think. Our present course may be like Russian roulette with a bullet in all but one chamber. But any really meaningful program for man's maintenance of an acceptable ecologic balance with the environment seemingly will require first a major change in attitudes toward nature that date back several centuries. Aldo Leopold's eloquent plea for a new *land ethic* characterized by an ecological conscience with man behaving more as steward than exploiter seems even more relevant today than when he wrote it in 1949.

In Western industrial societies, both religious and economic philosophies long have been invoked to rationalize the contention that man was the ultimate earthly being (the perfect creation) destined to rule over all else, a tradition that can be traced at least back to medieval Europe (see Chap. 2). Seemingly nature existed for the benefit of man—a commodity to be arbitrarily exploited. From the practical necessity for frontier societies to conquer nature, a subtle moral justification for exploitation was an easy outgrowth. It became virtuous to cut down the forest and plow up the ground, that is, to make the land "useful." Wilderness was wasteland, and anything judged useless also was regarded as "bad." Therefore, the world was seen to be improved by human efforts to make nature "useful." Such an attitude is still conspicuous today, for example in the presumptive value judgments incorporated in elaborate justifications for some grandiose reclamation schemes. As Western societies became more industrialized, a self-justifying morality was invented, which preached that those who would develop and thereby improve nature were somehow superior to other, "primitive" peoples. Thus were the subservience of certain peoples as well as of nature itself conveniently rationalized in the nineteenth century. In our century, industry's increasing ability to produce goods has led to an *ethic of consumption*—"more is better." We need to replace this with an *ethic of conservation* so as to achieve an equilibrium or *steady state* between population demand and resource supply, especially of food and energy. Chancellor and Goss have analyzed computer models and shown that if near-zero population growth, reduced resource consumption, and near-zero energy growth can be realized by early in the twenty-first century, such a food and energy equilibrium could be sustained even with an acceptable standard of living. Their model presupposes an increasing reliance upon alternate energy sources such as the sun, however.

Fundamental overhaul on a social level of some of the most cherished tenets of industrial society seems to be the prerequisite to future welfare of the earth's surface, which is largely in the hands of man. Inasmuch as history reveals, above all else, the inevitability of change, it also suggests that thinking man has the choice of how to influence future changes in his environment. Can we accept more difficult and uncertain times? Can we innovate creatively to deal with them? Most organisms are at their best when challenged. Let us hope that we are. Surely an appreciation of geologic time, the inevitability of change, and of the operation of natural selection in the past should help to prepare anyone to cope better with the future. Ultimately all of our problems converge upon the same great question: *Will this continue to be the best of all possible worlds?*

Key Concepts

best of all possible worlds
capital formation
diversity
ecologic replacement
ecosystem
energy efficiency
ethic of conservation (of Leopold)
ethic of consumption
exponential increase of resource consumption
exponential population growth
gene pool
inorganic evolution
Leopold's land or conservation ethic
man-caused extinctions
mass extinctions
monoculture
nature as a commodity
overspecialization
prediction of consequences of evolution
renewable versus nonrenewable resources
soft versus hard technology
steady state
three maxims:
   geologic time and rates
   certainty of evolutionary change
   evolutionary interaction of living and nonliving

Readings

Alvarez, L. W., Alvarez, W., Asaro, F., and Michel, H. V., 1980, Extraterrestrial cause for the Cretaceous-Tertiary extinction, Science, vol. 208, no. 4448, pp. 1905–1108.

American Association for the Advancement of Science, 1974, Energy: Science, v. 184, (special issue), pp. 211–402.

Bryson, R. A., and Murray, T. J., 1979, Climates of hunger: Madison, University of Wisconsin Press.

Cain, S., 1967, The problem of people: Geoscience News, v. 1, p. 6.

Cameron, E. N. (ed.), 1973, The mineral position of the United States, 1975–2000: Madison, Univ. of Wisconsin Press.

Chancellor, W. J., and Goss, J. R., 1976, Balancing energy and food production, 1975–2000: Science, v. 192, pp. 213–218.

Ehrlich, P., Holdren, J. P., and Holm, R. W., 1971,

Man and the ecosphere: Scientific American Readings: San Francisco, Freeman.

Flawn, P. T., 1966, Geology and the new conservation movement: Science, v. 151, pp. 409–412.

Gardner, J. W., 1963, Self-renewal: the individual and the innovative society: New York, Harper & Row. (Colophon Paperback CN 54)

Hardin, G., 1968, The tragedy of the commons: Science, v. 162, pp. 1243–1248.

Harman, W. W., 1976, An incomplete guide to the future: San Francisco, San Francisco Book Co.

Hoyle, F., 1964, Of men and galaxies: Seattle, Univ. of Washington Press. (Washington Paperbacks WP–1)

Hubbert, M. K., 1969, Energy resources, in Resources and man: National Academy of Sciences, San Francisco, W. H. Freeman.

Kesler, S. E., 1975, Our finite mineral resources: New York, McGraw-Hill. (Paperback)

Leopold, A., 1949, A Sand County almanac: Oxford.

(Paperback edition GB263, 1968)

Lovins, A. B., 1977, Soft energy paths—toward a durable peace: San Francisco, Friends of the Earth International.

National Academy of Sciences Committee on Resources and Man, 1969, Resources and man: San Francisco, Freeman.

Newell, N. D., 1963, Crises in the history of life: Scientific American. (Reprint 867)

Orwell, G., 1949, 1984: New York, Harcourt, Brace and Co.

Stavrian, L. S., 1976, The promise of the coming dark age: San Francisco, W. H. Freeman.

Steinhart, J. S., and Steinhart, C. E., 1974, Energy: sources, use, and role in human affairs: North Scituate, Duxbury.

Watt, K. E. F., 1974, The Titanic effect: Stamford, Sinauer.

White, L., Jr., 1967, The historical roots of our ecological crisis: Science, v. 155, pp. 1203–1207.

APPENDIX I
A SYNOPTIC CLASSIFICATION OF PLANTS AND ANIMALS

Any classification of organisms is a progress report on the status of knowledge of different groups of organisms. Several things must be kept in mind regarding this report. For one thing, it is partly a subjective concoction. Even though there is a general agreement among specialists, synthesizers who scan the whole field of classification may differ with those specializing on individual groups. Thus, some classifications will show the graptolites as a Phylum Graptolithina, while others may place them as orders of either the Phyla Coelenterata or Protochordata. Another point to keep in mind is that some groups have been worked on more intensively than others so that we have more modern concepts in the systematics of the Mammalia than, say, the Porifera (sponges). The classification presented here is abbreviated; categories rarely or never found in the fossil record have been omitted, such as various "worm" phyla. Conversely, some groups are expanded so that important subcategories are outlined.

The following classification is a compromise between new and somewhat older systems in order to present as simplified a classification as possible.

KINGDOM ANIMALIA

Phylum Protozoa: Single-celled or unicelled animals subdivided on the basis of locomotive devices. (Prepaleozoic-Recent)

Class Mastigophora: Flagellated forms which may have calcareous or siliceous skeletons; includes coccoliths. Many students believe this group belongs to the plant kingdom. (Fig. 16.42)

Class Sarcodina: Forms with pseudopods used for feeding and locomotion.

Order Foraminiferida: Sarcodinids with calcareous or siliceous skeletons, the most important and common fossil Protozoa; includes fusulinids, globigerinids, miliolids, and nodosarids. (Cambrian-Recent) (Fig. 17.41)

Phylum Porifera: A primitive multicellular group generally consisting of vase-shaped body with three wall layers and with ciliated canals that carry out digestive and other functions; there are no organs. The skeleton is composed of calcite, silica, or spongin (a chitinous, flexible substance). The skeleton uniquely is composed of discrete particles called spicules which are manufactured in the middle body layer (includes the "bath" and glass sponges). (Cambrian-Recent)

Order Stromatoporoidea: A group having a fibrous skeleton of pillars and laminae; important reef formers in the Silurian and Devonian. (Ordovician-Recent)

Phylum Archaeocyatha: An extinct corallike group. The calcareous skeleton has a double wall with connecting and supporting plates. (Early-Middle Cambrian) (Fig. 9.9)

Phylum Coelenterata: Solitary or colonial animals with a three-layered wall and a body cavity called the coelenteron in which most body functions occur, such as circulation, digestion, and reproduction. Forms are either naked or have a calcareous or chitinophosphatic skeleton. They are polymorphic; that is, during the reproductive cycle there may be an alteration of form. The two principal types are the medusa, a jellyfish stage, and the polyp, a stage attached to a hard substrate. Both forms possess tentacles for food gathering. (Prepaleozoic-Recent)

Class Scyphozoa: Exclusively medusoid forms; includes many of the common jellyfish. They are preserved rarely as impressions. (Prepaleozoic-Recent)

Class Anthozoa: The largest and most important class; includes sea anemones and all corals. They are polypoid only and have the coelenteron subdivided by numerous bladelike tissues which, if calcified, are called septa. Greatest majority of genera had calcareous skeletons. (Ordovician-Recent)

Order Rugosa: Solitary and colonial forms with a basic fourfold symmetry of septa; important reef formers during the middle Paleozoic. (Ordovician-Permian) (Fig. 9.20)

Order Scleractinia: Solitary or colonial corals with a basic sixfold symmetry of septa. They formed reefs from Triassic times onward, and are the most important reef formers today. (Triassic-Recent) (Fig. 16.32)

Order Tabulata: Exclusively colonial forms related to the corals; reduced septa but with well-developed horizontal plates (tabulae) subdividing the elongated corallites. The walls have characteristic pores. (Late Cambrian-Jurassic)

Phylum Bryozoa: Exclusively colonial, very minute forms that build a variety of colonies. They possess a stomach and gastrointestinal tract, which serves several functions; reproduction is accomplished in a separate chamber. The calcareous skeleton, if present, is external, and the animals can completely draw into it for protection. These were important animals during the early Paleozoic and Tertiary, both in numbers and distribution. They are used for correlation and ecological interpretations. (Ordovician-Recent)

Phylum Brachiopoda: Forms with two valves serving as exoskeletons; they are bilaterally symmetrical. The bryozoans and brachiopods share a respiratory organ or gill which is a tube with a series of hairlike cilia. The brachiopods have a primitive heart and kidney, which is housed with the stomach and reproductive glands in a visceral sac. The anterior cavity houses the circulatory vessels and sinuses. Shell may be attached to bottom by a fleshy stalk (pedicle). (Cambrian-Recent)

Class Inarticulata: A conservative group in evolution having chitinophosphatic or calcareous shells. They lack hinging structures; articulation (opening and closing the valves) was accomplished by complex muscles. (Cambrian-Recent) (Fig. 9.10)

Class Articulata: The dominant, highly evolved, and most common group with calcareous, hinged shells. (Cambrian-Recent)

Order Orthida: With wide hinge line or oval-shaped biconvex shells. Most common during lower Paleozoic. (Cambrian-Permian) (Fig. 9.18)

Order Pentamerida: Short hinge line, large internal platform for muscles and other structures. Important index fossils in the Silurian. (Cambrian–Late Devonian)

Order Strophomenida: One of the very common lower and middle Paleozoic groups. Has a pseudo-punctate shell and a wide hinge line. One valve is plane or convex. Pedicle opening very small or absent. (Ordovician-Recent) (Fig. 9.18)

Order Productoidea: A large, varied, and sometimes bizarre group most common and important in the upper Paleozoic. Interarea (between valves) small or absent, hinge line is moderately long. The shell is pseudopunctate and generally covered by spines. Attachment to hard substrate with spines, or spines used as stilts to raise shell off a muddy bottom. (Ordovician-Permian)

Order Rhynchonellida: Convex shells with sharp ribbing. The *beaks* may be large; hinge line is short. A relatively conservative group. (Ordovician-Recent) (Fig. 14.31)

Order Spiriferida: Typically with a long hinge line and with the interarea on pedicle valve only. Internally there is a spiral gill. An important group particularly in the Devonian. (Ordovician-Jurassic) (Fig. 13.3)

Order Terebratulida: Biconvex, punctate shells with a short hinge line and an interarea as in the spiriferids. Internally there is a "looped" gill support. (Silurian-Recent)

Phylum Mollusca: A highly variable group of forms with a two- or three-chambered heart (rarely with more), kidneys, gills or lungs, and with a highly developed nervous system that may include eyes. Single, coiled shells (as in the gastropods and cephalopods), or two-valved shells (as in the pelecypods), or eight-plated shells as in the amphineurans (chitons). One of the most common invertebrate forms living today. (Cambrian-Recent)

Class Monoplacophora: A segmented bilaterally symmetrical form with a single cap-shaped shell. Fairly common in certain lower Paleozoic strata, not known after the Devonian until Recent. One of the best examples of a "living fossil" is *Neopilinia*, which was dredged up from deep water near Central America. (Cambrian-Recent) (Fig. 9.11)

Class Polyplacophora: This group also is called Amphineura or chitons. Its forms have bilateral symmetry; most possess a shell consisting of eight separate calcareous plates. They are adapted for living on rocks in the surf zone or hard substrate. They are rare in the fossil record, being represented only by isolated plates. (Late Cambrian–Recent)

Class Gastropoda: Single-shelled, generally coiled forms, which may have an operculum for protection. Distinct from all other molluscs by having the gastrointestinal tract contorted into a figure-8 pattern so that the anus and mouth are close together. One of the most successful of all invertebrate classes. (Cambrian-Recent) (Figs. 10.12 and 14.29)

Class Cephalopoda: Molluscs that have evolved into highly efficient swimmers. Because of their mobility, they have developed excellent eyes and a jet-propulsion system of locomotion in addition to having tentacles for feeding and locomotion. Includes the living octopus, squid, and chambered nautilus. Fossil cephalopods include the subclass Nautiloidea, dominant in the lower Paleozoic, and the subclass Ammonoidea, dominant in Mesozoic seas. (Late Cambrian–Recent) (Figs. 14.30 and 16.31)

Class Bivalvia (Pelecypoda): Molluscs having two shells that are mirror images in contrast to the brachiopods in which the left and right sides of each shell reflect bilateral symmetry. Clams share with gastropods the honor of being among the most diverse of the invertebrates in today's seas. (Middle Cambrian-Recent) (Fig. 16.33)

Phylum Arthropoda: The largest group of invertebrates includes the insects (probably numbering close to a million species compared to the molluscs with 80,000). The chief characteristic is a chitinous, jointed exoskeleton and specialized, jointed appendages used for locomotion, capturing prey, respiration, and reproduction. The appendages are most important in classification of many groups. (Cambrian-Recent)

Class Trilobita: The earliest appearing class. Most trilobites had *cephalon*, *thorax*, and *pygidium* segments, and were bilaterally symmetrical. The name comes from the fact that there is an axial lobe which separates two lateral lobes of the body. The exoskeleton frequently was reinforced by granules of calcite. Classification is based on characters of the exoskeleton rather than the appendages, because they are so rarely preserved. Trilobites were very common in the early Paleozoic when they underwent considerable evolution. They became progressively rarer until their extinction in the Permian. (Cambrian-Permian) (Figs. 9.7 and 9.8)

Class Crustacea: A varied group which may have a tough calcite-impregnated exoskeleton consisting of a fused cephalothorax and an abdomen. Appendages are highly specialized and are used in classification. The class includes barnacles, crabs, shrimp, copepods (the most important food source for many animals in the sea), and the minute bivalved ostracods, which are very common and important in the fossil record. (Late Cambrian–Recent) (Fig. 14.28)

Class Merostomata: Another large group which has four pairs of appendages and whose exoskeleton is more flexible than that of the Crustacea. They also have appendages modified into antennae as do the trilobites and crustaceans. The respiratory organs are specialized, enabling many of the animals to live on land. The class includes the spiders, scorpions, eurypterids, and the horseshoe crab (*Limulus*). (Cambrian-Recent) (Fig. 13.6)

Class Insecta: The most diverse and common of all living invertebrate classes. Insects are six-legged air-breathing forms with flexible exoskeletons and a body divided into head, thorax, and abdomen. Practically all bear paired wings on the second and third thoracic segments. Although there are about 12,000 described fossil species, they are relatively unimportant and rare in the record because of their soft exoskeletons. The class includes mosquitoes, cockroaches, wasps, boll weevils, locusts, and the common housefly. (Silurian-Recent) (Fig. 2.2)

Phylum Echinodermata: A large group of organisms having a skeleton composed of polygonal plates and having a dominantly radial symmetry with a predominant fivefold structure. Circulation is by a system of tubes and canals associated with food-gathering appendages (isolated in feeding-ambulacral areas having perforated plates). Groups are based on symmetry, ontogeny, and mobility. (Prepaleozoic-Recent)

Subphylum Echinozoa: Globose echinoderms that do not develop arms; most forms are unattached.

Class Helicoplacophora: A unique Lower Cambrian group that is the earliest echinozoa known. They are free living, top-shaped with their plates arranged as helical spirals. (Early Cambrian) (Fig. 9.13)

Class Edrioasteroidea: A Paleozoic group, may be attached or free living. From three to five feeding areas are twisted into sigmoid shapes resembling starfish. (Prepaleozoic-Pennsylvanian) (Fig. 9.12)

Class Holothuroidea: The sea cucumbers, sediment feeders having a reduced skeleton consisting of calcareous spicules imbedded in a tough skin. Radial symmetry is observable internally. Known mainly from curiously shaped disarticulated spicules in the fossil record. (Ordovician-Recent)

Class Echinoidea: The largest group of echinoderms. Dominantly with a bun-shaped exoskeleton composed of fused or articulated polygonal plates. The anus is on the upper surface, the mouth on the lower. Typically with protective spines on the intraambulacral plates. The echinoids tend to be sessile or move very slowly by moving their spines. One group is more mobile and tends to develop a secondary bilateral symmetry. When this occurs, the anus and mouth migrate to posterior-anterior positions. (Ordovician-Recent)

Subphylum Homalozoa: A small group of lower Paleozoic forms with flattened, asymmetrical bodies. Another name for the group is Carpoidea; representatives are rare. (Cambrian-Devonian)

Subphylum Crinozoa: Forms attached by means of a calcified jointed stem.

Class Cystoidea: Globular to pear-shaped forms attached directly to the substrate by means of a stem. The porous head (calyx) plates are not stabilized in numbers or shape. (Ordovician-Devonian)

Class Blastoidea: A relatively small class of Paleozoic attached echinoderms having rather short stems. The calyx has prominent ambulacra and the 13 interambulacral plates are arranged in three circles. The top of the calyx has five openings leading to complex, calcified internal circulatory structures. (Ordovician-Permian)

Class Crinoidea: The most important class of Paleozoic echinoderms, which has gradually diminished in numbers and kinds until today. The majority of the crinoids were attached by a well-developed and jointed stem and a root system. The viscera are contained in a calyx, which is variable in structure but with a regular system of plates, and a system of food-gathering arms. (Ordovician-Recent) (Fig. 14.33)

Subphylum Asterozoa: A group of stemless echinoderms that are vagrant benthonic. Mobility is achieved by arms and tube feet.

Class Somasteroidea: Starfish with flattened, petal-shaped arms, and lacking an anus. The tube feet are nonextendable (Ordovician-Recent; quite rare in the fossil record)

Class Asteroidea: The common starfish with a water vascular system carrying water by radial tubes from near the mouth to the tips of the arms. The mouth is on the underside and anus on the central dorsal surface; water for the circulatory system enters a sieve plate near the anus. (Ordovician-Recent)

Class Ophiuroidea: Arms set off from a central disc containing all organs. Arms are jointed and the organism is highly mobile; the common name is the brittle starfish. (Ordovician-Recent)

Phylum Protochordata: A group of sessile forms and burrowers that includes acorn worms. The most important feature is the presence of a notochord (dorsal nerve chord) sometime during the life history.

Class Graptolithina: A group of colonial genera having a chitinous exoskeleton. Found most commonly as carbon films in black shales of the lower Paleozoic (primarily in the Ordovician). The graptolites were apparently planktonic forms, either as floaters or attached to some floating or swimming organisms. As a result, genera are found worldwide and are very important for intercontinental correlation. (Cambrian-Mississippian) (Fig. 9.21)

Phylum Chordata: A highly organized group of animals with a protected central nerve cord located dorsally; gill slits are usually present, at least in the embryo. (Upper Cambrian-Recent)

Class Agnatha: Fishlike forms without paired appendages or jaws. These were precursors of the fish and the early genera had a heavy bony armor covering the head and anterior part of the body; includes the lamprey eel of today. (Upper Cambrian-Recent) (Fig. 13.7)

Class Placodermii: Primitive fish with diamond-shaped scales and, in some, spines. Jaws were either bony or cartilaginous and appendages were paired. They dominated Devonian seas, and gradually were replaced by more efficient and advanced bony fishes. (Devonian) (Fig. 13.8)

Class Chondrichthyes: The sharks long have been successful marine predators. They have cartilaginous skeletons, two pairs of appendages, and lack lungs or air bladders. (Devonian-Recent) (Fig. 13.7)

Class Osteichthyes: These are the bony fishes and have well-developed skeletons, complicated skulls, and paired appendages. Most modern fishes belong to this class. (Devonian-Recent) (Fig. 13.7)

Subclass Actinopterygii: The ray-finned fish. "Primitive" forms had diamond-shaped scales. Nearly all common fish, such as the salmon, tuna, and sardine, belong to this group. (Devonian-Recent)

Subclass Choanichthyes: A very important fish group that gave rise to the amphibians. They have internal nostrils and their fins are supported by bony rods rather than rays. The subclass includes the coelacanths represented by the very rare living fossils, *Latimeria* and *Malania.* (Devonian-Recent) (Fig. 13.9)

Class Amphibia: The most primitive tetrapods and the first of the terrestrial vertebrates. They have internal nostrils, lungs, and paired bony limbs. They are dependent on being near water, for the eggs must hatch and early larval stages must live in water. Much "experimentation" in the ossification of the vertebrate skeleton occurred in this group in response to the body-weight problem engendered by life on land. (Devonian-Recent) (Fig. 14.41)

Class Reptilia: A large and varied group of vertebrates with an advanced, sturdy, bony skeleton. They evolved from the Amphibia and are independent of water in the sense that their amniote eggs can be laid on land. The eggs are protected by a shell which prevents drying out. Includes the dinosaurs, crocodiles, snakes, and lizards. (Pennsylvanian-Recent) (Fig. 16.35)

Subclass Anapsida: Includes the cotylosaurs, which are the earliest reptiles, as well as the turtles. (Pennsylvanian-Recent)

Subclass Synapsida: Includes the pelycosaurs and the Therapsida (mammallike reptiles). (Permian-Jurassic) (Figs. 14.42, 15.22, and 15.18)

Subclass Ichthyopterygia: The Mesozoic ichthyosaurs. These forms, which superficially resemble sharks or porpoises, are good examples of convergent organisms. (Triassic-Cretaceous) (See Fig. 16.41)

Subclass Archosauria: Includes the thecodonts, phytosaurs, pterosaurs (Mesozoic flying reptiles), the plesiosaurs, and the dinosaurs. (Permian-Recent) (See Figs. 16.35 and 16.41)

Class Aves: The birds are very efficiently developed flying animals. The breastbone is enlarged for emplacement of large muscles that operate the anterior limbs, which have been modified into wings. Hollow bones reduce weight and feathers retain body heat and aid in flight. Examples include the dove and the hawk. (Jurassic-Recent) (Fig. 16.40)

Class Mammalia[1] (Triassic-Recent)

This is a diverse group of vertebrates that underwent the most rapid and spectacular period of adaptive radiation of any animal class. Within a short span during early Cenozoic time mammals came to dominate the majority of vertebrate land habitats. They are warm-blooded, have mammary glands and hair as distinguishing features. Includes the rat, the skunk, and man. (Jurassic-Recent)

[1]Mammalian groups are based on types of placental or nonplacental development on the higher levels (subclass). Orders and smaller units of fossils are based on skeletal characters, particularly skull and tooth features. This classification is abbreviated, listing only those groups discussed in the text or of general interest. This classification is really just a list of selected mammalian orders rather than the latest and highly complicated one used by paleontologists. There is, at the time of this writing, a controversy over the methods of higher taxonomic ranking. To illustrate, we include the McKenna (1975) higher classification of the single order, Carnivora.

Subdivision Mammalia
Infradivision Theria
Supercohort Eutheria
Subcohort Epitheria
Infracohort Preptotheria
Supersection Tokotheria
Section Ferae sedis mutabilis
Order Carnivora.

Subclass Prototheria
  Order Monotremata   Pleistocene-Recent. Includes egg-laying duck-billed platypus
Subclass Allotheria
  Order Multituberculata   Jurassic-Eocene
Subclass Theria
  Infraclass Pantotheria
    Order Pantotheria   Jurassic. Includes the trituberculates
  Infraclass Metatheria
    Order Marsupialia   Cretaceous-Recent. Includes the opossum, kangaroo, etc.
  Infraclass Eutheria
    Order Insectivora   Cretaceous-Recent. Includes the shrews of today
    Order Chiroptera   Eocene-Recent. Bats
    Order Primates   Cretaceous-Recent. Includes apes and man
    Order Carnivora   Paleocene-Recent. Includes cats and dogs
    Order Perissodactyla   Paleocene-Recent. Odd-toed hoofed mammals, such as the horse
    Order Artiodactyla   Eocene-Recent. Even-toed hoofed mammals, such as pigs
    Order Condylarthra   Cretaceous-Miocene
    Order Pantodonta   Paleocene-Oligocene. Includes the coryphodonts
    Order Hyracoidea   Paleocene-Recent
    Order Embrithopoda   Paleocene-Recent
    Order Astrapotheria ⎫
    Order Pyrotheria ⎬ Paleocene to Recent This group of orders is restricted to South America and includes hoofed forms
    Order Litopterna ⎪
    Order Notoungulata ⎭
    Order Proboscidea   Eocene-Recent. Includes the elephant
    Order Cetacea   Eocene-Recent. Includes the whales
    Order Rodentia   Paleocene-Recent. Includes the mice
    Order Lagomorpha   Paleocene-Recent. Includes the rabbits

KINGDOM PLANTAE

Phylum Thallophyta: The most primitive group of plants, which includes such single-celled forms as bacteria and diatoms, and such multicelled types as fungi, lichens, and algae. The earliest-known living organisms, bacteria and calcareous algae, indicate the long and varied history of this important phylum. (Prepaleozoic-Recent) (Figs. 9.2 and 9.3)

Phylum Bryophyta: The bryophytes have developed special reproductive cycles and have differentiated tissues, which permit them to live on land. They include the mosses and liverworts, neither of which have left any significant fossil record. (Devonian-Recent)

Phylum Psilopsida: The most primitive and earliest appearing of the vascular phyla, which developed a method of raising water from the substrate to the upper portions of the plant to make possible a nonaquatic habitat. There were no true roots or leaves. (Silurian-Devonian) (Fig. 14.35)

Phylum Lycopsida: Scale trees represented today by a few genera such as the club mosses and the ground pine; they had true roots and leaves, and dominated the Paleozoic both in abundance and size, some reaching 100 feet in height. Most commonly found in Carboniferous coal forests; includes *Lepidodendron* and *Sigillaria*. (Devonian-Recent) (Fig. 14.36)

Phylum Sphenopsida: (Also called arthrophytes.) A moderately common upper Paleozoic group characterized by vertical ribbing and jointed stems; *Calamites* of the late Paleozoic is most typical. The modern *Equisetum* or horsetail found along railroad tracks is the only living member and can be said to be a "living fossil." (Devonian-Recent) (Fig. 14.36)

Phylum Filicineae: The true ferns, which have a very well-developed vascular system and a unique spore-bearing reproductive cycle. (Devonian-Recent) (Fig. 14.36)

Phylum Gymnospermae: The most primitive seed-bearing plants. Members are quite different-appearing; includes forms that dominated the land during the late Paleozoic and early Mesozoic. The earliest order, the cycadofilicales, had fernlike leaves and are difficult to separate from the true ferns in the fossil record. The cycadofilicales became extinct in the Mesozoic. The group includes the important Permian Southern Hemisphere form *Glossopteris*. (Devonian-Mesozoic) (Figs. 14.37 and 14.39)
Other important orders include the cycadales (Mesozoic-Recent) and the cycadeoidales (Mesozoic). These look like palms in having a long, unbranched trunk with a crown of palmlike leaves.
The conifers are the most common gymnosperm of today. They have blade or needlelike leaves and possess cones with exposed seeds. They began in the Late Carboniferous, but it was not until the Jurassic that they reached their climax. Competition with the angiosperms is responsible for their subsequent decline. (late Paleozoic-Recent)
Another group is the order Ginkgoales (Permian-Recent) which have peculiar fan-shaped leaves. This group is another example of a "living fossil" because they have existed virtually unchanged since the Permian. This resistance to change and tolerance of grossly unfavorable environment is reflected in the observation that the ginkgo happily survives along the streets of New York City.

Phylum Angiospermae (Magnoliophyta): The largest, most diverse and most successful phylum of plants, characterized by internal, covered seeds, and true flowers. The very rapid adaptive radiation is parallel to that of the Mammalia, which synchroneity is interesting. The phylum arose from the cycadeoids in the Jurassic, and by Cretaceous time dominated the world flora. Examples include all hardwood trees, the rose, and the Venus's-flytrap. (Cretaceous-Recent)

APPENDIX II
ENGLISH EQUIVALENTS OF METRIC MEASURES

Units of Length
1 millimeter (mm) = 0.1 centimeter (cm) = 0.039 inch (in.)
1,000 mm = 100 cm = 1 meter (m) = 39.37 in. = 3.28 feet (ft) = 1.09 yard (yd)
1,000 m = 1 kilometer (km) = 0.62 mile (mi)
1 in. = 2.54 cm
1 mi = 1.61 km

Units of Area
1 m^2 = 10.76 ft^2
1 km^2 = 0.39 mi^2
1 ft^2 = 0.09 m^2
1 mi^2 = 2.58 km^2

Units of Volume
1 m^3 = 35.32 ft^3
1 km^3 = 0.24 mi^3
1 ft^3 = 0.03 m^3
1 mi^3 = 4.17 km^3

Units of Weight
1 gram (g) = 0.001 kilogram (kg) = 0.0022 pound (lb)
1 kg = 1,000 g = 2.2 lb

Temperature Scales
1°F = 0.56°C
1°C = 1.8°F
0°C = 32°F (freezing point of water)
100°C = 212°F (boiling point of water)

GLOSSARY

"When I use a word, it means just what I choose it to mean—nothing more nor less."

Humpty Dumpty in Lewis Carroll,
Through the Looking Glass

Activity. The process of emission of particles and/or radiation from the nuclei of unstable atoms during radioactive decay (see Emissions). Rate of activity or decay provides a basis for calculating isotopic dates.

Actualism. The assumption that present laws of science would apply at all times; thus, presently known processes are presumed to have always acted in the same way but with greatly varying intensities.

Adaptation. The modification of specific features in organisms in response to environmental change.

Adaptive Radiation. Diversification by means of speciation into a new (not previously occupied) environment and subsequent fragmentation of that environment as speciation occurs.

Aerobic. Free oxygen present in an environment; also applied to any organism that requires free oxygen.

Albedo. Reflectivity of surfaces (e.g., water, forest, snow) measured by percent of incoming solar radiation that is reflected; an important factor in global climate. (See Table 18.3.)

Allopatric model. The hypothesis that new species originate in small, isolated populations on the periphery of an ancestral species' geographic range.

Amino Acids. Relatively simple organic compounds that form proteins.

Anaerobes. Microorganisms that live without oxygen to process for energy as opposed to *Aerobes* that can.

Anaerobic. No free oxygen present in an environment; also a type of organism that cannot use free oxygen in its metabolism.

Andean-type plate margin. Margin like that of western South America with oceanic crust being subducted beneath continental crust to produce a volcanic arc (Andes volcanoes) on the edge of a continent.

Andesite. Volcanic rock of intermediate composition like that of coarser diorite (see endpapers) characteristic of volcanic arcs above subduction zones.

Angiosperm. Vascular plants that produce flowers containing ovaries in which capsuled seeds are formed.

Arch. Large-scale cratonic structure that has a broad anticlinal form; total deposition was less and was interrupted by more unconformities than in its adjacent basins.

Atoll. A ring-shaped group of islands composed of coral reefs and coral sands built on a submerged volcano.

Atomic number. Number of protons in an atomic nucleus (see Mass number).

Aulacogen. Fault-bounded trough (graben) generally extending into a craton margin and filled with thick strata; thought to form when a supercontinent begins to break apart.

Autotroph. A plant that can feed on inorganic material and that manufactures food.

Axis of spreading. Hypothetical axis that defines the rotation of one or a pair of lithosphere plates; not the same as geographic spin axis of the earth. (See Fig. 7.10.)

Basalt. Fine-grained, dark (mafic) igneous rock composed chiefly of calcium-rich feldspar and pyroxene. (See endpapers.)

Basin. Large-scale structure with synclinal form and especially (though not exclusively) important in cratons; deposition thicker and with fewer unconformities than in adjacent areas, such as arches.

Benthonic. Organisms that live on the bottom of lakes, streams or oceans. They are either sessile (stationary) or vagrant (can move around) and may live on the bottom (epifaunal) or beneath the surface (infaunal).

Biotic. Pertaining to plants *and* animals.

Block faulting. Fragmentation or rifting of the crust due to tension, which results in parallel fault blocks dropped down varying amounts relative to one another. (See Figs. 17.15 and 17.25.)

Carnivore. An animal (rarely a plant) which eats animals for its basic food.

Catastrophism. Explanation of phenomena, such as great unconformities and extinction of fossil species by appeal to supernatural causes acting rapidly and violently.

Chemical sediments. Accumulated by precipitation of grains from chemically saturated water (e.g., the evaporite deposits, salt and gypsum).

Chromosome. A structure in a cell nucleus carrying genes. During cell division the chromosome splits into two sections so that identical genetic information is passed on to the daughter cells.

Classification. A grouping of similar organisms into categories such as species based upon the number of characteristics that they share in common. The categories are further grouped into a higher category containing members with fewer shared characters which are more general in nature. *Genera* have fewer shared characters than *species*, *families* fewer than *genera*, etc.

Clastic sediments. Fragmental sediments formed of broken grains of older minerals, rocks, shells, or other skeletal materials. *Terrigenous clastic* sediments were derived by erosion of older rocks exposed in land areas; *nonterrigenous clastic* materials were derived from skeletons formed in the sea but then washed about on the sea floor.

Clastic wedge. Prism of clastic sediments with wedge-shaped cross section, which was deposited adjacent to rising mountains. As the mountains rose, their debris was deposited on adjacent lowlands.

Conformable. Strata lacking any discontinuities (unconformities), thus representing an apparently complete record of continuous deposition.

Continental accretion. An increase through time of the volume and area of the continental crust by the formation of new granitic and andesitic rocks within mobile belts as well as by the collision of volcanic arcs and microcontinents with the continents. (See Fig. 16.2)

Continental drift. Theory of Taylor, Wegener, and others that continents may be broken and displaced large distances; where they split apart, new ocean basins are formed and where they collide, mountains are formed.

Convergence (Evolutionary). Two or more organisms, living in a similar environment but not closely related, having similar characteristics (e.g., the body streamlining and fins in the sharks and dolphins).

Convergent plate margin. Margin along which two lithosphere plates move toward each other. In some cases an oceanic plate is subducted (or consumed) beneath the other, causing andesitic vol-

canism and granite intrusion, whereas, in other cases, two continents or arcs may converge and collide.

Coriolis effect. Deflection of wind or ocean currents due to spin of the earth, so that such currents follow curved rather than straight paths (Fig. 11.25).

Correlation. Comparison of strata at two or more localities to establish a similarity of age. Most commonly correlation is established by comparing index fossils, but it may also be based upon similar isotopic dates. (See Fig. 3.16)

Cosmic particles. High-energy particles from extraterrestrial space, which may shatter nuclei of oxygen and nitrogen atoms in the upper atmosphere, releasing neutrons that then produce ^{14}C by collision with other nitrogen atoms.

Cosmogonists. Those who developed all-encompassing theories for the origin of the earth and solar system; e.g., Buffon.

Craton. Large tectonically stable nucleus of a continent which has relatively subdued topography. Most existing continental cratons have been relatively stable throughout most of Phanerozoic time (past 700 m.y.).

Cross-cutting age relationships. Any igneous intrusion must be younger than all rocks penetrated by it but older than any that cut it. Inference of relative age from such relationships is an extension of the *superposition principle*.

Cross stratification. Smaller-scale lamination at an angle (typically 20° to 30°) to that of the major stratification, and produced by the migration of ripples or dunes. (See Figs. 8.8 and 8.15.)

Daughter-isotope dating. Comparison of ratios among different stable daughter isotopes to determine rock ages. Because different daughter isotopes are generated at different rates, their ratios to one another change progressively through time, providing a basis for dating. (See Fig. 4.5.)

Decay series. Radiogenic parent

isotope and all daughter isotopes derived therefrom by decay; e.g., ^{40}K to ^{40}Ar, ^{87}Rb to ^{87}Sr, and ^{238}U to ^{206}Pb series.

Diluvialists. People who held that most strata and fossils were deposited by the biblical Flood.

Diorite. Coarse-grained igneous rock of intermediate composition; contains feldspar and pyroxene, but little or no quartz. (See endpapers.)

Discordant dates. Two (or more) differing isotopic dates obtained by different methods for the same rock, generally due to selective resetting of one decay series that is more susceptible to heating (e.g., K–Ar); a valuable clue to complex metamorphic histories.

Disjunct floras and faunas. Portions of the same flora or fauna now isolated but originally spread over all the area between present separated outposts.

Divergence (evolutionary). The process of the evolution of organisms away from the general adaptive mode of the ancestral group, such as the amphibians becoming progressively land-oriented by developing strengthened backbones and limbs.

Divergent plate margin. Margin of lithosphere plate that is moving away from a spreading zone; characterized initially by tensional faulting and basaltic volcanism followed by stable tectonic behavior.

DNA (Deoxyribonucleic acid). A complex double-stranded, helical, protein molecule composed of four amino acid bases arranged in groups with a phosphate, a sugar, and a purine.

Dolomite. Sedimentary rock composed of calcium-magnesium carbonate $[CaMg(CO_3)_2]$; most commonly formed by replacement of limestone $(CaCO_3)$ through introduction of magnesium ions carried in solution in pore water.

Ecologic niche. A habitat occupied by a species or group of organisms especially adapted for existing in the habitat.

Ecologic replacement. The filling

of a vacated ecologic niche by the evolution of a new species or the migration of a different species.

Ecologic zone. A group of ecologic niches related by such physical factors as temperature, depths of water, salinity, etc.

Ecosystem. All living and nonliving things in a given area that have a relationship that is self-renewing.

Emissions. Spontaneous expulsions from an atomic nucleus during radioactive decay of one or more of the following: *alpha particle* (He nucleus), *beta particle* (electron) or *gamma rays* (similar to x-rays). Emission rate reflects decay rate (see Activity).

Endemic species. A species known only from one fauna or ecologic zone.

Epeiric sea. Shallow seas that flooded continental cratons frequently through geologic time; also called *epicontinental seas* (epi = above, thus "seas above or over continents").

Epifauna. Animals living on the bottom of aqueous bodies, such as oceans, lakes, and streams, but not beneath the sediment surface.

Equilibrium. A balance between two or more processes; e.g., uplift and erosion, or new snowfall and glacier melting, or rise of sea level and upward growth of reefs.

Eucaryote. All cells that have a nucleus and certain organelles. (See procaryote.)

Evaporite intrusions (salt domes). Upward intrusions by plastic flow of salt or gypsum from deeply buried evaporite layers, which become isostatically unstable because they are less dense than overlying strata.

Evaporites. Sedimentary rocks, such as salt (NaCl) and gypsum ($CaSO_4$) formed by the evaporation of large volumes of sea water.

Evolution. The continuous adaptation of organisms to a changing environment by natural selection and mutation.

Exponential change. Change at an increasing (or decreasing) rather than a constant rate (e.g., population growth).

Facies fossils. Types of fossils that tend to be restricted to a single lithology or facies, such as graptolites in black shales, so they are not useful for correlation from that facies into other facies.

Facies map. Shows distribution of different sedimentary facies (i.e., lithologies) over some geographic area for some specified moment or interval of geologic time. Important for interpreting ancient depositional environments and paleogeography.

Fauna. All the animals in a given region or time period or all the species of a phylum in a given region or time period.

Faunal province or realm. A group of interrelated faunas that are widespread and distributed according to some broad control, such as temperature. For example, the recent Floral province or realm, Arctic faunal province.

Fermentation. Bacterial action causing the breakdown of large organic molecules usually under anaerobic (oxygen-depleted) conditions.

Filter feeders. Organisms that can strain water to remove food particles.

Fission tracks. Imperfections in minerals and volcanic glass caused by spontaneous fission of an unstable atomic nucleus, which propels energetic particles through surrounding material. Density of tracks is a function of numbers of atoms that have undergone fission, thus also of age.

Five-hundred-year flood. Concept of violent natural phenomena such as floods that are rare on the human time scale, but not on the geologic scale.

Flora. All the plants in a given region or time period.

Floral province or realm. The same as faunal province but involving plants.

Flysch. Collective term for very evenly layered, alternating thin sandstones and shales; most characteristic of sequences in which the sandstones were deposited by turbidity currents. (See Fig. 16.26.)

Focus of earthquakes. Actual location of an earthquake source beneath the surface; *shallow-focus* events characterize ocean ridges and *deep focus* only characterize volcanic arcs. (See Fig. 7.6.)

Food web. A diagram of different levels of energy consumption by organisms.

Formation. The most fundamental local rock division of stratigraphic classification, which has some distinctive homogeneity of color, texture, fossil content, or the like; generally named formally for some geographic locality (e.g., Morrison Formation for Morrison, Colorado).

Fossil. Any record of past life, such as actual bones, shells, teeth, but also including impressions, footprints, burrows, etc.

Fossil zone. Restricted thickness of strata characterized by a distinctive *index fossil*, which constitutes a biologic datum for correlation.

Gene. The fundamental inheritance unit that carries a characteristic from parent to offspring composed of a linear segment of the double stranded DNA molecule.

Gene pool. The actively breeding portion of a population in which genes are exchanged as a result of reproduction.

Genetics. The study of heredity and the causes of variation of organisms.

Genus. The formal name of a group of related species in a classification.

Geosyncline. Relatively thick sequence of sedimentary and/or volcanic rocks that was deposited within a subsiding linear zone of the crust (generally a mobile belt) and that ultimately was upheaved to form mountains.

Gondwanaland. Southern supercontinent of Paleozoic time composed of South America, Africa, India, Australia, and Antarctica.

Graded bedding. A regular gradation from coarser to finer grain size upward through a clastic stratum; implies settling of grains at rates proportional to their size or mass. (See Fig. 8.7.)

Granite (granitic). Coarse-grained, light-colored (silicic) igneous rock with quartz and potassium feldspar predominating over plagioclase feldspar, therefore relatively rich in K, A1, and SiO_2. (See end papers.)

Graywacke. A heterogeneous poorly sorted (i.e., texturally immature) dark sandstone. (See Fig. 8.18.)

Greenhouse effect. Heating of air because some solar radiation reaching the earth's surface is radiated back as long-wavelength infrared radiation, most of which is trapped in the atmosphere by H_2O and CO_2 molecules (or by glass in a greenhouse). (See Fig. 5.11.)

Greenstone belts. Zones of the earth's crust (chiefly Prepaleozoic) characterized by mildly metamorphosed volcanic (greenstones) and associated immature clastic sedimentary rocks.

Half-life. Time required for decay of one-half of the parent isotope present in a radioactive sample. After one half-life, 50% of the parent is left; after two half-lives only 25%, and so on. (See Fig. 4.4.)

Herbivore. Animals that eat plants for their primary source of food.

Heterotroph. A consumer of food, either plants and/or animals.

Inclination angle. Vertical angle of inclination of rock remanent magnetization (*"fossil magnets"*) relative to the surface of the earth, which indicates paleolatitude of a specimen because steep inclination of the earth's magnetic field is characteristic of high latitudes and low inclination is characteristic of low latitudes. (See Fig. 6.17.)

Included-fragment age relationships. Any fragment included within a clastic sediment or an igneous rock must be older than the enclosing rock; an extension of the *superposition principle* for relative age.

Index fossils. Fossils unique to a limited thickness of strata, widespread geographically, and easily recognized. As a result, they are important for correlation of strata.

Infauna. Organisms living in sediment of the bottoms of aqueous bodies, such as oceans, lakes, or streams.

Inheritance of acquired characters. The origin of new species by passing on to offspring new habits (hence ultimately new characteristics) acquired during the lifetime of an adult because of changing (new) needs.

Inorganic evolution. Irreversible evolutionary change in the nonliving realm, such as change from an oxygen-free early atmosphere to an oxygen-rich later one, and from no initial continental crust to the present configuration.

Iron formation. Evenly laminated iron-rich layers alternating with either chert or carbonate layers; unique to Prepaleozoic sequences.

Isolation. The separation of one unit of population of a species from other populations so that there no longer is gene flow between them.

Isostasy. Gravitational equilibrium such that large parts of the brittle crust "float" on the plastic mantle; their relative elevations are functions of thickness and density of the crustal blocks.

Isostatic rebound. Rise of the crust in response to removal of a load that had previously depressed the crust as in Scandinavia and around Hudson Bay. (See Fig. 1.8.)

Isotope. A nuclear species of a chemical element having a different number of neutrons in the atomic nucleus than other isotopes of the same element. Unstable or radioactive isotopes decay to stable ones, producing a spontaneous change in the number of neutrons.

Isotopic dating. All mineral or rock dating using radioactive isotopes. In the usual case, the ratio of radiogenic daughter isotope to its parent unstable isotope contained in a mineral is multiplied by the rate of decay of the parent to determine how long the parent isotope has been incorporated in that mineral (see also Total-Rock Dating and Daughter-Isotope Dating).

Land bridge. A narrow strip of land connecting two continents, permitting land organisms previously restricted to one continent to migrate to the other.

Lateral fault. see Transcurrent Fault.

Laurasia. Northern supercontinent of late Paleozoic time composed of North America, Europe, and Asia.

Limestone. Sedimentary rock composed of calcium carbonate ($CaCO_3$); generally made up in considerable part of invertebrate fossil skeletal debris.

Lithology. The total characteristics of a rock, including texture composition, fossil constituents, color, etc.

Lithosphere. Rigid outer portion of the earth (60 to 100 kilometers thick) above a low-seismic-velocity zone; includes continental and oceanic crust in its top part.

Low-seismic-velocity zone. Zone in upper mantle (below 60 or 100 kilometers) that defines base of lithosphere and seems to be zone in which most lithosphere plate motion is concentrated. This zone has a seismic velocity lower than those above or below, indicating less rigidity.

Magnetic Anomalies. Any local deviations from normal (either greater or lesser) in the intensity of the earth's magnetic field, such as the narrow positive and negative anomaly strips parallel to ocean ridges.

Mantle plume. A hypothetical plume of partially molten material rising up through the mantle and doming the crust above; may cause initiation of rifts, such as the Red Sea.

Mass number. Total sum of number of protons and neutrons in any atomic nucleus; generally written at top left of chemical symbol for an element, thus ^{14}C is the isotope of carbon with a mass number of 14.

Meteorites. Small solid objects believed to be fragments of asteroids moving in space; occasionally they collide with planets or their moons. Two major types include *metallic* (Fe–Ni–Co-rich) and *stony* (Ca–Mg–Fe-rich).

Microcontinent. Any crustal block smaller than a continent with a continental-type crust (e.g., Japan).

Milankovitch effect. Cyclic change of temperature of earth due to combined effects of variations of tilt of axis, wobble of axis (precession), and ellipticity of orbit. (See Table 18.2.)

Mineral. Naturally formed chemical element or compound with a definite composition, a characteristic crystal form, and other distinctive physical properties. (See Table 5.4 and endpapers for major rock-forming minerals.)

Mobile (or orogenic) belt. Linear or arcuate zones of the crust that experienced unusually severe compressional tectonics for long spans of time. They are sites of convergence and subduction of lithosphere plates, of andesitic volcanism, granitic batholiths, and mountain ranges.

Moho (Mohorovicic discontinuity). Discontinuity in seismic velocity long used to define the base of the crust (see Fig. 5.7.) Seismic velocity increases rather abruptly below the Moho from about 6 or 7 km per sec to 8 km per sec (P-wave velocities).

Molasse. Collective term for coarse, clastic sediments with large-scale cross stratification, channel structures, coal, and in some cases red beds; generally nonmarine deposits. (See Fig. 12.20.)

Mutation. An abrupt heritable difference of offspring from parent caused by a change in the genes.

Natural selection. The process which results in the survival of those individuals best adapted for particular changes in the environment.

Nektonic. Organisms that swim.

Neptunism. Archaic theory championed by Werner, which held that all rocks (including granite and basalt) were deposited from an early, universal ocean.

Normal Fault. Fault along which the upper block has been moved down relative to the lower block along a nearly vertical surface; characteristic wherever crust has been subjected to tension (e.g., Red Sea rift, Midocean ridge crests, Basin and Range region).

Ontogeny. The historical change in developmental features from the fertilized egg to death of an organism.

Oölite. Sedimentary rock commonly of limestone composition consisting of spheres called oölites with microscopic concentric layers, which were precipitated from shallow, supersaturated waters as the grains were agitated vigorously by waves or currents. (See Fig. 11.18.)

Ophiolite suite. European term for sequences found in many mountain belts that consist of peridotite overlain successively by basalt, chert, and black shale. Ophiolite rocks represent uplifted deep sea floor. (See Fig. 12.20.)

Organelles. Structures within a cell that process food, eliminate waste, trigger reproduction, and perform many other functions.

Organic evolution. Cumulative change of organisms in response to environmental change through time.

Original horizontality. Strata are originally deposited in horizontal layers because they accumulate grain-by-grain under the influence of gravity. Inclined strata imply post-depositional deformation.

Original lateral continuity. Inference of original extent of strata now partially eroded. Such extent must be estimated in order to infer past extent of land and sea or to reconstruct deeply eroded structures, such as cratonic arches.

Orogenic belt. see Mobile Belt.

Orogeny. Mountain building, especially a mountain building event in history, such as the *Taconian orogeny.*

Overspecialization. An outdated idea that a character had evolved beyond its functional utility; e.g., the antlers of the Irish Elk.

Oxygen-isotope analysis. Ratio of ^{18}O to ^{16}O in shells composed of $CaCO_3$ provides indication of paleotemperature of sea water when shell was formed. (See Fig. 16.28.)

Paleoecology. The study of fossil organisms in relation to past environments.

Paleogeography. The restoration of geography of any past time, including land and sea distributions, and such things as mountains and lowlands, paleocurrents and paleowinds, hot tropics and cold areas, etc.

Paleolatitude. Latitude of a continent or any other part of the crust at some past time. Due to continental drift, a continent may have changed paleolatitude significantly.

Paleontology. The study of past life.

Pan-African orogeny. Widespread metamorphism and mountain building between 400 and 600 m.y. ago in all southern continents except India. (See Figs. WM.9 and 15.9.)

Pangea. Single supercontinent of Permian and Triassic time formed by collision of Gondwanaland with Laurasia.

Parathethys Sea. Relict of northern part of Tethys seaway now represented by Black Sea, Caspian Sea, and Aral Sea. (See Fig. 17.33.)

Partial melting. Only a portion of the mantle or crust is melted to form magma; e.g., andesitic magma seems to represent selective, partial melting of sodium, aluminum and silica from an iron-magnesium-rich source.

Peridotite. Dark, ultramafic rock composed of pyroxene and olivine; characteristic of mantle. (See endpapers.)

Photic zone. The upper 100 meters or so of water in which sufficient sunlight can penetrate to allow photosynthesis.

Photosynthesis. The method by which plants chemically transform carbon dioxide and water into simple carbohydrates (the basic food) utilizing the sun's visible radiation. Free oxygen (O_2) is released.

Phylogeny. The evolutionary history of a group of related organisms.

Pillow (or ellipsoidal) lava. Lava flows with distinctive internal ellipsoidal structures resembling a pile of pillows; forms when lava is erupted in or flows into water. (See Figs. 8.4 and 12.2.)

Plankton. Organisms that float in water.

Plate tectonics. Arrangement of the outer earth in 10 or 12 large, rigid plates of lithosphere all in motion and "floating" upon plastic mantle material. Plate margins are defined by zones of seismicity. Movement of the plates is away from one another along *divergent spreading ridges* and toward one another at *convergent subduction zones.*

Plateau basalts. Basalts formed by eruptions of basaltic lava flows from many fissures, resulting in great thicknesses of flow-upon-flow to produce large plateaus.

Plutonism. Hutton's theory of the origin of all granitic rocks (and mountains) from molten magmas forcibly intruded upward into the crust due to subterranean heat.

Polarity reversals. Nonperiodic reversals of polarity of the earth's magnetic field (last of which occurred about 700,000 years ago). Provide basis for a magnetic reversal time scale. (See Fig. 7.13.)

Population density. The number of individuals occupying a given area or stratum.

Prepaleozoic borderland. Hypothetical high land composed of Prepaleozoic metamorphic and igneous rocks postulated by Americans before 1940 to be the source of all Paleozoic and younger clastic sediments deposited in geosynclines. (See Tectonic land).

Primordial lead. The ratios among four lead isotopes found in meteorites. Because neither parent Th nor U are present in the meteorites, these ratios are assumed to represent the primary or primordial ratios in the solar system 4.5 to 4.7 billion years ago. (See Fig. 4.5.)

Procaryotes. Primitive single-celled plants lacking a cell nucleus. (See also *Eucaryotes*).

Protein. Large organic molecules constructed of amino acids, the basic molecule in all cells.

Radioactive decay. Spontaneous change of an unstable atomic nucleus to a stable one. By emission of particles or radiation an unsta-ble parent isotope changes to a stable daughter isotope.

Red beds. Any sedimentary rocks that are red-colored due to disseminated, fully oxidized iron coatings on grains.

Regressive facies. Retreat of the sea away from a land area results in deposition of facies that characteristically become coarser upward at any one locality and generally have an unconformity at the top. (See Fig. 3.11.)

Remanent magnetism. Component of a rock's magnetism induced by the earth's magnetic field when that rock formed. Measurement of its orientation provides a clue to paleolatitude of the sample (thus of continents) at time of formation.

Repetitive (cyclic) sedimentation. Cyclically changing conditions, such as fluctuations of sea level or climatic changes implied by a distinct succession of sediment types is repeated many times in a vertical sequence. Especially conspicuous in upper Paleozoic strata.

Resetting of isotopic clocks. Loss of daughter isotopes from minerals generally due to heating by metamorphism (e.g., loss of ^{40}Ar gas), which results in apparent ages younger than true age. Valuable in dating metamorphic events. (See *Discordant dates* and Fig. 4.8.)

Respiration. The metabolic process by which organisms supply their cells with oxygen needed for energy processing (usually called breathing in forms having lungs).

Restored cross section. Cross section that shows thickness and facies of strata with any later structural deformation and erosion ignored (i.e., restored); valuable for portraying inferred conditions at some past time. (See Fig. 3.23B.)

Rift. Long, narrow zone of normal faulting due to tension of the crust, as in *aulacogens*, ocean-ridge crests, etc. (See Fig. 8.23.)

Rock sequences. Very widespread, relatively thick, formally designated intervals of strata definable chiefly on cratons. They are bounded by profound, widespread unconformities and include many formations (see Fig. 12.5.)

Rudistid clams. Large, irregular, and conical organisms that formed corallike reefs in the Cretaceous.

Samfrau belt. Permo-Triassic geosyncline or mobile belt extending across Gondwanaland from South America along southern Africa and Antarctica to eastern Australia. (See Fig. 15.5.)

Sandstone maturity. Relative purity of a sandstone in terms of *compositional maturity* or of relative uniformity of texture in terms of *textural maturity.*

Sea-level fluctuation ("Vail") curve. Graph of apparent worldwide changes of sea level through Phanerozoic time of 200 to 300 meters above or below present level, which resulted in half a dozen profound unconformities that seem to be present on several continents. (See endpapers.)

Sedimentary facies. Overall lithology of strata reflecting environment of deposition; facies characteristic of one environment, such as beach sand, grade laterally into facies of another environment, such as offshore mud. (See Fig. 3.9.)

Sorting. The range of grain sizes in clastic sediments; a restricted range of sizes, as in beach sands and dune sands, indicates good sorting.

Specialized organisms. Organisms that have become adapted to narrowly defined ecological niches.

Speciation. An evolutionary event in which a new species arises by mutation forming a new character and which is reproductively isolated from any closely related populations.

Steady state. A dynamic system (such as a beach) is said to be in a steady state when energy is being dissipated, but the system appears the same from time to time; also called *dynamic equilibrium.*

Stratification. Layering characteristic of sedimentary deposits, which reflects differences of color, texture, or composition.

Strike-slip fault. See Transcurrent Fault.

Stromatolites. Complex microbial ecosystem in the form of mounds and columns of layered anaerobic and aerobic bacteria, algae and fine sediments. (See Fig. 8.19).

Subduction. Process whereby one lithosphere plate is thrust beneath another along a convergent plate margin. Generally a deep-ocean trench and an adjacent volcanic arc result.

Subduction complex. Intensely sheared and fractured assemblages of rocks with many overthrust faults, characteristic of lithosphere plate margins above subduction zones. (See Figs. 12.19 and 16.4.)

Subsidence. Sinking of a large part of the earth's crust due to such causes as loading by ice or thick sediments, gradual increase of density as by cooling of young oceanic crust or downward convective flow in the mantle beneath.

Superposition. Relationship among strata such that the lowest layer in a sequence is the oldest; basis of relative age of strata. For deformed strata, original superposition must be determined with fossils, isotopic dating, or features like graded bedding that indicate the original up direction (see Fig. 8.7.)

Suspect terrane. A tectonic region (terrane) whose original location is uncertain (suspect); some examples have moved thousands of kilometers and collided with a continent or another suspect terrane.

Suture zones. The zones of collision between continental plates, which are generally characterized by intense deformation, metamorphism, and (especially) strips of mafic and ultramafic oceanic rocks.

Sweepstakes routes. Dispersal routes across hostile regions so that migration is rare and difficult. Chance plays a major role in determining which organisms will make it across (e.g., an insect blown across a mountain range).

Synthetic theory of evolution. The modernized version of Darwin's theory incorporating genetics, systematic zoology, ecology, and paleontology.

Taxonomy. The grouping together of related organisms into categories arranged in a classification.

Tectonic. The structural behavior of large parts of crust through significant geologic time (e.g., mobile tectonic belts, rifted tectonic zones, stable tectonic areas).

Tectonic land. Large, mountainous land raised within a mobile belt by an orogeny and composed of only slightly older sedimentary, volcanic and plutonic rocks; principal source of so-called geosynclinal sediments. (See Fig. 12.15.)

Tethys sea. Permian through Eocene seaway between Gondwanaland and Laurasia with a distinctive, very diverse tropical fauna. (See Fig. WM.6.)

Thermal convection. Slow circulation or turnover of mantle material by plastic flow due to differential heating just as air rises from a radiator, cools, and then sinks on the opposite side of a room.

Thrust fault. Surface or zone inclined less than 45° along which the upper block moved over the lower one perhaps for tens of kilometers. It is characteristic of portions of the crust subjected to compression. (See Fig. 14.24.)

Tillite. Lithified glacial till characterized by heterogeneous composition, unsorted texture, and a lack of stratification. (See Fig. 15.12.)

Total-rock dating. Isotopic dating using an entire rock specimen rather than a single mineral. Thus, parent-daughter-isotope ratios are determined for a specimen containing several different minerals.

Trace fossils. Indirect evidence of life, such as footprints, trails, and burrows.

Transcurrent fault. A lateral or strike-slip fault with horizontal displacement, such that one side moves laterally past the other (e.g., San Andreas fault).

Transform fault. A type of transcurrent or strike-slip fault characteristic of oceanic spreading ridges, which offsets the ridge axes as spreading progresses. (See Fig. 7.9.)

Transgressive facies. Advance of the sea over a land area results in deposition of *transgressive facies,* which overlie an unconformity and are characterized by deposits that become finer upward at any one locality. (See Fig. 3.10.)

Triple junction. Junction of three linear tectonic elements, such as three spreading ridges or three subduction zones.

Tsunamis. Japanese name for giant breaking waves generated by submarine earthquakes; very damaging to coastal structures.

Turbidity currents. Currents whose movement is driven by the action of gravity pulling moredense, sediment-laden, turbulent water masses downslope beneath less-dense, clear water. (See Fig. 8.16.)

Unconformity. Break or discontinuity in a stratigraphic sequence, which implies an interval of erosion. (See Fig. 3.20.)

Uniformitarianism. In the strict sense of Lyell, this was an assumption that present processes acted the same way through all time *and with about the same intensity.* (See Actualism).

Uplift. Raising of large regions by structural forces, such as upward convective flow in the mantle, rise of masses of magma through the crust, or compression of the crust as by subduction or continental collision.

Variability (organic). Morphological differences between individuals or groups which may be sequential or random.

Walther's principle. Observation that, when environments change position, adjacent sedimentary facies will succeed each other in vertical depositional sequences. Vertical changes of lithology due to transgression or regression are mirrored by similar lateral changes (e.g., coarsening laterally as well as upward). (See Fig. 3.10.)

INDEX

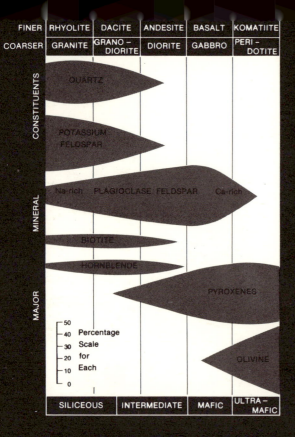

The Major Igneous-Rock-Forming Minerals (see also Table 5.4).

Name	Composition	Color
Quartz	SiO_2	colorless
Potassium feldspars: orthoclase and microcline	K–Al silicate	white to pink
Plagioclase feldspar series:	Na–Ca–Al silicate	
albite, oligoclase, andesine	Na > Ca	white to light gray
labradorite, bytownite, anorthite	Na < Ca	dark gray
Iron-magnesium silicates:		
biotite mica	K–Mg–Fe–Al silicate	black or brown
hornblende (amphibole group)	Ca–Mg–Fe–Al silicate	black to dark green
pyroxene group	Ca–Mg–Fe silicates	green, brown, black
olivine	Mg–Fe silicate	green